Photon-Involving Purification of Water and Air

Special Issue Editor

Pierre Pichat

MDPI • Basel • Beijing • Wuhan • Barcelona • Belgrade

MDPI

Special Issue Editor
Pierre Pichat
Photocatalyse et Environnement, CNRS/Ecole Centrale de Lyon (STMS)
France

Editorial Office
MDPI AG
St. Alban-Anlage 66
Basel, Switzerland

This edition is a reprint of the Special Issue published online in the open access
journal *Molecules* (ISSN 1420-3049) in 2017 (available at:
http://www.mdpi.com/journal/molecules/special_issues/photon-involv_purif).

For citation purposes, cite each article independently as indicated on the article
page online and as indicated below:

Lastname, F.M.; Lastname, F.M. Article title. *Journal Name.* **Year**. *Article number,*
page range.

First Edition 2018

ISBN 978-3-03842-700-1 (Pbk)
ISBN 978-3-03842-699-8 (PDF)

Table of Contents

Section 1: Sun-Driven Processes in Natural and Treated Waters

Section 2: Ultraviolet and Solar Homogeneous Processes to Decontaminate Waters

Section 3: Assisted Photocatalytic Treatment of Water

Section 4: Photocatalysts: Modeling; Efficacy Effects of Composition, Characteristics, Supports and Modifications

Section 5: Modeling and Testing Photocatalytic Reactors for Air Purification

About the Special Issue Editor

Pierre Pichat (pierre.pichat@ec-lyon.fr), as "Directeur de Recherche de 1ère classe" (first-class) with the CNRS (National Center for Scientific Research, France), has been active in heterogeneous photocatalysis for many years. He has founded a laboratory dealing with both basic investigations on this field and applications regarding self-cleaning materials and purification of air or water. He has published numerous research papers and several reviews of the domain. He has edited two books and special issues. He is a frequent invited lecturer at Conferences on photon-involving Advanced Oxidation Processes. He is a member of the International Scientific Committees of most of the International Conferences whose one of the topics is photocatalysis. Over the years, he has served on CNRS-related Committees on diverse aspects of chemistry; he has been the coordinator or advisor of European Community projects on photocatalysis; he has evaluated projects on environmental chemistry for various countries. He has received an International Appreciation Award acknowledging his pioneering contributions to heterogeneous photocatalysis.

Preface to "Photon-Involving Purification of Water and Air"

In the framework of the new section on Photochemistry launched by the Editorial Office of Molecules in September 2015, I proposed a feature paper issue titled "Photon-involving purification of water and air". A series of reviews and articles were published in this issue of Molecules from March 2017 to October 2017 after rigorous peer-review. These reviews and articles are freely accessible online. Nevertheless, it was thought that a printed book gathering them in an organized manner would be very useful. The book format allows one to browse through the articles in a much easier way. Anybody in a laboratory can have the printed book at hand for consulting at any time. Attention of potential readers to the existence of a book can be drawn readily in libraries and online. A book is also more appropriate for storage than a pile of copies!

This book contains six reviews and twelve articles written by distinguished experts on the various photon-driven processes, either natural or man-made, that can change the quality of water and air. Its publication will allow the community of senior scientists and students interested in these domains to possess a book of great significance for a low price. According to their topic, the reviews and articles are arranged into five consecutive sections, three of which concern photocatalysis over semiconductors. The contents of each section are summarized hereafter.

The book begins (**Section 1**) with a review by Y. Yang and J.J. Pignatello, which is essential, since it details the multiple implications (122 references) of halide ions—which are ubiquitous in natural and wastewaters—in the fate of chemical compounds in the "natural" environment and in the treated waters. An article by L. Carena and D. Vione, in this Section dealing with "natural" conditions, addresses the modelling, as a function of solar irradiation, pH and the availability of oxidizing species, of the oxidation of As (III), a crucial pollutant, particularly in waters where rice is grown.

Section 2 is devoted to homogeneous processes using UV-lamps or solar irradiation for decontaminating waters. The efficacy of several of these processes, especially the photo-Fenton one, for the elimination of representative chemicals and pathogens in municipal and hospital wastewaters is reviewed by C. Pulgarin and co-workers, taking into account the varying conditions that must be faced depending on the development level of the country. Using 2-aminobenzoate as an example of chemical compound dispersed into the environment because of anthropogenic activities (including the protection of some agriculture facilities), the article by D. Vione and co-workers reports on the fate of this compound under UV-C irradiation, and when H_2O_2/UV or S_2O_8/UV processes are used. It also considers the effects of Cl^- and CO_3^{2-} anions.

The review and the two articles gathered in **Section 3** refer to the possibilities of improving, chemically or electrically, the efficacy of photocatalysis over semiconductors. Thus, the detailed review by F. Beltran and A. Rey deals with the addition of ozone to a photocatalyst excited by sunlight (natural or simulated). It considers all the aspects from the photocatalysts to the reactors. It concludes that this combination may be viable, especially if solar energy can be used to completely operate the water purification device, including the production of ozone. J.A. Byrne and co-workers, in cooperation with two other teams, report a substantial increase in the inactivation of *E. coli* when aligned TiO_2 nanotubes (pristine or doped with N or both N and F) were immobilized on a conducting support and an external electrical bias was utilized; this increase was presumably due to electrostatic attraction of the negatively charged bacteria. N-doping increased the efficiency under UV-visible irradiation, though it had no effect under visible light only, drawing attention on possible wavelength-dependent disinfection mechanisms. The article by J. Krysa and co-workers describes the fabrication of transparent TiO_2 nanotube arrays by anodization of Ti thin layers sputtered on fluorine doped tin oxide glass. The photoelectrochemical and photocatalytic activities were measured and compared with those of layers obtained via oxidation of Ti foils. The interest of these transparent nanotube arrays is to enable back-side irradiation, which can be useful for some photoelectrochemical applications.

Section 4 encompasses many domains of semiconductor photocatalysis from calculations to practical aspects. Both the materials and the degradation effects on chemicals and microorganisms are considered. The review by Z. Cinar may be regarded as an introduction to this Section. It clearly presents, by use of examples, the interest that molecular modeling methods can present to explain and even predict, on one hand, the effects of various doping and surface modifications of TiO$_2$, and, on the other hand, the degradabilities of typical molecules both in liquid water or the gas phase. An article by Z. Cinar and co-workers illustrates the use of these calculations to help determine the electronic levels and dopant locations in the case of Se/N co-doped TiO$_2$. Many of the modifications of TiO$_2$ explored aim at extending light absorption to the visible spectral range in order to use LED lamps or possibly sunlight. That was the objective of A. Zelinska-Medynska and co-workers who depict the preparation of TiO$_2$-V$_2$O$_5$ (or MnO$_2$) nanotube arrays in a duo of papers. The efficacy of these materials for the removal of gaseous toluene under visible irradiation was measured and attributed to V$_2$O$_5$ (or MnO$_2$) species; the effects of some characteristics of the nanotubes were determined. The combination of semiconductors with natural or synthetic polymeric supports which can affect the adsorption of pollutants, the absorption of light and possibly the lifetime of the photo-produced charge carriers, is reviewed (114 references) by J.C. Colmenares and E. Kuna. In particular, the potential use of non-expensive and easily available polymers is emphasized. The review by C. Pulgarin and co-workers considers some antimicrobial coatings to combat the spread of infections in hospitals. The emphasis is on the use of magnetron sputtering deposition of Cu or Fe-oxides, alone or in combination with TiO$_2$, on polymeric substances. Two microorganisms were employed to assess the antimicrobial activity in the dark or under visible light, comparatively. Though most of the papers in this Section involve TiO$_2$, two other semiconductors are also considered. Unlike TiO$_2$, Cu$_2$O absorbs in the visible spectral range, but it photocorrodes. The article by W.C.J. Ho, Z. Chen and co-workers underlines the importance of the exposed facets of Cu$_2$O for both adsorption and degradation of a dye under visible light. Photocorrosion can be decreased by appropriate hole scavengers. Graphitic carbon nitride has been reported to be an attractive photocatalyst, active in the visible spectral range. In their article, X. Wang and co-workers detail a soft-templating synthesis of a series of S-doped C$_3$N$_4$ samples and report the efficacy of these materials under visible light to remove Rhodamine B or reduce Cr(VI). They show the decisive impact of the synthesis whose numerous, interrelated parameters can be adjusted.

The potential of semiconductor photocatalysis to purify air is questionable. For gaseous effluents, the possibilities of passing the obstacles related to low removal rates and interferences between pollutants depend on each case. For outdoor air, a significant impact is limited to confined spaces. For indoor air, the process appears not viable until now, because of the progressive deactivation of photocatalysts and the formation of degradation toxic by-products. In **Section 5** of this book, a review and two articles address this latter problem by considering the expected improvements of reactors through proper modeling and testing O. Alfano and co-workers review their modeling studies that can be used to scale-up and optimize the design of photocatalytic wall reactors by computing the local superficial rate of photon absorption. They show that their approach can be applied to model a corrugated wall reactor. In a duo of articles, E. Dumont, V Héquet and co-workers show the interest of using a recirculating close-loop reactor to determine the clean air delivery rate of diverse photocatalytic devices incorporated in the reactor. This reactor was modeled and then used with one selected photocatalytic device to illustrate the effects of various parameters, including the concentration of toluene chosen as the pollutant. The authors conclude that this reactor is a good tool to compare the efficacy of various photocatalytic devices.

In conclusion, this book will be helpful to the beginners who would like to learn more about the diverse aspects of the environment that are covered, as well as to the senior scientists who will find reviews and articles allowing them to refresh or update their knowledge of some aspects of this multidisciplinary field.

I sincerely thank the contributors for their response to my solicitation. Initially, none of them knew who would respond positively and they were just confident in me for being able to accomplish this venture. I think they do not regret their decision, given the group of eminent scientists from many countries who authored these articles published in Molecules and now gathered in this book. I am also very grateful to the many reviewers who accepted to evaluate the manuscripts and to write constructive comments. Obviously, hearty thanks are also due to Ms. Layla Zhang and Dr. Yu Wang, Editors at MDPI, who took care of the publishing of the Molecules issue and the book. I always had excellent and efficient email relationships with them. I also thank the Assistant Editors who helped with high competence to speed up the reviewing process.

Pierre Pichat
Special Issue Editor

Section 1:
Sun-Driven Processes in Natural and Treated Waters

Review

Participation of the Halogens in Photochemical Reactions in Natural and Treated Waters

Yi Yang and Joseph J. Pignatello *

Department of Environmental Sciences, The Connecticut Agricultural Experiment Station, 123 Huntington St., P.O. Box 1106, New Haven, CT 06504-1106, USA; yangyihit@hotmail.com
* Correspondence: joseph.pignatello@ct.gov; Tel.: +1-203-974-8518

Received: 18 September 2017; Accepted: 4 October 2017; Published: 13 October 2017

Abstract: Halide ions are ubiquitous in natural waters and wastewaters. Halogens play an important and complex role in environmental photochemical processes and in reactions taking place during photochemical water treatment. While inert to solar wavelengths, halides can be converted into radical and non-radical reactive halogen species (RHS) by sensitized photolysis and by reactions with secondary reactive oxygen species (ROS) produced through sunlight-initiated reactions in water and atmospheric aerosols, such as hydroxyl radical, ozone, and nitrate radical. In photochemical advanced oxidation processes for water treatment, RHS can be generated by UV photolysis and by reactions of halides with hydroxyl radicals, sulfate radicals, ozone, and other ROS. RHS are reactive toward organic compounds, and some reactions lead to incorporation of halogen into byproducts. Recent studies indicate that halides, or the RHS derived from them, affect the concentrations of photogenerated reactive oxygen species (ROS) and other reactive species; influence the photobleaching of dissolved natural organic matter (DOM); alter the rates and products of pollutant transformations; lead to covalent incorporation of halogen into small natural molecules, DOM, and pollutants; and give rise to certain halogen oxides of concern as water contaminants. The complex and colorful chemistry of halogen in waters will be summarized in detail and the implications of this chemistry for global biogeochemical cycling of halogen, contaminant fate in natural waters, and water purification technologies will be discussed.

Keywords: hydroxyl radical; sulfate radical; photocatalysis; atmospheric aerosols; reactive oxygen species; reactive halogen species; advanced oxidation processes; dissolved natural organic matter; halogenation; reclaimed waters

---

1. Introduction

Halide ions are ubiquitous in natural waters. Ordinary levels of halides in seawater are 540 mM chloride, 0.8 mM bromide, and 100–200 nM iodide [1,2]. Halide levels range downward in estuaries and upward in saltier water bodies relative to typical seawater levels. Surface fresh water and groundwater may contain up to 21 mM chloride and 0.05 mM bromide [1], with higher levels in some places. Even though the halides themselves do not absorb light in the solar region, in nature they provide far more than just background electrolytes—they participate in a rich, aqueous-phase chemistry initiated by sunlight that has many implications for dissolved natural organic matter (DOM) processing, fate and toxicity of organic pollutants, and global biogeochemical cycling of the halogens.

Advanced oxidation processes (AOPs) employing solar, visible, or ultraviolet light have been used or are under study for removal of organic pollutants from reclaimable waters, such as industrial wastewater, petrochemical produced waters, municipal wastewater, and landfill leachates, in order to meet agricultural, residential, business, industrial, or drinking water standards. While generalizations are difficult, such waters often contain moderate-to-very-high halide ion concentrations, as well as high concentrations of other photochemically important solutes like carbonate that can impact halogen chemistry [1].

This review aims to summarize the reactions of halides and their daughter products and offer insight into their effects on photochemical transformations taking place in water. Halides can undergo sensitized photolysis and react with many secondary photoproducts to produce reactive halogen species (RHS) that can participate in a variety of reactions with DOM and anthropogenic compounds, including oxidation and incorporation of halogen. These reactions are described and discussed. Extensive tabulations of rate constants for relevant reactions or RHS generation and decay have been collected for the convenience of the reader in Supplementary Section Table S1. Halides, and the RHS derived from them, affect the concentrations of photogenerated reactive oxygen species (ROS) and other reactive species; influence the photobleaching of DOM; alter the rates and products of pollutant transformations; lead to covalent incorporation of halogen into small natural molecules, dissolved natural organic matter, and pollutants; and give rise to certain halogen oxides of concern as water contaminants. The concentrations of halides is an important consideration in water treatment because halides can scavenge desired reactive oxidants and lead to unwanted halogenated byproducts. The identity of the halogen substituent(s) is critical because toxicity ordinarily increases in the order $Cl < Br < I$ for compounds of similar structure [3,4].

Halogen reactions in the atmosphere have been well studied in relation to ozone chemistry [5]. This article will not discuss gas phase reactions or surface reactions in the atmosphere, a topic recently addressed in a comprehensive review [5]; however, it will cover relevant reactions that occur in the liquid phase or at the air-liquid interface of atmospheric aerosols. A number of important reactions that take place on snow, ice, and solid microparticles actually occur on or within a surface liquid layer that is often rich in salts [6]. Compared to bulk natural waters, aerosol liquid phases can reach lower pH, and the evidence supports altered rates and/or unique chemical reactions close to the air-liquid interface.

2. Sources and Speciation of RHS Produced from Halide Ions

Reactive halogen species are generated by sensitized photochemical reactions or by reaction of halides with other oxidants of a photochemical origin. Halogen interconversion reactions are dealt with in detail. Scheme 1 provides an overview.

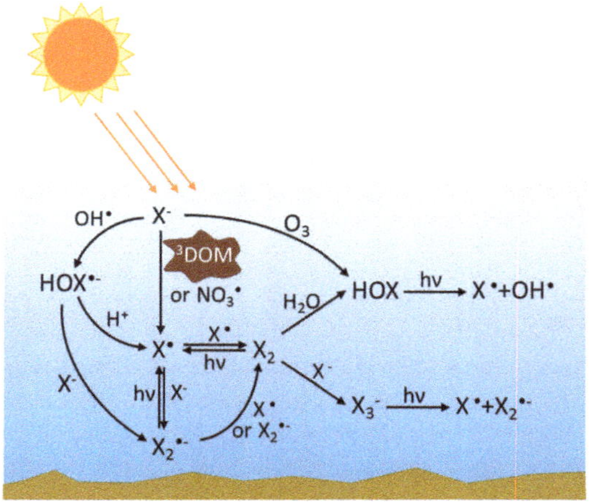

Scheme 1. Generation of RHS in waters through the action of sunlight.

2.1. Sensitized Photolysis

Halide ions in aqueous solution have absorption edges below ~260 nm and therefore do not photolyze at solar wavelengths. However, recent studies indicate that photo-sensitization by DOM may be an important source of RHS in natural waters [7,8]. Irradiation of DOM with solar light generates a short-lived excited singlet state (^{1}DOM*) that can relax to the ground state or intersystem crosses (ISC) to a much longer-lived excited triplet state (^{3}DOM*). ^{3}DOM* is a mixture of excited triplet states of diverse structures with energies ranging from 94 kJ·mol^{-1} to above 250 kJ·mol^{-1} [9]. While the nature of the chromophoric groups of DOM giving rise to triplet states is not known for certain, it has been said that aromatic ketones and other carbonyl-containing groups (e.g., coumarin and chromone moieties) are candidates for production of the high-energy triplet states of DOM [10]. The steady-state concentration of ^{3}DOM* is estimated to be 10^{-14} to 10^{-12} M, depending on light intensities, [DOM] and [O$_2$] [10] and, undoubtedly, the nature of DOM in the water parcel.

^{3}DOM* is a known precursor of photochemically-produced reactive oxygen species (ROS) such as singlet oxygen (^{1}O$_2$) and hydroperoxyl/superoxide (HO$_2$$^\bullet$/O$_2$$^{-\bullet}$, p$K_a$ = 4.88), and is a suspected precursor of hydroxyl (HO$^\bullet$). In addition, ^{3}DOM* also can engage in triplet energy transfer or oxidation reactions with itself and with other solutes. It has been shown that ^{3}DOM* can oxidize or reduce various organic compounds [11], and that model triplet ketone sensitizers with similar reactivity as ^{3}DOM* can oxidize CO$_3$$^{2-}$ to CO$_3$$^{-\bullet}$, NO$_2$$^-$ to NO$_2$$^\bullet$ [12], etc.

The question arises whether ^{3}DOM* can oxidize halide ions. The standard reduction potential of ^{3}DOM* obtained in different studies of terrestrial and freshwater NOM reference standards is estimated to be "centered near 1.64 V" [10] and about 1.6–1.8 V [8]. The estimated one-electron reduction potentials of the halogens E$°_{X·/X^-}$ are 2.59 V (Cl), 2.04 V (Br), and 1.37 V (I) in water [13]. These values are about 0.4–0.5 V lower in polar organic solvents—an important consideration because DOM exists as supramolecular aggregates and colloids, in which the electric field in the vicinity of the chromophoric site may be somewhere in between water and polar organic solvents. It thus appears that bromide and iodide, and possibly chloride, are potentially susceptible to one-electron oxidation by ^{3}DOM*.

Jammoul et al. [7] found that the triplet excited state of benzophenone, which can be regarded as a surrogate for aromatic carbonyl compounds in seawater DOM, can oxidize halide ions to X$_2$$^{-\bullet}$, Reaction (1):

$$[(C_6H_5)_2C=O]^{3*} + 2X^- \xrightarrow{hv(355 \text{ nm})} [(C_6H_5)_2C-O]^{-\bullet} + X_2^{-\bullet} \qquad (1)$$

The rate constant for Reaction (1) follows the order, I$^-$ (~8 × 10^9) > Br$^-$ (~3 × 10^8) > Cl$^-$ (<1 × 10^6 M^{-1} s^{-1}) which is consistent with the order in their reduction potential. The triplet state of anthraquinone derivatives was observed to oxidize bromide and chloride [12,14].

Building on previous theory [15], Loeff et al. [12] modeled reactions sensitized by simple organic compounds according to Scheme 2.

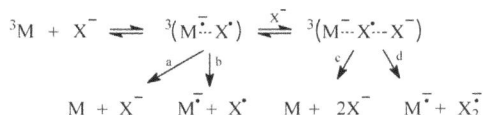

Scheme 2. Proposed pathways of sensitized oxidation of halide ions in water.

According to this model, halide ion reacts with the triplet excited state (^{3}M) to form a charge-transfer binary exciplex, ^{3}(M$^-$ --- X), or, at higher halide concentrations, the ternary exciplex, ^{3}(M$^-$---X --- X$^-$). Both the binary and ternary exciplexes can decay to the ground state (paths a or c) or dissociate to the radical pair (paths b or d). The ternary exciplex has a lower tendency than the binary exciplex to decay to the ground state because it has weaker spin-orbit coupling of the incipient radical. Therefore, the ternary exciplex more favorably dissociates to the radical products, M$^{-\bullet}$ and X$_2$$^{-\bullet}$.

In seawater, the mixed dihalogen radical anion, BrCl$^{-\bullet}$, is expected to predominate, since bromide is more readily oxidized [16], while chloride is more abundant.

Comparing artificial seawater with ionic strength controls (NaClO$_4$), Parker and Mitch [8] report that ^{3}DOM* contributes to RHS formation, which, in turn, affects the oxidation of certain added organic compounds. Using a series of radical quenching agents, they found a strong linear correlation between the observed rate constant for degradation of the marine algal toxin domoic acid sensitized by a DOM reference standard, and the same rate constant sensitized by bromoacetophenone which generates Br$^\bullet$ upon photolysis. In support of Scheme 2 for DOM, the researchers found that chloride enhances bromination in samples containing bromide.

In summary, Scheme 2 has been able to rationalize the behavior of simple sensitizer molecules. Even though the scant data available on DOM is consistent with it, it is far from being "established" for DOM and further studies are called for.

2.2. Oxidation of Halide Ions by Secondary Photo-Products

Sunlight directly or indirectly produces OH$^\bullet$, ozone (O$_3$), ^{1}O$_2$, HO$_2$$^\bullet$/O$_2$$^{-\bullet}$, and hydrogen peroxide (H$_2$O$_2$) in natural waters. Such ROS are important in many AOPs, as well. Halide ions are susceptible to oxidation by several of these ROS.

One of the most important is HO$^\bullet$. Hydroxyl originates from direct photolysis of H$_2$O$_2$, NO$_3$$^-$, NO$_2$$^-$, DOM, and dissolved iron species, and can also be produced by (dark) Fenton-type reactions of H$_2$O$_2$ catalyzed by redox-switchable transition metal ions, especially Fe. Which of these sources are most important depends on local conditions and is difficult to ascertain in most situations. The exact mechanism of HO$^\bullet$ generation from DOM has been the subject of debate for many years, without consensus [17–19]. Hydroxyl reacts with halides via the adduct HOX$^{-\bullet}$ to form the corresponding halogen and dihalogen radicals:

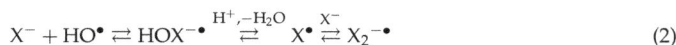

$$X^- + HO^\bullet \rightleftarrows HOX^{-\bullet} \xrightarrow{H^+, -H_2O} X^\bullet \overset{X^-}{\rightleftarrows} X_2^{-\bullet} \tag{2}$$

Reaction (2) is fast, reversible, and dependent on [X$^-$] and [H$^+$] [20]. Reactions with bromide and iodide lie far to the right at any normal environmental pH, while the oxidation of chloride to Cl$^\bullet$ and Cl$_2$$^{-\bullet}$ is favorable only under acidic conditions and comparatively high halide concentrations. For example, at pH 3, oxidation of chloride is significant whenever [Cl$^-$] is much above a few millimolar [21]. However, oxidation of chloride can be important in aerosols, where the pH can be as low as 2. Bromide and iodide are important OH$^\bullet$ scavengers in seawater [17]. Scavenging of OH$^\bullet$ does not necessarily protect other solute molecules from oxidation, as the resulting RHS are themselves strong oxidants, albeit more selective (see Section 4).

Ozone is an important component of the troposphere due to the action of sunlight on nitrogen oxides and organic vapors. Ground-level ozone concentrations can be appreciable especially in urban and industrial areas [22–24]. The reaction of ozone with halide initially produces hypohalite or hypohalous acid (XO$^-$/XOH; p$K_{a,HOCl}$ = 7.82 (0 °C), 7.54 (25 °C); p$K_{a,HOBr}$ = 8.55; p$K_{a,HOI}$ = 10.5), via a transient halo-ozonide intermediate [25]:

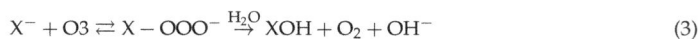

$$X^- + O3 \rightleftarrows X - OOO^- \xrightarrow{H_2O} XOH + O_2 + OH^- \tag{3}$$

The observed rate constants for overall Reaction (3) differ by more than twelve orders of magnitude among the halogens (k_{Cl-} < 3 × 10^{-3} M^{-1} s^{-1}; k_{Br-} = 258 M^{-1} s^{-1}; k_{I-} = 1.2 × 10^9 M^{-1} s^{-1}) [25,26]. Given normal seawater halide concentrations, the ratio of *rates* for ozone oxidation of iodide, bromide, and chloride is thus approximately 2300:130:1.

Reaction with O$_3$ is suggested to be a principal source of bromo- and iodo-RHS in seawater [27]. Since HOI can react with Br$^-$ and Cl$^-$ to form molecular bromine and chlorine species and regenerate I$^-$ (see Section 2.4), iodide has been implicated as a catalyst for volatilization of bromine and chlorine from marine aerosol microdroplets [28].

Halides react only slowly with $^1O_2$; second order rate constants are 1×10^3 for Cl^- (in D_2O); $< 1 \times 10^6$ for Br^- (in acetone/bromobenzene solution), and 8.7×10^5 M^{-1} s^{-1} for I^- (in water)—too slow to compete with physical quenching of $^1O_2$ by water (2.5×10^5 s^{-1} [29]). The halides do not react with $HO_2^\bullet/O_2^{-\bullet}$ in water at environmentally significant rates. A few other oxidation reactions of halides are important in aerosol systems (Section 2.3).

2.3. Heterogeneous Reactions Leading to RHS

Halides also participate in both dark and actinic heterogeneous chemistry in or on atmospheric particles [30,31]. Atmospheric aerosols broadly encompass polar stratospheric cloud particles of nitric and sulfuric acid hydrates; cloud particles of water ice; soil dusts; marine boundary layer aerosols consisting of sea salts; secondary organic aerosols resulting from oxidation of biogenic compounds in the troposphere; combustion aerosols of fuels and biomass; and inorganic ammonium salt aerosols. Many of these types of particles are relevant here, either because they are aqueous liquids, or because their surfaces are coated with aqueous films that exist due to the high salt levels which attract water.

Reactions of halides in aerosol liquids can be qualitatively and quantitatively different from reactions in terrestrial waters owing to their small size and the significance of gas-particle interfacial phenomena [30]:

(i)   The pH is often more acidic in the bulk liquid phase of aerosols than in terrestrial water bodies. By contrast, the air-liquid interface can be significantly more basic than the bulk aerosol phase; for example, it is known that the pH is 7 at the surface of bulk water at pH 3 [32].

(ii) The heavier halide ions (Br^-, I^-) concentrate at the air-liquid interface. Evidence exists for unique chemical reactions close to the air-liquid interface [33].

(iii) Particles may become depleted in bromide and iodide with respect to chloride, so that the chemistry can change over time.

(iv) Reactions may be sensitive to humidity which governs film thickness.

Halide conversion to RHS on atmospheric aerosols is initiated mainly by reactions with $HO^\bullet$, O_3, nitrate radical ($NO_3^\bullet$), and N_2O_5. Their reactions with $HO^\bullet$ and O_3 are given in Reactions (2) and (3) above. Pratt et al. [6] found that Br_2 is generated on arctic fresh snow by oxidation of Br^- by $HO^\bullet$ formed by photolysis of NO_2^- or H_2O_2 within the quasi-brine layer on the snow surfaces. The volatilized Br_2 is postulated to get pumped by the wind into the troposphere where it contributes to the episodic depletion of tropospheric ozone during the Arctic springtime.

Nitrate radical, which originates from oxidation of nitrogen dioxide (NO_2) by ozone [34], is an important atmospheric free radical, especially at night. It rapidly oxidizes aerosol halides (Reaction 4) [35,36]:

$$X^- + NO_3^\bullet \rightarrow X^\bullet + NO_3^-\ \ k_{Cl-} = 3.5 \times 10^8 M^{-1}s^{-1}; k_{Br-} = 4 \times 10^9 M^{-1}s^{-1} \qquad (4)$$

The nitrate radical interconverts with dinitrogen pentoxide if a suitable surface is available ($NO_3^\bullet + NO_2 \rightleftharpoons N_2O_5$) [34]. In water N_2O_5 dissociates to NO_3^- and NO_2^+; the latter pairs with a halide to form XNO_2, which reacts with a second halide to give X_2 [37,38]:

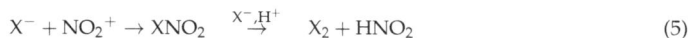

$$X^- + NO_2^+ \rightarrow XNO_2 \ \overset{X^-, H^+}{\rightarrow}\ X_2 + HNO_2 \qquad (5)$$

For chloride, Reaction (5) occurs only below pH 2 [38].

2.4. Speciation and Interconversion of RHS in Waters

Radical and non-radical RHS (rRHS and nrRHS) undergo well-known species and interconversion reactions in aqueous solutions. Unfortunately, rate constants are not available for iodine speciation in most cases.

Halogen atoms react rapidly and reversibly with halide ion to form the dihalogen radical anion:

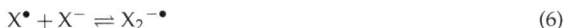

$$X^\bullet + X^- \rightleftharpoons X_2^{-\bullet} \tag{6}$$

The equilibrium constants are large (on the order of 10^5 M^{-1}, Supplementary Table S1) and the equilibria lie far to the right in both seawater and freshwater containing typical levels of halides. When I$^\bullet$ and Br$^\bullet$ are generated, the mixed dihalogen radical anion ClX$^{-\bullet}$ can form, as chloride is normally predominant. The reverse of Reaction (6) preferentially gives Cl$^-$ and the other halogen atom because chlorine is the most electronegative of the pair.

Kinetic modeling for seawater containing phenol in which reactions were initiated with OH$^\bullet$ indicates that the sum of all $X_2^{-\bullet}$ concentrations is more than 1000-times greater than the sum of all $X^\bullet$ concentrations, and that [Br$_2$$^{-\bullet}$] is about 2.7 times greater than [BrCl$^{-\bullet}$] [1].

Interconversion of halogen is possible among the rRHS. Some pertinent reactions and their equilibrium constants are given in Reactions (7) and (8):

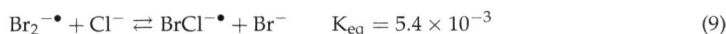

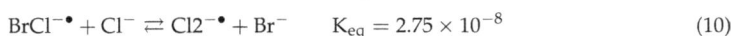

$$HOBr^{-\bullet} + Cl^- \rightleftharpoons BrCl^{-\bullet} + OH^- \qquad K_{eq} = 9.5 \tag{7}$$

$$HOCl^{-\bullet} + Br^- \rightleftharpoons BrCl^{-\bullet} + OH^- \qquad K_{eq} = 330 \tag{8}$$

$$Br_2^{-\bullet} + Cl^- \rightleftharpoons BrCl^{-\bullet} + Br^- \qquad K_{eq} = 5.4 \times 10^{-3} \tag{9}$$

$$BrCl^{-\bullet} + Cl^- \rightleftharpoons Cl2^{-\bullet} + Br^- \qquad K_{eq} = 2.75 \times 10^{-8} \tag{10}$$

rRHS dimerize or disproportionate to give the nrRHS:

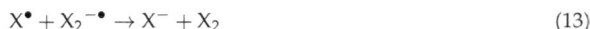

$$X^\bullet + X^\bullet \to X_2 \tag{11}$$

$$2X_2^{-\bullet} \to X_2 + 2X^- \tag{12}$$

$$X^\bullet + X_2^{-\bullet} \to X^- + X_2 \tag{13}$$

Molecular halogen reacts reversibly with halide to form the trihalide ion Reaction (14). For example,

$$BrCl + Cl^- \rightleftharpoons BrCl_2^- \qquad K_{eq} = 5.88 M^{-1} \tag{14}$$

Pertinent to aerosol chemistry, the reactions of Cl$_2$ and Br$_2$ with bromide and iodide are much faster at the air-microdroplet interface than in bulk aqueous solution presumably due to differences in solvation [39]; the same is likely true for chloride but it was not included in the study.

Molecular halogen and trihalide ions hydrolyze to hypohalous acid or the hypohalite ion [40]. Some relevant reactions are:

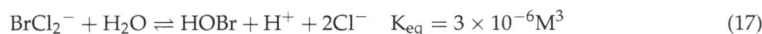

$$XCl + H_2O \rightleftharpoons HOX + H^+ + Cl^- \tag{15}$$

$$XCl + OH^- \rightleftharpoons HOCl + X^- \tag{16}$$

$$BrCl_2^- + H_2O \rightleftharpoons HOBr + H^+ + 2Cl^- \qquad K_{eq} = 3 \times 10^{-6} M^3 \tag{17}$$

Reactions (15) and (16) lie far to the right and are complete within seconds.

We may consider speciation of nrRHS in different hypothetical waters (Table 1). One represents seawater (540 mM Cl$^-$, 0.8 mM Br$^-$, 2.3 mM carbonates, pH 8.1) [1], the other a wastewater (141 mM Cl$^-$, 0.05 mM Br$^-$, 11.5 mM carbonates, pH 7.0). Modeling was performed with 163 reactions using Kintecus V6.01 [41], with an OH$^\bullet$ generation rate of 1×10^{-9} M^{-1} s^{-1}, no organic matter present, and a total simulation time of 5 or 60 min. Iodide was not included because many rate constants are unknown.

Table 1. Simulated speciation of nrRHS in different waters. Molar ratio relative to Cl_2 after 5 min except where noted.

RHS/Cl$_2$	Br$_2$	BrCl	Cl$_3$$^-$	BrCl$_2$$^-$	Br$_2$Cl$^-$	Br$_3$$^-$	HOBr/OBr$^-$	HOCl/OCl$^-$
Wastewater	4.01×10^3	2.27	0.0257	33.5	417	3.5	0.95×10^9 (1.74×10^{10}) *	2.57×10^5 (5.92×10^5) *
Seawater	1.04×10^4	24.7	0.0982	533	3800	145	6.42×10^9 (7.08×10^9) *	1.79×10^5 (6.2×10^5) *

* After 60 min.

It can be seen from Table 1 that the principal X_2 species is Br_2 and the principal $X_3$$^-$ species is Br_2Cl^-. Among all the molecular halogen species, between 87% (seawater) and 92% (wastewater) exist as Br_2Cl^- and the remainder mostly as $BrCl_2^-$. Nevertheless, the vast majority of the nrRHS are HOX/OX$^-$ species, with HOBr/OBr$^-$ dominating over HOCl/OCl$^-$ by more than a factor of 10^3 (wastewater) or 10^4 (seawater). While the concentrations of all X_2 and $X_3$$^-$ stay constant between 5 and 60 min, the concentrations of HOX/OX$^-$ continue to increase during this interval because there is no sink for them and the starting concentrations of all reactants and the pH are held constant during the simulations. Interestingly, in seawater where chloride is at much higher concentration than in the wastewater, HOCl/OCl$^-$ increases at a faster rate than HOBr/OBr$^-$ between 5 and 60 min. This suggests that Br^0 species are partially converted to Cl^0 species over time. The most likely explanation is a series of reactions that converts HOBr to HOCl, beginning with (and probably rate-limited by) substitution of Br for Cl in HOBr:

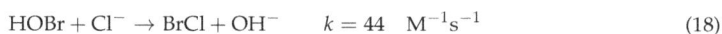

$$HOBr + Cl^- \rightarrow BrCl + OH^- \qquad k = 44 \quad M^{-1}s^{-1} \tag{18}$$

Following Reaction (18) would be, in sequence: (i) Reaction (14) to give $BrCl_2^-$; (ii) the reverse of Reaction (14) which gives Cl_2 rather than Br_2 about 5% of the time; and (iii) hydrolysis of Cl_2 to HOCl (via Reactions (15) or (16)).

Both HOCl and HOBr readily oxidize iodide [42,43]:

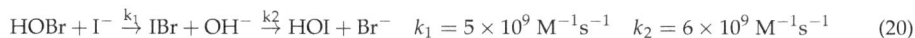

$$HOCl + I^- \rightarrow HOI + Cl^- \qquad k = 4.3 \times 10^8 \, M^{-1}s^{-1} \tag{19}$$

$$HOBr + I^- \xrightarrow{k1} IBr + OH^- \xrightarrow{k2} HOI + Br^- \quad k_1 = 5 \times 10^9 \, M^{-1}s^{-1} \quad k_2 = 6 \times 10^9 \, M^{-1}s^{-1} \tag{20}$$

Reactions (19) and (20) will therefore generate a lot of HOI regardless of which RHS is initially formed. In water, HOI is slowly converted to iodate (IO_3^-) [44]. Iodate can be an appreciable fraction of total iodine in the sea [45,46].

Since Reactions (14)–(17) are reversible, and X_2 species are volatile, atmospheric aerosols can become depleted in bromide and iodide relative to chloride [30].

3. Reactions of RHS

3.1. Photolysis of nrRHS (X_2, $X_3$$^-$, HOX)

Molecular halogens, X_2 and $X_3$$^-$, all absorb at wavelengths in the solar UV and into the visible. Photolysis of X_2 yields two $X^\bullet$ atoms [47,48] ($\Phi_{Br_2,500nm} = 0.85$ [49]; $\Phi_{IBr,500nm} = 0.73$ [49]), while photolysis of $X_3$$^-$ yields $X_2$$^{-\bullet}$ and $X^\bullet$ [50–52] ($\Phi_{Br_3^-,260nm} = 0.15$ [52]). However, molecular halogens are transient and their concentrations so small that photolysis is not likely an important fate mechanism.

The absorption spectra of the hypohalites partially overlap the UV solar emission spectrum, and the molar absorption of OX$^-$ is greater than that of HOX. Solar UV cleaves the O–X bond homolytically or heterolytically [53–57] to give halogen atoms, halide ions, and a variety of ROS, including hydroxyl radical OH$^\bullet$/O$^{-\bullet}$ (pK_a, 11.5 [58]), singlet-state atomic oxygen O(^{1}D), ground-state atomic oxygen O(^{3}P), ozone, and hydrogen peroxide (Scheme 3). For OCl$^-$, as wavelength increases the quantum yield of homolytic cleavage decreases while that of heterolytic cleavage increases [53,59].

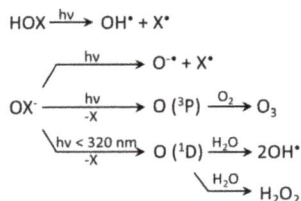

$$HOX \xrightarrow{h\nu} OH^\bullet + X^\bullet$$

Scheme 3. Photolysis of hypohalites.

The absorption spectra of HOBr and HOI are red-shifted in the gas phase compared to the aqueous phase [60–62]. Thus, the quantum efficiency of HOBr and HOI reactions may be different in aerosols than in bulk solution due to gas-liquid interfacial effects.

3.2. Reactions of RHS with Inorganic Species

Radical and nrRHS exhibit a complex chemistry with inorganic water constituents. Potentially important scavengers include carbonates, hydrogen peroxide, nitrite, and ozone. Hydrogen peroxide is a common component of natural waters owing to biological and photochemical processes. An overview of the reactions is given in Scheme 4. As strong oxidants, RHS may also oxidize metal ions that are present at low concentrations in natural waters, such as Fe^{II}, As^{III}, and Mn^{II}. Reactions of RHS with metal ions are covered elsewhere [63].

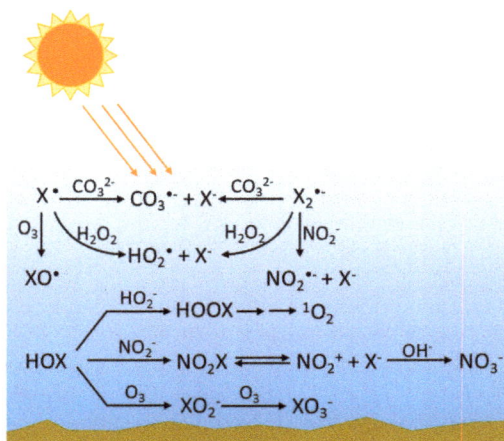

Scheme 4. Reactions of RHS with inorganic species.

3.2.1. Radical RHS

rRHS can be scavenged by carbonate and bicarbonate ions to form carbonate radicals, which, like rRHS, are strong oxidants of organic compounds:

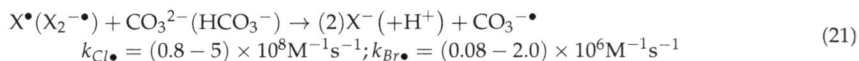

$$X^\bullet (X_2^{-\bullet}) + CO_3^{2-} (HCO_3^-) \rightarrow (2)X^- (+H^+) + CO_3^{-\bullet}$$
$$k_{Cl\bullet} = (0.8 - 5) \times 10^8 M^{-1}s^{-1}; k_{Br\bullet} = (0.08 - 2.0) \times 10^6 M^{-1}s^{-1} \tag{21}$$

Carbonates also affect RHS levels indirectly by scavenging OH$^\bullet$. Kinetic modeling shows that under OH$^\bullet$-generating conditions, addition of 2.3 mM carbonates to a solution containing 0.54 M chloride and 0.8 mM bromide steeply reduces rRHS [64]. Conversely, addition of halide ions to carbonate solutions boosts [CO$_3$$^{-\bullet}$] [1,64] and increases the contribution of CO$_3$$^{-\bullet}$ to transformation of phenol [1].

rRHS species oxidize H$_2$O$_2$ to HO$_2$$^\bullet$/O$_2$$^{-\bullet}$ Reactions (22) and (23). Rate constants are 2×10^9 M^{-1} s^{-1} for Cl$^\bullet$ and 4×10^9 M^{-1} s^{-1} for Br$^\bullet$, but are much smaller for X$_2$$^{-\bullet}$ ($k_{Cl_2^-} = 1.4 \times 10^5$M^{-1}s^{-1}; $k_{Br_2^-} = 5.0 \times 10^2$M^{-1}s^{-1}. The products HO$_2$$^\bullet$/O$_2$$^{-\bullet}$ are not very reactive in water toward most organic compounds:

$$X^\bullet + H_2O_2 \rightarrow HO_2^\bullet + X^- + H^+ \tag{22}$$

$$X_2^{-\bullet} + H_2O_2 \rightarrow HO_2^\bullet + 2X^- + H^+ \tag{23}$$

Nitrite reduces X$_2$$^{-\bullet}$ to the halide and nitrite radical, NO$_2$$^\bullet$:

$$X_2^{-\bullet} + NO_2^- \rightarrow NO_2^\bullet + X^- \quad (k_{Cl_2^-} = 2.5 \times 10^8 M^{-1}s^{-1}; \; k_{Br_2^-} = 2 \times 10^7 M^{-1}s^{-1}) \tag{24}$$

Ozone reacts rapidly with Br$^\bullet$ to form XO$^\bullet$. Data are unavailable for Cl$^\bullet$ and I$^\bullet$:

$$Br^\bullet + O_3 \rightarrow BrO^\bullet + O_2 \quad k = 1.5 \times 10^8 M^{-1}s^{-1} \tag{25}$$

Ozone also reacts with X$_2$$^{-\bullet}$ ($k_{Cl_2^-} = 9 \times 10^7 M^{-1}s^{-1}$). The ClO$^\bullet$ radical appears to be much less reactive than X$^\bullet$ and X$_2$$^{-\bullet}$ toward organic compounds [65].

Since in most waters carbonates will be at millimolar concentrations, whereas H$_2$O$_2$, NO$_2$$^-$, and O$_3$ will seldom exceed micromolar concentrations, scavenging of the rRHS by carbonates will usually predominate over the others. For their scavenging rates to be equal, [scavenger] = [carbonate]$\cdot k_{carbonate}/k_{scavenger}$. Thus, for example, ozone would have to be >~1 mM for it to out-compete 1 mM CO$_3$$^{2-}$ for scavenging of Cl$_2$$^{-\bullet}$.

3.2.2. Non-Radical RHS

Hypohalites can oxidize H$_2$O$_2$ to give the halide and ^{1}O$_2$ (Reactions (26) and (27)). The highest rate constants are observed when the acidic form of one reactant is paired with the basic form of the other—namely, OX$^-$ + H$_2$O$_2$ or HOX + HO$_2$$^-$. The reaction proceeds by nucleophilic attack of H$_2$O$_2$/HO$_2$$^-$ on the electrophilic halogen atom of HOX/OX$^-$ to give initially H-O-O-X and then H-O-O-O-H [66], which decomposes spontaneously to ^{1}O$_2$ [67]. Singlet oxygen is reactive towards many compounds, but physical quenching by water severely limits this reactivity (Yang et al. manuscript in preparation).

$$HOX + HO_2^- \rightarrow HOOX + OH^- \tag{26}$$

$$H_2O + HOOX \xrightarrow{-X^-, -H^+} HOOOH \xrightarrow{-H_2O} {}^1O_2 \tag{27}$$

Nitrite attacks hypohalites nucleophilically to generate NO$_2$X Reaction (28) [68,69]. Hypochlorites are more reactive than hypobromites. At typically low NO$_2$$^-$ concentrations, the principal decomposition pathway of NO$_2$X is reversible dissociation to X$^-$ and NO$_2$$^+$, followed by hydrolysis of NO$_2$$^+$ to nitrate (Reactions (29) and (30)):

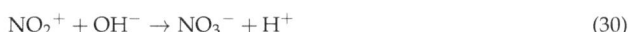

$$HOX + NO_2^- \rightarrow NO_2X + OH^- \tag{28}$$

$$NO_2X \rightleftharpoons NO_2^+ + X^- \tag{29}$$

$$NO_2^+ + OH^- \rightarrow NO_3^- + H^+ \tag{30}$$

In acidic aerosols, it is also possible to regain the nrRHS through Reaction (5). Hypohalites react with O_3 giving XO_2^- and eventually XO_3^- [25,26,70]. Bromate (BrO_3^-) is of concern in drinking water as a carcinogen [71]:

$$XOH + O_3 \rightarrow XO_2^- + O_2 + H^+ \tag{31}$$

$$XO_2^- + O_3 \rightarrow XO_3^- + O_2 + H^+ \tag{32}$$

4. Involvement of Halogen Species in Organic Matter Processing and Transformations of Organic Compounds

Organic matter entering natural waters is processed in part by its own photo-excitation. Photo-excitation of DOM can lead to bleaching, molecular fragmentation, and mineralization (to CO_2). DOM can also sensitize the photolysis of dissolved compounds such as pollutants, either through direct reaction between the solute and the $^3DOM^*$ (via either triplet energy transfer or electron transfer [10,11]), or through reactions of the solute with secondary photoproducts of DOM such as $^1O_2$, $OH^\bullet$, or $HO_2^\bullet/O_2^{-\bullet}$. Halides can affect photoexcitation and photobleaching of DOM, and give rise to RHS that can react with DOM and other organic compounds.

4.1. Impact of Halide Ions on Photoexcitation and Photobleaching of DOM

DOM-sensitized photolysis is an important mechanism for attenuation of organic contaminants in natural waters [10,11]. Increasing halide concentrations up to seawater levels decreased the DOM-sensitized photolysis rate of the female sex hormone, 17β-estradiol, by 90% [72]. About four fifths of the rate decrease was due to a general ionic strength effect, with the remainder to halide-specific effects, especially for bromide. The halide-specific effect was attributed to halide enhancement of DOM photobleaching, which reduced the concentration of chromophoric groups acting as sensitizers [72]. There have been a few other studies on the effects of halides on DOM-sensitized photodegradation, but they have either neglected to include ionic strength controls, or have attributed the observed effects to unrelated causes (see [72]).

Halide ions have been shown to influence the yield and lifetime of $^3DOM^*$, parameters that can be measured by a sorbic acid isomerization probe method [73]. Two studies independently report substantial increases in the steady-state $^3DOM^*$ concentration in photolyzed water as the halide concentration increases to seawater levels [74,75]. One study [74] attributed it to a general ionic strength enhancement of $^3DOM^*$ lifetime by slowing intra-organic matter electron transfer, which is known to be an important decay pathway for $^3DOM^*$. In the other [75], it was proposed that halides quench $^3DOM^*$, but at the same time increase the rate of singlet-to-triplet intersystem crossing ($^1DOM^* \rightarrow {}^3DOM^*$). Exactly how halide affects $^3DOM^*$ lifetime and intersystem crossing rates remain to be resolved.

Photobleaching has important implications for the depth of the photic zone, the processing of DOM itself, and the ability of DOM to photosensitize transformations of other chemical species. The fundamental mechanisms accounting for photobleaching of DOM are not well understood, but halide ions may have an effect on rate. Using either a terrestrial DOM reference standard or an algal exudate representing seawater DOM, Mitch and co-workers [76] found that seawater levels of Cl^- and Br^- enhanced DOM photobleaching rates, independent of ionic strength. About 12% of the rate increase was attributed to the formation of RHS (from reaction with $OH^\bullet$) that target electron-rich chromophores more selectively than does $OH^\bullet$. The rest was unresolved. Studies on environmental grab samples are mixed; some report no consistent effect, others rate enhancement, and still others rate suppression with increasing salinity (see [76,77]).

4.2. Reactions of RHS with Organic Compounds

It is useful to summarize what is known generally about the reactions of RHS with organic compounds. The reactivities of rRHS ($X^\bullet$, $X_2^{-\bullet}$) in aqueous solution have not received a great deal of attention, and rate constants are far more plentiful for $X_2^{-•}$ than $X^•$. rRHS are commonly generated by pulse radiolysis or flash photolysis, and rate constants are calculated from the decay or growth of the UV/visible signal. As mentioned above, the major nrRHS in aqueous solution are the hypohalites, HOX and OX$^-$. A sizable literature on these reactions exists due to their importance in chlorine disinfection chemistry [63,78]. To stay relevant to natural waters we will focus here mainly on initial reactions in dilute solutions.

4.2.1. Radical RHS

Three major pathways for reactions of rRHS with organic compounds have been identified: H-atom abstraction from C-H groups Reaction (33); one electron removal from heteroatoms (Z = N, O, or S; Reaction (34); and addition to unsaturated bonds Reaction (35). Rate constants range from 10^4 to 10^9 M^{-1} s^{-1} [79,80] (http://kinetics.nist.gov/solution/):

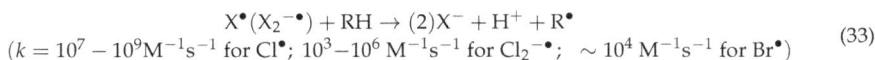

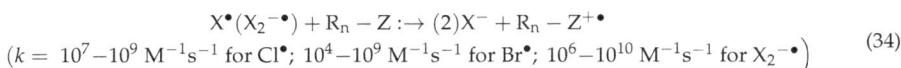

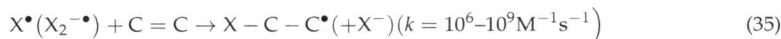

$$X^•(X_2^{-•}) + RH \rightarrow (2)X^- + H^+ + R^•$$
$$(k = 10^7 - 10^9 M^{-1}s^{-1} \text{ for } Cl^•; \; 10^3 - 10^6 \text{ M}^{-1}s^{-1} \text{ for } Cl_2^{-•}; \; \sim 10^4 \text{ M}^{-1}s^{-1} \text{ for } Br^•) \quad (33)$$

$$X^•(X_2^{-•}) + R_n - Z : \rightarrow (2)X^- + R_n - Z^{+•}$$
$$(k = 10^7 - 10^9 \text{ M}^{-1}s^{-1} \text{ for } Cl^•; \; 10^4 - 10^9 \text{ M}^{-1}s^{-1} \text{ for } Br^•; \; 10^6 - 10^{10} \text{ M}^{-1}s^{-1} \text{ for } X_2^{-•}) \quad (34)$$

$$X^•(X_2^{-•}) + C = C \rightarrow X - C - C^•(+X^-)(k = 10^6 - 10^9 M^{-1}s^{-1}) \quad (35)$$

As expected from their reduction potentials (Section 2.3), reactivity of rRHS generally follows the order: $Cl^• > Br^•$; $Cl_2^{-•} > Br_2^{-•}$; and $X^• >> X_2^{-•}$. For many organic compounds, the rate constants for reaction with $Br^•$ and $Cl^•$ rival those with $OH^•$. While rate constants for $X^•$ may exceed $X_2^{-•}$ by several orders of magnitude, the steady-state concentrations of the latter can exceed those of the former by several orders. Thus, both $X^•$ and $X_2^{-•}$ must be considered. The reactivity of $BrCl^{-•}$ with organic compounds is essentially unknown. The reduction potential of $BrCl^{-•}$ ($E^°_{BrCl^-/2Cl^-} = 1.85V$) lies in between that of $Cl_2^{-•}$ (2.30 V) and $Br_2^{-•}$ (1.66 V) [81], suggesting intermediate reactivity.

H-abstraction [65] seems to occur only for aliphatic C-H groups and the rate constant increases with decreasing C-H bond dissociation energy [79,80]). Molecules containing amino, hydroxyl, ether, keto, and sulfide groups preferentially undergo one-electron oxidation, as in Reaction (34).

rRHS add to the double bond alkenes reversibly Reaction (35). Rate constants for $Cl_2^{-•}$ and $Br_2^{-•}$ increase with electron-donating ability of the alkene substituents [80]. The resulting β-substituted organoradical can react with oxygen ($10^8 - 10^9$ M^{-1} s^{-1}) to give β-halo organoperoxyl radicals (X-C-C-OO$^•$) that decompose through various pathways to give such products as halohydrins, haloketones or haloaldehydes, ketones/aldehydes, epoxides, and diols.

Reactions of $Br^•$, $Cl_2^{-•}$, and $Br_2^{-•}$ with simple aromatic compounds depend on substituents [80]. In general, if the substituent bears an electropositive atom with an electron pair, (e.g., -OH, OR, -NH$_2$), reaction proceeds by electron transfer as in Reaction (34); whereas, if the substituent is H, alkyl, -Cl, -NO$_2$, etc., the radical can add to the aromatic ring.

Mitch and co-workers [1] kinetically modeled the transformation of phenol in artificial saline media employing 180 different elementary reactions with known rate constants. They assumed that the process was initiated by photoproduction of $OH^•$, and used short times to minimize involvement of nrRHS. In solutions containing just 540 mM Cl$^-$ and 0.02 mM Br$^-$, the contributions to phenol transformation were 74.4% by $OH^•$, 21.9% by $BrCl^{-•}$, 3.3% by $Cl_2^{-•}$, and 0.4% by $Br_2^{-•}$. In simulated seawater that included 2.3 mM carbonates, they were 52.6% by $CO_3^{-•}$, 6% by $OH^•$, 21.5% by $ClBr^{-•}$, 0.1% by $Cl_2^{-•}$, and 19.7% by $Br_2^{-•}$. The lessons to be learned from this with respect to phenol transformation are that, (a) carbonates divert oxidative power away from $OH^•$ toward $CO_3^{-•}$ and $X_2^{-•}$; (b) $X^•$ seems to play no significant role; and (c) $BrCl^{-•}$, is an important rRHS.

4.2.2. Non-Radical RHS

Hypohalites react with organic compounds by electrophilic substitution, electrophilic addition, or oxidation. Known apparent second-order rate constants for HOCl reactions with organic compounds range widely from 10^{-2} to 10^7 M^{-1} s^{-1} [63]. The most reactive functional groups are amino, keto/aldehyde, phenolic, and low-valent sulfur.

The neutral form of amines reacts rapidly with HOCl (primary > secondary >> tertiary) to form chloramines Reaction (36). α-Amino acid groups undergo further decarboxylation and deamination reactions [63]:

$$R - NH_2 + HOCl \rightarrow R - NHCl + H_2O \tag{36}$$

α-Amino acid groups undergo further decarboxylation and deamination reactions [63]. Aromatic compounds react with HOX by electrophilic substitution of halogen. HOBr and HOI are more reactive than HOCl [42,82]. Ring substituents increase the rate constant in the approximate order, R– < RO– < HO– < $(HO-)_n > 1$. Phenols give *o*- and *p*-X substituted products. The phenoxide ion is ~10^5-times more reactive than the free phenol, and reactivity correlates with electron donor character of the substituents. *Ortho*- and *para*-substituted dihydroxyaromatics undergo oxidation to the corresponding quinone [83].

Above pH ~5 ketones and aldehydes are halogenated by electrophilic substitution at the α carbon Reaction (37), an important reaction in disinfection chemistry because it leads to hazardous byproducts:

$$R - C(=O) - CH_3 \overset{OH^-}{\rightleftharpoons} R - C(-O^-) = CH_2 \overset{HOX}{\rightarrow} R - C(=O) - CH_2Cl \tag{37}$$

Alkenes are attacked electrophilically by the halogen atom of HOX at the least-substituted end of the double bond to form the halohydrin Reaction (38).

$$C = C + HOX \rightarrow X - C - C - OH \tag{38}$$

The halohydrin can undergo internal or solvolytic elimination of halide to form, respectively, the epoxide or the α,β-dihydroxy compound.

Hypohalites can also oxidize some functional groups—for example, primary and secondary alcohols to aldehydes and ketones, respectively; aldehydes to carboxylic acids; and sulfides to sulfones, sulfoxides, or sulfonic acids [78]. Some of those reactions may go through halogenated intermediates via electrophilic pathways.

4.3. Photo-Initiated Incorporation of Halogen into Organic Compounds under Natural Conditions

Given that RHS are photochemically produced in natural waters, the question arises as to whether this could lead to incorporation of halogen into natural compounds and water contaminants. Of the nearly five thousand natural organohalogen compounds identified to date [84], only a few have been assigned an abiotic origin. It is important to understand abiotic halogenation pathways because organohalogen compounds are known to play important roles in climate warming, ozone depletion, and toxicity. In addition, abiotic halogenation may play a role in pollutant fate.

4.3.1. Incorporation of Halogen into Simple, Defined Molecules

A number of volatile gases of importance in stratospheric ozone chemistry and climate warming are thought to originate from abiotic reactions in the oceans. Table 2 lists examples of these compounds and their origins. Moreover, sunlight illumination of natural and artificial saline samples has been observed to halogenate specific organic probe compounds. Table 2 also lists these compounds, which include natural compounds, lignin-like model compounds representing NOM, and pollutants. The mechanism of halogenation is unambiguously established in few of these cases.

A noteworthy example of abiotic halogenation of a natural compound with potentially important ramifications for animal and human exposure was recently reported by Kumar et al. [27].

They found that ozonation of seawater samples in the dark stimulated polyhalogenation (X = Br, Cl) of 1,1-dimethyl-2,2′-bipyrrole and 1′-methyl-1,2′-bipyrrole. The polyhalogenated derivatives of these two bipyrroles are widely distributed among oceanic sealife and found in air samples and human breastmilk, but a satisfactory biological explanation for their existence has been elusive. Under laboratory ozonation conditions, bromination predominated over chlorination and there was no detectable iodine incorporation. Kumar et al. [27] proposed that tropospheric ozone exchanging with seawater generates HOCl and HOBr (see Reaction (3)), which halogenates the bipyrroles. The presence of chlorine RHS is unexpected because, at seawater halide levels, bromide reacts ~130 times faster than chloride with O_3 (see Section 2.2). However, since rather high ozone concentrations were employed (0.2–2.2 mmol/L), it is possible that bromide became depleted in solution. Alternative explanations for the chlorinated products are conversion of some bromine RHS to chlorine RHS, or nucleophilic displacement of bromide by chloride on intermediate products. Interestingly, in kinetic modeling of hydroxyl radical-generating systems in water containing 540 mM NaCl and 0.02 mM NaBr, 3.3% of phenol degradation was due to reaction with $Cl_2^{-\bullet}$, which could only come from oxidation of chloride [1]. The $Cl_2^{-\bullet}$ may originate from chloride displacement of hydroxide from the reversibly-formed species, $ClOH^{-\bullet}$:

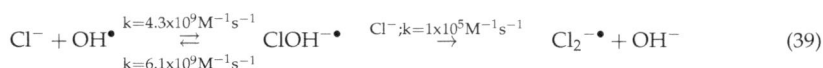

$$Cl^- + OH^\bullet \underset{k=6.1\times10^9 M^{-1}s^{-1}}{\overset{k=4.3\times10^9 M^{-1}s^{-1}}{\rightleftarrows}} ClOH^{-\bullet} \overset{Cl^-;k=1\times10^5 M^{-1}s^{-1}}{\rightarrow} Cl_2^{-\bullet} + OH^- \qquad (39)$$

Or it could come from chloride displacement of bromide in $BrCl^{-\bullet}$ Reaction (10; $k = 1.1 \times 10^2$ M^{-1} s^{-1}).

Table 2. Some examples of halogenation reactions of specific organic compounds in illuminated salty water systems.

Compound	Proposed Origin	References
CH_3Cl	(a) nucleophilic displacement by chloride on CH_3I and/or CH_3Br in seawater; (b) is produced on irradiation of lignin-like DOM model compounds (4-methoxy-1-naphthol; syringic acid; 2-methoxyphenol; 3,4,5-trimethoxy benzoic acid; and 2-methoxyhydroquinone) in chloride solution	(a) [85] (b) [86]
CH_3I	formed after simulated solar irradiation of filtered seawater; production was enhanced when samples were degassed or iodide was added; proposed origin is recombination of $CH_3^\bullet$ and $I^\bullet$ radicals.	[87]
CH_2I_2, CHI_3, and CHI_2Cl	formed by reactions of DOM with HOI generated via oxidation of I^- by O_3	[88]
CH_2ICl	photolysis product of CH_2I_2 in seawater	[89]
$Cl-CH_2CH(OH)CH_2OH$ and $Br-CH_2CH(OH)CH_2OH$	$CH_2=CHCH_2OH$ reaction with reactive halogen species	[90]
3-Cl and 3,3-diCl bisphenol A	solar irradiation of bisphenol A in coastal seawater and saline solution containing 0.13–0.66 mM Fe(III) and fulvic acid; $Cl_2^{-\bullet}$ was detected by its absorption spectrum, and $OH^\bullet$ as its DMPO adduct by EPR spectroscopy; proposed source of halogen radicals: $Fe^{III}Cl^- \rightarrow Fe^{II} + Cl^\bullet$ or $Fe^{III}OH^- \rightarrow Fe^{II} + OH^\bullet$, followed by $OH^\bullet + Cl^- \rightarrow Cl^\bullet$.	[91]
5-bromo-and 3,5-dibromosalicylic acids	solar irradiation of salicylic acid in artificial seawater and brackish lagoon water	[92]
mixed poly-brominated/ chlorinated bipyrroles	irradiation of 1,1-dimethyl-2,2′-bipyrrole and 1′-methyl-1,2′bipyrrole in ozonated seawater; proposed oxidation of Br^- and I^- by O_3 to form HOX/X_2.	[27]
halogenated dicarboxylic acids	isolated from arctic aerosols; unclear whether transformations occurred in the liquid phase	[93]

Mitch and co-workers [1] investigated halogen incorporation into phenol both in artificial seawater and wastewater concentrate (141 mM NaCl, 0.05 mM NaBr, and 11.5 mM carbonates at pH 7.0) spiked with H_2O_2 and irradiated with UV for 35 min. Phenol can be considered a model compound for terrestrial DOM and some pollutants. Both chloro- and bromophenols were produced, with bromophenols constituting the majority of products. However, the total yields based on initial phenol were only 0.52% in seawater and 0.03% in wastewater concentrate. The yields were unaffected by eliminating the carbonate component, despite carbonate's ability to scavenge rRHS Reaction (19). While not established by these results, it is more likely that phenol was halogenated by nrRHS, given the greater reactivity of nrRHS than rRHS toward phenolic compounds (Section 4.2).

4.3.2. Incorporation of Halogen into Bulk DOM

Recent studies [94,95] show convincingly that bulk natural organic matter is photo-halogenated under natural or simulated natural conditions. In the study by Mitch, Dodd, and co-workers [94], organo-Br and organo-I were quantified by solid-phase extraction and silver-form cation exchange filtration to remove the high background of halide ions, followed by non-specific quantification of Br and I by inductively-coupled plasma mass spectrometry (ICP-MS) (the method was insensitive for Cl). In the study by Hao et al. [95], the organohalogen compounds were identified at the formula level by ultra-high resolution electrospray ionization Fourier transform ion cyclotron resonance mass spectrometry (ESI-FT ICR MS).

Native organobromine and organoiodine compounds were found in a variety of seawater samples at concentrations ranging $(3.2\text{–}6.4) \times 10^{-4}$ mol Br/mol C and $(1.1\text{–}3.8) \times 10^{-4}$ mol I/mol C (or 19–160 nmol Br L^{-1} and 6–36 nmol I L^{-1}) [94,95], and diminishing with ocean depth [94]. Simulated and natural solar irradiation of terrestrial NOM spiked to artificial and natural seawaters led to halogenation that increased with light fluence [94]. With added NaI, iodination increased at the expense of bromination [94]. Addition of the probe, 3,5-dimethyl-1*H*-pyrazole (DMP), to irradiated natural seawater samples generated 4-Br and 4-I DMP. Since rRHS oxidize rather than halogenate DMP, this result verifies production of nrRHS in these systems [94] and points to nrRHS as the most likely source of halogenated DOM.

Control experiments indicated that some of the native and photo-generated organobromine and organoiodine compounds are photolabile [94,95]. This indicates that the prevailing levels of organohalogen found in environmental samples likely reflect a balance between formation and decomposition. Experiments in artificial seawater showed that chloride ion stimulates organobromine production [94,95]. This implies that chloride facilitates oxidation of bromide. If ^{3}DOM* is the active oxidant species, one may postulate a mechanism involving the formation and subsequent decay of a ternary exciplex, as previously discussed (Section 2.1):

$$^3\text{DOM}^* + \text{Br}^- + \text{Cl}^- \rightarrow {}^3\left[\text{DOM}^- - - - \text{Br}^\bullet - - - \text{Cl}^-\right] \rightarrow {}^3\text{DOM}^{-\bullet} + \text{BrCl}^{-\bullet} \tag{40}$$

The ESI-FT ICR-MS study provided a wealth of information on the types of reactions that occur [95]. Most native and photo-produced organohalogen compounds were mono- or di-Br or I molecules (a few contained Cl) of 250–700 Daltons in size, and there was considerable overlap among the natively-present and photo-produced compounds. Some products could be attributed to simple H-for-X substitution or X-addition reactions, but most were the result of multiple processes, often accompanied by photooxidation. Most brominated compounds fell in regions of the van Krevelin diagram indicating derivation from unsaturated aliphatic compounds and saturated fatty acids and carbohydrates, while smaller numbers were derived from polycyclic aromatic and polyphenol moieties. Most iodo compounds appeared to be derived mainly from lignin- or tannic-like structures.

In summary, the results suggest that sunlight-driven reactions of RHS with DOM play an important role in bromine and iodine geochemical cycling in marine environments. It has been estimated [94] that photochemical halogenation of terrestrial DOM in estuaries could generate 30 Gg of organobromine and 70 Gg of organoiodine annually worldwide. Those values do not even include RHS-driven halogenation of marine DOM in the open ocean.

5. Impacts of Halides on Water Treatment Processes

Photo-driven AOPs using oxygen, ozone, and peroxides as bulk oxidants are frequently used to destroy pollutants in drinking water and wastewaters. Semiconductor materials are often used as photocatalysts. While earlier work employed UV light, recent emphasis has been on reactions that are viable in the visible or solar spectral regions to reduce energy costs. Wastewaters such as landfill leachates, production waters, industrial wastewaters, and reverse osmosis brines intended for reuse or safe disposal often contain high levels of halide ions. Application of AOPs for treating salty waters is challenging due, among other things, to the conversion of ROS to RHS, which can affect the efficiency of organic compound degradation and generate unwanted halogenated byproducts. It should be noted that solutions irradiated with UV wavelengths below the absorption edges of halide ions (~260 nm) may generate rRHS from direct photolysis of halides Reaction (41):

$$X^- \overset{UV}{\rightleftarrows} (X^\bullet, e^-) \overset{H^+ (H_2O)}{\rightarrow} X^\bullet + H^\bullet (+OH^-) \tag{41}$$

This reaction proceeds through a reversible halogen atom-electron solvent cage complex that produces free halogen and hydrogen atoms upon reaction with water [96,97]. The hydrogen atoms are normally rapidly scavenged by O_2 to produce $HO_2^\bullet/O_2^{-\bullet}$. One study [97] reports that Reaction (41) can be driven in the bay of a diode array spectrophotometer (!), and cautions about the potential for analytical interference.

5.1. Hydroxyl Radical-Based AOPs

Numerous AOPs generate $OH^\bullet$, including H_2O_2/UV, Fenton reactions, and heterogeneous photocatalysis (e.g., TiO_2), among others. There are several reports of decreased rates and organohalogen formation when $OH^\bullet$-based AOPs were conducted in the presence of elevated halide concentrations. A few examples are given. One study involving a Fenton-based AOP to destroy dyes indicated that dye removal decreased while total halogenated organic compounds (AOX) increased when 57 mM Cl^- was present at pH 2.8 and 1 [98]. Another observed auto-inhibition of 1,2-dibromoethane oxidation in a (dark) Fenton-based AOP due to the generation of bromide ions during the reaction that scavenged $OH^\bullet$ [99]. Machulek observed that chloride inhibited mineralization of organic compounds by the photo-Fenton reaction, both in a synthetic phenol wastewater and an extract of gasoline [100]. In Fenton reactions at pH 2.8, the impact of chloride scavenging on organic compound transformation rate was noticeable above 0.01 M Cl^- [21].

Kinetic modeling of phenol oxidation by an $OH^\bullet$-generating reaction (H_2O_2/UV) in phosphate-buffered water containing 0.8 mM NaBr showed that $OH^\bullet$ accounted for most (74%) of phenol transformation, $Br_2^{-\bullet}$ for 24%, and $Br^\bullet$ for 0.8% [1]. In a synthetic wastewater (141 mM chloride, 0.05 mM bromide, 11.5 mM carbonates) at pH 7, $OH^\bullet$ was still the most important radical (67%), followed by $CO_3^{-\bullet}$ (31%), $BrCl^{-\bullet}$ (2.1%), and $Br_2^{-\bullet}$ (0.3%).

It has been proposed that halide ions can be oxidized on the surfaces of semiconductor photocatalysts such as TiO_2, either by surface-associated hydroxyl radicals or valence band holes [101,102]. It is known that chloride forms an inner-sphere complex with Ti at the surface [103].

5.2. The UV/Chlorine AOP

The UV photolysis of HOCl has been proposed as an alternative AOP. The photolysis of $HOCl/OCl^-$ at 254 nm yields $OH^\bullet$ and $Cl^\bullet$ (Section 3.1; Scheme 3). Some $OH^\bullet$ and $Cl^\bullet$ are scavenged by $HOCl/OCl^-$, however, to give $ClO^\bullet$, which is a less-reactive radical towards organics [65,104]:

$$OH^\bullet/Cl^\bullet + HOCl/OCl^- \rightarrow ClO^\bullet \tag{42}$$

In bromide-containing waters, $HOCl/OCl^-$ rapidly converts to $HOBr/OBr^-$ [105]. Photolysis of $HOBr/OBr^-$ generates $OH^\bullet$ and $Br^\bullet$ (Scheme 3) [106]. HOBr is also known to oxidize HOI to

IO_3^-. [107,108] When HOCl is in excess, the oxidation of I^- to IO_3^- is catalyzed by Br^-. Obviously, the use of UV/chlorine AOP has the potential to generate high levels of halogenated byproducts.

5.3. Sulfate Radical-Based AOPs

Sulfate radical ($SO_4^{-\bullet}$)-based AOPs are attractive alternatives to hydroxyl radical-based AOPs. The sulfate radical is nearly as reactive toward organic compounds as hydroxyl. UV/peroxydisulfate ($S_2O_8^{2-}$) has a higher quantum yield of $SO_4^{-\bullet}$ from $S_2O_8^{2-}$ (1.4) at 254 nm [109] than does $OH^\bullet$ from H_2O_2 (1.0) [110]. The photolysis of peroxymonosulfate at 254 nm generates $OH^\bullet$ and $SO_4^{-\bullet}$ simultaneously [111]. Sulfate radical can also be generated from peroxysulfates by various non-photolytic means, as well. Sulfate radicals convert to hydroxyl radicals in water above pH 7.

Sulfate radical reacts directly and rapidly with the halide ions:

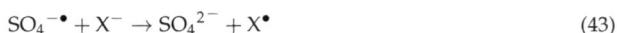

$$SO_4^{-\bullet} + X^- \rightarrow SO_4^{2-} + X^\bullet \tag{43}$$

While oxidation of Cl^- by $OH^\bullet$ is important only in acidic solution, oxidation of Cl^- by $SO_4^{-\bullet}$ Reaction (43) is pH-independent and therefore impacts water treatment over a much broader pH range. Several studies have shown that halide ions can strongly affect $SO_4^{-\bullet}$-based oxidation rates [64] and lead to halogenated byproducts [112].

Experiment and kinetic modeling show that $SO_4^{-\bullet}$-based processes are more strongly affected by halides than are $OH^\bullet$-based processes [64]. In the presence of halides and carbonates, the steady-state concentrations of $X_2^{-\bullet}$ and $CO_3^{-\bullet}$ are much higher than those of $SO_4^{-\bullet}$ and $OH^\bullet$ [80,113]. This, combined with the fact that $X_2^{-\bullet}$ and $CO_3^{-\bullet}$ are typically less reactive and more selective toward organic compounds than $SO_4^{-\bullet}$ and $OH^\bullet$, means that oxidation efficiency can be significantly impacted [80,113]. However, the impact depends on molecular structure. Benzoic acid transformation by UV/$S_2O_8^{2-}$ was strongly suppressed in 0.54 M chloride solution compared to phosphate-buffered water, while cyclohex-3-ene carboxylic acid was hardly affected at all [64]. This is because the major rRHS that formed, $Cl_2^{-\bullet}$, is poorly reactive toward benzoic acid, but highly reactive toward the double bond in cyclohex-3-ene carboxylic acid [64]. A similar reason was offered to explain the effects of halides and carbonates on the UV/$S_2O_8^{2-}$ reactivity of different pharmaceuticals in reverse-osmosis brine compared to water—namely, that $X_2^{-\bullet}$ and $CO_3^{-\bullet}$ were more reactive toward some than others [114].

Sulfate radical AOPs can yield bromate (BrO_3^-) as a final product under some conditions (Scheme 5) [115,116]. Bromate is a suspected human carcinogen with a drinking water standard of 10 μg/L as set by U.S. EPA and the World Health Organization [117]. Both experiment and modeling indicate that HOBr/OBr^- is a required intermediate in the production of bromate (Scheme 5) [115,116]. The yield of bromate is pH-dependent, as HOBr is about 2 orders of magnitude less reactive than OBr^- toward $Br^\bullet$ [116,118]. Organic solutes, DOM, and generated superoxide can scavenge $Br^\bullet$, $Br_2^{-\bullet}$, and HOBr. This has the effect of significantly reducing or eliminating bromate formation, as well as recycling bromine back into the bromide form [116].

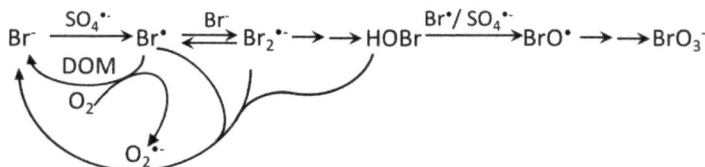

Scheme 5. The mechanism of BrO_3^- formation by $SO_4^{-\bullet}$.

Halides can also react directly with peroxymonosulfate. The bimolecular rate constant follows the order $I^- > Br^- > Cl^-$ [119]. The evidence is consistent with nucleophilic attack of halide on the peroxy group. The product HOX (except HOCl) is further oxidized by peroxymonosulfate [120]. Oxidation of HOI to IO_2^- is strongly pH dependent due to speciation effects [121] (Scheme 6.). The reaction of IO_2^- to IO_3^- is very fast. The oxidation of HOBr to BrO_3^- is much slower [120]:

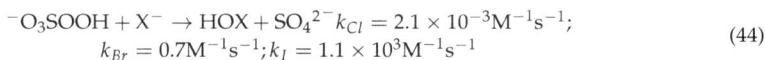

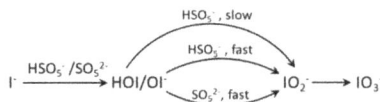

$$^-O_3SOOH + X^- \rightarrow HOX + SO_4{}^{2-} \quad k_{Cl} = 2.1 \times 10^{-3}M^{-1}s^{-1};$$
$$k_{Br} = 0.7M^{-1}s^{-1}; k_I = 1.1 \times 10^3M^{-1}s^{-1} \tag{44}$$

Scheme 6. The mechanism of IO_3^- formation by peroxymonosulfate oxidation of iodide.

In the presence of iodide, peroxymonosulfate reactions can also lead to incorporation of iodine into DOM and form byproducts of concern derived from DOM, namely, iodoform (CHI_3) and iodoacetic acid [121].

6. Concluding Remarks

Halogen plays an important and colorful role in environmental photochemical processes in natural waters and in chemical reactions taking place during photochemical water purification. In the environment, halides can be oxidized to rRHS ($X^\bullet$, $X_2^{-\bullet}$) principally through DOM-sensitized photolysis and reactions with ROS of photochemical origin, especially hydroxyl radical, but also ozone and nitrate radical in atmospheric aerosols. Much more work needs to be done to establish the mechanism and importance of DOM-sensitized photolysis of halide ions with respect to generation of RHS. The nature of the chromophoric groups and the quantum yields of initial RHS products as a function of DOM type need to be established. It is noteworthy that chloride enhances bromination and further work is needed to establish whether the cause is formation of a ternary exciplex like the one in Scheme 2. DOM photosensitization of RHS formation is diminished with photobleaching of DOM, a process that itself is affected by halide ions.

rRHS dimerize or disproportionate to give nrRHS, chiefly the hypohalites (HOX) interconverting with smaller amounts of molecular halogen species, X_2 and X_3^-. nrRHS can photolyze to regenerate halogen atoms and produce hydroxyl radical, ozone, or hydrogen peroxide, depending on pH and wavelength. Hypohalites can react with hydrogen peroxide, nitrite, and ozone to give singlet oxygen, nitrate, and oxyhalide anions, respectively—products that oxidize organic compounds less efficiently than the hypohalites. Rate constants are lacking for many speciation reactions and reactions of RHS with other photo-generated species, especially in the case of I. Usually the most important inorganic scavenger of rRHS will be the carbonates.

Halide ions at relatively high concentrations can apparently increase the steady-state concentration of excited triplet-state dissolved organic matter (^{3}DOM*). Exactly how this happens is not entirely clear and deserves further research; one study attributed it to a general ionic strength effect, while another to an increase in the rate of singlet-to-triplet intersystem crossing. Some studies report that high halide concentration accelerates photobleaching of DOM. The mechanism has not been established with confidence. An increase in the steady-state ^{3}DOM* concentration could lead to a loss of chromophoric groups through intra-DOM reactions, or an increase in the generation of RHS that can oxidize DOM.

Reactions of $X^\bullet$, $X_2^{-\bullet}$, and HOX with organic compounds in water have been characterized, although rate constants are sparse for iodine compounds and for $X^\bullet$ relative to $X_2^{-\bullet}$. Rate constants for the mixed species, $BrCl^{-\bullet}$, are unavailable. Depending on structure, rRHS can react by H-atom abstraction, one-electron oxidation, and addition to double bonds and aromatic rings. The addition reactions can lead to incorporation of halogen into the products. Hypohalites react principally by non-radical electrophilic reactions, including halogen incorporation into amines, ketones, alkenes, and aromatic rings. Incorporation of halogen seems to be more likely with nrRHS than rRHS. Hypohalites can also oxidize alcohols, aldehydes, and sulfides without halogen incorporation. Limited interconversion of halogen within and between rRHS and nrRHS is possible, and can lead to incorporation of all halogens into organic molecules, regardless of which RHS species are generated initially. Some naturally-occurring halogenated compounds are known to form abiotically, including ozone-depleting gases. Many of these are thought to have a photolytic origin. Evidence has appeared for abiotic incorporation of halogen into water contaminants and model compounds representing natural organic matter initiated by photochemical processes. Evidence has also appeared for the photochemical incorporation of halogen into natural compounds creating products toxic to oceanic sealife. More examples of natural abiotic incorporation of halogen atoms into natural compounds and water contaminants are likely to appear in the future.

Recent studies show that DOM from both oceanic and terrestrial sources is halogenated under simulated or natural conditions of irradiation. The mechanisms of halogen incorporation have not been identified precisely. Likewise, the scope of such reactions and the effects of water chemistry are as yet poorly characterized. Complicating matters is the finding that photo generation and decomposition of halogenated DOM seem to be taking place simultaneously. The results so far indicate that sunlight-driven oxidation by RHS and halogenation reactions may play important roles in halogen geochemical cycling in marine and estuarine environments, especially in regard to bromine and iodine. It is possible that the presence of natural halogenated compounds has contributed to the evolution of enzymatic dehalogenation pathways of halogenated molecules.

Oxidation and halogen incorporation are of demonstrated importance in AOPs for salty water treatment that use light. Rates can be markedly slowed (or sometimes accelerated) and undesirable byproducts can be formed. There is much to be learned about the influence of halide salts. It is noteworthy that halides can be photolyzed by UV below 260 nm. Whether halides are oxidized on photocatalyst surfaces is largely an open question. The rate constants of RHS with many organic compounds are unknown, but important for evaluating the contribution of RHS to organic compound degradation. The yields of halogenated byproducts depend strongly on the parent compound and the solution conditions. An insufficient database exists to predict precisely where rRHS will attack in a complex molecule. The halogenated byproducts may be of greater toxicity than the original contaminants. Although many studies reported the appearance of halogenated products, quantitative yields are often not reported [101,104,122]. For these reasons the impact of halides on toxicity of the treated waters should routinely accompany investigations, and such information should be used to judge the suitability of the AOP.

Supplementary Materials: The following are available online. Table S1: Rate constants for relevant reactions of halides and reactive halogen species.

Acknowledgments: The authors thank the Chinese International Postdoctoral Exchange Fellowship Program (No. 20160074) for support for Y.Y.

Author Contributions: Both Y.Y. and J.J.P. contributed to writing the manuscript.

Conflicts of Interest: The authors declare no conflict of interest. The founding sponsors had no role in the writing of the manuscript, nor in the decision to publish it.

References

1. Grebel, J.E.; Pignatello, J.J.; Mitch, W.A. Effect of halide ions and carbonates on organic contaminant degradation by hydroxyl radical-based advanced oxidation processes in saline waters. *Environ. Sci. Technol.* **2010**, *44*, 6822–6828. [CrossRef] [PubMed]
2. Luther, G.W., III; Swartz, C.B.; Ullman, W.J. Direct determination of iodide in seawater by cathodic stripping square wave voltammetry. *Anal. Chem.* **1988**, *60*, 1721–1724. [CrossRef]

3. Plewa, M.J.; Muellner, M.G.; Richardson, S.D.; Fasano, F.; Buettner, K.M.; Woo, Y.-T.; McKague, A.B.; Wagner, E.D. Occurrence, synthesis, and mammalian cell cytotoxicity and genotoxicity of haloacetamides: An emerging class of nitrogenous drinking water disinfection byproducts. *Environ. Sci. Technol.* **2007**, *42*, 955–961. [CrossRef]

4. Richardson, S.D.; Plewa, M.J.; Wagner, E.D.; Schoeny, R.; DeMarini, D.M. Occurrence, genotoxicity, and carcinogenicity of regulated and emerging disinfection by-products in drinking water: A review and roadmap for research. *Mutat. Res. Rev. Mutat. Res.* **2007**, *636*, 178–242. [CrossRef] [PubMed]

5. Simpson, W.R.; von Glasow, R.; Riedel, K.; Anderson, P.; Ariya, P.; Bottenheim, J.; Burrows, J.; Carpenter, L.J.; Frieß, U.; Goodsite, M.E.; et al. Halogens and their role in polar boundary-layer ozone depletion. *Atmos. Chem. Phys.* **2007**, *7*, 4375–4418. [CrossRef]

6. Pratt, K.A.; Custard, K.D.; Shepson, P.B.; Douglas, T.A.; Pohler, D.; General, S.; Zielcke, J.; Simpson, W.R.; Platt, U.; Tanner, D.J.; et al. Photochemical production of molecular bromine in Arctic surface snowpacks. *Nat. Geosci.* **2013**, *6*, 351–356. [CrossRef]

7. Jammoul, A.; Dumas, S.; D'Anna, B.; George, C. Photoinduced oxidation of sea salt halides by aromatic ketones: A source of halogenated radicals. *Atmos. Chem. Phys.* **2009**, *9*, 4229–4237. [CrossRef]

8. Parker, K.M.; Mitch, W.A. Halogen radicals contribute to photooxidation in coastal and estuarine waters. *Proc. Natl. Acad. Sci. USA* **2016**, *113*, 5868–5873. [CrossRef] [PubMed]

9. Zepp, R.G.; Schlotzhauer, P.F.; Sink, R.M. Photosensitized transformations involving electronic energy transfer in natural waters: Role of humic substances. *Environ. Sci. Technol.* **1985**, *19*, 74–81. [CrossRef]

10. McNeill, K.; Canonica, S. Triplet state dissolved organic matter in aquatic photochemistry: Reaction mechanisms, substrate scope, and photophysical properties. *Environ. Sci. Process Impacts* **2016**, *18*, 1381–1399. [CrossRef] [PubMed]

11. Canonica, S. Oxidation of aquatic organic contaminants induced by excited triplet states. *Chim. Int. J. Chem.* **2007**, *61*, 641–644. [CrossRef]

12. Loeff, I.; Rabani, J.; Treinin, A.; Linschitz, H. Charge transfer and reactivity of $n\pi^*$ and $\pi\pi^*$ organic triplets, including anthraquinonesulfonates, in interactions with inorganic anions: A comparative study based on classical Marcus theory. *J. Am. Chem. Soc.* **1993**, *115*, 8933–8942. [CrossRef]

13. Isse, A.A.; Lin, C.Y.; Coote, M.L.; Gennaro, A. Estimation of standard reduction potentials of halogen atoms and alkyl halides. *J. Phys. Chem. B* **2011**, *115*, 678–684. [CrossRef] [PubMed]

14. De Laurentiis, E.; Minella, M.; Maurino, V.; Minero, C.; Mailhot, G.; Sarakha, M.; Brigante, M.; Vione, D. Assessing the occurrence of the dibromide radical (Br_2^-) in natural waters: Measures of triplet-sensitised formation, reactivity, and modelling. *Sci. Total Environ.* **2012**, *439*, 299–306. [CrossRef] [PubMed]

15. Hurley, J.K.; Linschitz, H.; Treinin, A. Interaction of halide and pseudohalide ions with triplet benzophenone-4-carboxylate: Kinetics and radical yields. *J. Phys. Chem.* **1988**, *92*, 5151–5159. [CrossRef]

16. Wardman, P. Reduction potentials of one-electron couples involving free radicals in aqueous solution. *J. Phys. Chem. Ref. Data* **1989**, *18*, 1637–1755. [CrossRef]

17. Mopper, K.; Zhou, X. Hydroxyl radical photoproduction in the sea and its potential impact on marine processes. *Science* **1990**, *250*, 661–664. [CrossRef] [PubMed]

18. Vaughan, P.P.; Blough, N.V. Photochemical formation of hydroxyl radical by constituents of natural waters. *Environ. Sci. Technol.* **1998**, *32*, 2947–2953. [CrossRef]

19. Sun, L.; Qian, J.; Blough, N.V.; Mopper, K. Insights into the photoproduction sites of hydroxyl radicals by dissolved organic matter in natural waters. *Environ. Sci. Technol. Lett.* **2015**, *2*, 352–356. [CrossRef]

20. Jayson, G.G.; Parsons, B.J.; Swallow, A.J. Some Simple, Highly reactive, inorganic chlorine derivatives in aqueous solution. *J. Chem. Soc., Faraday Trans. I* **1973**, *69*, 1597–1607. [CrossRef]

21. Pignatello, J. Dark and photoassisted Fe^{3+}-catalyzed degradation of chlorophenoxy herbicides by hydrogen peroxide. *Environ. Sci. Technol.* **1992**, *26*, 944–951. [CrossRef]

22. Ryerson, T.; Trainer, M.; Angevine, W.; Brock, C.; Dissly, R.; Fehsenfeld, F.; Frost, G.; Goldan, P.; Holloway, J.; Hübler, G. Effect of petrochemical industrial emissions of reactive alkenes and NO_x on tropospheric ozone formation in Houston, Texas. *J. Geophys. Res. Atmos.* **2003**, *108*, 4249. [CrossRef]

23. Zhang, R.; Lei, W.; Tie, X.; Hess, P. Industrial emissions cause extreme urban ozone diurnal variability. *Proc. Natl. Acad. Sci. USA* **2004**, *101*, 6346–6350. [CrossRef] [PubMed]

24. Zheng, J.; Shao, M.; Che, W.; Zhang, L.; Zhong, L.; Zhang, Y.; Streets, D. Speciated VOC emission inventory and spatial patterns of ozone formation potential in the Pearl River Delta, China. *Environ. Sci. Technol.* **2009**, *43*, 8580–8586. [CrossRef] [PubMed]

25. Liu, Q.; Schurter, L.M.; Muller, C.E.; Aloisio, S.; Francisco, J.S.; Margerum, D.W. Kinetics and mechanisms of aqueous ozone reactions with bromide, sulfite, hydrogen sulfite, iodide, and nitrite ions. *Inorg. Chem.* **2001**, *40*, 4436–4442. [CrossRef] [PubMed]

26. Haag, W.R.; Hoigne, J. Ozonation of bromide-containing waters: Kinetics of formation of hypobromous acid and bromate. *Environ. Sci. Technol.* **1983**, *17*, 261–267. [CrossRef]

27. Kumar, A.; Borgen, M.; Aluwihare, L.I.; Fenical, W. Ozone-activated halogenation of mono- and dimethylbipyrrole in seawater. *Environ. Sci. Technol.* **2017**, *51*, 589–595. [CrossRef] [PubMed]

28. Enami, S.; Vecitis, C.D.; Cheng, J.; Hoffmann, M.R.; Colussi, A.J. Global inorganic source of atmospheric bromine. *J. Phys. Chem. A* **2007**, *111*, 8749–8752. [CrossRef] [PubMed]

29. Haag, W.R.; Gassman, E. Singlet oxygen in surface waters-Part I: Furfuryl alcohol as a trapping agent. *Chemosphere* **1984**, *13*, 631–640. [CrossRef]

30. Finlayson-Pitts, B.J. The tropospheric chemistry of sea salt: A molecular-level view of the chemistry of NaCl and NaBr. *Chem. Rev.* **2003**, *103*, 4801–4822. [CrossRef] [PubMed]

31. Rossi, M.J. Heterogeneous reactions on salts. *Chem. Rev.* **2003**, *103*, 4823–4882. [CrossRef] [PubMed]

32. Mishra, H.; Enami, S.; Nielsen, R.J.; Stewart, L.A.; Hoffmann, M.R.; Goddard, W.A.; Colussi, A.J. Brønsted basicity of the air–water interface. *Proc. Natl. Acad. Sci. USA* **2012**, *109*, 18679–18683. [CrossRef] [PubMed]

33. Enami, S.; Hoffmann, M.R.; Colussi, A.J. Halogen radical chemistry at aqueous interfaces. *J. Phys. Chem. A* **2016**, *120*, 6242–6248. [CrossRef] [PubMed]

34. Wayne, R.P.; Barnes, I.; Biggs, P.; Burrows, J.P.; Canosa-Mas, C.E.; Hjorth, J.; Le Bras, G.; Moortgat, G.K.; Perner, D.; Poulet, G.; et al. The nitrate radical: Physics, chemistry, and the atmosphere. *Atmos. Environ. Part A* **1991**, *25*, 1–203. [CrossRef]

35. Poskrebyshev, G.A.; Huie, R.E.; Neta, P. The rate and equilibrium constants for the reaction $NO_3^\bullet + Cl^- \rightleftharpoons NO_3^- + Cl^\bullet$ in aqueous solutions. *J. Phys. Chem. A* **2003**, *107*, 1964–1970. [CrossRef]

36. Neta, P.; Huie, R.E. Rate constants for reactions of NO_3 radicals in aqueous solutions. *J. Phys. Chem.* **1986**, *90*, 4644–4648. [CrossRef]

37. Schweitzer, F.; Mirabel, P.; George, C. Multiphase chemistry of N_2O_5, $ClNO_2$, and $BrNO_2$. *J. Phys. Chem. A* **1998**, *102*, 3942–3952. [CrossRef]

38. Roberts, J.M.; Osthoff, H.D.; Brown, S.S.; Ravishankara, A.R. N_2O_5 Oxidizes chloride to Cl_2 in acidic atmospheric aerosol. *Science* **2008**, *321*, 1059. [CrossRef] [PubMed]

39. Hu, J.H.; Shi, Q.; Davidovits, P.; Worsnop, D.R.; Zahniser, M.S.; Kolb, C.E. Reactive Uptake of $Cl_2(g)$ and $Br_2(g)$ by aqueous surfaces as a function of Br^- and I^- ion concentration: The effect of chemical reaction at the interface. *J. Phys. Chem.* **1995**, *99*, 8768–8776. [CrossRef]

40. Wang, T.X.; Kelley, M.D.; Cooper, J.N.; Beckwith, R.C.; Margerum, D.W. Equilibrium, kinetic, and UV-spectral characteristics of aqueous bromine chloride, bromine, and chlorine species. *Inorg. Chem.* **1994**, *33*, 5872–5878. [CrossRef]

41. Ianni, J.C. Kintecus V6.01. Available online: www.kintecus.com (accessed on 30 July 2016).

42. Bichsel, Y.; von Gunten, U. Formation of iodo-trihalomethanes during disinfection and oxidation of iodide containing waters. *Environ. Sci. Technol.* **2000**, *34*, 2784–2791. [CrossRef]

43. Troy, R.C.; Margerum, D.W. Non-metal redox kinetics: Hypobromite and hypobromous acid reactions with iodide and with sulfite and the hydrolysis of bromosulfate. *Inorg. Chem.* **1991**, *30*, 3538–3543. [CrossRef]

44. Bichsel, Y.; von Gunten, U. Hypoiodous acid: Kinetics of the buffer-catalyzed disproportionation. *Water Res.* **2000**, *34*, 3197–3203. [CrossRef]

45. Barkley, R.A.; Thompson, T.G. The total Iodine and Iodate-iodine content of sea-water. *Deep Sea Res.* **1960**, *7*, 24–34. [CrossRef]

46. Chen, Z.; Megharaj, M.; Naidu, R. Speciation of iodate and iodide in seawater by non-suppressed ion chromatography with inductively coupled plasma mass spectrometry. *Talanta* **2007**, *72*, 1842–1846. [CrossRef] [PubMed]

47. Kurylo, M.J.; Ouellette, P.A.; Laufer, A.H. Measurements of the pressure dependence of the hydroperoxy (HO_2) radical self-disproportionation reaction at 298 K. *J. Phys. Chem.* **1986**, *90*, 437–440. [CrossRef]

48. Barnes, R.J.; Lock, M.; Coleman, J.; Sinha, A. Observation of a new absorption band of HOBr and its atmospheric implications. *J. Phys. Chem.* **1996**, *100*, 453–457. [CrossRef]

49. Haugen, H.K.; Weitz, E.; Leone, S.R. Accurate quantum yields by laser gain vs absorption spectroscopy: Investigation of Br/Br* channels in photofragmentation of Br_2 and IBr. *J. Phys. Chem.* **1985**, *83*, 3402–3412. [CrossRef]

50. Gershgoren, E.; Banin, U.; Ruhman, S. Caging and geminate recombination following photolysis of triiodide in solution. *J. Phys. Chem. A* **1998**, *102*, 9–16. [CrossRef]

51. Callow, A.; Griffith, R.; McKeown, A. The photo-reaction between bromine and hydrogen peroxide in aqueous solution. *Trans. Faraday Soc.* **1939**, *35*, 412–420. [CrossRef]

52. Treinin, A.; Hayon, E. Charge transfer spectra of halogen atoms in water. Correlation of the electronic transition energies of iodine, bromine, chlorine, hydroxyl, and hydrogen radicals with their electron affinities. *J. Am. Chem. Soc.* **1975**, *97*, 1716–1721. [CrossRef]

53. Forsyth, J.E.; Zhou, P.; Mao, Q.; Asato, S.S.; Meschke, J.S.; Dodd, M.C. Enhanced inactivation of bacillus subtilis spores during solar photolysis of free available chlorine. *Environ. Sci. Technol.* **2013**, *47*, 12976–12984. [CrossRef] [PubMed]

54. Jenkin, M.; Cox, R.; Hayman, G. Kinetics of the reaction of IO radicals with HO_2 radicals at 298 K. *Chem. Phys. Lett.* **1991**, *177*, 272–278. [CrossRef]

55. Francisco, J.S.; Hand, M.R.; Williams, I.H. Ab initio study of the electronic spectrum of HOBr. *J. Phys. Chem.* **1996**, *100*, 9250–9253. [CrossRef]

56. Minaev, B.F. The singlet-triplet absorption and photodissociation of the HOCl, HOBr, and HOI molecules calculated by the MCSCF quadratic response method. *J. Phys. Chem. A* **1999**, *103*, 7294–7309. [CrossRef]

57. Biedenkapp, D.; Hartshorn, L.G.; Bair, E.J. The O (^{1}D)+ H_2O reaction. *Chem. Phys. Lett.* **1970**, *5*, 379–381. [CrossRef]

58. Poskrebyshev, G.A.; Neta, P.; Huie, R.E. Temperature dependence of the acid dissociation constant of the hydroxyl radical. *J. Phys. Chem. A* **2002**, *106*, 11488–11491. [CrossRef]

59. Buxton, G.; Subhani, M. Radiation chemistry and photochemistry of oxychlorine ions. Part 2.— Photodecomposition of aqueous solutions of hypochlorite ions. *J. Chem. Soc. Faraday Trans. 1* **1972**, *68*, 958–969. [CrossRef]

60. Orlando, J.J.; Burkholder, J.B. Gas-phase UV/Visible absorption spectra of HOBr and Br_2O. *J. Phys. Chem.* **1995**, *99*, 1143–1150. [CrossRef]

61. Rowley, D.M.; Mössinger, J.C.; Cox, R.A.; Jones, R.L. The UV-visible absorption cross-sections and atmospheric photolysis rate of HOI. *JAtC* **1999**, *34*, 137–151.

62. Palmer, D.A.; Van Eldik, R. Spectral characterization and kinetics of formation of hypoiodous acid in aqueous solution. *Inorg. Chem.* **1986**, *25*, 928–931. [CrossRef]

63. Deborde, M.; von Gunten, U. Reactions of chlorine with inorganic and organic compounds during water treatment—Kinetics and mechanisms: A critical review. *Water Res.* **2008**, *42*, 13–51. [CrossRef] [PubMed]

64. Yang, Y.; Pignatello, J.J.; Ma, J.; Mitch, W.A. Comparison of halide impacts on the efficiency of contaminant degradation by sulfate and hydroxyl radical-based advanced oxidation processes (AOPs). *Environ. Sci. Technol.* **2014**, *48*, 2344–2351. [CrossRef] [PubMed]

65. Minakata, D.; Kamath, D.; Maetzold, S. Mechanistic insight into the reactivity of chlorine-derived radicals in the aqueous-phase UV–chlorine advanced oxidation process: Quantum mechanical calculations. *Environ. Sci. Technol.* **2017**, *51*, 6918–6926. [CrossRef] [PubMed]

66. Von Gunten, U.; Oliveras, Y. Kinetics of the reaction between hydrogen peroxide and hypobromous acid: Implication on water treatment and natural systems. *Water Res.* **1997**, *31*, 900–906. [CrossRef]

67. Cerkovnik, J.; Plesničar, B. Recent Advances in the chemistry of hydrogen trioxide (HOOOH). *Chem. Rev.* **2013**, *113*, 7930–7951. [CrossRef] [PubMed]

68. Johnson, D.W.; Margerum, D.W. Non-metal redox kinetics: A reexamination of the mechanism of the reaction between hypochlorite and nitrite ions. *Inorg. Chem.* **1991**, *30*, 4845–4851. [CrossRef]

69. Lahoutifard, N.; Lagrange, P.; Lagrange, J.; Scott, S.L. Kinetics and mechanism of nitrite oxidation by $HOBr/BrO^-$ in atmospheric water and comparison with oxidation by $HOCl/ClO^-$. *J. Phys. Chem. A* **2002**, *106*, 11891–11896. [CrossRef]

70. Von Gunten, U. Ozonation of drinking water: Part II. Disinfection and by-product formation in presence of bromide, iodide or chlorine. *Water Res.* **2003**, *37*, 1469–1487. [CrossRef]

71. Environmental Protection Agency, USA. National drinking water regulations: Disinfectants and disinfection byproducts. *Fed. Regist.* **1998**, *63*, 69390–69476.

72. Grebel, J.E.; Pignatello, J.J.; Mitch, W.A. Impact of halide ions on natural organic matter-sensitized photolysis of 17-β-estradiol in saline waters. *Environ. Sci. Technol.* **2012**, *46*, 7128–7134. [CrossRef] [PubMed]

73. Grebel, J.E.; Pignatello, J.J.; Mitch, W.A. Sorbic acid as a quantitative probe for the formation, scavenging and steady-state concentrations of the triplet-excited state of organic compounds. *Water Res.* **2011**, *45*, 6535–6544. [CrossRef] [PubMed]

74. Parker, K.M.; Pignatello, J.J.; Mitch, W.A. Influence of salinity on triplet-state natural organic matter loss by energy transfer and electron transfer pathways. *Environ. Sci. Technol.* **2013**, *47*, 10987–10994. [CrossRef] [PubMed]

75. Glover, C.M.; Rosario-Ortiz, F.L. Impact of halides on the photoproduction of reactive intermediates from organic matter. *Environ. Sci. Technol.* **2013**, *47*, 13949–13956. [CrossRef] [PubMed]

76. Grebel, J.E.; Pignatello, J.J.; Song, W.; Cooper, W.J.; Mitch, W.A. Impact of halides on the photobleaching of dissolved organic matter. *Mar. Chem.* **2009**, *115*, 134–144. [CrossRef]

77. Song, G.; Li, Y.; Hu, S.; Li, G.; Zhao, R.; Sun, X.; Xie, H. Photobleaching of chromophoric dissolved organic matter (CDOM) in the Yangtze River estuary: Kinetics and effects of temperature, pH, and salinity. *Environ. Sci. Process Impacts* **2017**, *19*, 861–873. [CrossRef] [PubMed]

78. Larson, R.A.; Weber, E.J. *Reaction Mechanisms in Environmental Organic Chemistry*; Lewis Publishers: Boca Ratan, FL, USA, 1994.

79. Wicktor, F.; Donati, A.; Herrmann, H.; Zellner, R. Laser based spectroscopic and kinetic investigations of reactions of the Cl atom with oxygenated hydrocarbons in aqueous solution. *Phys. Chem. Chem. Phys.* **2003**, *5*, 2562–2572. [CrossRef]

80. Hasegawa, K.; Neta, P. Rate constants and mechanisms of reactions of Cl_2^- radicals. *J. Phys. Chem.* **1978**, *82*, 854–857. [CrossRef]

81. Ershov, B.G.; Kelm, M.; Gordeev, A.V.; Janata, E. A pulse radiolysis study of the oxidation of Br- by dichloro radical anion in aqueous solution: Formation and properties of chlorobromo radical anion. *Phys. Chem. Chem. Phys.* **2002**, *4*, 1872–1875. [CrossRef]

82. Lee, Y.; Yoon, J.; von Gunten, U. Kinetics of the oxidation of phenols and phenolic endocrine disruptors during water treatment with ferrate (Fe(VI)). *Environ. Sci. Technol.* **2005**, *39*, 8978–8984. [CrossRef] [PubMed]

83. Criquet, J.; Rodriguez, E.M.; Allard, S.; Wellauer, S.; Salhi, E.; Joll, C.A.; von Gunten, U. Reaction of bromine and chlorine with phenolic compounds and natural organic matter extracts—Electrophilic aromatic substitution and oxidation. *Water Res.* **2015**, *85*, 476–486. [CrossRef] [PubMed]

84. Gribble, G.W. *Naturally Occurring Organohalogen Compopunds–A Comprehensive Update*; Springer: Wien, Austria; New York, NY, USA, 2010.

85. Jeffers, P.M.; Wolfe, N.L. On the degradation of methyl bromide in sea water. *Geophys. Res. Lett.* **1996**, *23*, 1773–1776. [CrossRef]

86. Moore, R.M. A photochemical source of methyl chloride in saline waters. *Environ. Sci. Technol.* **2008**, *42*, 1933–1937. [CrossRef] [PubMed]

87. Moore, R.M.; Zafiriou, O.C. Photochemical production of methyl iodide in seawater. *J. Geophys. Res. Atmos.* **1994**, *99*, 16415–16420. [CrossRef]

88. Martino, M.; Mills, G.P.; Woeltjen, J.; Liss, P.S. A new source of volatile organoiodine compounds in surface seawater. *Geophys. Res. Lett.* **2009**, *36*. [CrossRef]

89. Jones, C.E.; Carpenter, L.J. Solar Photolysis of CH_2I_2, CH_2ICl, and CH_2IBr in water, saltwater, and seawater. *Environ. Sci. Technol.* **2006**, *40*, 1372. [CrossRef]

90. Anastasio, C.; Matthew, B.M. A chemical probe technique for the determination of reactive halogen species in aqueous solution: Part 2—Chloride solutions and mixed bromide/chloride solutions. *Atmos. Chem. Phys.* **2006**, *6*, 2439–2451. [CrossRef]

91. Liu, H.; Zhao, H.; Quan, X.; Zhang, Y.; Chen, S. Formation of chlorinated intermediate from bisphenol a in surface saline water under simulated solar light irradiation. *Environ. Sci. Technol.* **2009**, *43*, 7712–7717. [CrossRef] [PubMed]

92. Tamtam, F.; Chiron, S. New insight into photo-bromination processes in saline surface waters: The case of salicylic acid. *Sci. Total Environ.* **2012**, *435*, 345–350. [CrossRef] [PubMed]

93. Narukawa, M.; Kawamura, K.; Hatsushika, H.; Yamazaki, K.; Li, S.-M.; Bottenheim, J.W.; Anlauf, K.G. Measurement of halogenated dicarboxylic acids in the arctic aerosols at polar sunrise. *J. Atmos. Chem.* **2003**, *44*, 323–335. [CrossRef]

94. Méndez-Díaz, J.D.; Shimabuku, K.K.; Ma, J.; Enumah, Z.O.; Pignatello, J.J.; Mitch, W.A.; Dodd, M.C. Sunlight-driven photochemical halogenation of dissolved organic matter in seawater: A natural abiotic source of organobromine and organoiodine. *Environ. Sci. Technol.* **2014**, *48*, 7418–7427. [CrossRef] [PubMed]

95. Hao, Z.; Yin, Y.; Cao, D.; Liu, J. Probing and comparing the photobromination and photoiodination of dissolved organic matter by using ultra-high-resolution mass spectrometry. *Environ. Sci. Technol.* **2017**, *51*, 5464–5472. [CrossRef] [PubMed]

96. Jortner, J.; Ottolenghi, M.; Stein, G. On the photochemistry of aqueous solutions of chloride, bromide, and iodide ions. *J. Phys. Chem.* **1964**, *68*, 247–255. [CrossRef]

97. Kalmar, J.; Doka, E.; Lente, G.; Fabian, I. Aqueous photochemical reactions of chloride, bromide, and iodide ions in a diode-array spectrophotometer. Autoinhibition in the photolysis of iodide ions. *Dalton Trans.* **2014**, *43*, 4862–4870. [CrossRef] [PubMed]

98. Kiwi, J.; Lopez, A.; Nadtochenko, V. Mechanism and kinetics of the OH-radical intervention during Fenton oxidation in the presence of a significant amount of radical scavenger (Cl$^-$). *Environ. Sci. Technol.* **2000**, *34*, 2162–2168. [CrossRef]

99. Pignatello, J.J.; Oliveros, E.; MacKay, A. Advanced oxidation processes for organic contaminant destruction based on the Fenton reaction and related chemistry. *Crit. Rev. Env. Sci. Technol.* **2006**, *36*, 1–84. [CrossRef]

100. Machulek, A.; Moraes, J.E.F.; Vautier-Giongo, C.; Silverio, C.A.; Friedrich, L.C.; Nascimento, C.A.O.; Gonzalez, M.C.; Quina, F.H. Abatement of the inhibitory effect of chloride anions on the photo-fenton process. *Environ. Sci. Technol.* **2007**, *41*, 8459–8463. [CrossRef] [PubMed]

101. Yuan, R.; Ramjaun, S.N.; Wang, Z.; Liu, J. Photocatalytic degradation and chlorination of azo dye in saline wastewater: Kinetics and AOX formation. *Chem. Eng. J.* **2012**, *192*, 171–178. [CrossRef]

102. Yamazaki, S.; Tanimura, T.; Yoshida, A.; Hori, K. Reaction mechanism of photocatalytic degradation of chlorinated ethylenes on porous TiO$_2$ pellets: Cl radical-initiated mechanism. *J. Phys. Chem. A* **2004**, *108*, 5183–5188. [CrossRef]

103. Kormann, C.; Bahnemann, D.; Hoffmann, M.R. Photolysis of chloroform and other organic molecules in aqueous titanium dioxide suspensions. *Environ. Sci. Technol.* **1991**, *25*, 494–500. [CrossRef]

104. Wu, Z.; Fang, J.; Xiang, Y.; Shang, C.; Li, X.; Meng, F.; Yang, X. Roles of reactive chlorine species in trimethoprim degradation in the UV/chlorine process: Kinetics and transformation pathways. *Water Res.* **2016**, *104*, 272–282. [CrossRef] [PubMed]

105. Heeb, M.B.; Criquet, J.; Zimmermann-Steffens, S.G.; von Gunten, U. Oxidative treatment of bromide-containing waters: Formation of bromine and its reactions with inorganic and organic compounds—A critical review. *Water Res.* **2014**, *48*, 15–42. [CrossRef] [PubMed]

106. Benter, T.; Feldmann, C.R.; Kirchner, U.; Schmidt, M.; Schmidt, S.; Schindler, R. UV/VIS-absorption Spectra of HOBr and CH$_3$OBr; Br(^{2}P$_{3/2}$) atom yields in the photolysis of HOBr. *Ber. Bunsenges. Phys. Chem.* **1995**, *99*, 1144–1147. [CrossRef]

107. Criquet, J.; Allard, S.; Salhi, E.; Joll, C.A.; Heitz, A.; von Gunten, U. Iodate and iodo-trihalomethane formation during chlorination of iodide-containing waters: Role of bromide. *Environ. Sci. Technol.* **2012**, *46*, 7350–7357. [CrossRef] [PubMed]

108. Allard, S.; Tan, J.; Joll, C.A.; von Gunten, U. Mechanistic study on the formation of Cl$^-$/Br$^-$/I$^-$ trihalomethanes during chlorination/chloramination combined with a theoretical cytotoxicity evaluation. *Environ. Sci. Technol.* **2015**, *49*, 11105–11114. [CrossRef] [PubMed]

109. Mark, G.; Schuchmann, M.N.; Schuchmann, H.-P.; von Sonntag, C. The photolysis of potassium peroxodisulphate in aqueous solution in the presence of tert-butanol: A simple actinometer for 254 nm radiation. *J. Photochem. Photobiol. A Chem.* **1990**, *55*, 157–168. [CrossRef]

110. Baxendale, J.; Wilson, J. The photolysis of hydrogen peroxide at high light intensities. *Trans. Faraday Soc.* **1957**, *53*, 344–356. [CrossRef]

111. Guan, Y.-H.; Ma, J.; Li, X.-C.; Fang, J.-Y.; Chen, L.-W. Influence of pH on the formation of sulfate and hydroxyl radicals in the UV/peroxymonosulfate system. *Environ. Sci. Technol.* **2011**, *45*, 9308–9314. [CrossRef] [PubMed]

112. Beitz, T.; Bechmann, W.; Mitzner, R. Investigations of reactions of selected azaarenes with radicals in water. 2. Chlorine and bromine radicals. *J. Phys. Chem. A* **1998**, *102*, 6766–6771. [CrossRef]

113. Canonica, S.; Kohn, T.; Mac, M.; Real, F.J.; Wirz, J.; von Gunten, U. Photosensitizer method to determine rate constants for the reaction of carbonate radical with organic compounds. *Environ. Sci. Technol.* **2005**, *39*, 9182–9188. [CrossRef] [PubMed]

114. Yang, Y.; Pignatello, J.J.; Ma, J.; Mitch, W.A. Effect of matrix components on UV/H_2O_2 and $UV/S_2O_8{}^{2-}$ advanced oxidation processes for trace organic degradation in reverse osmosis brines from municipal wastewater reuse facilities. *Water Res.* **2016**, *89*, 192–200. [CrossRef] [PubMed]

115. Fang, J.Y.; Shang, C. Bromate formation from bromide oxidation by the UV/Persulfate process. *Environ. Sci. Technol.* **2012**, *46*, 8976–8983. [CrossRef] [PubMed]

116. Lutze, H.V.; Bakkour, R.; Kerlin, N.; von Sonntag, C.; Schmidt, T.C. Formation of bromate in sulfate radical based oxidation: Mechanistic aspects and suppression by dissolved organic matter. *Water Res.* **2014**, *53*, 370–377. [CrossRef] [PubMed]

117. Fawell, J.; Walker, M. Approaches to determining regulatory values for carcinogens with particular reference to bromate. *Toxicology* **2006**, *221*, 149–153. [CrossRef] [PubMed]

118. Kläning, U.K.; Wolff, T. Laser flash photolysis of HCIO, CIO^-, HBrO, and BrO^- in aqueous solution. reactions of Cl-and Br-atoms. *Ber. Bunsenges. Phys. Chem.* **1985**, *89*, 243–245. [CrossRef]

119. Fortnum, D.H.; Battaglia, C.J.; Cohen, S.R.; Edwards, J.O. The kinetics of the oxidation of halide ions by monosubstituted peroxides. *J. Am. Chem. Soc.* **1960**, *82*, 778–782. [CrossRef]

120. Lente, G.; Kalmár, J.; Baranyai, Z.; Kun, A.; Kék, I.; Bajusz, D.; Takács, M.; Veres, L.; Fábián, I. One-versus two-electron oxidation with peroxomonosulfate ion: Reactions with iron(II), vanadium(IV), halide ions, and photoreaction with cerium(III). *Inorg. Chem.* **2009**, *48*, 1763–1773. [CrossRef] [PubMed]

121. Li, J.; Jiang, J.; Zhou, Y.; Pang, S.-Y.; Gao, Y.; Jiang, C.; Ma, J.; Jin, Y.; Yang, Y.; Liu, G.; et al. Kinetics of oxidation of iodide (I^-) and hypoiodous acid (HOI) by peroxymonosulfate (PMS) and formation of iodinated products in the PMS/I^-/NOM system. *Environ. Sci. Technol. Lett.* **2017**, *4*, 76–82. [CrossRef]

122. Pan, Y.; Cheng, S.; Yang, X.; Ren, J.; Fang, J.; Shang, C.; Song, W.; Lian, L.; Zhang, X. UV/chlorine treatment of carbamazepine: Transformation products and their formation kinetics. *Water Res.* **2017**, *116*, 254–265. [CrossRef] [PubMed]

Sample Availability: Not available.

![molecules logo] *molecules*

MDPI

Article

A Model Study of the Photochemical Fate of As(III) in Paddy-Water

Luca Carena [1] and Davide Vione [1,2,*]

[1]   Dipartimento di Chimica, Università di Torino, Via Pietro Giuria 5, 10125 Torino, Italy; luca.carena@unito.it
[2]   Centro Interdipartimentale NatRisk, Università di Torino, Largo Paolo Braccini 2,
      10095 Grugliasco (TO), Italy
*   Correspondence: davide.vione@unito.it; Tel.: +39-011-670-5296

Academic Editor: Pierre Pichat
Received: 27 January 2017; Accepted: 6 March 2017; Published: 11 March 2017

Abstract: The APEX (Aqueous Photochemistry of Environmentally-occurring Xenobiotics) software previously developed by one of us was used to model the photochemistry of As(III) in paddy-field water, allowing a comparison with biotic processes. The model included key paddy-water variables, such as the shielding effect of the rice canopy on incident sunlight and its monthly variations, water pH, and the photochemical parameters of the chromophoric dissolved organic matter (CDOM) occurring in paddy fields. The half-life times ($t_{1/2}$) of As(III) photooxidation to As(V) would be ~20–30 days in May. In contrast, the photochemical oxidation of As(III) would be much slower in June and July due to rice-canopy shading of radiation because of plant growth, despite higher sunlight irradiance. At pH < 8 the photooxidation of As(III) would mainly be accounted for by reaction with transient species produced by irradiated CDOM (here represented by the excited triplet states $^3CDOM^*$, neglecting the possibly more important reactions with poorly known species such as the phenoxy radicals) and, to a lesser extent, with the hydroxyl radicals ($HO^\bullet$). However, the carbonate radicals ($CO_3^{\bullet-}$) could be key photooxidants at pH > 8.5 provided that the paddy-water $^3CDOM^*$ is sufficiently reactive toward the oxidation of CO_3^{2-}. In particular, if paddy-water $^3CDOM^*$ oxidizes the carbonate anion with a second-order reaction rate constant near (or higher than) 10^6 $M^{-1} \cdot s^{-1}$, the photooxidation of As(III) could be quite fast at pH > 8.5. Such pH conditions can be produced by elevated photosynthetic activity that consumes dissolved CO_2.

Keywords: arsenic contamination; paddy-field floodwater; sunlight-induced reactions

1. Introduction

Arsenic (As) contamination of paddy fields is an important pollution problem in several regions around the world, such as the Bengal Basin area [1]. Contamination by As is usually caused by irrigation of paddy fields with groundwater containing elevated As levels [1,2]. In a paddy field, As can be found both in the soil and in the flooding water, and its concentration range can be quite wide [3]. Moreover, several types of rice plants are able to uptake soil As in the roots and to accumulate it, leading to human exposure through food [3–5]. There are many processes governing As chemistry in paddy fields, which depend on the physicochemical and biological features of paddies and may affect As speciation. The main As redox states are As(V) and As(III), usually occurring as inorganic compounds (arsenate and arsenite, respectively) and, to a minor extent, as organic species, such as the methylated As forms [6]. Under aerobic soil conditions that occur before the flooding of rice crops, As is sorbed as As(V) because the oxyanion arsenate has high affinity for soil minerals and, particularly, for Fe (hydr)oxides. Anaerobic regimes develop into the paddy soil after flooding, and Fe(III) (hydr)oxides and As(V) undergo fast microbial reduction. Arsenic is, thus, released as As(III) into the soil pore water, because the more mobile (and toxic) As(III) has lesser affinity than As(V) towards soil minerals [7].

The results reported by Takahashi et al. [7] suggest that the half-life time ($t_{1/2}$) of As(V) is ~30 days. The produced aqueous As(III) could subsequently undergo oxidation to As(V) in the upper soil layers, where oxygen occurs in higher amounts. In particular, biotic oxidation takes place into the oxygenated bulk soil and in the rhizosphere of the rice plant roots, where Fe (hydr)oxides form an iron plaque and As is re-sorbed as As(V) [2,8]. The biotic oxidation of As(III) occurs mainly as a detoxification process, but As(III) can also be used as an electron/energy source during the growth of microorganisms. Biotic detoxification of As(III) to As(V) may be fast, and apparent $t_{1/2}$ values have been found to be <1 day depending on experimental conditions (although the fastest processes might not always be representative of actual paddies) [9]. As(V) can, in turn, undergo biotic reduction to As(III) in both aerobic and anaerobic environments, by dissimilatory arsenate-reduction processes [10], which may be fast under anaerobic conditions [11]. As(V) reduction is also fast in aerobic environments, and the $t_{1/2}$ of As(V) has been found to be <3 days [9,12]. Therefore, microbial oxidation/reduction processes are key phenomena of As speciation in paddy soil and in surface water environments [13]. They could also be important in the paddies floodwater, where light is an important factor as well.

Photochemistry plays a key role in the transformation of inorganic and organic substances in surface water. The photochemical reactions occurring in sunlit natural waters can generally be divided into direct and indirect phototransformation pathways. The direct photolysis of a substrate takes place after absorption of sunlight, if the absorption spectrum of the substance overlaps with the sunlight spectrum. In contrast, indirect photochemistry is triggered by compounds called photosensitizers, which occur naturally in surface waters and are able to absorb sunlight. The main photosensitizers are the chromophoric moieties of dissolved organic matter (chromophoric dissolved organic matter or CDOM), nitrate and nitrite. The irradiated photosensitizers produce high-energy transient species, such as hydroxyl radicals (HO$^{\bullet}$), carbonate radicals (CO$_3^{\bullet-}$), singlet oxygen (^{1}O$_2$), and the excited triplet states of CDOM (^{3}CDOM*). These transient species react with dissolved substrates and cause their indirect phototransformation [14]. As(III) is well known to react with both HO$^{\bullet}$ and ^{3}CDOM* [15–18], whereas its reactivity with carbonate radicals has been studied only under alkaline conditions [15]. As(III) reactivity toward ^{1}O$_2$ was found to be negligible in solutions containing humic substances under irradiation [18]. Direct As(III) photolysis could occur significantly at λ = 254 nm [19], but it is not an important photochemical pathway under sunlight [16,17].

It should be pointed out that the oxidation of As(III) in the presence of irradiated CDOM likely involves radical species (presumably phenoxy radicals) to a higher extent than ^{3}CDOM* [18]. Unfortunately, little to nothing is known about the formation and reactivity of these radicals, which prevents photochemical modelling. In the present study, the photochemical fate of As(III) in rice-field water was modelled using the available literature values of the second-order rate constants of the reactions between As(III) and several transients ($^{\bullet}$OH, ^{3}CDOM*, CO$_3^{\bullet-}$), as well as the photochemical parameters of paddy floodwater. This approach provides a lower limit for the photochemical oxidation kinetics of As(III), by neglecting the reactions involving the phenoxy radicals.

2. Results and Discussion

The half-life time ($t_{1/2}$) of As(III), in the form of H$_3$AsO$_3$, in sunlit rice field water was modelled for the months of May, June, and July. This period follows the flooding of paddies in late April–early May. The details of the photochemical model are reported in the Methods section. Figure 1a,b show the half-life times of H$_3$AsO$_3$ in May and June as a function of nitrate concentration and of the dissolved organic carbon (DOC) content of paddy water. The light transmittance through the rice plant canopy was taken into account in the model, considering that the canopy shading effect increases as the rice plants grow higher. Consequently, during the rice growing season there can be considerable variations in the sunlight irradiance that reaches the paddy-water surface [20]. Sunlight transmittance through the rice canopy has been measured in the La Albufera coastal lagoon (Spain, latitude ~39°N) during the months of May, June, and July, 1989 [21]. The reported monthly values of the average transmittance were ~53%, ~8% and ~6%, respectively. These values were used to model As(III) photochemistry in the present work.

It can be seen from Figure 1 that $t_{1/2}$ undergoes important variations with nitrate concentration and DOC. In particular, at low DOC values (< 1 mg·C·L^{-1}) the increase of nitrate causes a significant $t_{1/2}$ decrement, because in such conditions HO$^{\bullet}$ that is partially photogenerated by nitrate plays a major role in As(III) photochemical oxidation (see also Figure 2a, which reports the fraction of H$_3$AsO$_3$ oxidized by HO$^{\bullet}$ as a function of DOC and nitrate). Low-DOC conditions are scarcely representative of the rice field floodwater environment, however, because a significant amount of dissolved organic matter (DOM, not necessarily chromophoric) is released by soil/sediment and by the rice plants. When the DOC is higher (>6 mg·C·L^{-1}, more representative of paddy floodwater), nitrate has a limited influence on the rate of As(III) photooxidation. Under these circumstances CDOM is the main chromophore, it limits the sunlight absorption by nitrate and it is also the main photochemical HO$^{\bullet}$ source. Moreover, DOM (note that high-DOC waters are DOM-rich by definition [22]) is a major HO$^{\bullet}$ scavenger and causes the steady-state [HO$^{\bullet}$] to be low. In contrast, [3CDOM*] is higher at high DOC [14] and, in these circumstances, the triplet-sensitised processes can play an important role in As(III) photooxidation (see Figure 2b). The maximum values of $t_{1/2}$ occur at DOC = 1–2 mg·C·L^{-1} (Figure 1). In these conditions, there is already enough DOM to scavenge HO$^{\bullet}$ significantly, but not yet enough CDOM to produce elevated [3CDOM*].

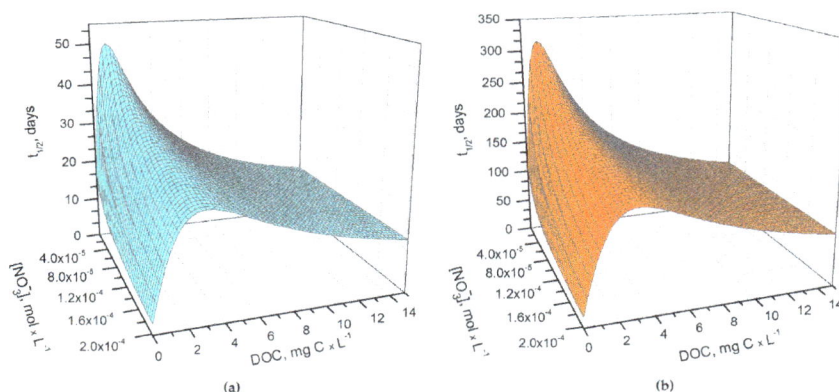

Figure 1. (**a**) Photochemical half-life time ($t_{1/2}$) of H$_3$AsO$_3$ in rice field water in May. Water conditions: 5 cm depth; 1 µM NO$_2^-$, 1 mM bicarbonate, 10 µM carbonate, and 0.53 sunlight transmittance; (**b**) Photochemical half-life time of H$_3$AsO$_3$ in rice field water in June. Water conditions: 5 cm depth; 1 µM NO$_2^-$, 1 mM bicarbonate, 10 µM carbonate and 0.08 sunlight transmittance. The day units of $t_{1/2}$ are referred to average mid-latitude irradiance conditions occurring in mid-May and mid-June, respectively.

It is interesting to note that the $t_{1/2}$ values are lower in May than in June, despite the significantly higher sunlight irradiance in June [14]. The reason is that the rice plants are higher in June, and their canopy produces a considerable decrease of the irradiance over the water surface. The shading by the rice canopy deeply affects all the photochemical processes as shown in Figure 3a, which reports the $t_{1/2}$ of As(III) as a function of water DOC and canopy transmittance. Figure 3b shows the overall As(III) photooxidation kinetics in May through July, which is affected both by sunlight irradiance and by shading from the rice canopy. It is clear that photooxidation is faster in May, while limited differences can be observed between June and July. Photochemical processes are predicted to be quite slow in June and July due to canopy shading. In contrast, the value $t_{1/2}$ ~20 days in May is, interestingly, of the same order of magnitude as the As(III) level fluctuations observed in a laboratory experiment that used paddy water and soil [7]. As(III) photooxidation may, thus, potentially play a role in As processing in paddy fields in the month of May, but in definite circumstances it has been found that the As redox processes in paddy soil and in the rice rhizosphere may be considerably faster [8,9,12].

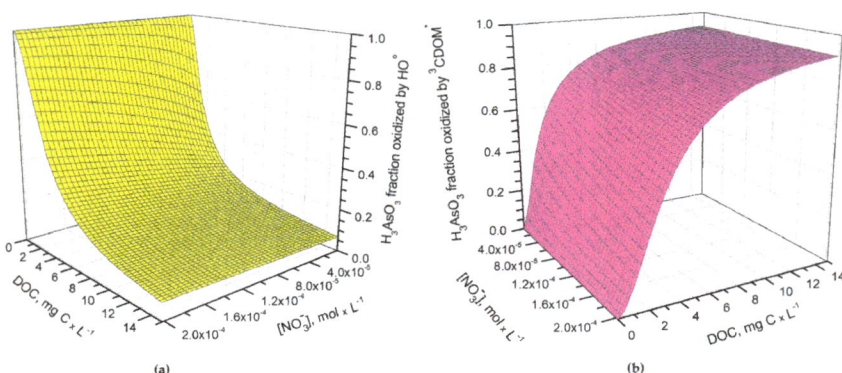

Figure 2. Fraction of H_3AsO_3 transformed by (**a**) hydroxyl radicals and (**b**) CDOM triplet states ($^3CDOM^*$) in May, as a function of nitrate and dissolved organic carbon (DOC). Water conditions: 5 cm depth, 1 μM NO_2^-, 1 mM bicarbonate, and 10 μM carbonate. Note that direct photolysis, carbonate radicals and singlet oxygen reactions were not taken into account in the model.

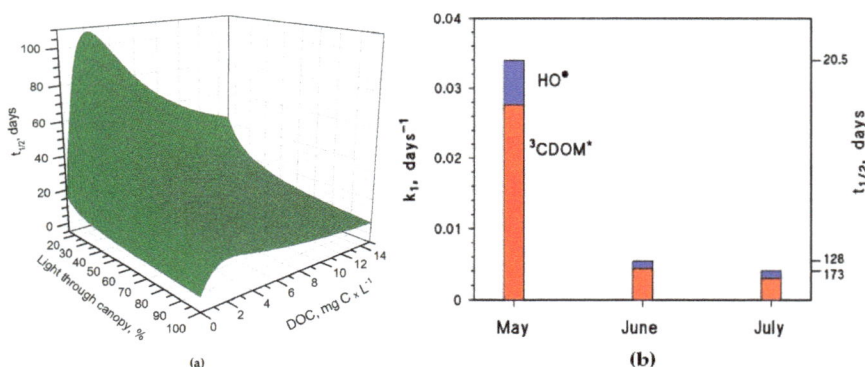

Figure 3. (**a**) Photochemical half-life time of H_3AsO_3 in rice-field water, as a function of the DOC and of light transmittance through the rice canopy. Other conditions: 5 cm depth, 0.1 mM NO_3^-, 1 μM NO_2^-, 1 mM bicarbonate, 10 μM carbonate, and May sunlight. (**b**) Pseudo-first-order rate constant of H_3AsO_3 photooxidation in the period from May to July. The corresponding half-life times ($t_{1/2} = \ln2$ $(k_1)^{-1}$) are also reported in the right Y-axis. Water conditions: 5 cm depth, 7 mg$\cdot$C$\cdot$L^{-1} DOC, 0.1 mM NO_3^-, 1 μM NO_2^-, 1 mM bicarbonate, and 10 μM carbonate. Transmittance values through the rice canopy are reported in the main text. The days are referred to average irradiance conditions occurring at mid latitude in, where relevant, mid-May, mid-June, and mid-July.

Under DOC conditions that are typically representative of paddy water, such as those assumed in the modelling of Figure 3b, $^3CDOM^*$ would be the main photooxidant for As(III) and account for ~80% of its transformation, while the remaining ~20% could be ascribed to HO$^\bullet$. This finding is consistent with the experimental results reported by Buschmann et al. in irradiation experiments of As(III) solutions containing humic substances [18], where the role of HO$^\bullet$ in As(III) phototransformation was a minor one. Note that the relative importance of ^{3}CDOM* and HO$^\bullet$ in As(III) photooxidation does not depend on the light transmittance through the rice canopy, but it is rather a function of the water chemistry. Moreover, it should be recalled here that the reaction between As(III) and ^{3}CDOM* is only a lower limit for the reaction kinetics between As(III) and irradiated CDOM, because the main oxidative process is actually carried out by additional transients (possibly phenoxy radicals) [18].

The above discussion refers to photochemical scenarios in which HO$^\bullet$ and ^{3}CDOM* are assumed to be the main photooxidants of As(III). However, H_3AsO_3 is in equilibrium with its conjugate base $As(OH)_2O^-$ (pK_a = 9.2), for which the photoreactivity has been studied in basic conditions only [15]. To carry out a more complete description of As(III) indirect photochemistry, pH was included into the model (see Methods for the mathematical treatment). By considering pH as a master variable and introducing the photoreactivity of $As(OH)_2O^-$, the role of $CO_3^{\bullet-}$-induced oxidation gains importance as a photochemical process. The radical $CO_3^{\bullet-}$ is produced by bicarbonate and carbonate oxidation (Reactions 1–3), and the corresponding second-order reaction rate constants are $k^{HO^\bullet}_{HCO_3^-}$ = 8.5·10^6 M^{-1} s^{-1}, $k^{HO^\bullet}_{CO_3^{2-}}$ = 3.9·10^8 M^{-1} s^{-1}, as well as $k^{3CDOM^*}_{CO_3^{2-}}$. In particular, we found that the reactivity of ^{3}CDOM* with the carbonate anions (expressed by the parameter $k^{3CDOM^*}_{CO_3^{2-}}$, see Reaction 3) is a key variable of the model. The value of $k^{3CDOM^*}_{CO_3^{2-}}$ has been found to vary widely depending on the proxy molecules (triplet sensitizers) used to study the reactivity between CO_3^{2-} and ^{3}CDOM* [23].

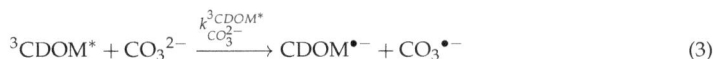

$$HO^\bullet + HCO_3^- \xrightarrow{k^{HO^\bullet}_{HCO_3^-}} H_2O + CO_3^{\bullet-} \tag{1}$$

$$HO^\bullet + CO_3^{2-} \xrightarrow{k^{HO^\bullet}_{CO_3^{2-}}} HO^- + CO_3^{\bullet-} \tag{2}$$

$$^3CDOM^* + CO_3^{2-} \xrightarrow{k^{3CDOM^*}_{CO_3^{2-}}} CDOM^{\bullet-} + CO_3^{\bullet-} \tag{3}$$

Figure 4a shows how the pH and the $k^{3CDOM^*}_{CO_3^{2-}}$ rate constant values affect the half-life time of As(III) in paddy floodwater in May. For pH < 8, $k^{3CDOM^*}_{CO_3^{2-}}$ and pH have a scarce influence on As(III) photochemistry and one has $t_{1/2}$ ~20 days, in analogy with the results reported in Figures 1 and 3. In this scenario, As(III) photooxidation would be mainly accounted for by the reactions between H_3AsO_3 and ^{3}CDOM*/HO$^\bullet$ (see also Figure 4c, which reports the fractions of As(III) photooxidation accounted for by HO$^\bullet$, $CO_3^{\bullet-}$, and ^{3}CDOM*, for different values of pH and of $k^{3CDOM^*}_{CO_3^{2-}}$). Indeed, the molar fraction of $As(OH)_2O^-$ ($\alpha_{As(OH)_2O^-}$, see the Methods section) and the steady-state $[CO_3^{\bullet-}]$ (Figure 4b) are too low at pH < 8 for $CO_3^{\bullet-}$ to be involved significantly in As(III) oxidation.

At pH > 8–8.5 one has different trends depending on the value of $k^{3CDOM^*}_{CO_3^{2-}}$. If $k^{3CDOM^*}_{CO_3^{2-}}$ is below 10^6 M^{-1}·s^{-1}, $t_{1/2}$ has some increase with increasing pH (Figure 4a), because $[CO_3^{\bullet-}]$ is too low in these conditions (see Figure 4b) to significantly affect the photooxidation of As(III), and because $As(OH)_2O^-$ is less reactive than H_3AsO_3 towards HO$^\bullet$ and ^{3}CDOM* [15,18]. In this case, the photooxidation of H_3AsO_3 prevails as the As(III) transformation process (see Figure 4c). In contrast, if $k^{3CDOM^*}_{CO_3^{2-}}$ is around 10^6 M^{-1}·s^{-1} or higher, one has an important decrease of $t_{1/2}$ with increasing pH. The main reason is the reaction between $As(OH)_2O^-$ and $CO_3^{\bullet-}$ (see Figure 4c), which is enhanced by the parallel increase with pH of both $\alpha_{As(OH)_2O^-}$ and $[CO_3^{\bullet-}]$.

(a)

(b)

(c)

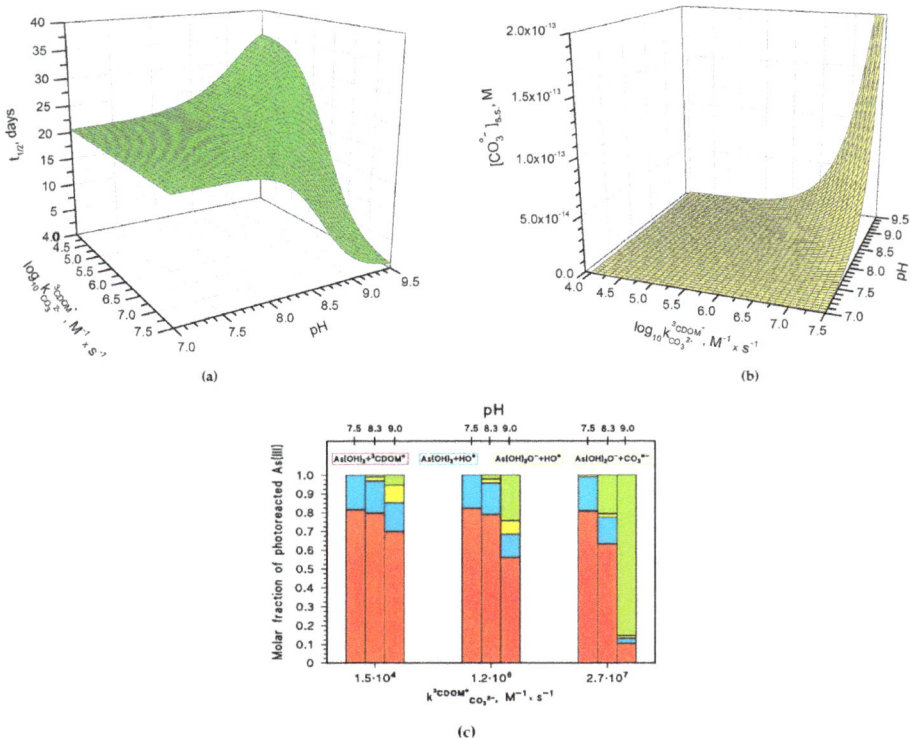

Figure 4. Photochemical parameters of As(III) photooxidation computed for the month of May. Water conditions: 5 cm depth, 0.1 mM NO_3^-, 1 μM NO_2^-, inorganic carbon (IC) = 1.1 mmol·C·L^{-1}, and DOC = 7 mg·C·L^{-1}. (**a**) Half-life time of As(III) and (**b**) steady-state concentrations of the carbonate radicals, as a function of pH and of the decimal logarithm of $k_{CO_3^{2-}}^{3CDOM^*}$. The days are referred to average irradiance conditions occurring at mid-latitude in mid-May. (**c**) Roles of the main reactions accounting for the photooxidation of dissolved As(III). The $k_{CO_3^{2-}}^{3CDOM^*}$ values used in (**c**) are those reported for the three CDOM proxy molecules 3'-methoxyacetophenone (1.5·10^4 M^{-1}·s^{-1}), benzophenone (1.2·10^6 M^{-1}·s^{-1}) and duroquinone (2.7·10^7 M^{-1}·s^{-1}) [23].

The pH increase of $[CO_3^{\bullet -}]$ shown in Figure 4b is mainly due to the reaction between CO_3^{2-} and $^3CDOM^*$, and it is expected to occur only if $k_{CO_3^{2-}}^{3CDOM^*}$ ~10^6 M^{-1}·s^{-1} or higher. The important role of $k_{CO_3^{2-}}^{3CDOM^*}$ is additionally highlighted in Figure 5a, which reports the relative weight of the various $CO_3^{\bullet -}$ generation processes for different values of pH and $k_{CO_3^{2-}}^{3CDOM^*}$. The value of $k_{CO_3^{2-}}^{3CDOM^*}$ has a clear correlation with the triplet-state reduction potential, as shown in Figure 5b. In the presence of triplet states with sufficiently high oxidizing power (e.g., $E°(^3CDOM^*/CDOM^{\bullet -}) > 1.7$ V, causing significant $CO_3^{\bullet -}$ production) and at basic pH, $t_{1/2}$ could reach very low values (below one week) and make photochemistry a very significant factor in the As redox cycling.

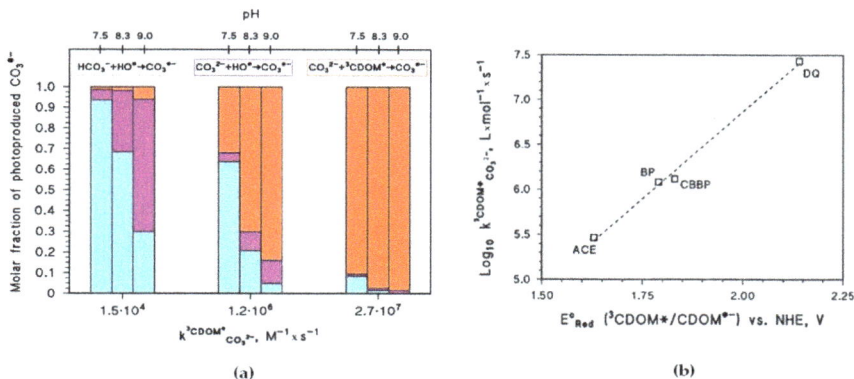

Figure 5. (**a**) Relative contributions of Reactions 1–3 to the production of $CO_3^{\bullet-}$ in the scenario described by Figure 4. (**b**) Correlation plot between the reduction potentials of some excited triplet states and their reactivity toward the carbonate anion, expressed as second-order reaction rate constants: ACE = acetophenone; BP = benzophenone; CBBP = 4-carboxybenzophenone; DQ = duroquinone. The $k_{CO_3^{2-}}^{3CDOM^*}$ values used are those reported in [23].

In the presence of sufficiently reactive $^3CDOM^*$, favourable pH conditions for As(III) photooxidation could be produced by algal photosynthesis that consumes dissolved CO_2 and may considerably increase the water pH. The reaction rate constants between paddy-field $^3CDOM^*$ and CO_3^{2-} are presently unknown and the topic deserves further investigation, but the occurrence of humic substances in paddy water [24] may suggest a non-negligible photoreactivity of the relevant CDOM [25].

3. Methods (Photochemical Modelling)

The assessment of As(III) photooxidation was carried out with the APEX Aqueous Photochemistry of Environmentally-occurring Xenobiotics) software, which is freely available as electronic supplementary information of [26]. APEX allows for the modelling of the photochemistry of surface-water environments (in the form of steady-state concentrations of photogenerated transients) and of the photochemical fate of xenobiotics. APEX computes the xenobiotics pseudo-first-order transformation rate constants (k_1) and half-life times ($t_{1/2}$ = ln2 $(k_1)^{-1}$) [26] as a function of sunlight irradiance, water (photo)chemistry, and depth. For this purpose, APEX has been used successfully to investigate the photochemical behaviour of several xenobiotics and the formation of their phototransformation intermediates [27–30]. APEX requires input data concerning water depth, chemical composition (nitrate, nitrite, dissolved organic carbon, and inorganic carbon species) and photochemistry (absorption spectrum and formation quantum yields of transient species by CDOM). Moreover, second-order reaction rate constants with the photogenerated transients and photolysis quantum yields are key parameters to model the photochemical fate of xenobiotics.

APEX uses as standard sunlight irradiance and standard sunlight spectrum the quantities relative to fair-weather 15 July at mid latitude (45° N), at 9 a.m. or 3 p.m. solar time. It also allows an approximate assessment of photochemistry throughout the year thanks to the *APEX_season* function, which operates corrections in order to compute transient steady-state concentrations and photoreaction kinetics in different months. However, APEX does not automatically take into account the effect of sunlight shielding by the rice plant canopy. The photochemistry of paddy floodwater occurs during the rice-growing season from May to July, and the shielding effect was included in the model by considering the transmittance (*T*) of sunlight through the plant canopy in each month (i.e., literature-available T^{month}, where month is May, June or July). Be k_1^{month} the APEX-modelled photooxidation kinetics (first-order decay constant) of As(III) without taking into account the shielding effect. The correction was carried out by computing $(k_1^{month})\prime = k_1^{month} \times T^{month}$. It must be pointed out that T^{month} was considered to be wavelength-independent, which is justified by the fact that the size of the shielding medium (plant leaves and stems) is much longer than the light wavelengths. Another key input parameter in the model is the water depth, d. Depth, together with the solar zenith angle (z), influences the optical path length of sunlight into the water. Here, a value of 5 cm was assumed for the water depth, which is typical of paddy fields in May and undergoes some non-substantial increase in the following months. By considering a constant water depth one slightly overestimates the photochemistry in June and July, but calculations showed that photochemical reactions would, in any case, be negligible in these months due to light shielding by the rice canopy. The light optical path length l was obtained as $l = d \left[\sqrt{(1 - (n^{-1} \times \sin z)^2)} \right]^{-1}$, where $n = 1.34$ is the refraction index of water and z is the zenith angle. As a result, $l > d$ was 6 cm in May and 5.7 cm in June and July [26].

The chemical and photochemical parameters of paddy water have been set in analogy with the results of previous work [24], carried out on paddy-water samples from three rice farms located in the province of Vercelli (Piedmont region, northwestern Italy). In particular, the average formation quantum yields of the transient species i from irradiated CDOM (Φ_i^{CDOM}) had the following values: $\Phi_{HO^\bullet}^{CDOM} = 1.7 \cdot 10^{-5}$; $\Phi_{^3CDOM^*}^{CDOM} = 2.9 \cdot 10^{-2}$, and $\Phi_{^1O_2}^{CDOM} = 9 \cdot 10^{-3}$. The chemical analysis of the paddy water samples yielded on average DOC ~7 mg·C·L^{-1} and inorganic carbon (IC) ~13 mg·C·L^{-1} (namely, ~10^{-3} mol·C·L^{-1}). Nitrate and nitrite were under the limit of detection in two out of three samples, the third one having 1.7 mg·N·L^{-1} nitrate and 18 µg·N·L^{-1} nitrite (namely, 1.2 mmol·L^{-1} nitrate and 1.3 µmol·L^{-1} nitrite). Similar values of nitrate concentration have been reported for flooded paddy fields in Spain [21].

The modelling of As(III) photochemistry was initially carried out by considering the reactions of H_3AsO_3 with $HO^\bullet$ and $^3CDOM^*$, for which the following second-order rate constants were used: $k_{H_3AsO_3}^{HO^\bullet} = 8.5 \cdot 10^9$ M^{-1}·s^{-1} [15,16], and $k_{H_3AsO_3}^{CDOM} = 1.6 \cdot 10^7$ M^{-1}·s^{-1} [18]. The reaction of As(III) with $HO^\bullet$ forms an As(IV) species that is quite unstable under aerated conditions and is quickly oxidized to As(V) upon reaction with dissolved oxygen, following different possible pathways [15,16]. Since paddy water is oxygen-rich during the day due to photosynthesis [31,32], we considered the reaction between As(IV) and O_2 to be very fast and assumed the oxidation of As(III) by $HO^\bullet$ to be the rate-determining step of As(V) production.

The reactivity of As(III) with $CO_3^{\bullet-}$ has been investigated in the literature only in basic conditions (pH > 9), and the second-order rate constant has been reported as $k_{As(OH)_2O^-}^{CO_3^{\bullet-}} = 1.1 \cdot 10^8$ M^{-1}·s^{-1} [15]. Since H_3AsO_3 has pK_a = 9.2, the reaction between $CO_3^{\bullet-}$ and As(OH)$_2$O$^-$ should be important only at basic pH where As(OH)$_2$O$^-$ is a significant As(III) species. Singlet oxygen has been found to react slowly with As(III) [18], and it was thus neglected in our model. It must be pointed out that phenoxy radicals, produced by CDOM irradiation, could be key photooxidants of As(III) [18]. Unfortunately, these reactive species are very difficult to take into account in a photochemical model because of insufficient knowledge concerning the amount of phenolic domains in DOM and their possible variability in different environments. Moreover, the formation quantum yields of phenoxy radicals and their second-order rate constants with xenobiotics are not known. For this reason, the present photochemical model did not take into account the oxidation of As(III) by phenoxy radicals. Therefore, the photochemistry of As(III) in rice field water might be underestimated.

The photochemical pseudo-first order rate constants of As(III) oxidation were also modelled as a function of water pH, by assuming that the total concentration of dissolved As(III) (namely, [As(III)]$_{tot}$) is the sum of [H$_3$AsO$_3$] and [As(OH)$_2$O$^-$]. Each concentration can be expressed as a function of [As(III)]$_{tot}$ and pH, by using the molar fractions α_x (where x = H$_3$AsO$_3$ or As(OH)$_2$O$^-$):

$$[H_3AsO_3] = [As(III)]_{tot} \, \alpha_{H_3AsO_3} = [As(III)]_{tot} \frac{[H^+]}{[H^+] + K_a} \tag{4}$$

$$[As(OH)_2O^-] = [As(III)]_{tot}\, \alpha_{As(OH)_2O^-} = [As(III)]_{tot}\frac{K_a}{[H^+] + K_a} \tag{5}$$

The photochemical reactions taken into account are listed below:

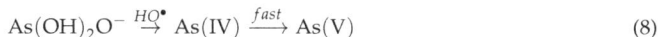

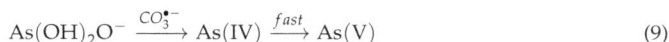

$$H_3AsO_3 \xrightarrow{HO^\bullet} (IV) \xrightarrow{fast} As(V) \tag{6}$$

$$H_3AsO_3 + {}^3CDOM^* \to As(V) \tag{7}$$

$$As(OH)_2O^- \xrightarrow{HO^\bullet} As(IV) \xrightarrow{fast} As(V) \tag{8}$$

$$As(OH)_2O^- \xrightarrow{CO_3^{\bullet-}} As(IV) \xrightarrow{fast} As(V) \tag{9}$$

Most of the relevant second-order reaction rate constants are available in the literature but, because the rate constant of $As(OH)_2O^- + HO^\bullet$ is not known, we assumed it to be equal to the corresponding H_3AsO_3 rate constant. This assumption is supported by the fact that, according to Kim et al. [16], the phototransformation rate of As(III) by $HO^\bullet$ in the 4.5–12 pH interval is mostly affected by $[HO^\bullet]$ variations and much less by the speciation of As(III). The total phototransformation rate of As(III) ($R^{tot}_{As(III)}$) can be written as the sum of the separate rates of Reactions (6–9), as follows:

$$R^{tot}_{As(III)} = R^{HO^\bullet}_{H_3AsO_3} + R^{{}^3CDOM^*}_{H_3AsO_3} + R^{HO^\bullet}_{As(OH)_2O^-} + R^{CO_3^{\bullet-}}_{As(OH)_2O^-} \tag{10}$$

where $R^i_{As(III)} = k^i_{As(III)}\,[As(III)][i]$, $i = HO^\bullet$, ${}^3CDOM^*$ or $CO_3^{\bullet-}$, and As(III) = H_3AsO_3 or $As(OH)_2O^-$. By introducing Equations (4) and (5) in Equation (10) and by dividing for $[As(III)]_{tot}$, one obtains:

$$k^{tot}_{As(III)} = \alpha_{H_3AsO_3}(k^{HO^\bullet}_{H_3AsO_3}[HO^\bullet] + k^{{}^3CDOM^*}_{H_3AsO_3}[{}^3CDOM^*]) + \alpha_{As(OH)_2O^-}(k^{HO^\bullet}_{As(OH)_2O^-}[HO^\bullet] + k^{CO_3^{\bullet-}}_{As(OH)_2O^-}[CO_3^{\bullet-}]) \tag{11}$$

where $k^{tot}_{As(III)} = R^{tot}_{As(III)}([As(III)]_{tot})^{-1}$ is the overall pseudo-first order rate constant of As(III) oxidation. Since $k^{tot}_{As(III)}$ contains the terms $\alpha_{H_3AsO_3}$ and $\alpha_{As(OH)_2O^-}$, it is a function of pH as suggested by Equations (4) and (5).

The steady-state $[CO_3^{\bullet-}]$ obviously depends on the value of $k^{{}^3CDOM^*}_{CO_3^{2-}}$, for which different estimates are available [23]. The take the different possibilities into account, $[CO_3^{\bullet-}]$ was calculated as follows:

$$[CO_3^{\bullet-}] = \frac{[HO^\bullet]\,(k^{HO^\bullet}_{HCO_3^-}[HCO_3^-] + k^{HO^\bullet}_{CO_3^{2-}}[CO_3^{2-}]) + k^{{}^3CDOM^*}_{CO_3^{2-}}[{}^3CDOM^*][CO_3^{2-}]}{k^{CO_3^{\bullet-}}_{DOM}\,DOC} \tag{12}$$

where $[HO^\bullet]$ and $[3CDOM^*]$ were provided by APEX, $k^{HO^\bullet}_{HCO_3^-} = 8.5 \times 10^6$ $M^{-1}\cdot s^{-1}$, and $k^{HO^\bullet}_{CO_3^{2-}} = 3.9 \times 10^8$ $M^{-1}\cdot s^{-1}$ [33]. Moreover, $k^{CO_3^{\bullet-}}_{DOM} = 10^2$ L·(mg·C)$^{-1}\cdot s^{-1}$ is the second-order rate constant of $CO_3^{\bullet-}$ scavenging by DOM.

4. Conclusions

Photochemical modelling suggests that As(III) would undergo photooxidation in paddy floodwater and, at pH < 8.5, mainly upon reaction with ${}^3CDOM^*$ and $HO^\bullet$. Note that As(III) oxidation by irradiated CDOM likely involves phenoxy radicals or other species to a higher extent than ${}^3CDOM^*$; thus, our approach can underestimate the importance of photochemical reactions. The predicted half-life times $t_{1/2}$ were much lower in May compared to June and July because, in the latter months, the rice plant growth would produce an important sunlight-shielding effect by the canopy. Therefore, As(III) photooxidation can be neglected in June and July. When typical chemical composition data of paddy water in May are taken into account, one obtains As(III) half-life times in the range of 20–30 days. In this case, photochemical As(III) oxidation could be overcome by microbial As(V) reduction to As(III). However, the photooxidation of As(III) might be much faster at pH > 8.5 if the carbonate radicals, $CO_3^{\bullet-}$, were produced efficiently by the reaction between $^3CDOM^*$ and CO_3^{2-}. Much depends on the $k_{CO_3^{2-}}^{3CDOM^*}$ second-order rate constant: if it is around 10^6 $M^{-1} \cdot s^{-1}$ or higher, the production of $CO_3^{\bullet-}$ and the subsequent reaction between $CO_3^{\bullet-}$ and $As(OH)_2O^-$ can play a key role in As(III) photooxidation. Further work is needed to better understand the $CO_3^{\bullet-}$ formation processes in paddy water concerning, most notably, the $CO_3^{\bullet-}$ formation quantum yield at basic pH and the $k_{CO_3^{2-}}^{3CDOM^*}$ rate constant.

Acknowledgments: Financial support by Università di Torino—Ricerca Locale is gratefully acknowledged.

Author Contributions: D.V. conceived the study and corrected the paper; L.C. elaborated the model, carried out the model calculations, and wrote the first draft of the paper.

Conflicts of Interest: The authors declare no conflict of interest. The funding sponsors had no role in the design of the study; in the collection, analyses, or interpretation of data; in the writing of the manuscript, and in the decision to publish the results.

References

1. Ahmed, M.F.; Ahuja, S.; Alauddin, M.; Hug, S.J.; Lloyd, J.R.; Pfaff, A.; Pichler, T.; Saltikov, C.; Stute, M.; van Geen, A. Ensuring safe drinking water in Bangladesh. *Science* **2006**, *314*, 1687–1688. [CrossRef] [PubMed]
2. Garnier, J.-M.; Travassac, F.; Lenoble, V.; Rose, J.; Zheng, Y.; Hossain, M.S.; Chowdhury, S.H.; Biswas, A.K.; Ahmed, K.M.; Cheng, Z.; et al. Temporal variations in arsenic uptake by rice plants in Bangladesh: The role of iron plaque in paddy fields irrigated with groundwater. *Sci. Total Environ.* **2010**, *408*, 4185–4193. [CrossRef] [PubMed]
3. Otero, X.L.; Tierra, W.; Atiaga, O.; Guanoluisa, D.; Nunes, L.M.; Ferreira, T.O.; Ruales, J. Arsenic in rice agrosystems (water, soil and rice plants) in Guayas and Los Ríos provinces, Ecuador. *Sci. Total Environ.* **2016**, *573*, 778–787. [CrossRef] [PubMed]
4. Abedin, M.J.; Feldmann, J.; Meharg, A.A. Uptake kinetics of arsenic species in rice plants. *Plant Physiol.* **2002**, *128*, 1120–1128. [CrossRef] [PubMed]
5. Zhao, F.-J.; Zhu, Y.-G.; Meharg, A.A. Methylated arsenic species in rice: Geographical variation, origin, and uptake mechanisms. *Environ. Sci. Technol.* **2013**, *47*, 3957–3966. [CrossRef] [PubMed]
6. Campbell, K.M.; Nordstrom, D.K. Arsenic speciation and sorption in natural environments. *Rev. Mineral. Geochem.* **2014**, *79*, 185–216. [CrossRef]
7. Takahashi, Y.; Minamikawa, R.; Hattori, K.H.; Kurishima, K.; Kihou, N.; Yuita, K. Arsenic behavior in paddy fields during the cycle of flooded and non-flooded periods. *Environ. Sci. Technol.* **2004**, *38*, 1038–1044. [CrossRef] [PubMed]
8. Zheng, R.L.; Sun, G.X.; Zhu, Y.G. Effects of microbial processes on the fate of arsenic in paddy soil. *Chin. Sci. Bull.* **2013**, *58*, 186–193. [CrossRef]
9. Macur, R.E.; Jackson, C.R.; Botero, L.M.; McDermott, T.R.; Inskeep, W.P. Bacterial populations associated with the oxidation and reduction of arsenic in an unsaturated soil. *Environ. Sci. Technol.* **2004**, *38*, 104–111. [CrossRef] [PubMed]
10. Oremland, R.S.; Stolz, J.F. The ecology of arsenic. *Science* **2003**, *300*, 939–944. [CrossRef] [PubMed]
11. Soda, S.O.; Yamamura, S.; Zhou, H.; Ike, M.; Fujita, M. Reduction kinetics of As(V) to As(III) by a dissimilatory arsenate-reducing bacterium, *Bacillus* sp. SF-1. *Biotechnol. Bioeng.* **2006**, *93*, 812–815. [CrossRef] [PubMed]
12. Macur, R.E.; Wheeler, J.T.; McDermott, T.R.; Inskeep, W.P. Microbial populations associated with the reduction and enhanced mobilization of arsenic in mine tailings. *Environ. Sci. Technol.* **2001**, *35*, 3676–3682. [CrossRef] [PubMed]
13. Hellweger, F.L.; Lall, U. Modeling the effect of algal dynamics on arsenic speciation in Lake Biwa. *Environ. Sci. Technol.* **2004**, *38*, 6716–6723. [CrossRef] [PubMed]

14. Vione, D.; Minella, M.; Maurino, V.; Minero, C. Indirect photochemistry in sunlit surface waters: Photoinduced production of reactive transient species. *Chem. Eur. J.* **2014**, *20*, 10590–10606. [CrossRef] [PubMed]

15. Klaning, U.K.; Bielski, B.H.J.; Sehested, K. Arsenic(IV). A pulse-radiolysis study. *Inorg. Chem.* **1989**, *28*, 2717–2724.

16. Kim, D.-H.; Lee, J.; Ryu, J.; Kim, K.; Choi, W. Arsenite oxidation initiated by the UV photolysis of nitrite and nitrate. *Environ. Sci. Technol.* **2014**, *48*, 4030–4037. [CrossRef] [PubMed]

17. Dutta, P.K.; Pehkonen, S.O.; Sharma, V.K.; Ray, A.K. Photocatalytic oxidation of arsenic(III): Evidence of hydroxyl radicals. *Environ. Sci. Technol.* **2005**, *39*, 1827–1834. [CrossRef] [PubMed]

18. Buschmann, J.; Canonica, S.; Lindauer, U.; Hug, S.J.; Sigg, L. Photoirradiation of dissolved humic acid induces arsenic(III) oxidation. *Environ. Sci. Technol.* **2005**, *39*, 9541–9546. [CrossRef] [PubMed]

19. Brockbank, C.I.; Batley, G.E.; Low, G.K.-C. Photochemical decomposition of arsenic species in natural waters. *Environ. Technol. Lett.* **1988**, *9*, 1361–1366. [CrossRef]

20. Kurasawa, H. The weekly succession in the standing crop of plankton and zoobenthos in the paddy field. Parts 1 and 2. *Bull. Res. Sci. Jpn.* **1956**, *41–42*, 86–98.

21. Quesada, A.; Leganés, F.; Fernàndez-Valiente, E. Environmental factors controlling N2 fixation in Mediterranean rice fields. *Microb. Ecol.* **1997**, *34*, 39–48. [CrossRef] [PubMed]

22. Graeber, D.; Goyenola, G.; Meerhoff, M.; Zwirnmann, E.; Ovesen, N.B.; Glendell, M.; Gelbrecht, J.; de Mello, F.T.; Gonzalez-Bergonzoni, I.; Jeppesen, E.; et al. Interacting effects of climate and agriculture on fluvial DOM in temperate and subtropical catchments. *Hydrol. Earth Syst. Sci.* **2015**, *19*, 2377–2394. [CrossRef]

23. Canonica, S.; Kohn, T.; Mac, M.; Real, F.J.; Wirz, J.; von Gunten, U. Photosensitizer method to determine rate constants for the reaction of carbonate radical with organic compounds. *Environ. Sci. Technol.* **2005**, *39*, 9182–9188. [CrossRef] [PubMed]

24. Carena, L.; Minella, M.; Barsotti, F.; Brigante, M.; Milan, M.; Ferrero, A.; Berto, S.; Minero, C.; Vione, D. Phototransformation of the herbicide propanil in paddy field water. *Environ. Sci. Technol.* **2017**, *51*, 2695–2704. [CrossRef] [PubMed]

25. Coelho, C.; Guyot, G.; ter Halle, A.; Cavani, L.; Ciavatta, C.; Richard, C. Photoreactivity of humic substances: Relationship between fluorescence and singlet oxygen production. *Environ. Chem. Lett.* **2011**, *9*, 447–451. [CrossRef]

26. Bodrato, M.; Vione, D. APEX (Aqueous Photochemistry of Environmentally occurring Xenobiotics): A free software tool to predict the kinetics of photochemical processes in surface waters. *Environ. Sci. Process. Impacts* **2014**, *16*, 732–740. [CrossRef] [PubMed]

27. Fabbri, D.; Minella, M.; Maurino, V.; Minero, C.; Vione, D. Photochemical transformation of phenylurea herbicides in surface waters: A model assessment of persistence, and implications for the possible generation of hazardous intermediates. *Chemosphere* **2015**, *119*, 601–607. [CrossRef] [PubMed]

28. Bianco, A.; Fabbri, D.; Minella, M.; Brigante, M.; Mailhot, G.; Maurino, V.; Minero, C.; Vione, D. New insights into the environmental photochemistry of 5-chloro-2-(2,4-dichlorophenoxy)phenol (triclosan): Reconsidering the importance of indirect photoreactions. *Water Res.* **2015**, *72*, 271–280. [CrossRef] [PubMed]

29. De Laurentiis, E.; Prasse, C.; Ternes, T.A.; Minella, M.; Maurino, V.; Minero, C.; Sarakha, M.; Brigante, M.; Vione, D. Assessing the photochemical transformation pathways of acetaminophen relevant to surface waters: Transformation kinetics, intermediates, and modelling. *Water Res.* **2014**, *53*, 235–248. [CrossRef] [PubMed]

30. Marchetti, G.; Minella, M.; Maurino, V.; Minero, C.; Vione, D. Photochemical transformation of atrazine and formation of photointermediates under conditions relevant to sunlit surface waters: Laboratory measures and modelling. *Water Res.* **2013**, *47*, 6211–6222. [CrossRef] [PubMed]

31. Saito, M.; Watanabe, I. Organic matter production in rice field flood water. *Soil Sci. Plant Nutr.* **1978**, *24*, 427–440. [CrossRef]

32. Roger, P.A. *Biology and Management of the Floodwater Ecosystem in Rice Fields*; International Rice Research Institute: Manila, Philippines, 1996; pp. 7–14.

33. Buxton, G.V.; Greenstock, C.L.; Helman, W.P.; Ross, A.B. Critical review of rate constants for reactions of hydrated electrons, hydrogen atoms and hydroxyl radicals ($^{\bullet}$OH/O$^{-\bullet}$) in aqueous solution. *J. Phys. Chem. Ref. Data* **1988**, *17*, 513–886. [CrossRef]

Sample Availability: No samples were used in this work.

Section 2:
Ultraviolet and Solar Homogeneous Processes to Decontaminate Waters

molecules

MDPI

Review

Light-Assisted Advanced Oxidation Processes for the Elimination of Chemical and Microbiological Pollution of Wastewaters in Developed and Developing Countries

Stefanos Giannakis * , Sami Rtimi and Cesar Pulgarin *

SB, ISIC, Group of Advanced Oxidation Processes (GPAO), École Polytechnique Fédérale de Lausanne (EPFL), Station 6, CH-1015 Lausanne, Switzerland; sami.rtimi@epfl.ch
* Correspondence: stefanos.giannakis@epfl.ch (S.G.); cesar.pulgarin@epfl.ch (C.P.);
  Tel.: +41-21-693-03-66 (S.G.); +41-21-693-47-20 (C.P.)

Received: 4 May 2017; Accepted: 23 June 2017; Published: 26 June 2017

Abstract: In this work, the issue of hospital and urban wastewater treatment is studied in two different contexts, in Switzerland and in developing countries (Ivory Coast and Colombia). For this purpose, the treatment of municipal wastewater effluents is studied, simulating the developed countries' context, while cheap and sustainable solutions are proposed for the developing countries, to form a barrier between effluents and receiving water bodies. In order to propose proper methods for each case, the characteristics of the matrices and the targets are described here in detail. In both contexts, the use of Advanced Oxidation Processes (AOPs) is implemented, focusing on UV-based and solar-supported ones, in the respective target areas. A list of emerging contaminants and bacteria are firstly studied to provide operational and engineering details on their removal by AOPs. Fundamental mechanistic insights are also provided on the degradation of the effluent wastewater organic matter. The use of viruses and yeasts as potential model pathogens is also accounted for, treated by the photo-Fenton process. In addition, two pharmaceutically active compound (PhAC) models of hospital and/or industrial origin are studied in wastewater and urine, treated by all accounted AOPs, as a proposed method to effectively control concentrated point-source pollution from hospital wastewaters. Their elimination was modeled and the degradation pathway was elucidated by the use of state-of-the-art analytical techniques. In conclusion, the use of light-supported AOPs was proven to be effective in degrading the respective target and further insights were provided by each application, which could facilitate their divulgation and potential application in the field.

Keywords: urban and hospital wastewater; urine treatment; pathogen microorganisms; emerging contaminants; UV/H_2O_2; photo-Fenton; *E. coli*; MS2 coliphage; saccharomyces cerevisiae; micropollutants

---

1. Problems Related with Presence of Microorganisms and Micropollutants

Wastewater treatment plants (WWTPs) have been built, transformed and updated through the years to effectively prevent solids, organic and inorganic compounds (carbon, nitrogen, phosphorus, etc.) and more, which enter the environment. The challenge posed by the pollutants is a matter that the majority of the WWTPs are not equipped to handle. The micropollutants are (in a high percentage) invulnerable to biological treatment; the transfer from source to the environment is therefore facilitated, leading to further accumulation in the environment; their presence has been associated with minor and major health risks, toxicity and more [1,2].

The risk of microorganisms' presence in natural water bodies is more explored compared to micropollutants [3,4]. Water scarcity has led to reuse concepts and many countries worldwide have included legislation concerning the removal thresholds according to the subsequent water reuse, e.g., in Italy, [5]. However, chlorination is still the most widespread technique, with its inherent problems, such as disinfection by-products (e.g., trihalomethanes) formation, and where funds are available UV has been applied.

The problem of microorganisms' occurrence in water sources has been identified for many decades. Water related diseases plague the developing countries using compromised drinking water sources [6], or reusing wastewater for food production [7], but also many outbreaks have occurred by cross contamination of public access waters in the developed states [8]. Apart from the notorious diseases such as cholera, gastro-enteritis or dengue fever, the contamination of water sources poses risks to populations which have incorporated their use in their economical-related activities, such as fishing or other economic activities.

The main problem associated with the occurrence of micropollutants (MPs) in the environment is the lack of information, concerning the side-effects of their presence in the receiving matrix [9]. For instance, little is known on the long term effects of pollutants, classified as "potentially not harmful"; there are no studies on the accumulation or the chronic toxicity of such substances in plants, animals, or humans. There is a relatively long list of MPs actually found in the environment, which derived from the various anthropogenic activities [10,11]. To date, not all substances have been assessed on their potential actions against the environment.

Furthermore, toxicologically speaking, most of the substances that have been assessed for their risks, have been directed through single-compound investigations, when the reality differs significantly [12]. The environment contains a mixture of a vast number of compounds, which could even react with each other, affecting their mode of action [13]. The result has been demonstrated to be severe, with cases reporting compounds that had no prior effect on species, to demonstrate harmful properties when found in mixtures [14–16]. Apart from the toxicity problems, other critical drug-related issues that have emerged over the years are: (i) the enhancement of antibiotic resistance, by the presence of antibiotics and their metabolites in the environment [17]; (ii) the problematic identification of the transformed metabolites of drugs [18]; and (iii) the bioaccumulation of pollutants in living organisms in the environment.

To better suggest treatment goals and methods, this work will address the use of Advanced Oxidation Processes (AOPs) as a greener and sustainable disinfection and decontamination method of hospital and urban wastewater in both developed and developing countries. Each context will be treated with regard to its specific issues, dealing with the economic feasibility and technical applicability of the various light-assisted AOPs. The UV-based ones will form the basis of treatment for developed countries whereas the photo-Fenton process will be presents as a viable solution for the sunny developing countries around the globe. After contextualization of the problems micropollutants (MPs) and microorganisms (MOs) cause to the environment and to humans, details will be provided on their presence, the contribution of hospitals in aggravating the actual situation and an overview of the results obtained in the different environments by the various AOPs.

2. Hospital Wastewater Effluents and Types and Pollutants

2.1. Hospital Wastewater and Micropollutants

Hospital wastewater (HWW) is the result of the residue collection from the various water-consuming activities taking place within its premises. These services include [19]:

(i)   Human sewage
(ii)   Kitchen and laundry
(iii)   Heating and cooling processes
(iv)   Laboratorial discharge (clinics and research centers)
(v)   Wards and outpatients contribution

The first categories are common also in municipal wastewater (MWW), which is the reason that led practitioners to suggest co-treatment in MWWTPs, sometimes with only a pre-treatment (e.g., chlorination) [20], only to limit the microbiological-related risk. The reality suggests that within HWW there are many substances, such as disinfectants, organic compounds, therapeutic metals, rare microbial agents or antibiotic-resistant ones [21–23], often in high concentrations that modify significantly the composition of HWW compared to MWW. For this reason, many authors have openly objected to the co-treatment practice so far [24–26].

The presence of Pharmaceutically Active Compounds (PhACs) in hospital wastewater is a hot topic, with a variety of works dedicated to the characterization of their nature [19,27–30] and their importance in the overall load [18]. The characteristics of HWW are influenced by a variety of factors, such as the size of the hospital, the range of services and activities, the season of the year, the time of the day [19,31], and more. Table 1 presents an overview comparing HWW with MWW, in terms of a single unit (patient/inhabitant).

It is obvious that the composition of either micro- or macropollutants in each case is significantly different (see Table 1), indicating one of the reasons for failure of treatment by conventional WWTPs. The micropollutants arriving in WWTPs are of a range of μg or ng, and also are reported to affect the nature of the WW in the treatment plant (solubility, volatility, adsorbability, absorbability, biodegradability, polarity, and stability) [31].

For the pre-mentioned reasons, HWW should be treated as a separate entity [31]. The economic and overall risks should be assessed [25], on-site treatment should be implemented as close as possible to the source [32,33], the consideration of no-mix toilets for urine separation must be taken into account [34] and reducing the quantities can considerably mitigate the effluent amounts if direct discharge in surface water is expected.

Table 1. Comparison between indicative MWW and HWW effluents characteristics [31,35,36] and references therein.

Parameter	Details	Municipal WW	Hospital WW	Units
BOD		60	160	mg/L
COD		110	280	mg/L
SS		80	135	mg/L
pH		7.5	8	
TKN		20–70	33	mg/L
Chlorides		50	200	mg/L
Total P		4	7	mg/L
Bacteria	*Total Coliforms*	10^6	7.7×10^9	CFU/mL
Viruses	*Norovirus*	1.6×10^2	2.4×10^6	PFU/mL
	Hepatitis A virus	10^2	10^4	PFU/mL
	Adenovirus	1.6×10^2	2.8×10^6	PFU/mL

2.2. Hospital Wastewaters and Microorganisms

Similar to MPs, HWW are carriers of higher microbiological agent loads, compared to their urban counterparts. In principle, the MWW are subjected to higher dilution and the intrinsic properties of HWW imply the presence of higher and possibly more infectious agents. In a recent review [35] the differences between MWW and HWW effluents in various countries have been presented, with compelling variations in the distribution of microorganisms' species and quantities. This microorganism load is usually led to co-treatment with MWW in MWWTPs and their removal efficiency is a function of the existing treatment. Among others mentioned before, some authors [36] stressed the importance of treating HWW separately, on-site, to effectively reduce micro- and macropollutants, but also stressing the need for microorganisms' elimination. Problematic treatment or inexistent treatment can lead to the problems mentioned before (antibiotic resistance, bioaccumulation, etc.) as well as directly jeopardize the drinking water sources in developing countries [37].

3. Light-Assisted AOPs and Their Action in Chemical and Microbiological Pollutants' Degradation

Due to the recalcitrant nature of many existing MPs, the existing biological and physicochemical treatment methods have been proven unable to efficiently degrade them in WWTPs [2]. Subsequently, their elimination relies on further, "quaternary" treatment, such as advanced oxidation treatment. These processes have successfully mineralized or converted the persistent MPs to less harmful forms [33]. AOPs can be used as pre-treatment or post-biological treatment processes. Depending on the target, they can achieve conversion of recalcitrant pollutants to biodegradable ones, or act as a polishing step. In the respective cases, the residence time in biological treatment is reduced or the residual pollutant content can be eliminated [38–41].

Ozonation and AOPs are redox technologies with main characteristic the non-selectivity on the target and share the production of the highly oxidative hydroxyl radical ($HO^\bullet$) [2]. After fluorine, the $HO^\bullet$ is the second most powerful oxidant (3.03 eV, compared to 2.80), with reaction rates ranging from 10^{-6} to $10^{-9}\ M^{-1}\,s^{-1}$ [42].

The AOPs typically involve chemical agents (metals, ozone or hydrogen peroxide) and an assistive energy source, such as UV or visible light, current, ultrasound or γ-irradiation [43]. Some examples of AOPs are:

- **Ozone-based**: O_3/H_2O_2, O_3/UV, $O_3/UV/H_2O_2$
- **UV-based**: UV, UV/H_2O_2
- **Fenton-related**: (Fe/H_2O_2), including photo-Fenton, electro-Fenton, etc.
- **Heterogeneous photocatalysis**, such as (TiO_2/hv)
- **γ-radiolysis**
- **Ultrasound-based**: sonolysis, ultrasound-supported Fenton, etc.

Although the hydroxyl radical is the main oxidizing agent in these processes, their application often induces the production and participation of other reactive oxygen species (ROS), such as superoxide radical anions, hydroperoxyl radicals, singlet and triplet oxygen, etc. [33,43]. Another main advantage of the AOPs application is the characteristic versatility with which the method can be achieved. For instance, photolysis acts directly or indirectly, by absorption of energy and excitation or photosensitizing agents, typically dissolved organic matter (DOM) [42].

3.1. UV-Based Processes (UV, UV/H₂O₂)

UV treatment consists on the direct photo-transformation of organic compounds. In UV direct photolysis, the micropollutant must absorb the incident radiation and undergo degradation starting from its excited state. This treatment has been the most known and widely used irradiation method in initiating oxidative degradation processes. Some organic pollutants effectively absorb UV-C light directly, and absorption of this high-energy can cause destruction of the chemical bonds and subsequent breakdown of the contaminant.

The main factors which will affect the degradation of MPs in the UVC light assisted UV/H_2O_2 processes are the UV absorption and its quantum yield [44]. The molar absorption coefficient, i.e., UV absorption is an indication of the strength with which a molecule absorbs UV, and consequently, cause its degradation [42,45,46]. In principle, reaction kinetics of the organic substrate with the oxidant are described by second order law, as follows Equation (1):

$$r(-M) = \frac{d[M]_t}{dt} = k_{oxidant,M}[M][oxidant],$$

(1)

where r(-M) represents the rate of degradation of the MP. At the same time, direct photolysis contributes in the dual manner mentioned before, provided that other WW constituents and physicochemical characteristics are present, such as pH conditions [33]. Other important factors include the concentration of the target compound, the pH of the matrix, the amount of H_2O_2, the presence/absence of scavenging compounds (e.g., bicarbonates) and the reaction time.

UV/H_2O_2 treatment is considered an advanced oxidation process because it involves the generation of hydroxyl radicals (HO•) produced by homolytic cleavage of hydrogen peroxide. Photolysis of H_2O_2 yields two HO• radicals per photon absorbed. The hydroxyl radicals are strong oxidants ($E° = 2.8$ V) with fast reactivity due to their non-selectiveness. The efficiency of the process will depend strongly on the HO• production velocity. The most basic initiation, propagation and termination reactions with main components in water are as follows (Equations (2)–(12)) [47]:

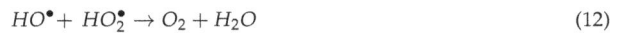

$$H_2O_2 \overset{h\eta}{\rightarrow} 2\,HO^{\bullet} \tag{2}$$

$$H_2O_2 + HO^{\bullet} \rightarrow HO_2^{\bullet} + H_2O \tag{3}$$

$$H_2O_2 + O_2^{\bullet-} \rightarrow HO^{\bullet} + O_2 + OH^- \tag{4}$$

$$HO^{\bullet} + O_2^{\bullet-} \rightarrow O_2 + OH^- \tag{5}$$

$$HO^{\bullet} + targets \rightarrow Products \tag{6}$$

$$HO^{\bullet} + DOC/NOM \rightarrow Products \tag{7}$$

$$HO^{\bullet} + CO_3^- \rightarrow CO_3^{\bullet-} + OH^- \tag{8}$$

$$HO^{\bullet} + HCO_3^- \rightarrow CO_3^{\bullet-} + H_2O2 \tag{9}$$

$$2\,HO^{\bullet} \rightarrow H_2O_2 \tag{10}$$

$$2\,HO_2^{\bullet} \rightarrow H_2O_2 + O_2 \tag{11}$$

$$HO^{\bullet} + HO_2^{\bullet} \rightarrow O_2 + H_2O \tag{12}$$

The molar absorption coefficient of H_2O_2 is only 18.7 $M^{-1} \cdot cm^{-1}$ at 254 nm [48]. Hence, the efficiency of UV/H_2O_2 process decreases drastically with the presence of strong photon absorbers or when the UV absorbance of the target pollutant is high.

3.2. Fenton-Related Reactions (Fenton, Photo-Fenton, Solar Light)

The Fenton process is an attractive oxidative system for water and wastewater treatment, due to iron abundance in nature and low inherent toxicity, as well as the fact that hydrogen peroxide is easy to handle and environmentally safe, decomposing spontaneously to H_2O and O_2.

It has been demonstrated that Fenton's reagent is able to destroy toxic compounds in wastewater [49]. Production of $HO^{\bullet}$ by the Fenton reagent takes place by addition of H_2O_2 to Fe^{2+} salts trough the following reactions [50,51]. "R" is used to describe the reacting organic compound (Equations (13)–(34)) [52]:

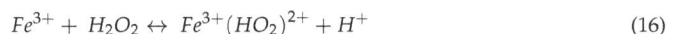

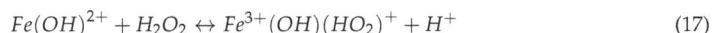

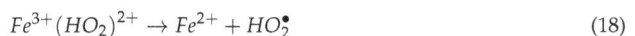

$$Fe^{3+} + H_2O \leftrightarrow Fe(OH)^{2+} + H^+ \tag{13}$$

$$Fe^{3+} + 2H_2O \leftrightarrow Fe(OH)_2^+ + 2H^+ \tag{14}$$

$$2Fe^{3+} + 2H_2O \leftrightarrow Fe_2(OH)_2^{4+} + 2H^+ \tag{15}$$

$$Fe^{3+} + H_2O_2 \leftrightarrow Fe^{3+}(HO_2)^{2+} + H^+ \tag{16}$$

$$Fe(OH)^{2+} + H_2O_2 \leftrightarrow Fe^{3+}(OH)(HO_2)^+ + H^+ \tag{17}$$

$$Fe^{3+}(HO_2)^{2+} \rightarrow Fe^{2+} + HO_2^{\bullet} \tag{18}$$

$$Fe^{3+}(OH)(HO_2)^+ \rightarrow Fe^{2+} + HO_2^\bullet + OH^- \tag{19}$$

$$Fe^{2+} + H_2O_2 \rightarrow Fe^{3+} + HO^\bullet + OH^- \tag{20}$$

$$Fe^{2+} + HO^\bullet \rightarrow Fe^{3+} + OH^- \tag{21}$$

$$HO^\bullet + H_2O_2 \rightarrow HO_2^\bullet + H_2O \tag{22}$$

$$Fe^{2+} + HO_2^\bullet \rightarrow Fe^{3+}(HO_2)^{2+} \tag{23}$$

$$Fe^{2+} + O_2^{\bullet-} + H^+ \rightarrow Fe^{3+}(HO_2)^{2+} \tag{24}$$

$$Fe^{3+} + HO_2^\bullet \rightarrow Fe^{2+} + O_2 + H^+ \tag{25}$$

$$Fe^{3+} + O_2^{\bullet-} \rightarrow Fe^{2+} + O_2 \tag{26}$$

$$HO_2^\bullet \rightarrow O_2^{\bullet-} + H^+ \tag{27}$$

$$O_2^{\bullet-} + H^+ \rightarrow HO_2^\bullet \tag{28}$$

$$HO_2^\bullet + HO_2^\bullet \rightarrow H_2O_2 + O_2 \tag{29}$$

$$HO_2^\bullet + O_2^{\bullet-} + H_2O \rightarrow H_2O_2 + O_2 + OH^- \tag{30}$$

$$HO^\bullet + HO_2^\bullet \rightarrow H_2O + O_2 \tag{31}$$

$$HO^\bullet + O_2^{\bullet-} \rightarrow O_2 + OH^- \tag{32}$$

$$HO^\bullet + HO^\bullet \rightarrow H_2O_2 \tag{33}$$

$$HO^\bullet + RH \rightarrow H_2O + R^\bullet \tag{34}$$

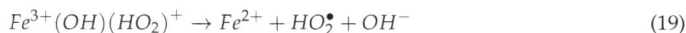

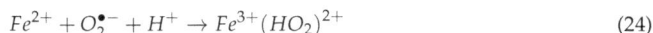

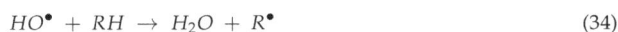

However, exposure to light enhances the Fenton reaction by the photo-regeneration of Fe (II), when reducing Fe (III) or via a ligand to metal charge transfer (LMCT). Hence, there is a double production of hydroxyl radicals (L is an organic ligand) [53] (Equations (35) and (36)):

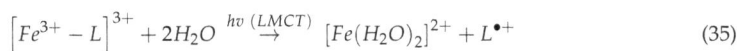

$$\left[Fe^{3+} - L\right]^{3+} + 2H_2O \xrightarrow{hv\ (LMCT)} [Fe(H_2O)_2]^{2+} + L^{\bullet+} \tag{35}$$

$$[Fe(H_2O)_5(OH)]^{3+} \xrightarrow{hv} Fe^{2+} + H^+ + HO^\bullet \tag{36}$$

Thus, photo-Fenton is a process that is able to use solar radiation as input taking advantage not only of the UV portion contained in solar radiation but also because of the ability of some compounds such as Fe^{3+}-hydroxy and Fe^{3+}-acid to absorb energy in the visible spectra.

The use of solar light as source of radiation for activating the hydroxyl radicals is not a new concept: several researches have proven the efficiency of solar light as an activating agent for the Fenton reaction. In this process, the Fe^{2+} is continuously recycled, reducing the amount of iron salts required (and their further disposal) for the Fenton's reaction. This feature makes the photo-Fenton process more applicable and attractive for application in sunny regions around the globe, as reviewed extensively in [54,55].

4. Problem Identification and Contextualization: Micropollutants and Microorganisms in Developed and Developing Countries

Since micropollutants have been identified in many cases as high risk compounds, many works have been initiated to identify their presence in the environment [2,56]. Moving in backward steps, the presence in environmental water matrices is a result of a variety of pathways. One of the main sources, which will be further analyzed later on, are the MWWTPs, due to the collection of urban and sometimes, (pre-treated) industrial effluents [57].

Although the treatment in WWTPs is followed by natural processes, such as sorption, photolysis and biodegradation, that can reduce the contaminant loads up to 10 times [58,59], the MPs' presence is still unambiguous. In a research conducted among many countries, the most frequently encountered drugs were the non-steroidal anti-inflammatory drugs (NSAIDs), Sulfamethoxazole, Carbamazepine and Triclosan [2]. Generally, the occurrence of MPs was less frequent in summer (probably due to elevated, temperature-driven biodegradation), and even though winter rain promoted dilution, sometimes the contribution in natural water was important [60]. Finally, the concentrations found in surface waters were well correlated with the population distribution, linked with the massive utilization of parent chemicals by a bigger number of users [2].

Concerning drinking water, the studies are relatively few, because the occurrence is sometimes below the detection limit [60,61]. However, this is often a limitation of the experimental capabilities of the analytical laboratories. Kummerer has discussed this problem, in the appearance of "new" compounds, which could have been normally encountered (for pollutants in ng or μg scale) [62], if the technology allowed so. In addition, as far as long-term side-effects are concerned, the presence of certain compounds or their intermediated in drinking water has not been under study (yet). In overall, in the review published by Luo et al. [2] it is mentioned that most of the countries investigated (France, USA, Spain, and Canada) were capable of removing the presence of some MPs in drinking water. This is a critical step, considering that drinking water treatment is literally the last line of defense among end-users and micropollutants [33].

Although in developed countries water is considered a de facto supply in each household, in developing countries, the reality is sometimes far from this state. Water acquisition can be an everyday struggle for many families. If in this scenario, one thinks of the potential problems that could appear if MP pollution is high, the risks are more imminent. The quality of life of the affected population is considerably endangered, and more specifically not by chronic or potential problems, but from the harsh reality of raw, untreated wastewater in the water supplies.

In many developing countries, the combination of rampant population growth and the lack of financial means, have led to insufficient (up to inexistent) sanitation facilities. Therefore, the collection and the treatment of wastewater is problematic. The poorest fractions of the population, who employ themselves in handcraft, fishing and agricultural activities suffer the most, since the situation in centralized, capital areas is slightly better. However, these areas have demonstrated unacceptable treatment, especially in (semi)industrial or hospital effluents, with cases describing direct untreated water being discharged in rivers and sea.

"Fortunately", the risk of MPs is relatively lower, when compared with developed countries. The availability of drugs and the capability of purchase restrict the widespread use and the diffuse pollution. The main areas expected to provide major MP flows are the hospitals and similar facilities. Recent research that has been performed in the framework of the "Treatment of the Hospital wastewaters in Ivory Coast and in Colombia by advanced oxidation processes" (unpublished data) indicated that even in this case, the majority of administered drugs are biodegradable and the MP content is limited in isolated hospitals in Colombia, but in the University Hospital in Abidjan, the situation is quite problematic.

On the other hand, even when the amounts of MPs is not alarming, the presence of microorganisms in WW is an important matter, which becomes top priority, since no disinfection process is applied in the effluents. Therefore, the focus should be directed at least to mitigating microbial pollution [63,64], when it comes to discharged WW in developing countries, which poses direct and acute illness risks. Hospitals are an identified contributor to fecal and overall pathogen microorganisms in surface waters, and the lack of treatment is directly jeopardizing their use [37]. The current practices in agriculture for instance, include the use of contaminated water for crops irrigation, and the transfer of pathogens is highly possible. Therefore, treatment designed taking account the local particular conditions and monitoring of total coliform bacteria and aerobic mesophilic bacteria, as representative of fecal and non-fecal routes, respectively, should be monitored and their elimination should precede discharge in natural water bodies [65,66].

5. Treatment Strategy and Research Results

As described in the previous sections, the issue of hospital wastewater treatment has multiple contextual, application and engineering extensions. The relevant literature has successful applications of various AOPs both as pre- and as post- biological treatment methods for hospital wastewater [67–74]. Our main goal here is to address the difference that the context of application can make in the choice and application of AOPs for hospital contaminants. The necessary strategies need to be specifically addressed towards developed or developing countries, where HWW is channeled in MWW or is directly discharged, respectively; a summary of the strategies is presented in Figure 1. The context differs significantly; the developed countries have more or less under control the problem of MOs and focus on MPs, while developing countries' priority should be the acute risk caused by MOs presence. Furthermore, the AOP chosen has to be a function of the technical and economic status of the place of application, with the UV-based methods being more prominent in developed countries and the solar based ones more suitable for developing countries. As a result, one should take into account the aforementioned constraints when focusing on HWW treatment by AOPs in developed and developing countries, emphasizing both on the application point of view, as well as the underlying mechanisms governing micropollutant degradation and microorganism inactivation.

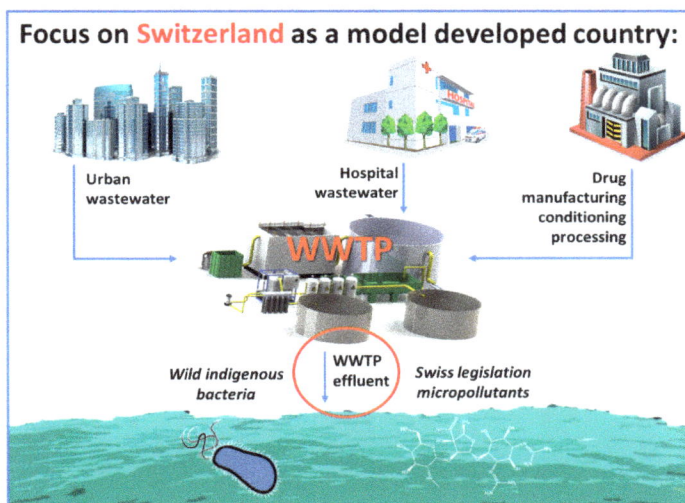

Figure 1. Schematic representation: Treatment focus, strategy and targets of WWTP by light-assisted AOPs in developing countries. Switzerland and its motion of upgrading the WWTPs is studied, presenting solutions based on the application of mainly UV-based AOPs, compared with the solar photo-Fenton. The targets are the micropollutants chosen by the Swiss evolution of control policy and the indigenous population of bacterial microorganisms present in WW.

5.1. Developed Countries and Municipal WWTPs: Treatment of MPs and MOs by Light-Assisted AOPs

In Switzerland (as an example of developed country), the wastewater effluents are already treated, and the implementation of the relative AOPs promoted by the Federal Office for the Environment focuses on the elimination of the chronic risk caused by the presence of micropollutants in natural waters, and less for the acute risk of microbial infection due to the pathogens carried within the flows.

As such, the micropollutants chosen derive from the modifications in the Swiss legislation, namely Carbamazepine, Clarithromycin, Diclofenac, Metoprolol, Venlafaxine, Benzotriazole and Mecoprop. Concerning microorganisms, the bacterial populations contained in urban effluents of the WWTP of Vidy, Lausanne were selected as microbial targets.

Figure 2 summarizes the results achieved by treating the effluents of Vidy WWTP by UV/H_2O_2 and the photo-Fenton process, where Figure 2a focuses on MPs and Figure 2b on bacterial microorganisms. The different color traces correspond to the different secondary pretreatment process applied in the plant before the lab-treatment by AOPs.

The UV-based process is a far more energetic and therefore more efficient process for the removal of microbiological and chemical contaminants. The difference is in orders of magnitude higher, as a maximum 20% MP removal was attained by the photo-Fenton process in 30 min of exposure and at 10 min the corresponding removal by the UV/H_2O_2 process was 100% for activated sludge and moving bed bioreactor effluents (AS, MBBR) and 30 min were necessary for the coagulation-flocculation effluents (CF). Similar trends were found in microorganisms' elimination, with 5 min exposure having completely eliminated microorganisms in all effluents, while more than 3 h were demanded by the photo-Fenton process. However, this process does not reflect the actual difference of the two processes, as increasing the Fenton reagents, for instance, would have a dramatic increase in efficiency, or, similarly, the H_2O_2 content in the UV-based process. Nevertheless, here we present only the mild conditions tested in lab-scale, but the literature has offered many successful applications of MP removals by the photo-Fenton process [75–77]. Apart from the reduction of micropollutants and microorganisms, a significant amount of effluent organic matter (EfOM) was eliminated from the bulk during treatment, reducing the overall charge carried before disposal [78].

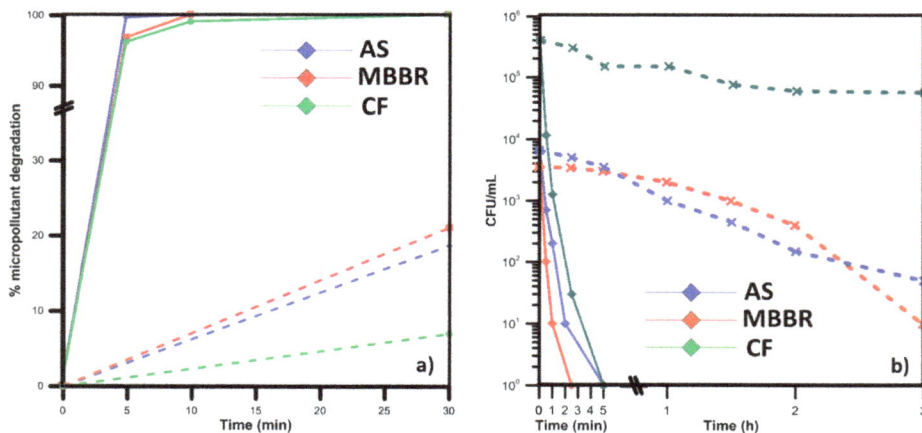

Figure 2. **Degradation of chemical and microbiological contaminants in MWW.** (**a**) Weighted average removal of seven selected micropollutants by the UV/H_2O_2 process (ng/L initial MP content and 25 mg/L H_2O_2); and (**b**) reduction of bacteria contained in WW by the UV/H_2O_2 (continuous trace) and the solar photo-Fenton process (dashed line). The colors correspond to the previous secondary treatment (blue: Activated Sludge, red: Moving Bed BioReactors, green: Coagulation-Flocculation). More details on the micropollutants' initial content, experimental configuration and wastewater characteristics can be found in [78,79].

The key in the application of either method is the economical and geographical context. For instance, there are very few countries that could support the application of such costly AOPs for domestic wastewater treatment; the estimated cost for the treatment plant of Vidy reaches 0.16–0.18 € m^3 of treated waste for the application of activated carbon and ozonation [80].

The electrical cost and reagent (H_2O_2) supplies were calculated in a pilot plant tested in the same premises [81] and is of the same order. Nevertheless, this option is viable, considering that many plants in the USA operate chlorination basins, which are ideal for conversion to UV streams, and part of the cost is already administered and later will be recovered.

Concerning photo-Fenton, prolonging the treatment inflicts further degradation of contaminants and microorganisms, and a plant design with residence time of day(s) could be considered as an option. However, the main constraints under consideration are the land cost and the latitude of the site; beyond a certain point the clear/sunny days are decreased dramatically. The combination of these two factors on the other hand, apart from the developing countries who would greatly benefit from this process, already indicate USA and Australia as excellent candidates for application, especially due to high number of sunlit hours in Western USA and thoroghout Australia. Even more, the application of maturation ponds is already an existing solution in the aforementioned regions, hence an addition of a polishing step aided by the photo-Fenton reagents could complete an already successful existing practice [82,83].

Finally, it was found that one of the most important concerns of wastewater treatment, the bacterial regrowth was sufficiently hindered [79]. For the UV-based processes, the complete elimination of microorganisms was attained by combined extensive mutations and hydroxylation of the cell membrane, which cannot be repaired by the enzymatic mechanisms possessed by microorganisms. The photo-Fenton process, even when it was not concluded, it did not present bacterial regrowth, owing to the presence of the Fenton reagents in the bulk; the continuous (dark) Fenton process ensures limitation of rampant bacterial growth and maintaining a good quality effluent in already inactivation was achieved. The design of the process should take into account both chemical and microbiological targets, since MPs require more time to degrade than bacteria, but bacteria engulf the regrowth risk, while the extended treatment times allow possible adaptation and growth in the rich WW medium.

5.2. Developing Countries and Microorganisms Disinfection by Photo-Fenton

In Ivory Coast and Colombia, WWTPs are virtually inexistent. The practice of organized WW treatment is present only in major cities (e.g., Medellin), or in special establishments (hospitals in Ivory Coast; University Hospital in Abidjan). Furthermore, the treatment either stops at primary/secondary space, or has malfunctioned and not working properly since its construction. Therefore, the application of solar photo-Fenton as a feasible AOP is implied only after the construction of basic treatment before (primary-secondary), or as a possible implementation of a barrier of microbiological-related problems. The acute risk of microbial infections is prioritized over MPs, as it poses an immediate threat to the populations, and the assessment was made by taking a viral and a yeast pathogen model into study. A summary of the strategy is presented in Figure 3.

Viruses are a special category of pathogens, responsible for a high number of disease outbreaks in developing countries. Their diversity and polymorphic nature make them an important target to consider when treating wastewaters. The MS2 bacteriophage model virus was chosen as it is particularly resistant to UV irradiation, hence its removal by solar-assisted processes could be delayed. As such, a step-wise construction was attempted, starting from solar light up to the photo-Fenton process.

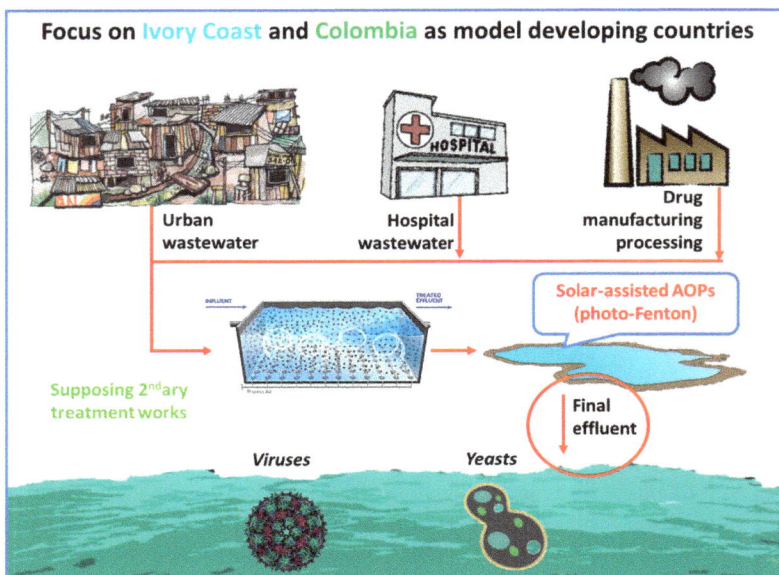

Figure 3. Schematic representation: Using the photo-Fenton as a disinfection method in developing countries. The application of Fenton reagents, aided by the numerous sunny days in the circum-Equatorial regions, which coincide with the geographical distribution of developing countries, was assessed on their virucidal and fungal inactivation capacities, using bacteriophages and common yeasts as proxies, while assessing the important role of dissolved organic matter and iron.

Figure 4 presents the aforementioned approach on dissociating the events that take place during photo-Fenton treatment of MS2 coliphage. The step-wise construction was as follows:

(i) Solar light alone is unable to inflict high removal of MS2, hence underlines the need for application of an oxidative process.

(ii) Adding 1 mg/L H_2O_2, which is a moderate amount for inactivation of microorganisms, but could simulate the in-situ generation of H_2O_2 by irradiation of Dissolved Organic Matter (DOM) [84], again has almost no effect. The mild oxidative potential of H_2O_2 is unable to inactivate more than 1 log of MS2.

(iii) Continuing with the components of the photo-Fenton process, iron, in salts form, Fe^{2+} or Fe^{3+} was tested. Natural waters in Africa have often been found to contain high amounts of iron, especially in the form of oxides [85] but also dissolved. Although iron has no oxidative actions, its complexation with the viral capsid enables, upon irradiation, a LMCT reaction, with the reduction of Fe^{3+} to Fe^{2+} and oxidation of the ligand. This oxidation damages the external capsid proteins, thus reducing the infectivity of the virus.

(iv) The voluntary addition of very low amounts of H_2O_2 in the bulk indicate that the photo-Fenton process is potentially a very efficient treatment technique in the elimination of viral pathogens; the inactivation is either very sharp, or at least faster than the Fe or solar alone. Fe^{2+} is readily oxidizable, generating $HO^\bullet$ species, reaching viral inactivation, and Fe^{3+} after an initial reduction step, participates in the photo-Fenton catalytic cycle.

(v) Finally, the presence of organic matter in all the experiments did not appear to significantly hinder the inactivation of viruses. In fact, in presence of organic matter, the quantity of dissolved iron was followed, and the precipitation due to the neutral operating pH was avoided. Hence, a sustainable catalytic cycle can be maintained, aided by the iron-DOM complexes in WW [86].

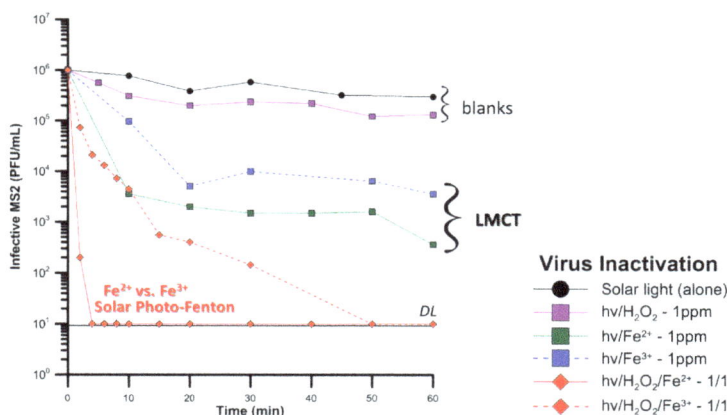

Figure 4. Stepwise construction of the solar photo-Fenton process. Effects of solar light, solar/H_2O_2, solar/Fe and solar photo-Fenton in viral (MS2) infectivity in simulated secondary wastewater, as in [87]. Note the difference inflicted by the initial speciation of iron in the solution.

For yeast-related experiments, *Saccharomyces cerevisiae* were used as a model of eukaryotic pathogen microorganisms. The strategy followed initially was similar to the virus inactivation, with stepwise construction of the inactivation process. On the other hand, the target was to study the internal events that take place during exposure to the photo-Fenton reagents. After initial experiments in simulated wastewater, where the inactivation of the yeast was normally achieved, the continuation involved tests only in pure water in order to avoid interferences with the genomic and proteomic measurements.

Figure 5 summarizes the results of the protein and DNA degradation experiments that took place during the photo-Fenton process. The events that take place during inactivation are inverse, compared to the expected series of events; one would suppose that since Fe and H_2O_2 are added in the bulk, the rapid reaction and generation of hydroxyl radicals would lead to the fast degradation of the (external) cell wall proteins. In fact this is partially true, as depicted in the first graph. The disappearance of the intensity of the bands during electrophoresis means that the proteins are degraded. However, if we correlate the damage taking place in 60 min, during which total disappearance of the DNA and heavy damages in the cytoplasmic proteins occurs, the slight fade in the cell wall protein bands is clearly not the main reason for yeast inactivation.

As it appears, even in a complex, evolved microorganism such as the yeast, who possess repairing enzymatic functions, and a fortified (compared to bacteria) external cell wall, the Achilles' heel is located in the internal part of the microorganism [88]. H_2O_2 can diffuse in to the cell, and with the lack of enzymes able to catalyze its dismutation to H_2O and O_2, due to sunlight exposure, the inactivation indeed takes place in the internal part. Even more, the iron that can be transported internally with various routes, or the already existing iron, released via the oxidative damage and the light-induced events, can facilitate an internal photo-Fenton process, able to inactivate yeasts.

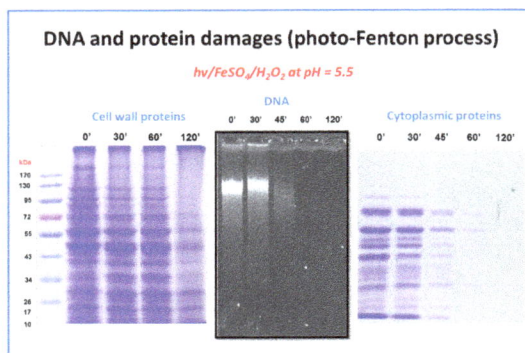

Figure 5. Internal and external damages during *Saccharomyces cerevisiae* inactivation by the photo-Fenton process at near-neutral pH. The panels indicate the external and two types of internal damage inflicted to the model yeast pathogen: cell wall proteins, genome and cytoplasmic proteins, respectively.

5.3. Hospital-Derived and Highly Concentrated Flows in Developed and Developing Countries: Iodinated Contrast Media and Drugs Treatment by Light-Assisted AOPs

In both developed and developing countries, as mass flows of special drugs derive from hospital and production sites, two pollutants (Iohexol and Venlafaxine) in high amounts have been chosen and their degradation by relevant AOPs was studied. These drugs can be encountered in the production wastewaters or in urine, due to the treatment of patients, and the (pre)treatment of concentrated flows at hospital or manufacturing level is desirable before dilution in the municipal wastewater streams. A summary of the strategies can be found in Figure 6.

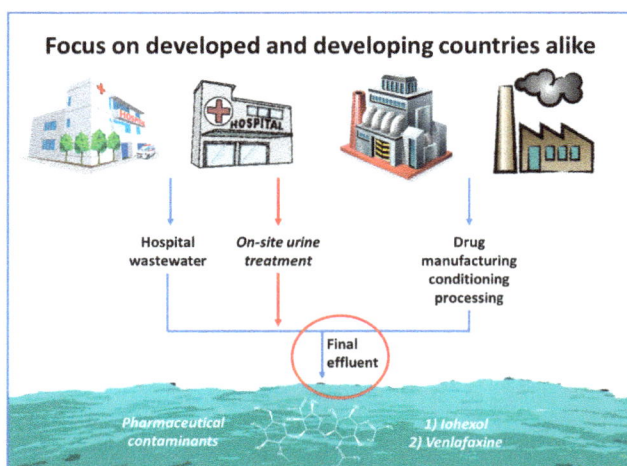

Figure 6. Schematic representation: Treatment of hospital and highly concentrated flows of PhACs. The treatment methods were associated with the concentration of the matrix (urine), hence UV-based solutions were sought, or the WW flows, allowing the assessment of photo-Fenton as a viable solution.

The Iodinated Contrast Medium (ICM) Iohexol is characterized by its recalcitrance to biological treatment, since its medical prescription aims to stay unchanged in the body for the duration of the tests of the patient. As such, it leaves the body with the same composition, making its discharge a problem; the doses for various exams may reach grams and was equally detected in Swiss and Ivorian HWW flows. In Figure 7, it is shown that applying UVC irradiation to spiked water, (simulated) wastewater and urine samples was very efficient in removing the parent compound, giving rise to de-iodinated intermediates that remained in the bulk, with no total organic carbon (TOC) reduction. Enhancing its action with H_2O_2 and by the production of $HO^\bullet$, did not significantly enhance the degradation kinetics, however moderately reduced the TOC of the solution, thus removing some of the recalcitrant intermediates. Similarly, adding iron to enhance the action of UV/H_2O_2 by the parallel UV-Fenton process only slightly enhanced the degradation kinetics, however the analysis of the intermediates revealed pathways that were not achieved during the degradation of the parent compound by UVC alone [89].

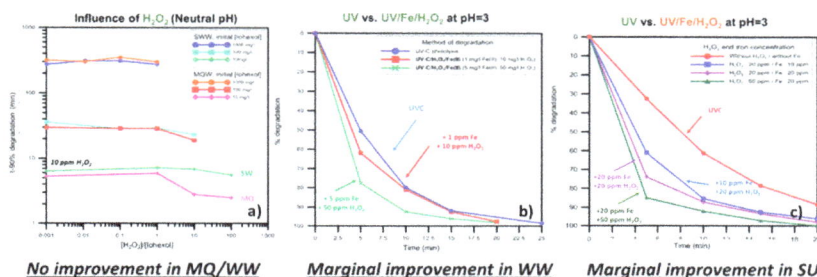

Figure 7. ICM Iohexol degradation by UV-based AOPs in: (a) water; (b) wastewater; and (c) urine. H_2O_2 addition and Fe-assisted experiments, aided by acidification of the matrix hold minor improvement in degradation kinetics.

Another highlight of the investigation was the opportunity presented to treat urine contaminated with Iohexol. By simulating the immediate collection of urine from the patient, it was shown that the contamination levels, can be significantly reduced before the discharge and subsequent dilution of the urine in HWW or UWW collection systems. The economic gain of such process would impact the treatment necessities of the final plant, e.g., AOPs, after secondary treatment. Hence, UV-based AOPs can play also a role of pre-treatment for Iodinated Contrast Media-containing effluents. Finally, although the use of iron was not justified by the kinetic investigation, the acute fractionation of the compound could be correlated with the opportunity to pre-treat such drugs prior to biological treatment, which is significantly cheaper from the AOPs treatment.

The other drug, Venlafaxine, is a compound that belongs in the broad family of serotonin-norepinephrine reuptake inhibitors, and is widely prescribed as an antidepressant. Although the amounts given are nothing like the ones encountered in the case of Iohexol, the results of the exposure of aquatic organisms in this drug has shown severe side-effects, such as alteration of the predation behavior of fish and disrupted locomotion of invertebrates [90–92]. Its treatment has been assessed by AOPs as TiO_2 photocatalysis and was further achieved by several AOPs, including UV-based and solar-assisted ones [93,94]. In Figure 8, the kinetic analysis showed a drug that can be removed without resolving to extreme conditions; moderate H_2O_2 addition (UVC light or few mg/L Fe, either combined with 10–25 mg/L) and reasonable residence time was necessary to remove the parent compound. However, the intermediate degradation was delayed revealing a recalcitrant nature of the drug. Nevertheless, its degradation was achieved in aqueous matrices such as MQ water, WW (real secondary and synthetic one), leaving promise for its efficient elimination in UWW or HWW. However, even in urine, which is a highly loaded matrix, with high UV absorbance, quantities as low as few µg/L were efficiently removed as followed by UPLC/MS analyses.

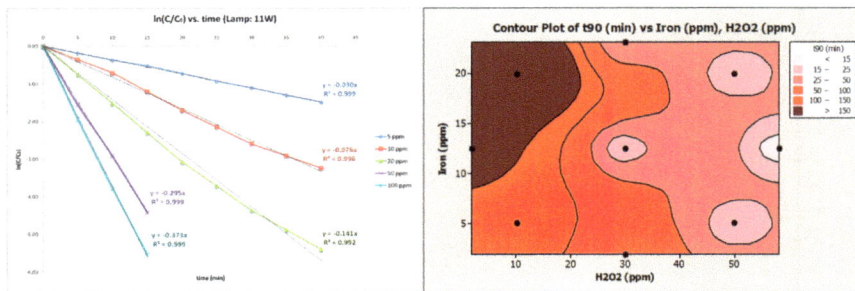

Figure 8. Antidepressant Venlafaxine degradation in MQ water. (**a**) UV/H_2O_2 mediated degradation and kinetic evaluation of H_2O_2 incremental addition; and (**b**) solar photo-Fenton at acidic pH, with the optimal degradation areas (light color contour).

Although the kinetics are important when its degradation is considered in MWTPs, other remarkable observations were made, with application in HWW or industrial-level (mg/L) contaminated waters. The pretreatment with UV-based or solar-assisted AOPs enhanced the biodegradability of the treated effluent, compared to the initial one. The gold standard of biodegradability tests, the Zahn–Wellens test revealed up to 20% increase in biodegradability of the treated effluents, by UV/H_2O_2 or solar photo-Fenton [95], in simulated WW containing mg/L amounts of the selected pharmaceutical. Hence, the restriction of dilution of HWW or manufacturing flows could be effectively pre-treated.

6. Conclusions

The use of light-assisted AOPs towards pollutant decontamination and disinfection of effluents, namely UV, UV/H_2O_2, solar light (shown to work as an AOP), Fenton and solar photo-Fenton are established as powerful allies in the ongoing task of wastewater purification. The key conclusions can be summarized as follows:

(1) UV-based AOPs are efficient for MP removal and MO inactivation. Although changing dynamically, the Swiss reality on hospital wastewater treatment dictates their discharge in the municipal collection network, and therefore implies their co-treatment with municipal wastes. The UV-based AOPs (UV and UV/H_2O_2) were found to be effective micropollutant removal strategies in ng/L level and bacterial inactivating processes, after biological secondary pre-treatment, as found in municipal wastewaters. When used in simulated hospital wastewaters and urine treatment, as alternative micropollutant elimination strategies, their efficiency was measured and established against a list of contaminants, with parallel elimination of the contained organic matter. The degradation was fast, and the reactants addition and necessary light doses were moderate.

(2) The solar photo-Fenton process and its constituents can be very effective in the proper context. Despite the lower apparent efficiency of this process when compared with its UV-based counterparts, photo-Fenton was found to effectively and non-selectively remove micropollutants and effluent organic matter. Furthermore, their application resulted in high bacterial removal, regrowth suppression, and yeasts and viruses inactivation from water and wastewater effluents. Most importantly, through systematic studies the mechanism and the key points of the process against the aforementioned targets were characterized. Special emphasis was given to the organic matter present in WW, as it is found to hinder the inactivation process but other benefits, such as iron complexation, also occur.

(3) The selected model hospital/industrial contaminants (Iohexol, Venlafaxine) helped elucidate the pitfalls and opportunities in HWW treatment by AOPs. The AOPs were found to work particularly well against the concentrated, (simulated) industrial wastewater, hospital flows and urine. Therefore, their application in hospitals and related industrial activities is promising. In addition, the structural deformation of the selected pollutants provided helpful insights on the operational and chemical constraints on applying the various AOPs; for instance the use of iron (when H_2O_2 is present) is strongly recommended for faster and more intense degradation of the contaminants in HWW. Finally, apart from the degradation point of view, the AOPs studied increased the biodegradability of the selected compounds treated solutions, which could allow their use as pre-treatment methods in HWWTPs.

In conclusion, more work is necessary to establish these methods as suitable for application in hospital environments. However, the initial results strongly support their further development, and future work stemming from the present research is encouraged to be sought.

Acknowledgments: The authors wish to thank the Swiss Agency for Development and Cooperation (SDC) and the Swiss National Foundation for the Research for Development (r4d) Grant, for the funding through the project "Treatment of the hospital wastewaters in Côte d'Ivoire and in Colombia by advanced oxidation processes" (Project No. 146919).

Conflicts of Interest: The authors declare no conflict of interest.

Abbreviations

Activated Sludge (AS), Advanced Oxidation Processes (AOPs), Biochemical Oxygen Demand (BOD), Chemical Oxygen Demand (COD), Coagulation-Flocculation (CF), Dissolved Organic Matter (DOM), Hospital wastewater (HWW), Iodinated Contrast Medium (ICM), Ligand-to-Metal Charge Transfer (LMCT), Microorganisms (MOs), Micropollutants (MPs), Moving Bed Bioreactor (MBBR), Municipal Wastewater (MWW), Non-Steroidal Anti-Inflammatory drugs (NSAIDs), Pharmaceutically Active Compounds (PhACs), Reactive Oxygen Species (ROS), Suspended Solids (SS), Total Organic Carbon (TOC), Wastewater (WW), Wastewater treatment plant (WWTP).

References

1. Fent, K.; Weston, A.A.; Caminada, D. Ecotoxicology of human pharmaceuticals. *Aquat. Toxicol.* **2006**, *76*, 122–159. [CrossRef] [PubMed]
2. Luo, Y.; Guo, W.; Ngo, H.H.; Nghiem, L.D.; Hai, F.I.; Zhang, J.; Liang, S.; Wang, X.C. A review on the occurrence of micropollutants in the aquatic environment and their fate and removal during wastewater treatment. *Sci. Total Environ.* **2014**, *473–474*, 619–641. [CrossRef] [PubMed]
3. Giannakis, S.; Darakas, E.; Escalas-Cañellas, A.; Pulgarin, C. Environmental considerations on solar disinfection of wastewater and the subsequent bacterial (re)growth. *Photochem. Photobiol. Sci.* **2015**, *14*, 618–625. [CrossRef] [PubMed]
4. Giannakis, S.; Merino Gamo, A.I.; Darakas, E.; Escalas-Cañellas, A.; Pulgarin, C. Monitoring the post-irradiation *E. coli* survival patterns in environmental water matrices: Implications in handling solar disinfected wastewater. *Chem. Eng. J.* **2014**, *253*, 366–376. [CrossRef]
5. Liberti, L.; Notarnicola, M.; Petruzzelli, D. Advanced treatment for municipal wastewater reuse in agriculture. UV disinfection: Parasite removal and by-product formation. *Desalination* **2003**, *152*, 315–324. [CrossRef]
6. McGuigan, K.G.; Conroy, R.M.; Mosler, H.J.; du Preez, M.; Ubomba-Jaswa, E.; Fernandez-Ibanez, P. Solar water disinfection (SODIS): A review from bench-top to roof-top. *J. Hazard. Mater.* **2012**, *235–236*, 29–46. [CrossRef] [PubMed]
7. Qadir, M.; Wichelns, D.; Raschid-Sally, L.; McCornick, P.G.; Drechsel, P.; Bahri, A.; Minhas, P. The challenges of wastewater irrigation in developing countries. *Agric. Water Manag.* **2010**, *97*, 561–568. [CrossRef]
8. Hoebe, C.J.; Vennema, H.; de Roda Husman, A.M.; van Duynhoven, Y.T. Norovirus outbreak among primary schoolchildren who had played in a recreational water fountain. *J. Infect. Dis.* **2004**, *189*, 699–705. [CrossRef] [PubMed]
9. Deblonde, T.; Cossu-Leguille, C.; Hartemann, P. Emerging pollutants in wastewater: A review of the literature. *Int. J. Hyg. Environ. Health* **2011**, *214*, 442–448. [CrossRef] [PubMed]

10. Focazio, M.J.; Kolpin, D.W.; Barnes, K.K.; Furlong, E.T.; Meyer, M.T.; Zaugg, S.D.; Barber, L.B.; Thurman, M.E. A national reconnaissance for pharmaceuticals and other organic wastewater contaminants in the United States—(II) Untreated drinking water sources. *Sci. Total Environ.* **2008**, *402*, 201–216. [CrossRef] [PubMed]
11. Schwarzenbach, R.P.; Escher, B.I.; Fenner, K.; Hofstetter, T.B.; Johnson, C.A.; von Gunten, U.; Wehrli, B. The Challenge of Micropollutants in Aquatic Systems. *Science* **2006**, *313*, 1072–1077. [CrossRef] [PubMed]
12. Gregorio, V.; Chèvre, N. Assessing the risks posed by mixtures of chemicals in freshwater environments: Case study of Lake Geneva, Switzerland. *Wiley Interdiscip. Rev. Water* **2014**, *1*, 229–247. [CrossRef]
13. Backhaus, T.; Altenburger, R.; Arrhenius, Å.; Blanck, H.; Faust, M.; Finizio, A.; Gramatica, P.; Grote, M.; Junghans, M.; Meyer, W.; et al. The BEAM-project: Prediction and assessment of mixture toxicities in the aquatic environment. *Cont. Shelf Res.* **2003**, *23*, 1757–1769. [CrossRef]
14. Deneer, J.W. Toxicity of mixtures of pesticides in aquatic systems. *Pest Manag. Sci.* **2000**, *56*, 516–520. [CrossRef]
15. Junghans, M.; Backhaus, T.; Faust, M.; Scholze, M.; Grimme, L.H. Application and validation of approaches for the predictive hazard assessment of realistic pesticide mixtures. *Aquat. Toxicol.* **2006**, *76*, 93–110. [CrossRef] [PubMed]
16. Rodney, S.I.; Teed, R.S.; Moore, D.R.J. Estimating the Toxicity of Pesticide Mixtures to Aquatic Organisms: A Review. *Hum. Ecol. Risk Assess. Int. J.* **2013**, *19*, 1557–1575. [CrossRef]
17. Rizzo, L.; Manaia, C.; Merlin, C.; Schwartz, T.; Dagot, C.; Ploy, M.C.; Michael, I.; Fatta-Kassinos, D. Urban wastewater treatment plants as hotspots for antibiotic resistant bacteria and genes spread into the environment: A review. *Sci. Total Environ.* **2013**, *447*, 345–360. [CrossRef] [PubMed]
18. Fatta-Kassinos, D.; Meric, S.; Nikolaou, A. Pharmaceutical residues in environmental waters and wastewater: Current state of knowledge and future research. *Anal. Bioanal. Chem.* **2011**, *399*, 251–275. [CrossRef] [PubMed]
19. Verlicchi, P.; Al Aukidy, M.; Galletti, A.; Petrovic, M.; Barcelo, D. Hospital effluent: Investigation of the concentrations and distribution of pharmaceuticals and environmental risk assessment. *Sci. Total Environ.* **2012**, *430*, 109–118. [CrossRef] [PubMed]
20. Emmanuel, E.; Keck, G.; Blanchard, J.-M.; Vermande, P.; Perrodin, Y. Toxicological effects of disinfections using sodium hypochlorite on aquatic organisms and its contribution to AOX formation in hospital wastewater. *Environ. Int.* **2004**, *30*, 891–900. [CrossRef] [PubMed]
21. Boillot, C.; Bazin, C.; Tissot-Guerraz, F.; Droguet, J.; Perraud, M.; Cetre, J.C.; Trepo, D.; Perrodin, Y. Daily physicochemical, microbiological and ecotoxicological fluctuations of a hospital effluent according to technical and care activities. *Sci. Total Environ.* **2008**, *403*, 113–129. [CrossRef] [PubMed]
22. Emmanuel, E.; Perrodin, Y.; Keck, G.; Blanchard, J.M.; Vermande, P. Ecotoxicological risk assessment of hospital wastewater: A proposed framework for raw effluents discharging into urban sewer network. *J. Hazard. Mater.* **2005**, *117*, 1–11. [CrossRef] [PubMed]
23. Hawkshead, J.J., III. Hospital wastewater containing pharmaceutically active compounds and drug-resistant organisms: A source of environmental toxicity and increased antibiotic resistance. *J. Residuals Sci. Technol.* **2008**, *5*, 51–60.
24. Altin, A.; Altin, S.; Degirmenci, M. Characteristics and treatability of hospital(medical) wastewaters. *Fresenius Environ. Bull.* **2003**, *12*, 1098–1108.
25. Pauwels, B.; Verstraete, W. The treatment of hospital wastewater: An appraisal. *J. Water Health* **2006**, *4*, 405–416. [PubMed]
26. Vieno, N.; Tuhkanen, T.; Kronberg, L. Elimination of pharmaceuticals in sewage treatment plants in Finland. *Water Res.* **2007**, *41*, 1001–1012. [CrossRef] [PubMed]
27. Kosma, C.I.; Lambropoulou, D.A.; Albanis, T.A. Occurrence and removal of PPCPs in municipal and hospital wastewaters in Greece. *J. Hazard. Mater.* **2010**, *179*, 804–817. [CrossRef] [PubMed]
28. Kümmerer, K. Drugs in the environment: Emission of drugs, diagnostic aids and disinfectants into wastewater by hospitals in relation to other sources—A review. *Chemosphere* **2001**, *45*, 957–969. [CrossRef]
29. Mahnik, S.; Lenz, K.; Weissenbacher, N.; Mader, R.; Fuerhacker, M. Fate of 5-fluorouracil, doxorubicin, epirubicin, and daunorubicin in hospital wastewater and their elimination by activated sludge and treatment in a membrane-bio-reactor system. *Chemosphere* **2007**, *66*, 30–37. [CrossRef] [PubMed]
30. Suarez, S.; Lema, J.M.; Omil, F. Pre-treatment of hospital wastewater by coagulation–flocculation and flotation. *Bioresour. Technol.* **2009**, *100*, 2138–2146. [CrossRef] [PubMed]

31. Verlicchi, P.; Galletti, A.; Petrovic, M.; Barceló, D. Hospital effluents as a source of emerging pollutants: An overview of micropollutants and sustainable treatment options. *J. Hydrol.* **2010**, *389*, 416–428. [CrossRef]

32. Jones, O.H.; Voulvoulis, N.; Lester, J. Human pharmaceuticals in wastewater treatment processes. *Crit. Rev. Environ. Sci. Technol.* **2005**, *35*, 401–427. [CrossRef]

33. Ikehata, K.; Jodeiri Naghashkar, N.; Gamal El-Din, M. Degradation of Aqueous Pharmaceuticals by Ozonation and Advanced Oxidation Processes: A Review. *Ozone Sci. Eng.* **2006**, *28*, 353–414. [CrossRef]

34. Lienert, J.; Burki, T.; Escher, B. Reducing micropollutants with source control: Substance flow analysis of 212 pharmaceuticals in faeces and urine. *Water Sci. Technol.* **2007**, *56*, 87–96. [CrossRef] [PubMed]

35. Carraro, E.; Bonetta, S.; Bertino, C.; Lorenzi, E.; Bonetta, S.; Gilli, G. Hospital effluents management: Chemical, physical, microbiological risks and legislation in different countries. *J. Environ. Manag.* **2016**, *168*, 185–199. [CrossRef] [PubMed]

36. El-Ogri, F.; Ouazzani, N.; Boaâm, F.; Mandi, L. A survey of wastewaters generated by a hospital in Marrakech city and their characterization. *Desalination Water Treat.* **2016**, *57*, 17061–17074.

37. Kilunga, P.I.; Kayembe, J.M.; Laffite, A.; Thevenon, F.; Devarajan, N.; Mulaji, C.K.; Mubedi, J.I.; Yav, Z.G.; Otamonga, J.-P.; Mpiana, P.T. The impact of hospital and urban wastewaters on the bacteriological contamination of the water resources in Kinshasa, Democratic Republic of Congo. *J. Environ.Sci. Health Part A* **2016**, *51*, 1–9. [CrossRef] [PubMed]

38. Gernjak, W.; Fuerhacker, M.; Fernández-Ibañez, P.; Blanco, J.; Malato, S. Solar photo-Fenton treatment—Process parameters and process control. *Appl. Catal. B Environ.* **2006**, *64*, 121–130. [CrossRef]

39. Mantzavinos, D.; Kalogerakis, N. Treatment of olive mill effluents: Part I. Organic matter degradation by chemical and biological processes—An overview. *Environ. Int.* **2005**, *31*, 289–295. [CrossRef] [PubMed]

40. Oller, I.; Malato, S.; Sanchez-Perez, J.A. Combination of Advanced Oxidation Processes and biological treatments for wastewater decontamination—A review. *Sci. Total Environ.* **2011**, *409*, 4141–4166. [CrossRef] [PubMed]

41. Ribeiro, A.R.; Nunes, O.C.; Pereira, M.F.; Silva, A.M. An overview on the advanced oxidation processes applied for the treatment of water pollutants defined in the recently launched Directive 2013/39/EU. *Environ. Int.* **2015**, *75*, 33–51. [CrossRef] [PubMed]

42. Michael, I.; Frontistis, Z.; Fatta-Kassinos, D. Removal of pharmaceuticals from environmentally relevant matrices by advanced oxidation processes (AOPs). *Compr. Anal. Chem.* **2013**, *62*, 345–407.

43. Oppenländer, T. *Photochemical Purification of Water and Air: Advanced Oxidation Processes (AOPs)—Principles, Reaction Mechanisms, Reactor Concepts*; John Wiley & Sons: Hoboken, NJ, USA, 2003.

44. Legrini, O.; Oliveros, E.; Braun, A. Photochemical processes for water treatment. *Chem. Rev.* **1993**, *93*, 671–698. [CrossRef]

45. Kim, I.; Yamashita, N.; Tanaka, H. Performance of UV and UV/H_2O_2 processes for the removal of pharmaceuticals detected in secondary effluent of a sewage treatment plant in Japan. *J. Hazard. Mater.* **2009**, *166*, 1134–1140. [CrossRef] [PubMed]

46. Kim, I.; Yamashita, N.; Tanaka, H. Photodegradation of pharmaceuticals and personal care products during UV and UV/H_2O_2 treatments. *Chemosphere* **2009**, *77*, 518–525. [CrossRef] [PubMed]

47. Alpert, S.M.; Knappe, D.R.; Ducoste, J.J. Modeling the UV/hydrogen peroxide advanced oxidation process using computational fluid dynamics. *Water Res.* **2010**, *44*, 1797–1808. [CrossRef] [PubMed]

48. Zapata, A.; Oller, I.; Rizzo, L.; Hilgert, S.; Maldonado, M.; Sánchez-Pérez, J.; Malato, S. Evaluation of operating parameters involved in solar photo-Fenton treatment of wastewater: Interdependence of initial pollutant concentration, temperature and iron concentration. *Appl. Catal. B Environ.* **2010**, *97*, 292–298. [CrossRef]

49. Andreozzi, R.; Caprio, V.; Insola, A.; Marotta, R. Advanced oxidation processes (AOP) for water purification and recovery. *Catal. Today* **1999**, *53*, 51–59. [CrossRef]

50. Neyens, E.; Baeyens, J. A review of classic Fenton's peroxidation as an advanced oxidation technique. *J. Hazard. Mater.* **2003**, *98*, 33–50. [CrossRef]

51. Stasinakis, A. Use of selected advanced oxidation processes (AOPs) for wastewater treatment—A mini review. *Glob. NEST J.* **2008**, *10*, 376–385.

52. De Laat, J.; Gallard, H. Catalytic decomposition of hydrogen peroxide by Fe (III) in homogeneous aqueous solution: Mechanism and kinetic modeling. *Environ. Sci. Technol.* **1999**, *33*, 2726–2732. [CrossRef]

53. Poyatos, J.; Muñio, M.; Almecija, M.; Torres, J.; Hontoria, E.; Osorio, F. Advanced oxidation processes for wastewater treatment: State of the art. *Water Air Soil Pollut.* **2010**, *205*, 187–204. [CrossRef]

54. Giannakis, S.; Polo López, M.I.; Spuhler, D.; Sánchez Pérez, J.A.; Fernández Ibáñez, P.; Pulgarin, C. Solar disinfection is an augmentable, in situ-generated photo-Fenton reaction—Part 1: A review of the mechanisms and the fundamental aspects of the process. *Appl. Catal. B Environ.* **2016**, *199*, 199–223. [CrossRef]

55. Giannakis, S.; Polo López, M.I.; Spuhler, D.; Sánchez Pérez, J.A.; Fernández Ibáñez, P.; Pulgarin, C. Solar disinfection is an augmentable, in situ-generated photo-Fenton reaction—Part 2: A review of the applications for drinking water and wastewater disinfection. *Appl. Catal. B Environ.* **2016**, *198*, 431–446. [CrossRef]

56. Kolpin, D.W.; Furlong, E.T.; Meyer, M.T.; Thurman, E.M.; Zaugg, S.D.; Barber, L.B.; Buxton, H.T. Pharmaceuticals, Hormones, and Other Organic Wastewater Contaminants in U.S. Streams, 1999–2000: A National Reconnaissance. *Environ. Sci. Technol.* **2002**, *36*, 1202–1211. [CrossRef] [PubMed]

57. Kasprzyk-Hordern, B.; Dinsdale, R.M.; Guwy, A.J. The removal of pharmaceuticals, personal care products, endocrine disruptors and illicit drugs during wastewater treatment and its impact on the quality of receiving waters. *Water Res.* **2009**, *43*, 363–380. [CrossRef] [PubMed]

58. Gros, M.; Petrović, M.; Barceló, D. Wastewater treatment plants as a pathway for aquatic contamination by pharmaceuticals in the Ebro river basin (Northeast Spain). *Environ. Toxicol. Chem.* **2007**, *26*, 1553–1562. [CrossRef] [PubMed]

59. Pal, A.; Gin, K.Y.-H.; Lin, A.Y.-C.; Reinhard, M. Impacts of emerging organic contaminants on freshwater resources: Review of recent occurrences, sources, fate and effects. *Sci. Total Environ.* **2010**, *408*, 6062–6069. [CrossRef] [PubMed]

60. Wang, C.; Shi, H.; Adams, C.D.; Gamagedara, S.; Stayton, I.; Timmons, T.; Ma, Y. Investigation of pharmaceuticals in Missouri natural and drinking water using high performance liquid chromatography-tandem mass spectrometry. *Water Res.* **2011**, *45*, 1818–1828. [CrossRef] [PubMed]

61. Vulliet, E.; Cren-Olivé, C.; Grenier-Loustalot, M.-F. Occurrence of pharmaceuticals and hormones in drinking water treated from surface waters. *Environ. Chem. Lett.* **2011**, *9*, 103–114. [CrossRef]

62. Kümmerer, K. Emerging Contaminants versus Micro-pollutants. *CLEAN Soil Air Water* **2011**, *39*, 889–890. [CrossRef]

63. Giannakis, S.; Darakas, E.; Escalas-Cañellas, A.; Pulgarin, C. Solar disinfection modeling and post-irradiation response of Escherichia coli in wastewater. *Chem. Eng. J.* **2015**, *281*, 588–598. [CrossRef]

64. Giannakis, S.; Merino Gamo, A.I.; Darakas, E.; Escalas-Cañellas, A.; Pulgarin, C. Impact of different light intermittence regimes on bacteria during simulated solar treatment of secondary effluent: Implications of the inserted dark periods. *Sol. Energy* **2013**, *98*, 572–581. [CrossRef]

65. Giannakis, S.; Darakas, E.; Escalas-Cañellas, A.; Pulgarin, C. The antagonistic and synergistic effects of temperature during solar disinfection of synthetic secondary effluent. *J. Photochem. Photobiol. A Chem.* **2014**, *280*, 14–26. [CrossRef]

66. Giannakis, S.; Darakas, E.; Escalas-Cañellas, A.; Pulgarin, C. Elucidating bacterial regrowth: Effect of disinfection conditions in dark storage of solar treated secondary effluent. *J. Photochem. Photobiol. A Chem.* **2014**, *290*, 43–53. [CrossRef]

67. Chong, M.N.; Jin, B. Photocatalytic treatment of high concentration carbamazepine in synthetic hospital wastewater. *J. Hazard. Mater.* **2012**, *199*, 135–142. [CrossRef] [PubMed]

68. Ganzenko, O.; Huguenot, D.; van Hullebusch, E.D.; Esposito, G.; Oturan, M.A. Electrochemical advanced oxidation and biological processes for wastewater treatment: A review of the combined approaches. *Environ. Sci. Pollut. Res.* **2014**, *21*, 8493–8524. [CrossRef] [PubMed]

69. Kajitvichyanukul, P.; Suntronvipart, N. Evaluation of biodegradability and oxidation degree of hospital wastewater using photo-Fenton process as the pretreatment method. *J. Hazard. Mater.* **2006**, *138*, 384–391. [CrossRef] [PubMed]

70. Kanakaraju, D.; Glass, B.D.; Oelgemöller, M. Titanium dioxide photocatalysis for pharmaceutical wastewater treatment. *Environ. Chem. Lett.* **2014**, *12*, 27–47. [CrossRef]

71. Köhler, C.; Venditti, S.; Igos, E.; Klepiszewski, K.; Benetto, E.; Cornelissen, A. Elimination of pharmaceutical residues in biologically pre-treated hospital wastewater using advanced UV irradiation technology: A comparative assessment. *J. Hazard. Mater.* **2012**, *239*, 70–77. [CrossRef] [PubMed]

72. Kovalova, L.; Siegrist, H.; von Gunten, U.; Eugster, J.; Hagenbuch, M.; Wittmer, A.; Moser, R.; McArdell, C.S. Elimination of Micropollutants during Post-Treatment of Hospital Wastewater with Powdered Activated Carbon, Ozone, and UV. *Environ. Sci. Technol.* **2013**, *47*, 7899–7908. [CrossRef] [PubMed]

73. Nielsen, U.; Hastrup, C.; Klausen, M.M.; Pedersen, B.M.; Kristensen, G.H.; Jansen, J.L.C.; Bak, S.N.; Tuerk, J. Removal of APIs and bacteria from hospital wastewater by MBR plus O(3), O(3) + H(2)O(2), PAC or ClO(2). *Water Sci. Technol.* **2013**, *67*, 854–862. [CrossRef] [PubMed]

74. Sprehe, M.; Geissen, S.U.; Vogelpohl, A. Photochemical oxidation of iodized X-ray contrast media (XRC) in hospital wastewater. *Water Sci.Technol. J. Int. Assoc. Water Pollut. Res.* **2001**, *44*, 317–323.

75. Carra, I.; Sánchez Pérez, J.A.; Malato, S.; Autin, O.; Jefferson, B.; Jarvis, P. Performance of different advanced oxidation processes for tertiary wastewater treatment to remove the pesticide acetamiprid. *J. Chem. Technol. Biotechnol.* **2016**, *91*, 72–81. [CrossRef]

76. Rivas, G.; Carra, I.; García Sánchez, J.L.; Casas López, J.L.; Malato, S.; Sánchez Pérez, J.A. Modelling of the operation of raceway pond reactors for micropollutant removal by solar photo-Fenton as a function of photon absorption. *Appl. Catal. B Environ.* **2015**, *178*, 210–217. [CrossRef]

77. Klamerth, N.; Malato, S.; Agüera, A.; Fernández-Alba, A.; Mailhot, G. Treatment of municipal wastewater treatment plant effluents with modified photo-Fenton as a tertiary treatment for the degradation of micro pollutants and disinfection. *Enviro. Sci. Technol.* **2012**, *46*, 2885–2892. [CrossRef] [PubMed]

78. Giannakis, S.; Gamarra Vives, F.A.; Grandjean, D.; Magnet, A.; De Alencastro, L.F.; Pulgarin, C. Effect of advanced oxidation processes on the micropollutants and the effluent organic matter contained in municipal wastewater previously treated by three different secondary methods. *Water Res.* **2015**, *84*, 295–306. [CrossRef] [PubMed]

79. Giannakis, S.; Voumard, M.; Grandjean, D.; Magnet, A.; De Alencastro, L.F.; Pulgarin, C. Micropollutant degradation, bacterial inactivation and regrowth risk in wastewater effluents: Influence of the secondary (pre) treatment on the efficiency of Advanced Oxidation Processes. *Water Res.* **2016**, *102*, 505–515. [CrossRef] [PubMed]

80. Margot, J.; Kienle, C.; Magnet, A.; Weil, M.; Rossi, L.; De Alencastro, L.F.; Abegglen, C.; Thonney, D.; Chèvre, N.; Schärer, M. Treatment of micropollutants in municipal wastewater: Ozone or powdered activated carbon? *Sci. Total Environ.* **2013**, *461*, 480–498. [CrossRef] [PubMed]

81. De la Cruz, N.; Esquius, L.; Grandjean, D.; Magnet, A.; Tungler, A.; de Alencastro, L.F.; Pulgarín, C. Degradation of emergent contaminants by UV, UV/H2O2 and neutral photo-Fenton at pilot scale in a domestic wastewater treatment plant. *Water Res.* **2013**, *47*, 5836–5845. [CrossRef] [PubMed]

82. Von Sperling, M. Performance evaluation and mathematical modelling of coliform die-off in tropical and subtropical waste stabilization ponds. *Water Res.* **1999**, *33*, 1435–1448. [CrossRef]

83. Von Sperling, M. Modelling of coliform removal in 186 facultative and maturation ponds around the world. *Water Res.* **2005**, *39*, 5261–5273. [CrossRef] [PubMed]

84. Canonica, S. Oxidation of aquatic organic contaminants induced by excited triplet states. *CHIMIA Int. J. Chem.* **2007**, *61*, 641–644. [CrossRef]

85. Ndounla, J.; Spuhler, D.; Kenfack, S.; Wéthé, J.; Pulgarin, C. Inactivation by solar photo-Fenton in pet bottles of wild enteric bacteria of natural well water: Absence of re-growth after one week of subsequent storage. *Appl. Catal. B Environ.* **2013**, *129*, 309–317. [CrossRef]

86. Giannakis, S.; Liu, S.; Carratalà, A.; Rtimi, S.; Bensimon, M.; Pulgarin, C. Effect of Fe (II)/Fe (III) species, pH, irradiance and bacterial presence on viral inactivation in wastewater by the photo-Fenton process: Kinetic modeling and mechanistic interpretation. *Appl. Catal. B Environ.* **2017**, *204*, 156–166. [CrossRef]

87. Muthukumaran, S.; Nguyen, D.A.; Baskaran, K. Performance evaluation of different ultrafiltration membranes for the reclamation and reuse of secondary effluent. *Desalination* **2011**, *279*, 383–389. [CrossRef]

88. Giannakis, S.; Ruales-Lonfat, C.; Rtimi, S.; Thabet, S.; Cotton, P.; Pulgarin, C. Castles fall from inside: Evidence for dominant internal photo-catalytic mechanisms during treatment of Saccharomyces cerevisiae by photo-Fenton at near-neutral pH. *Appl. Catal. B Environ.* **2016**, *185*, 150–162. [CrossRef]

89. Giannakis, S.; Jovic, M.; Gasilova, N.; Pastor Gelabert, M.; Schindelholz, S.; Furbringer, J.-M.; Girault, H.; Pulgarin, C. Iohexol degradation in wastewater and urine by UV-based Advanced Oxidation Processes (AOPs): Process modeling and by-products identification. *J. Environ. Manag.* **2017**, *195*, 174–185. [CrossRef] [PubMed]

90. Bisesi, J.H.; Bridges, W.; Klaine, S.J. Effects of the antidepressant venlafaxine on fish brain serotonin and predation behavior. *Aquat. Toxicol.* **2014**, *148*, 130–138. [CrossRef] [PubMed]
91. Fong, P.P.; Ford, A.T. The biological effects of antidepressants on the molluscs and crustaceans: A review. *Aquat. Toxicol.* **2014**, *151*, 4–13. [CrossRef] [PubMed]
92. Fong, P.P.; Molnar, N. Antidepressants cause foot detachment from substrate in five species of marine snail. *Mar. Environ. Res.* **2013**, *84*, 24–30. [CrossRef] [PubMed]
93. García-Galán, M.J.; Anfruns, A.; Gonzalez-Olmos, R.; Rodríguez-Mozaz, S.; Comas, J. UV/H_2O_2 degradation of the antidepressants venlafaxine and O-desmethylvenlafaxine: Elucidation of their transformation pathway and environmental fate. *J. Hazard. Mater.* **2016**, *311*, 70–80. [CrossRef] [PubMed]
94. Lambropoulou, D.; Evgenidou, E.; Saliverou, V.; Kosma, C.; Konstantinou, I. Degradation of venlafaxine using TiO_2/UV process: Kinetic studies, RSM optimization, identification of transformation products and toxicity evaluation. *J. Hazard. Mater.* **2017**, *323*, 513–526. [CrossRef] [PubMed]
95. Giannakis, S.; Hendaoui, I.; Jovic, M.; Grandjean, D.; De Alencastro, L.F.; Girault, H.; Pulgarin, C. Solar photo-Fenton and UV/H_2O_2 processes against the antidepressant Venlafaxine in urban wastewaters and human urine. Intermediates formation and biodegradability assessment. *Chem. Eng. J.* **2017**, *308*, 492–504. [CrossRef]

molecules

MDPI

Article

Degradation of Methyl 2-Aminobenzoate (Methyl Anthranilate) by H$_2$O$_2$/UV: Effect of Inorganic Anions and Derived Radicals

Grazia Maria Lanzafame [1,2,†], Mohamed Sarakha [1], Debora Fabbri [2] and Davide Vione [2,3,*]

1   Institut de Chimie de Clermont-Ferrand, Clermont Université, Université Blaise Pascal, F-63177 Aubière, France; grazialanzafame@gmail.com (G.M.L.); mohamed.sarakha@uca.fr (M.S.)
2   Dipartimento di Chimica, Università di Torino, Via Pietro Giuria 5, 10125 Torino, Italy; debora.fabbri@unito.it
3   Centro Interdipartimentale NatRisk, Università di Torino, Largo Paolo Braccini 2, 10095 Grugliasco (TO), Italy
*   Correspondence: davide.vione@unito.it; Tel.: +39-11-670-5296
†   Present address: Institut National de l'environnement Industriel et des Risques (INERIS), Rue Jacques Taffanel, F-60550 Verneuil-en-Halatte, France

Academic Editor: Pierre Pichat
Received: 24 February 2017; Accepted: 6 April 2017; Published: 12 April 2017

Abstract: This study shows that methyl 2-aminobenzoate (also known as methyl anthranilate, hereafter MA) undergoes direct photolysis under UVC and UVB irradiation and that its photodegradation is further accelerated in the presence of H$_2$O$_2$. Hydrogen peroxide acts as a source of hydroxyl radicals (·OH) under photochemical conditions and yields MA hydroxyderivatives. The trend of MA photodegradation rate vs. H$_2$O$_2$ concentration reaches a plateau because of the combined effects of H$_2$O$_2$ absorption saturation and ·OH scavenging by H$_2$O$_2$. The addition of chloride ions causes scavenging of ·OH, yielding Cl$_2$·$^-$ as the most likely reactive species, and it increases the MA photodegradation rate at high H$_2$O$_2$ concentration values. The reaction between Cl$_2$·$^-$ and MA, which has second-order rate constant $k_{Cl_2^{\bullet-}+MA} = (4.0 \pm 0.3) \times 10^8$ M^{-1}·s^{-1} (determined by laser flash photolysis), appears to be more selective than the ·OH process in the presence of H$_2$O$_2$, because Cl$_2$·$^-$ undergoes more limited scavenging by H$_2$O$_2$ compared to ·OH. While the addition of carbonate causes ·OH scavenging to produce CO$_3$·$^-$ ($k_{CO_3^{\bullet-}+MA} = (3.1 \pm 0.2) \times 10^8$ M^{-1}·s^{-1}), carbonate considerably inhibits the photodegradation of MA. A possible explanation is that the elevated pH values of the carbonate solutions make H$_2$O$_2$ to partially occur as HO$_2$$^-$, which reacts very quickly with either ·OH or CO$_3$·$^-$ to produce O$_2$·$^-$. The superoxide anion could reduce partially oxidised MA back to the initial substrate, with consequent inhibition of MA photodegradation. Fast MA photodegradation is also observed in the presence of persulphate/UV, which yields SO$_4$·$^-$ that reacts effectively with MA ($k_{SO_4^{\bullet-}+MA} = (5.6 \pm 0.4) \times 10^9$ M^{-1}·s^{-1}). Irradiated H$_2$O$_2$ is effective in photodegrading MA, but the resulting MA hydroxyderivatives are predicted to be about as toxic as the parent compound for aquatic organisms (most notably, fish and crustaceans).

Keywords: advanced oxidation processes; methyl anthranilate; methyl 2-aminobenzoate; hydrogen peroxide; photodegradation intermediates; emerging contaminants

1. Introduction

Methyl 2-aminobenzoate (MA, C$_8$H$_9$NO$_2$) is a clear liquid that occurs in many essential oils. It has a melting point of 24 °C, a boiling point of 256 °C, and a density of 1.17 g·mL^{-1} [1]. MA can be found in Concord grapes, jasmine, bergamot, lemon, orange, and strawberries, and it is used as bird repellent in the protection of food crops such as corn, sunflower, rice, and fruits [2]. MA is also employed to prevent birds from accessing oil spill–contaminated water or water pools near airports, in the latter case reducing the risk of collision with aircraft [3]. MA is used as well for the flavouring of candies, soft drinks, chewing gum, and medicines [4], which suggests that it has limited toxicity towards mammals, including humans. Its LD50 (lethal dose 50) values for rats and rabbits are in fact in the g/(Kg body weight) range for either oral or dermal uptake [5]. Despite these apparently favourable features, MA shows non-negligible toxicity for aquatic organisms, with LC50 (lethal concentration 50 in water, i.e., acute toxicity) values of 20–30 mg·L^{-1} for fish [6]. More importantly, it can cause chronic effects to both fish and crustaceans at 10–70 μg·L^{-1} levels [7–9]. Therefore, toxicity to fish is to be taken into account when using MA to protect aquaculture facilities from predation by birds [10]. MA is also poorly biodegradable, little volatile, and it undergoes limited partitioning to solids. Moreover, its predicted hydrolysis time is in the range of several months to some years [11]. Following ingestion as a food additive and excretion, this compound is unlikely to be eliminated from the aqueous phase of the wastewater treatment plants (estimates for MA elimination are in the range of a few percent, mostly accounted for by sludge adsorption) [11]. Therefore, MA could easily reach surface water environments. Unfortunately, concentration data of MA in surface waters are extremely difficult to be found in the literature, as are data concerning the elimination of MA from aqueous solutions including wastewater.

Poor elimination during wastewater treatment is a widespread feature of several emerging substances used as drugs, fragrances, fire extinguishers etc. These compounds can be found in surface waters at significant levels due to point-source emission at the treatment plant outlets [12,13]. A likely future development in wastewater treatment will be the update of the existing plants to enable the removal of emerging contaminants. As technological upgrade options, Advanced Oxidation Processes (AOPs) are among the best promising tools [14,15]. Most AOPs are based on the thermal, electrochemical, photochemical or sonochemical generation of hydroxyl radicals (·OH), in the homogeneous phase or in heterogeneous systems (e.g., heterogeneous photocatalysis, [16]). The ·OH radicals are very strong oxidants and they can react with a very wide variety of organic and inorganic molecules, including pollutants. The relevant reactions include electron or hydrogen transfer, as well as addition to double bonds and aromatic rings [17]. Among AOPs, the UV irradiation of hydrogen peroxide (hereafter, H_2O_2/UV) presents several advantages including the elevated quantum yield of ·OH photogeneration (the reaction $H_2O_2 + h\nu \rightarrow 2$ ·OH has $\Phi_{H_2O_2}$ ~0.5 and $\Phi_{\bullet OH}$ ~1 [17,18]), the low cost of H_2O_2, and the formation of rather innocuous reaction by-products from H_2O_2 itself (mostly H_2O and O_2) [19,20]. H_2O_2 is also water-miscible and it is relatively safe to store and transport pending some precautions, but its water solutions need to be added with stabilisers (e.g., stannate, pyrophosphate, nitrate, colloidal silicate) that will finally end up in the water undergoing treatment [18]. Moreover, toxic or otherwise harmful photodegradation intermediates may be formed when dealing with the treatment of certain organic pollutants [21]. Another issue is that ·OH reacts not only with the target contaminant, but also with natural organic matter (NOM) and with inorganic anions that occur in aqueous solution. While NOM is mostly a ·OH quencher, in the case of some anions the framework is more complicated because their oxidation by ·OH yields reactive radical species, which are less reactive than ·OH itself but could still initiate some degradation processes on their own. Examples are the ·OH-induced formation of the carbonate radical ($CO_3 \cdot^-$) from carbonate and bicarbonate, of dichloride ($Cl_2 \cdot^-$) from chloride, of dibromide ($Br_2 \cdot^-$) from bromide, and of nitrogen dioxide (·NO_2) from nitrite [22]. Moreover, one of the possibilities to reduce the consumption of reactive species by the organic and inorganic components of natural waters is to replace ·OH with the more selective sulphate radical, $SO_4 \cdot^-$, which reacts with several pollutants but undergoes less interference from NOM and inorganic anions compared to ·OH [23]. To produce $SO_4 \cdot^-$, it is often sufficient to replace H_2O_2 with the analogous peroxide persulphate ($S_2O_8{}^{2-}$) in comparable processes [23,24].

The goal of this work is to study the photoinduced degradation of MA under UV irradiation (direct UVC photolysis, here used as benchmark) and with H_2O_2/UV and persulphate/UV treatments, as well as to assess the effect on the process of common inorganic anions such as chloride and carbonate. To better assess the effect of the added anions, the reactivity of $CO_3 \cdot^-$ and $Cl_2 \cdot^-$ with MA was studied by using the nanosecond laser flash photolysis technique. Because MA is not harmless to aquatic environments, this study investigates the following: (i) whether and to what extent MA could be photodegraded under AOP conditions, also in the presence of inorganic anions such as chloride and carbonate; and (ii) the potential of MA photodegradation to produce intermediates that might have higher impact than the parent compound, and that could be formed during the AOP removal of MA and/or other contaminants.

2. Results and Discussion

2.1. MA Photodegradation by UV and H_2O_2/UV

The photoinduced degradation of 0.1 mM MA was first studied under UVC irradiation alone (lamp maximum emission at 254 nm) and under UVC irradiation in the presence of different concentration values of H_2O_2 (see Figure 1 for the absorption spectra of MA and H_2O_2). The MA time evolution under these conditions is reported in Figure 2A, while Figure 2B reports the trend of the photodegradation rate of MA (R_{MA}) as a function of the H_2O_2 concentration. Table 1 reports the pseudo-first order photodegradation rate constants of MA (k_{MA}) for this and other series of experiments. Note that $R_{MA} = k_{MA} [MA]_o$, where $[MA]_o$ = 0.1 mM is the initial concentration of MA.

Table 1. Pseudo-first order photodegradation rate constants of MA (k_{MA}) in the different irradiation experiments. The initial concentration values of hydrogen peroxide, persulphate, chloride and carbonate are also reported. The error bounds to the k_{MA} data represent the sigma-level uncertainty of the pseudo-first order kinetic model. In all the cases the initial concentration of MA was 0.1 mM.

[H_2O_2], mM	[$S_2O_8{}^{2-}$], mM	[Cl^-], mM	[$CO_3{}^{2-}$], mM	k_{MA}, min^{-1} ($\pm\sigma$)
0	/	/	/	0.081 ± 0.007
5	/	/	/	2.72 ± 0.12
10	/	/	/	2.03 ± 0.13
20	/	/	/	2.87 ± 0.09
5	/	100	/	2.29 ± 0.16
10	/	100	/	2.97 ± 0.06
20	/	100	/	5.30 ± 0.34
5	/	/	100	1.40 ± 0.08
10	/	/	100	0.652 ± 0.121
20	/	/	100	0.764 ± 0.049
/	0	/	/	0.106 ± 0.007
/	1	/	/	0.133 ± 0.031
/	5	/	/	0.598 ± 0.018
/	10	/	/	7.09 ± 0.16

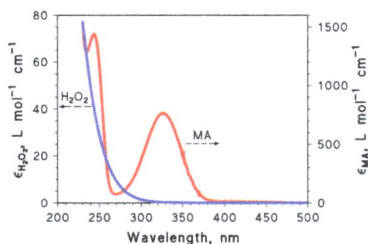

Figure 1. Absorption spectra (molar absorption coefficients) of methyl 2-aminobenzoate (MA) (right Y-axis) and H_2O_2 (left Y-axis).

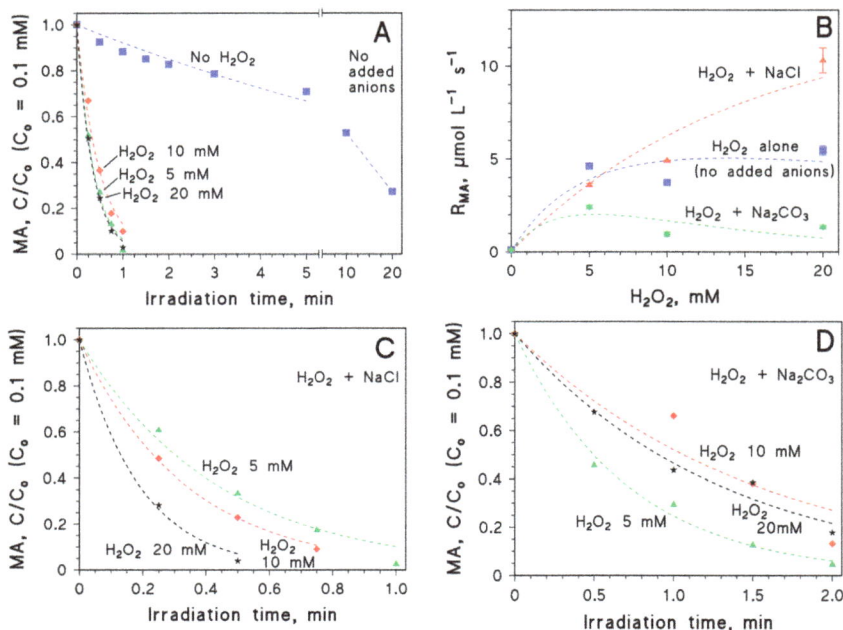

Figure 2. (**A**) Time trend of 0.1 mM MA under UVC irradiation, alone or in the presence of different concentration values of H_2O_2 (varied in the range of 0–20 mM); (**B**) Initial photodegradation rates of 0.1 mM MA (R_{MA}) as a function of H_2O_2 concentration, alone or in the presence of 0.1 M NaCl or 0.1 M Na_2CO_3; (**C**) Time trend of 0.1 mM MA in the presence of 0.1 M NaCl and different concentration values of H_2O_2 (varied in the range of 5–20 mM); (**D**) Time trend of 0.1 mM MA in the presence of 0.1 M Na_2CO_3 and different concentration values of H_2O_2 (varied in the range of 5–20 mM). The pH values of the studied systems were ~neutral, with the exception of the systems containing Na_2CO_3. The error bars shown in panel (**B**) represent the uncertainty associated to the calculation of the photodegradation rates by fitting the MA time trend data with a pseudo-first order kinetic model (intra-series variability). In several cases the error bars were smaller than the data points. The reproducibility between experimental replicas (inter-series variability) was in the range of 15–20%.

Some MA photodegradation with a half-life time of approximately 10 min took place in the absence of H_2O_2, due to MA direct photolysis. The direct photolysis quantum yield of MA was calculated as follows [25]:

$$\Phi_{MA} = \frac{R_{MA}}{\int\limits_{\lambda} p^{\circ}(\lambda)\left(1 - 10^{A_{MA}(\lambda)}\right)d\lambda} \tag{1}$$

where $p^{\circ}(\lambda)$ is the incident spectral photon flux density of the lamp and $A_{MA}(\lambda)$ is the absorbance of 0.1 mM MA. The photolysis quantum yield was measured in separate experiments under monochromatic irradiation (at 254 and 325 nm) and under broadband irradiation (see Supplementary Material for the detailed results). The Φ_{MA} values obtained in the different conditions are in the order of magnitude of 10^{-3}. The value at 254 nm ($\Phi_{MA}^{254nm} = 3.8 \times 10^{-3}$) is the most relevant to our steady irradiation experiments, and a decrease was observed in the values of Φ_{MA} as the irradiation wavelength increased. Therefore, when applying artificial irradiation, the UVC spectral range and in particular the radiation at 254 nm (very near the UVC absorption maximum of MA, see Figure 1) appears to be the most suitable option to induce MA direct photolysis.

The addition of H_2O_2 considerably accelerated the photodegradation of MA, and relatively similar kinetics were obtained in the H_2O_2 concentration range of 5–20 mM. A plateau trend of R_{MA} vs. $[H_2O_2]$ is apparent in Figure 2B, and in principle it might be accounted for by two different phenomena: (i) saturation of H_2O_2 absorption with increasing H_2O_2 concentration; and (ii) offset between photoinduced ·OH generation, and ·OH scavenging by H_2O_2 itself. The first effect depends on the absorbance of H_2O_2. Considering $\varepsilon_{H_2O_2,254nm}$ ~15 L·mol^{-1}·cm^{-1} and assuming b = 2 cm as the optical path length inside the irradiated solutions, the absorbance of the studied H_2O_2 solutions was approximately 0.15 (5 mM H_2O_2), 0.3 (10 mM), and 0.6 (20 mM). The absorbance of 0.1 mM MA at 254 nm is $A_{MA,254nm}$ ~0.2, and the fraction of radiation absorbed by H_2O_2 in the irradiated systems can be calculated as follows [25]:

$$\wp_{H_2O_2,254nm}^{H_2O_2+MA} = \frac{\varepsilon_{H_2O_2,254nm}\, b\, [H_2O_2]}{\varepsilon_{H_2O_2,254nm}\, b\, [H_2O_2] + A_{MA,254nm}} \left[1 - 10^{-\left(\varepsilon_{H_2O_2,254nm}\, b\, [H_2O_2] + A_{MA,254nm}\right)} \right] \quad (2)$$

On this basis, the values of $\wp_{H_2O_2,254nm}^{H_2O_2+MA}$ as a function of the H_2O_2 concentration are 0.24 (H_2O_2 5 mM), 0.41 (H_2O_2 10 mM), and 0.63 (H_2O_2 20 mM). Because the lamp radiation can be considered as monochromatic as a first approximation, these values are directly proportional to the formation rate of ·OH produced by the irradiation of H_2O_2 ($R_{\bullet OH}^{H_2O_2+h\nu} \propto \wp_{H_2O_2,254nm}^{H_2O_2+MA}$). The direct proportionality constant between $R_{\bullet OH}^{H_2O_2+h\nu}$ and $\wp_{H_2O_2,254nm}^{H_2O_2+MA}$ includes the formation quantum yield of ·OH by irradiated H_2O_2 and the incident photon flux in solution, which are constant values that can be included into a proportionality parameter α (as $R_{\bullet OH}^{H_2O_2+h\nu} = \alpha\, \wp_{H_2O_2,254nm}^{H_2O_2+MA}$). The photogenerated ·OH can react with either MA or H_2O_2, and in the latter case the second-order reaction rate constant is $k_{\bullet OH+H_2O_2}$ = 2.7 × 10^7 M^{-1}·s^{-1} [26]. By assuming $k_{\bullet OH+MA}$ as the (unknown) second-order reaction rate constant between ·OH and MA, the competition kinetics between MA and H_2O_2 yields the following results for the MA photodegradation rate (R_{MA}):

$$R_{MA} = R_{\bullet OH}^{H_2O_2+h\nu} \frac{k_{\bullet OH+MA}\,[MA]}{k_{\bullet OH+MA}\,[MA] + k_{\bullet OH+H_2O_2}[H_2O_2]} = \alpha\, \wp_{H_2O_2,254nm}^{H_2O_2+MA} \frac{1}{1 + \dfrac{k_{\bullet OH+H_2O_2}}{k_{\bullet OH+MA}}\dfrac{[H_2O_2]}{[MA]}} \quad (3)$$

With the known values of $\varepsilon_{H_2O_2,254nm}$ ~15 M^{-1}·cm^{-1}, $A_{MA,254nm}$ ~0.2, [MA] = 0.1 mM and $k_{\bullet OH+H_2O_2}$ = 2.7 × 10^7 M^{-1}·s^{-1}, it was possible to fit reasonably well the R_{MA} vs. $[H_2O_2]$ experimental data reported in Figure 2B (see dashed curve in the figure). The fit results suggested that $k_{\bullet OH+MA}$ would be about two orders of magnitude higher than $k_{\bullet OH+H_2O_2}$. This means that the reaction of ·OH with H_2O_2 is expected to prevail over that with 0.1 mM MA for $[H_2O_2]$ > 10 mM, which is right within the investigated range of H_2O_2 concentrations.

2.2. Effect of Inorganic Anions on MA Photodegradation

The effect of anions commonly occurring in surface waters, and most notably of chloride and carbonate, on the photodegradation of MA induced by H_2O_2/UV was studied upon UVC irradiation of MA, H_2O_2, and, where relevant, NaCl or Na_2CO_3.

The time evolution of 0.1 mM MA in the presence of 0.1 M NaCl and different concentration values of H_2O_2 is reported in Figure 2C, and the corresponding photodegradation rates are reported in Figure 2B. The figure shows that MA photodegradation became progressively faster as the H_2O_2 concentration increased and, differently from the previous case (MA + H_2O_2 + UV, without chloride), there was no obvious plateau trend. The experimental rate data could be fitted well with an equation of the form $R_{MA} = \beta\, \wp_{H_2O_2,254nm}^{H_2O_2+MA}$, where β is a constant proportionality factor (see the dashed curve in Figure 2B). In this case it seems that the observed trend just mirrored the photon absorption by H_2O_2, with no need to invoke an additional competition kinetics between MA, H_2O_2 and the reactive transient species. Moreover, at elevated H_2O_2 concentration the photodegradation of MA was considerably faster in the presence of 0.1 M NaCl than in the absence of chloride. These pieces of evidence suggest that the prevailing reactive species in the $MA/H_2O_2/Cl^-/UV$ system is very unlikely to be $\cdot OH$, which is expected to produce a plateau trend as per the above discussion. A different transient species should rather be involved, inducing competition kinetics between MA and H_2O_2 to a far lesser extent than $\cdot OH$. This reactive transient, provisionally indicated here as X, should react with MA and H_2O_2 in such a way that $k_{X+H_2O_2}(k_{X+MA})^{-1} << k_{\cdot OH+H_2O_2}(k_{\cdot OH+MA})^{-1}$. If this condition is met, one has $\frac{k_{X+H_2O_2}}{k_{X+MA}[MA]}[H_2O_2] < 1$ in Equation (4), which differs from Equation (3) in that the $\cdot OH$-based terms are replaced by X-based ones:

$$R_{MA} = R_X^{H_2O_2+h\nu} \frac{1}{1 + \frac{k_{X+H_2O_2}}{k_{X+MA}[MA]}[H_2O_2]} \tag{4}$$

In the presence of $\cdot OH + Cl^-$, the following reactions may take place [26–28]:

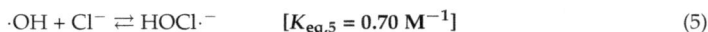

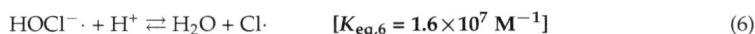

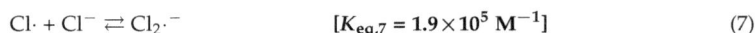

$$\cdot OH + Cl^- \rightleftarrows HOCl\cdot^- \qquad [\textbf{K}_{\textbf{eq,5}} = \textbf{0.70 M}^{-1}] \tag{5}$$

$$HOCl\cdot^- + H^+ \rightleftarrows H_2O + Cl\cdot \qquad [\textbf{K}_{\textbf{eq,6}} = \textbf{1.6}\times\textbf{10}^7 \textbf{ M}^{-1}] \tag{6}$$

$$Cl\cdot + Cl^- \rightleftarrows Cl_2\cdot^- \qquad [\textbf{K}_{\textbf{eq,7}} = \textbf{1.9}\times\textbf{10}^5 \textbf{ M}^{-1}] \tag{7}$$

Based on the above reactions, potential X-species in the system are $HOCl\cdot^-$, $Cl\cdot$, and $Cl_2\cdot^-$. The reactivity of $Cl_2\cdot^-$ can be studied by laser flash photolysis, thus one can check the possible involvement of $Cl_2\cdot^-$ in MA photodegradation by measuring $k_{Cl_2^{\bullet-}+MA}$.

In the $H_2O_2/Na_2CO_3/UV$ system with 0.1 M Na_2CO_3, the photodegradation of MA did not accelerate when increasing $[H_2O_2]$ above 5 mM (see Figure 2D for the MA time trends, and Figure 2B for the corresponding photodegradation rates). The $\cdot OH$ reactions with carbonate and bicarbonate are more straightforward than in the case of chloride and they lead to the unequivocal formation of $CO_3\cdot^-$ as additional reactive species [26,29]:

$$\cdot OH + HCO_3^- \rightarrow H_2O + CO_3\cdot^- \tag{8}$$

$$\cdot OH + CO_3^{2-} \rightarrow OH^- + CO_3\cdot^- \tag{9}$$

A comparison of the MA photodegradation rates in the systems "H_2O_2 alone" and "H_2O_2 + Na_2CO_3" in Figure 2B shows that the rates were lower in the presence of carbonate, coherently with the replacement of $\cdot OH$ with the less reactive species $CO_3\cdot^-$. Moreover, the ratio $R_{MA}^{H_2O_2\,alone}(R_{MA}^{H_2O_2+Na_2CO_3})^{-1}$ increased with increasing $[H_2O_2]$. A potential explanation for this phenomenon is that H_2O_2 competes more effectively with MA, for reaction with $CO_3\cdot^-$, than for reaction with $\cdot OH$. In other words, this hypothesis leads to the assumption that $k_{CO_3^{\bullet-}+H_2O_2}(k_{CO_3^{\bullet-}+MA})^{-1} > k_{\cdot OH+H_2O_2}(k_{\cdot OH+MA})^{-1}$. Considering that $k_{CO_3^{\bullet-}+H_2O_2} = 8 \times 10^5 \text{ M}^{-1}\cdot\text{s}^{-1}$ is known from the literature [30], the measurement of $k_{CO_3^{\bullet-}+MA}$ by laser flash photolysis is an appropriate test for this hypothesis.

2.3. MA Photodegradation by Persulphate/UV

The UV irradiation of persulphate yields the sulphate radical, $SO_4\cdot^-$ [31–33]. This radical has similar if not higher reduction potential compared to $\cdot OH$, but it tends to be preferentially involved in charge-transfer reactions while $\cdot OH$ often triggers hydrogen-transfer or addition processes in comparable conditions [17,34].

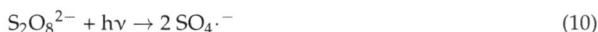

$$S_2O_8^{2-} + h\nu \rightarrow 2\,SO_4\cdot^- \tag{10}$$

The time trend of 0.1 mM MA upon UVC irradiation in the presence of varying concentration values of sodium persulphate (PS) is reported in Figure 3. The figure shows that PS above 1 mM concentration could considerably accelerate the photodegradation of MA, and that the photodegradation became considerably faster as the PS concentration was higher. Moreover, while there was limited difference between the MA time trends with 5 or 10 mM H_2O_2, the photodegradation of MA with 10 mM PS was considerably faster compared to 5 mM PS. This result suggests that the reaction between $SO_4 \cdot^-$ and PS interferes with MA photodegradation to a lesser extent than the reaction between $\cdot OH$ and H_2O_2.

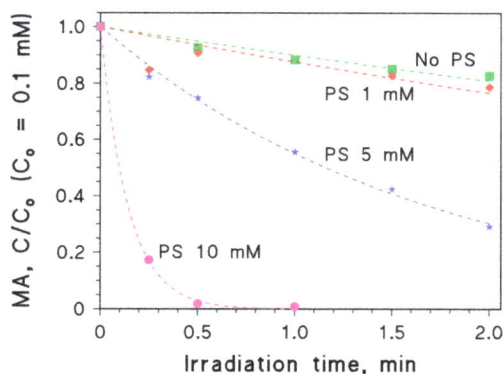

Figure 3. Time trend of 0.1 mM MA upon UVC irradiation, alone or in the presence of different concentration values of $Na_2S_2O_8$ (persulphate, PS). The PS concentration was varied in the range of 0–10 mM.

2.4. Second-Order Reaction Rate Constants of MA with $Cl_2 \cdot^-$, $CO_3 \cdot^-$ and $SO_4 \cdot^-$

The second-order reaction rate constants between MA and three reactive transient species ($Cl_2 \cdot^-$, $CO_3 \cdot^-$, and $SO_4 \cdot^-$) were measured by means of the laser flash photolysis technique. The radical $Cl_2 \cdot^-$ was produced by laser irradiation of H_2O_2 + NaCl (0.01 M chloride) at pH 3 by $HClO_4$, under which conditions the equilibria of reactions (4–6) are shifted towards the products and there is a consequent enhancement of the formation of $Cl_2 \cdot^-$ [26–28]. As far as the other transient species are concerned, $CO_3 \cdot^-$ was produced by laser irradiation of H_2O_2 + Na_2CO_3, and $SO_4 \cdot^-$ was produced by laser irradiation of $Na_2S_2O_8$. The actual occurrence of these radicals as the main transient species in the laser-irradiated solutions has been demonstrated in previous studies [35,36]. Figure 4A reports the absorption spectra of the studied solutions undergoing laser flash photolysis, obtained just after the laser pulse. Based on these results, in successive experiments the radical $Cl_2 \cdot^-$ was monitored at 350 nm, $CO_3 \cdot^-$ at 550 nm, and $SO_4 \cdot^-$ at 450 nm. Figure 4B reports the trends of the pseudo-first order decay constants k of each transient as a function of the MA concentration. Following the Stern-Volmer approach, the slopes of k vs. [MA] represent the second-order reaction rate constants of the transient species with MA. We obtained $k_{Cl_2^{\bullet-}+MA} = (4.0 \pm 0.3) \times 10^8$ $M^{-1} \cdot s^{-1}$ (the error bounds represent the σ-level uncertainty), $k_{CO_3^{\bullet-}+MA} = (3.1 \pm 0.2) \times 10^8$ $M^{-1} \cdot s^{-1}$, and $k_{SO_4^{\bullet-}+MA} = (5.6 \pm 0.4) \times 10^9$ $M^{-1} \cdot s^{-1}$. These values are consistent with the typical reactivity of the three transient species [30].

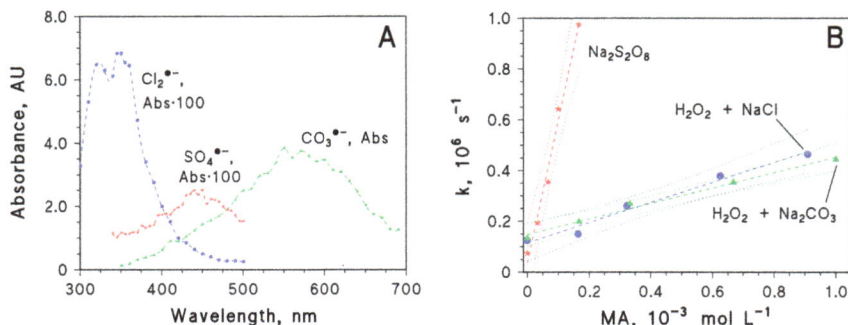

Figure 4. (**A**) Absorption spectra of the radicals $CO_3 \cdot^-$, $Cl_2 \cdot^-$, and $SO_4 \cdot^-$ produced by laser flash photolysis of 2.5 mM H_2O_2 + 0.1 M Na_2CO_3 ($CO_3 \cdot^-$), of 2.5 mM H_2O_2 + 0.01 M NaCl (pH 3, adjusted with $HClO_4$) ($Cl_2 \cdot^-$), and of 10 mM $Na_2S_2O_8$ ($SO_4 \cdot^-$). The absorbance signals were taken soon after the laser pulse. Laser irradiation at 266 nm, 35 mJ·pulse^{-1}; (**B**) First-order decay constants of the studied radical species ($SO_4 \cdot^-$, $Cl_2 \cdot^-$, $CO_3 \cdot^-$) as a function of MA concentration. The radical species were obtained by laser irradiation of 10 mM $Na_2S_2O_8$ ($SO_4 \cdot^-$), of 2.5 mM H_2O_2 + 0.01 M NaCl at pH 3 ($Cl_2 \cdot^-$), and of 2.5 mM H_2O_2 + 0.1 M Na_2CO_3 ($CO_3 \cdot^-$). The slopes of the lines give the second-order reaction rate constants between the relevant radicals and MA (Stern-Volmer approach).

The formation of $CO_3 \cdot^-$ and $SO_4 \cdot^-$ upon either laser-based or steady-state irradiation of, respectively, H_2O_2 + Na_2CO_3 and $Na_2S_2O_8$ is rather straightforward [35,36]. In the case of H_2O_2 + NaCl, the laser irradiation took place at pH 3 to ensure the formation of $Cl_2 \cdot^-$. In contrast, the corresponding steady irradiation experiments took place at the natural pH, where the involvement of $Cl_2 \cdot^-$ in MA photodegradation is less obvious.

To assess the actual involvement of $Cl_2 \cdot^-$ in the steady irradiation process one can check whether $k_{Cl_2^{\bullet -}+H_2O_2} (k_{Cl_2^{\bullet -}+MA})^{-1} \ll k_{\bullet OH+H_2O_2} (k_{\bullet OH+MA})^{-1}$, as suggested by the steady irradiation results where R_{MA} was directly proportional to $\wp_{H_2O_2,254nm}^{H_2O_2+MA}$ (see Figure 2B). Table 2 summarises the second-order reaction rate constants of $Cl_2 \cdot^-$, $CO_3 \cdot^-$ and $\cdot OH$ with MA, derived in this study, and those with H_2O_2 and HO_2^-, obtained from the literature [26,30].

Table 2. Second-order reaction rate constants (k_{X+Y}) between the transient species X = $Cl_2 \cdot^-$, $CO_3 \cdot^-$ or $\cdot OH$, and the compound Y = MA, H_2O_2, or HO_2^-.

k_{X+Y}, M^{-1} s^{-1}	MA	H_2O_2	HO_2^-
$Cl_2 \cdot^-$	4.0×10^8	1.4×10^5 [30]	n/a
$CO_3 \cdot^-$	3.1×10^8	8×10^5 [30]	3×10^7 [30]
$\cdot OH$	$\sim 10^9$	2.7×10^7 [26]	7.5×10^9 [26]

The experimental data of R_{MA} *vs.* $[H_2O_2]$ in the presence of H_2O_2 alone (see Figure 2B) are consistent with $k_{\bullet OH+H_2O_2} (k_{\bullet OH+MA})^{-1} \sim 0.01$. From this value and the condition $k_{Cl_2^{\bullet -}+H_2O_2} (k_{Cl_2^{\bullet -}+MA})^{-1} \ll k_{\bullet OH+H_2O_2} (k_{\bullet OH+MA})^{-1}$, one gets $k_{Cl_2^{\bullet -}+MA} \gg 1.4 \times 10^7$ M$^{-1} \cdot$s^{-1}. The laser flash photolysis experiments yielded $k_{Cl_2^{\bullet -}+MA} = (4.0 \pm 0.3) \times 10^8$ M$^{-1} \cdot$s^{-1}, in full agreement with the steady irradiation estimate. This means that $Cl_2 \cdot^-$ is a reasonable reactive species for the photodegradation of MA in the presence of H_2O_2 + Cl^- under irradiation in ~neutral solution.

The steady irradiation trend of R_{MA} *vs.* $[H_2O_2]$ in the presence of Na_2CO_3, when interpreted in the framework of a competition kinetics between MA and H_2O_2 for reaction with $CO_3 \cdot^-$, gives $k_{CO_3^{\bullet -}+H_2O_2} (k_{CO_3^{\bullet -}+MA})^{-1} > k_{\bullet OH+H_2O_2} (k_{\bullet OH+MA})^{-1} \sim 0.01$. From the literature datum $k_{CO_3^{\bullet -}+H_2O_2} = 8 \times 10^5$ M$^{-1} \cdot$s^{-1} [30] one gets $k_{CO_3^{\bullet -}+MA} < 8 \times 10^7$ M$^{-1} \cdot$s^{-1}, which is in stark contrast with the value $k_{CO_3^{\bullet-}+MA} = (3.1 \pm 0.2) \times 10^8$ M^{-1}·s^{-1} obtained by laser flash photolysis. The H_2O_2 + Na_2CO_3 solutions had more basic pH (~10) compared to those containing H_2O_2 + NaCl, thus H_2O_2 would in part occur as its conjugate base HO_2^- [37]. However, when considering that $k_{\bullet OH+HO_2^-} = 7.5 \times 10^9$ M^{-1}·s^{-1} [26] and $k_{CO_3^{\bullet-}+HO_2^-} = 3 \times 10^7$ M^{-1}·s^{-1} [30], from the condition $k_{CO_3^{\bullet-}+HO_2^-}\left(k_{CO_3^{\bullet-}+MA}\right)^{-1} > k_{\bullet OH+HO_2^-}\left(k_{\bullet OH+MA}\right)^{-1}$ one derives $k_{CO_3^{\bullet-}+MA} < 2 \times 10^7$ M^{-1}·s^{-1}, which is again not consistent with the laser flash photolysis results. A more reasonable explanation is that the reactions of HO_2^- with ·OH and $CO_3 \cdot^-$ are much faster than those of H_2O_2 (see Table 2), thereby causing a considerable production of $HO_2 \cdot / O_2 \cdot^-$ (reactions 11, 12; [26,30]). The superoxide radical anion that prevails at the pH conditions of the studied systems [37] is an effective reductant [38], and it could reduce the oxidised MA transients back to the initial compound (see e.g., reaction 13).

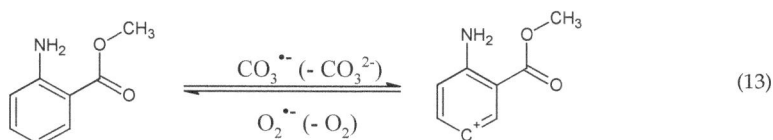

$$CO_3 \cdot^- + HO_2^- \rightarrow HCO_3^- + O_2 \cdot^- \tag{11}$$

$$\cdot OH + HO_2^- \rightarrow H_2O + O_2 \cdot^- \tag{12}$$

$$\tag{13}$$

The above reactions, ending up in an inhibition of MA photodegradation, might explain the trend of R_{MA} *vs.* $[H_2O_2]$ in the presence of carbonate, reported in Figure 2B.

2.5. MA Photodegradation Intermediates

The LC-MS analysis of the MA solutions irradiated in the presence of H_2O_2, with a conversion percentage of 32%, allowed the detection of MA at the retention time of 12.5 min and of several photodegradation intermediates, namely P1 (10.6 min), P2 (11.0 min), P3 (11.4 min), and P4 (13.1 min). Useful information was initially obtained from the MS spectrum of MA itself. A pattern of MA fragmentation, based on the information obtained in its MS^2 spectrum at 20 eV, is shown in Figure 5a. The spectrum shows the formation of a fragment ion with an accurate mass of m/z = 120.0499, which corresponds to an elemental composition of $C_7H_6ON^+$ (error = −17 ppm) and is formed by the loss of a CH_3OH group. This fragmentation is a peculiar behaviour of *ortho*-substituted esters [39]. Two additional fragment ions are also observed at m/z 92 and 65. The former with an accurate mass of 92.0500 ($C_6H_6N^+$, error = −1.3 ppm) arises from the loss of HCO_2CH_3 from the molecular ion, which is a common fragmentation process in the methyl esters of carboxylic acids [40]. The same fragment could also be produced by CO loss from the fragment ion at m/z 120.0499. The fragment with m/z = 65.0391 ($C_5H_5^+$, error = −3.1 ppm) is obtained from m/z = 92.0500 by loss of HCN.

As far as the intermediates P1, P2, and P3 are concerned, they were characterised by the molecular ion m/z = 168.0655. This is consistent with the elemental composition $C_8H_{10}O_3N^+$ (error = −3.4 ppm), corresponding to MA monohydroxy derivatives. Remarkably, despite the possibility to hydroxylate MA in four different positions, only three isomers were actually detected with P2 as the major one. The MS^2 product ions of these compounds are listed in Table 3, together with the LC retention times of the parent molecules.

In the case of P1, the most abundant product ion is 109.0515 m/z ($C_6H_7ON^+$, error = −11.6 ppm), which arises from the loss of $CH_3COO \cdot$ and is consistent with the presence of the -OH group in position 4 or 6 with respect to the ester functionality of MA. The fragment at 81.0590 m/z ($C_5H_7N^+$, error = +14.2 ppm) can be explained with the further loss of another CO group. The formation of the 141.0569 m/z fragment ($C_7H_9O_3^+$, error = +12.3 ppm) can be justified with the loss of HCN, whereas the detachment of a $CH_3O \cdot$ radical group would yield the fragment at 137.0470 m/z ($C_7H_7O_2N^+$, error = −5.0 ppm). Unfortunately, no further information is present in the spectrum that allows for the determination of the exact location of the OH group. For explanatory purposes, the fragmentation pathway of the 6-hydroxyderivative of MA is reported in Figure 5b. Remarkably, a totally similar fragmentation that yields fragment ions with the same *m/z* values could be proposed for the 4-hydroxyderivative.

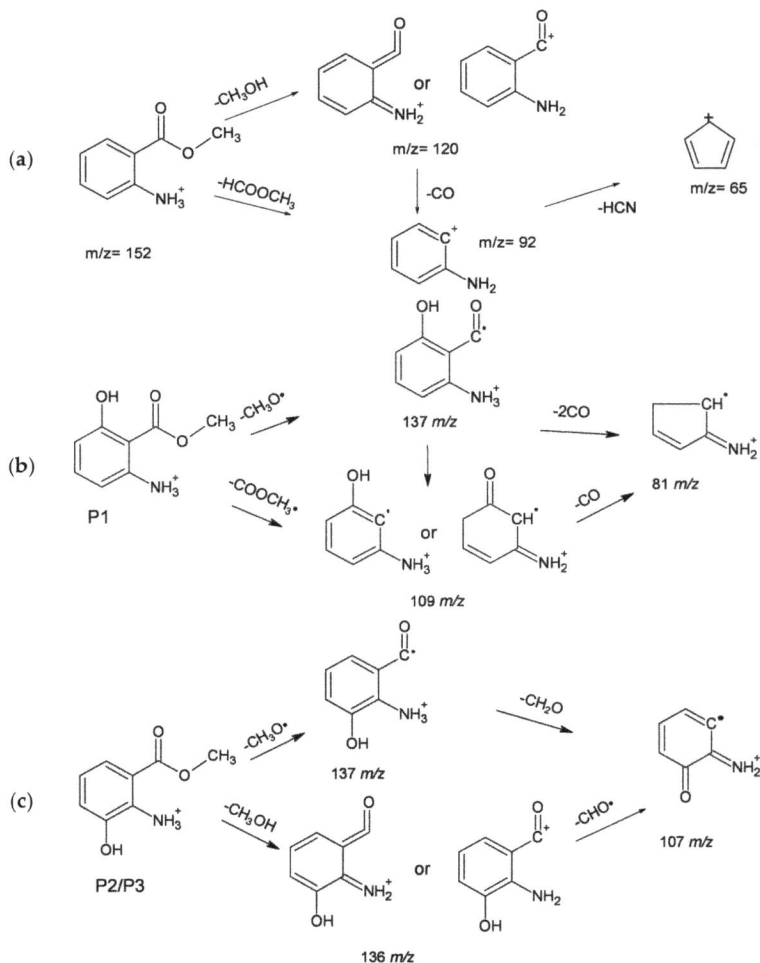

Figure 5. Proposed MS fragmentation pathways of: (**a**) MA; (**b**) a possible structure of P1; (**c**) a possible structure of P2/P3.

Table 3. Summary of the mass spectrometric data of the detected MA hydroxyderivatives.

Compound Acronym	LC Retention Time, (min)	MS^2 Fragments, *m/z* (% Relative Abundance)
P1	10.6	141 (8), 137 (7), 109 (100), 81 (38)
P2	11.0	136 (100), 137 (30), 107 (3)
P3	11.4	136 (100), 137 (25), 107 (25)

As far as P2 and P3 are concerned, the most abundant signal occurs at 136.0375 *m/z* ($C_7H_6O_2N^+$, error = -17.3 ppm) and, in analogy with the fragmentation of MA, it could arise from CH_3OH loss. As already seen for P1, one also observes the product ion at 137.0465 *m/z*. The occurrence of the product ion at 107.0358 *m/z* (H_2CO loss) suggests the presence of an OH group in *ortho* or *para* position with respect to the amino group (i.e., in position 3 or 5 with respect to the ester functionality). A possible fragmentation pathway for the 3-hydroxyderivative is shown in Figure 5c, but a fully similar pathway could be proposed for the 5-hydroxyderivative. From the available MS data it was unfortunately not possible to attribute uniquely each isomer to the corresponding signal. However, by assuming that P2 and P3 are the 3- and 5-hydroxyderivatives of MA (irrespective of which is which), one can tentatively conclude that both of them are anyway formed. In contrast, P1 may be either the 4- or the 6-hydroxyderivative. Therefore, one could tentatively assume that hydroxylation takes place in the 3 and 5 positions, plus 4 or 6 (in other words, either the 3-, 4-, and 5- or the 3-, 5-, and 6-hydroxyderivatives would be formed).

In the case of P4, the accurate mass of the molecular ion (*m/z* = 331.0915) corresponds to the elemental composition $C_{16}H_{15}N_2O_6^+$, with an error of -4.6 ppm. This indicates the possible presence of an oxidised dimeric structure. Unfortunately, based on the available MS data it was not possible to propose a univocal structure for this compound.

Based on ECOSAR predictions, the MA hydroxyderivatives would show comparable toxicity as the parent molecule [7]. In all the cases the major effects are predicted to be the acute and, most notably, the chronic toxicity towards fish and crustaceans.

3. Methods

3.1. Reagents and Materials

Methyl 2-aminobenzoate (MA, purity grade $\geq$98%), methanol (gradient grade), $HClO_4$ (70%), NaOH (1.0 M titrated solution), and Na_2CO_3 (99.9%) were purchased from Sigma-Aldrich (Saint-Quentin-Fallavier, France). Formic acid (98%), H_2O_2 (30%), and NaCl (99.5%) were purchased from Fluka (Saint-Quentin-Fallavier, France). The above chemicals were used as received. Ultra-pure water was prepared with a Millipore (Billerica, MA, USA) Milli-Q apparatus (resistivity $\geq$ 18.2 MΩ cm, TOC < 2 ppb).

3.2. Irradiation Experiments

The absorption spectra of the studied compounds (see Figure 1 for MA and H_2O_2) were taken with a Varian (Palo Alto, CA, USA) Cary 3 UV-vis spectrophotometer, using 1 cm quartz cuvettes. The solution pH was measured with a combined glass electrode connected to a Meterlab pH meter (Hach Lange, Loveland, CO, USA). Solutions containing 0.1 mM MA, and other components where relevant, were inserted inside a quartz tube (100 mL total volume), which was placed in the centre of an irradiation set-up consisting of six TUV Philips (Amsterdam, Netherlands) 15 W lamps with emission maximum at 254 nm. The lamp intensity was 7.6×10^{-9} Einstein cm$^{-2}\cdot$s^{-1}. The water solutions were magnetically stirred during irradiation. At scheduled irradiation times, 1.5 mL sample aliquots were withdrawn from the tube, placed into HPLC vials, and kept refrigerated until HPLC analysis. The time trend of MA was monitored by means of a high-performance liquid chromatograph interfaced to a photodiode-array detector (HPLC-PDA, model Nexera XR by Shimadzu, Kyoto, Japan), equipped with SIL20-AC autosampler, SIL-20AD pump module for low-pressure gradients, CT 0-10AS column oven (set at 40 °C), reverse-phase column Kinetex RP-C18 packed with Core Shell particles (100 mm $\times$ 2.10 mm $\times$ 2.6 μm) by Phenomenex (Torrance, CA, USA), and SPDM 20A photodiode array detector. The isocratic eluent was a A/B = 60/40 mixture of A = (0.5% formic acid in water, pH 2.3) and B = methanol, at a flow rate of 0.2 mL min^{-1}. In these conditions, the MA retention time was 7.3 min. The detection wavelength was set at 218 nm. A schematic of the experimental procedure is reported in Figure 6.

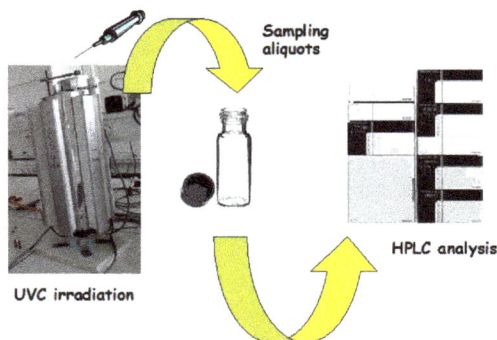

Figure 6. Schematic of the used experimental procedure.

The time evolution of MA concentration ([MA]) was fitted with the pseudo-first order equation [MA]$_t$ = [MA]$_o$ $e^{-k_{MA} t}$, where [MA]$_t$ is the concentration of MA at the time t, [MA]$_o$ the initial concentration of MA, and k_{MA} the pseudo-first order photodegradation rate constant of MA. The initial rate of MA photodegradation is $R_{MA} = k_{MA}$ [MA]$_o$.

3.3. Identification of Photodegradation Intermediates

The photodegradation intermediates of MA were identified by liquid chromatography interfaced with mass spectrometry (LC-MS). A Waters Alliance (Milford, MA, USA) instrument equipped with an electrospray (ESI) interface (used in ESI+ mode) and a Q-TOF mass spectrometer (Micromass, Manchester, UK) were used. Samples were eluted on a column Phenomenex Kinetex C18 (100 mm × 2.10 mm × 2.6 μm) with a mixture of acetonitrile (A) and 0.1% formic acid in water (B) at 0.2 mL·min^{-1} flow rate, with the following gradient: start at 5% A, then up to 95% A in 15 min, keep for 10 min, back to 5% A in 1 min, and keep for 5 min (post-run equilibration). The capillary needle voltage was 3 kV and the source temperature 100 °C. The cone voltage was set to 35 V. Data acquisition was carried out with a Micromass MassLynx 4.1 data system. Both MS and MS/MS experiments were carried out by using this chromatographic set-up.

3.4. Laser Flash Photolysis Experiments

The reactivity of the radicals Cl$_2\cdot^-$, CO$_3\cdot^-$, and SO$_4\cdot^-$ was studied by means of the nanosecond laser flash photolysis technique. Flash photolysis runs were carried out using the third harmonic (266 nm) of a Quanta Ray GCR 130-01 Nd:YAG laser system instrument, used in a right-angle geometry with respect to the monitoring light beam. The single pulses energy was set to 35 mJ unless otherwise stated. A 3 mL solution volume was placed in a quartz cuvette (path length of 1 cm) and used for a maximum of three consecutive laser shots. The transient absorbance at the pre-selected wavelength was monitored by a detection system consisting of a pulsed xenon lamp (150 W), monochromator, and a photomultiplier (1P28). A spectrometer control unit was used for synchronising the pulsed light source and programmable shutters with the laser output. The signal from the photomultiplier was digitised by a programmable digital oscilloscope (HP54522A). A 32 bits RISC-processor kinetic spectrometer workstation was used to analyse the digitised signal.

3.5. Model Assessment of Toxicity

The potential acute and chronic toxicity of the detected MA intermediates was assessed with the ECOSAR software (US-EPA, Washington DC, USA). ECOSAR uses a quantitative structure-activity relationship approach to predict the toxicity of a molecule of given structure. The relevant endpoints are the acute and chronic toxicity thresholds (LC50, EC50, chronic values ChV) for freshwater fish, daphnid, and algae. The values predicted by ECOSAR are apparently very precise but, as far as accuracy is concerned, a compound can be said to be more toxic than another only when the predicted values differ by at least an order of magnitude [7,8].

4. Conclusions

The H_2O_2/UV technique as photochemical $\cdot$OH source is a potentially effective tool to achieve MA photodegradation, and in fact the addition of hydrogen peroxide considerably accelerated the photodegradation of MA compared to UV irradiation alone. The addition of inorganic anions that act as $\cdot$OH scavengers, such as chloride and carbonate, did not necessarily quench MA photodegradation. The reason is the reactivity with MA itself of the generated radical species, i.e., $Cl_2\cdot^-$ produced from $Cl^- + \cdot OH$ and $CO_3\cdot^-$ produced from $CO_3^{2-} + \cdot OH$. In the case of chloride, there was even an acceleration of MA photodegradation at elevated $[H_2O_2]$, because $Cl_2\cdot^-$ competes more successfully than $\cdot$OH for reaction with MA in the presence of H_2O_2 (H_2O_2 behaves as a scavenger of $\cdot$OH and, to a lesser extent, of $Cl_2\cdot^-$ as well). The same effect was not observed with carbonate, possibly because the basic pH caused a considerable production of superoxide ($O_2\cdot^-$) upon oxidation of the H_2O_2 conjugated base, HO_2^-. The radical $O_2\cdot^-$ is a well-known reductant that could reduce the partially oxidised MA back to the starting compound. Effective MA photodegradation was also observed with persulphate/UV, probably because of the fast reaction between MA and photogenerated $SO_4\cdot^-$, and because of limited scavenging of $SO_4\cdot^-$ by persulphate itself.

Among the MA photodegradation intermediates detected in the H_2O_2/UV process, the hydroxyderivatives could be about as toxic as the parent compound. Therefore, decontamination is not yet achieved once MA has disappeared, and the H_2O_2/UV treatment of MA should at least ensure the photodegradation of the MA hydroxylated derivatives as well. Usually, the photodegradation of both the primary compound and its intermediates takes more time than the photodegradation of the starting compound alone.

Supplementary Materials: Supplementary materials are available online.

Acknowledgments: G.M.L. acknowledges the Erasmus Placement programme for financially supporting her stay in Clermont-Ferrand.

Author Contributions: M.S. conceived the study and corrected the paper. G.M.L. performed the experiments. D.F. interpreted the MS results. D.V. made kinetic calculations and wrote the paper.

Conflicts of Interest: The authors declare no conflict of interest. The founding sponsors had no role in the design of the study; in the collection, analyses, or interpretation of data; in the writing of the manuscript; or in the decision to publish the results.

Abbreviations

k_{MA}	pseudo-first order rate constant of MA photodegradation.
k_{X+Y}	second-order reaction rate constant between the species X and the compound Y.
R_{MA}	initial photodegradation rate of MA.
$\varepsilon_{Y,\lambda}$	molar absorption coefficient of the compound Y at the wavelength λ.
$p°(\lambda)$	incident spectral photon flux density of the used lamp at the wavelength λ.
$R_X^{Y+h\nu}$	formation rate of the species X in the presence of the compound Y under irradiation.
$\wp_{Y,\lambda}^S$	fraction of radiation absorbed by the compound Y at the wavelength λ, inside the solution S.

References

1. Sigma-Aldrich. Available online: http://www.sigmaaldrich.com/catalog/product/aldrich/w268208 (accessed on 3 March 2017).
2. Fraternale, D.; Ricci, D.; Flamini, G.; Giomaro, G. Volatiles profile of red apple from Marche region (Italy). *Rec. Nat. Prod.* **2011**, *5*, 202–207.

3.   Dolbeer, R.A.; Clark, L.; Woronecki, P.; Seamans, T.W. Pen tests of methyl anthranilate as a bird repellent in water. In Proceedings of the Eastern Wildlife Damage Control Conference, Ithaca, NY, USA, 6–9 October 1991; Volume 5, pp. 112–116.

4.   Brown, J.E.; Luo, W.T.; Lorne, I.M.; Pankow, J.F. Candy flavorings in tobacco. *N. Engl. J. Med.* **2014**, *370*, 2250–2252. [CrossRef] [PubMed]

5.   M&U International. Methyl Anthranilate Material Safety Data Sheet. Available online: http://www.mu-intel.com/upload/msds/20141119004116.pdf (accessed on 19 December 2016).

6.   Clark, L.; Cummings, J.; Bird, S.; Aronov, E. Acute toxicity of the bird repellent, methyl anthranilate, to fry of *Salmo salar*, *Oncorhynus mykiss*, *Ictalurus punctatus* and *Lepomis macrochirus*. *Pestic. Sci.* **1993**, *39*, 313–317. [CrossRef]

7.   Mayo-Bean, K.; Moran, K.; Meylan, B.; Ranslow, P. *Methodology Document for the ECOlogical Structure-Activity Relationship Model (ECOSAR) Class Program*; US-EPA: Washington DC, 2012; 46 pp.

8.   US-EPA, ECOSAR Software (Ecological Structure Activity Relationships Predictive Model). Available online: https://www.epa.gov/tsca-screening-tools/ecological-structure-activity-relationships-ecosar-predictive-model (accessed on 20 December 2016).

9.   Avery, M.L. Evaluation of methyl anthranilate as a bird repellent in fruit crops. In Proceedings of the Fifteenth Vertebrate Pest Conference, Newport Beach, CA, USA, 3–5 March 1992; Volume 15, pp. 115–129.

10. Dorr, B.; Clark, L.; Glahn, J.F.; Mezine, I. Evaluation of a methyl anthranilate-based bird repellent: Toxicity to channel catfish *Ictalurus punctatus* and effect on great blue heron *Ardea herodias* feeding behavior. *J. World Aquacult. Soc.* **1998**, *29*, 451–462. [CrossRef]

11. US-EPA EPI Suite™-Estimation Program Interface. Available online: https://www.epa.gov/tsca-screening-tools/epi-suitetm-estimation-program-interface (accessed on 15 December 2016).

12. Zuccato, E.; Castiglioni, S.; Bagnati, R.; Melis, M.; Fanelli, R. Source, occurrence and fate of antibiotics in the Italian aquatic environment. *J. Hazard. Mater.* **2010**, *179*, 1042–1048. [CrossRef] [PubMed]

13. Richardson, S.D.; Ternes, T.A. Water analysis: Emerging contaminants and current issues. *Anal. Chem.* **2014**, *86*, 2813–2848. [CrossRef] [PubMed]

14. Cooper, W.J.; Cramer, C.J.; Martin, N.H.; Mezyk, S.P.; O'Shea, K.E.; von Sonntag, C. Free radical mechanisms for the treatment of methyl tert-butyl ether (MTBE) via advanced oxidation/reductive processes in aqueous solutions. *Chem. Rev.* **2009**, *109*, 1302–1345. [CrossRef] [PubMed]

15. Brillas, E.; Sires, I.; Oturan, M.A. Electro-Fenton process and related electrochemical technologies based on Fenton's reaction chemistry. *Chem. Rev.* **2009**, *109*, 6570–6631. [CrossRef] [PubMed]

16. Pichat, P. (Ed.) *Photocatalysis and Water Purification*; Wiley-VCH: Weinheim, Germany, 2013.

17. Gligorovski, S.; Strekowski, R.; Barbati, S.; Vione, D. Environmental implications of hydroxyl radicals (•OH). *Chem. Rev.* **2015**, *115*, 13051–13092. [CrossRef] [PubMed]

18. Doré, M. *Chemistry of Oxidants and Water Treatments*; Wiley-VCH: Weinheim, Germany, 1995.

19. Ganiyu, S.O.; van Hullebusch, E.D.; Cretin, M.; Esposito, G.; Oturan, M.A. Coupling of membrane filtration and advanced oxidation processes for removal of pharmaceutical residues: A critical review. *Separ. Purif. Technol.* **2015**, *156*, 891–914. [CrossRef]

20. Wang, J.L.; Wang, S.Z. Removal of pharmaceuticals and personal care products (PPCPs) from wastewater: A review. *J. Environ. Manag.* **2016**, *182*, 620–640. [CrossRef] [PubMed]

21. Rozas, O.; Vidal, C.; Baeza, C.; Jardim, W.F.; Rossner, A.; Mansilla, H.D. Organic micropollutants (OMPs) in natural waters: Oxidation by UV/H_2O_2 treatment and toxicity assessment. *Water Res.* **2016**, *98*, 109–118. [CrossRef] [PubMed]

22. Minero, C.; Pellizzari, P.; Maurino, V.; Pelizzetti, E.; Vione, D. Enhancement of dye sonochemical degradation by some inorganic anions present in natural waters. *Appl. Catal. B Environ.* **2008**, *77*, 308–316. [CrossRef]

23. Avetta, P.; Pensato, A.; Minella, M.; Malandrino, M.; Maurino, V.; Minero, C.; Hanna, K.; Vione, D. Activation of persulfate by irradiated magnetite: Implications for the degradation of phenol under heterogeneous photo-Fenton-like conditions. *Environ. Sci. Technol.* **2015**, *49*, 1043–1050. [CrossRef] [PubMed]

24. Matzek, L.W.; Carter, K.E. Activated persulfate for organic chemical degradation: A review. *Chemosphere* **2016**, *151*, 178–188. [CrossRef] [PubMed]

25. Braslavsky, S.E. Glossary of terms used in photochemistry, 3rd edition (IUPAC Recommendations 2006). *Pure Appl. Chem.* **2007**, *79*, 293–465. [CrossRef]

26. Buxton, G.V.; Greenstock, C.L.; Helman, W.P.; Ross, A.B. Critical review of rate constants for reactions of hydrated electrons, hydrogen atoms and hydroxyl radicals ($^{\bullet}$OH/$^{\bullet}$O$^-$) in aqueous solution. *J. Phys. Chem. Ref. Data* **1988**, *17*, 1027–1284. [CrossRef]

27. Jayson, G.G.; Parsons, B.J.; Swallow, A.J. Some simple, highly reactive, inorganic chlorine derivatives in aqueous solution. *J. Chem. Soc., Faraday I* **1973**, 1597–1607.

28. Vione, D.; Maurino, V.; Minero, C.; Calza, P.; Pelizzetti, E. Phenol chlorination and photochlorination in the presence of chloride ions in homogeneous aqueous solution. *Environ. Sci. Technol.* **2005**, *39*, 5066–5075. [CrossRef] [PubMed]

29. Canonica, S.; Kohn, T.; Mac, M.; Real, F.J.; Wirz, J.; Von Gunten, U. Photosensitizer method to determine rate constants for the reaction of carbonate radical with organic compounds. *Environ. Sci. Technol.* **2005**, *39*, 9182–9188. [CrossRef] [PubMed]

30. Neta, P.; Huie, R.E.; Ross, A.B. Rate constants for reactions of inorganic radicals in aqueous solution. *J. Phys. Chem. Ref. Data* **1988**, *17*, 1027–1284. [CrossRef]

31. Hori, H.; Yamamoto, A.; Hayakawa, E.; Taniyasu, S.; Yamashita, N.; Kutsuna, S. Efficient decomposition of environmentally persistent perfluorocarboxylic acids by use of persulfate as a photochemical oxidant. *Environ. Sci. Technol.* **2005**, *39*, 2383–2388. [CrossRef] [PubMed]

32. Criquet, J.; Leitner, N.K.V. Degradation of acetic acid with sulfate radical generated by persulfate ions photolysis. *Chemosphere* **2009**, *77*, 194–200. [CrossRef] [PubMed]

33. Antoniou, M.G.; de la Cruz, A.A.; Dionysiou, D.D. Degradation of microcystin-LR using sul fate radicals generated through photolysis, thermolysis and e$^-$ transfer mechanisms. *Appl. Catal. B: Environ.* **2010**, *96*, 290–298. [CrossRef]

34. Herrmann, H.; Hoffmann, D.; Schaefer, T.; Braeuer, P.; Tilgner, A. Tropospheric aqueous-phase free-radical chemistry: Radical sources, spectra, reaction kinetics and prediction tools. *ChemPhysChem* **2010**, *11*, 3796–3822. [CrossRef] [PubMed]

35. Zuo, Z.H.; Cai, Z.L.; Katsumura, Y.; Chitose, N.; Muroya, Y. Reinvestigation of the acid-base equilibrium of the (bi)carbonate radical and pH dependence of its reactivity with inorganic reactants. *Radiat. Phys. Chem.* **1999**, *55*, 15–23. [CrossRef]

36. Wu, Y.L.; Bianco, A.; Brigante, M.; Dong, W.B.; de Sainte-Claire, P.; Hanna, K.; Mailhot, G. Sulfate radical photogeneration using Fe-EDDS: Influence of critical parameters and naturally occurring scavengers. *Environ. Sci. Technol.* **2015**, *49*, 14343–14349. [CrossRef] [PubMed]

37. Martell, A.E.; Smith, R.M.; Motekaitis, R.J. *Critically Selected Stability Constants of Metal Complexes Database, Version 4.0*; NIST: Gaithersburg, MD, USA, 1997.

38. Bielski, B.H.J.; Cabelli, D.E.; Arudi, R.L.; Ross, A.B. Reactivity of HO$_2$$^{\bullet}$/O$_2$$^{\bullet-}$ radicals in aqueous solution. *J. Phys. Chem. Ref. Data* **1985**, *14*, 1041–1100. [CrossRef]

39. Merritt, C.R.B. *Mass Spectrometry: Part A*; Merritt, C., Jr., McEwen, C.N., Eds.; Marcel Dekker: New York, NY, USA, 1979.

40. Barker, J. *Mass Spectrometry: Analytical Chemistry by Open Learning*, 2nd ed.; John Wiley and Sons: West Sussex, UK, 1999.

Sample Availability: Samples of the compounds are available from the authors.

Section 3:
Assisted Photocatalytic Treatment of Water

Review

Solar or UVA-Visible Photocatalytic Ozonation of Water Contaminants

Fernando J. Beltrán * and Ana Rey

Departamento de Ingeniería Química y Química Física, Instituto Universitario de Investigación del Agua, Cambio Climático y Sostenibilidad, Universidad de Extremadura, Av. Elvas s/n, 06006 Badajoz, Spain; anarey@unex.es
* Correspondence: fbeltran@unex.es; Tel.: +34-924-289-387

Received: 27 April 2017; Accepted: 4 July 2017; Published: 14 July 2017

Abstract: An incipient advanced oxidation process, solar photocatalytic ozonation (SPO), is reviewed in this paper with the aim of clarifying the importance of this process as a more sustainable water technology to remove priority or emerging contaminants from water. The synergism between ozonation and photocatalytic oxidation is well known to increase the oxidation rate of water contaminants, but this has mainly been studied in photocatalytic ozonation systems with lamps of different radiation wavelength, especially of ultraviolet nature (UVC, UVB, UVA). Nowadays, process sustainability is critical in environmental technologies including water treatment and reuse; the application of SPO systems falls into this category, and contributes to saving energy and water. In this review, we summarized works published on photocatalytic ozonation where the radiation source is the Sun or simulated solar light, specifically, lamps emitting radiation to cover the UVA and visible light spectra. The main aspects of the review include photoreactors used and radiation sources applied, synthesis and characterization of catalysts applied, influence of main process variables (ozone, catalyst, and pollutant concentrations, light intensity), type of water, biodegradability and ecotoxicity, mechanism and kinetics, and finally catalyst activity and stability.

Keywords: solar photocatalytic oxidation; ozonation; solar photocatalytic ozonation; water contaminants

1. Introduction

For wastewater to be released into a natural water environment or reused for social or industrial purposes, it first passes through the classical unit operations of a wastewater treatment plant (WWTP). It comes out as apparently clear water, with chemical and biochemical oxygen demand values (COD, BOD) normally below those allowed by law or by official environmental rules. However, such treatments often do not remove organic compounds, called micropollutants, at very low concentrations (μg to ng L^{-1}). These contaminants are generally due to human activities relating to agriculture, industries, or simply health and personal care. Thus, compounds such as pesticides, phenols, and pharmaceuticals, among others, are frequently found in the wastewater influent and effluent of WWTPs and in groundwater [1–4]. Many of these organics have well-defined maximum contaminant levels (MCL) (priority pollutants) [5], but others still do not have MCL and are called emerging contaminants. Thus, phenols and pesticides are priority pollutants while pharmaceuticals and personal care products are emerging contaminants. These compounds are a threat to water quality in different ways. For instance, antibiotics can increase the resistance some microorganisms have. A variety of pollutants can disrupt the endocrine system, causing tumors, sexual changes in animals, etc. [6]. In addition, water scarcity and/or drought increases the need for water reuse and makes necessary the effective elimination of biologically active compounds. Therefore, tertiary treatment methods such as adsorption, membrane, and chemical oxidation processes should be included in WWTPs to address these problems.

Adsorption and membrane operations only transfer contaminants from the treated water to a second phase (the adsorbent or the concentrate in the membrane process), while chemical oxidation can eliminate the pollutants.

Chemical oxidation can be accomplished with the use of individual chemicals such as ozone or hydrogen peroxide, or through combinations with other agents such as radiation and/or catalysts. These combinations are usually called advanced oxidation processes (AOPs), since they generate hydroxyl radicals (HO·). The strong oxidation potential of HO· can degrade or eliminate essentially all micropollutants except perhalogenated compounds. The rate constants of the reactions between HO· radicals and most organic compounds range from about 10^7 M^{-1} s^{-1} for the most recalcitrant organics (i.e., oxalic acid) up to 10^{10} M^{-1} s^{-1} for the most reactive compounds (i.e., phenol) [7].

Because of its high reactivity and potential combination with other agents, ozone forms an important group of AOPs. In fact, ozonation is an AOP alone since ozone may decompose into hydroxyl radicals or upon direct reactions with different organic compounds [8]. The reactivity of ozone is due to its electronic structure (see Figure 1), which has resonance forms with positively and negatively charged oxygen atoms.

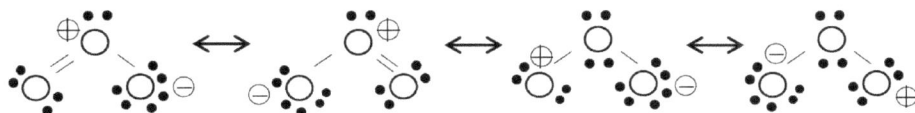

Figure 1. Electronic forms of an ozone molecule.

Ozone, which is an oxidant, reacts very fast with organic compounds with electron-rich systems such as double carbon bonds and aromatic rings. Pharmaceuticals often contain electron-rich functional groups in their structure and are very highly reactive to ozone attack (see Figure 2). The rate constants of the reactions of ozone with these compounds can also be very high, as seen in Table 1.

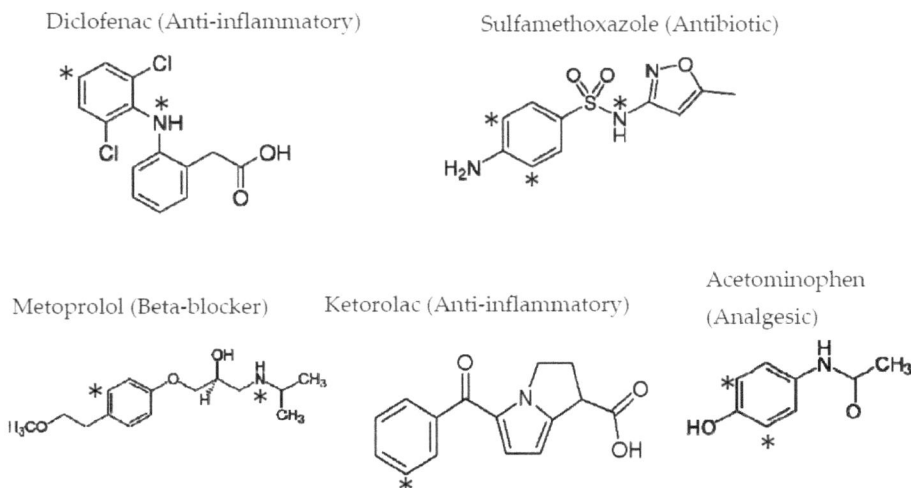

Figure 2. Examples of molecular structures of some pharmaceuticals usually found in urban wastewater. * Possible points of ozone attack.

Table 1. Rate constants of ozone reactions with some pharmaceuticals found in urban wastewater *.

Compound	Activity	Rate Constant ($M^{-1} s^{-1}$)	Reference
Sulfamethoxazole	Antibiotic	5.5×10^5	[9]
Diclofenac	Anti-inflammatory	10^5	[10]
Ketorolac	Analgesic	3.4×10^5	
Acetaminophen	Anti-inflammatory	2.7×10^5	
Metoprolol	Beta-blocker	2.5×10^3	
Carbamacepine	Analgesic	3×10^5	[11]
17α-ethinylestradiol	Hormone	3×10^6	
Tetracycline	Antibiotic	1.9×10^6	[12]
Fenoterol	Breath aider	2.8×10^6	[13]
Gemfibrozil	Lipid regulator	4.9×10^5	
Estriol	Hormone	10^5	[14]
Lyncomycine	Antibiotic	6.7×10^5	[15]

* At pH 7.

Consequently, in an ozonation process there are two types of reactions with the organics in water: the so-called direct reactions, and the reactions of hydroxyl radicals formed from the decomposition of ozone. The combinations of ozone with different agents condition the way hydroxyl radicals are formed and their concentration. For example, the peroxone process, that is, the combination between ozone and hydrogen peroxide at neutral pH, is a well-known AOP. In this case, the reaction of ozone with the hydroperoxide ion (the ionic form of hydrogen peroxide) initiates the formation of free radicals in a reaction with a high rate constant (2.8×10^6 $M^{-1} s^{-1}$) [16], while the reaction that initiates the formation of free radicals in ozonation alone has a value of only 70 $M^{-1} s^{-1}$ [17]. This is the reason why ozonation alone is, let us say, a poor AOP because of the low generation rate of free radicals. Other combinations of ozone with catalysts and/or radiation constitute other ozone AOPs where, in different ways [8,18–21], hydroxyl radicals are formed, as occurs with photocatalytic ozonation.

Photocatalytic ozonation is the combination of ozonation and photocatalytic oxidation, which is the application of radiation upon a semiconductor in the presence of oxygen or air. This process started in 1978, with the first published work dealing with the removal of contaminants in water [22]. Six years before, in 1972, it was reported that UV-irradiation of a TiO_2 anode produced hydrogen from water using an electrical bias [23]. Photocatalytic oxidation takes place when photons, with an energy equal to or higher than the band gap energy of the catalyst (or semiconductor) used, excite electrons from the highest occupied molecular orbital (HOMO) or the valence band to be transferred to the lowest unoccupied molecular orbital (LUMO) or conduction band of the catalyst. In this way, two charge carriers occur: an oxidizing point in the HOMO or hole, and electrons able to trigger reduction reactions in the LUMO [24–27]. The main problem with this process is that the electron-hole pair recombination inhibits the oxidation-reduction steps. Both oxidizing holes and electrons in the presence of adsorbed water and/or contaminants give rise to the formation of hydroxyl radicals or molecular oxidizing substances such as hydrogen peroxide [28].

From the beginning, there has been a great research interest in both ozonation and photocatalytic oxidation processes. In fact, a search in the Web of Science database from 2000 to the present day, with the keywords ozonation and water or photocatalytic oxidation and water, produces 5656 and 10,434 papers published on the two processes, respectively. Furthermore, their main characteristics, possible applications, and experimental and practical results (with regard to ozonation) have been the subject of different reviews [21,25–35].

Given the fact that ozone is a stronger oxidant than oxygen, the simultaneous application of ozone, radiation, and a catalyst with the abovementioned features has led to a new AOP called photocatalytic ozonation. Thus, Figure 3 shows the evolution of the numbers of publications on photocatalytic ozonation that have appeared in scientific literature from 2000 to 2016 according to the Web of Science database.

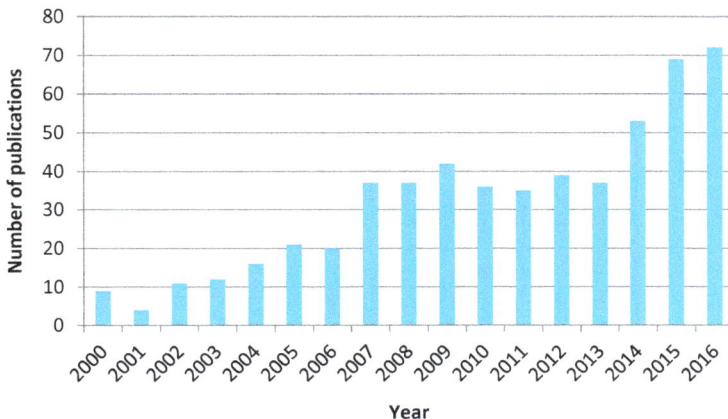

Figure 3. Evolution of the number of publications on water photocatalytic ozonation in the period 2000–2016.

Although the number is far lower than the corresponding figures of ozonation and photocatalytic oxidation as individual processes, the trend of publication is clearly increasing with time. In spite of its short history, this subject has already been the focus of some reviews where the characteristics, operating variables, mechanism, and kinetics of the process are dealt with from the works published so far [36–39].

In this work, however, the subject under study is not photocatalytic ozonation, but the logical consequence of this AOP to make it more environmentally sustainable: solar photocatalytic ozonation (SPO), or the use of solar or simulated solar radiation in the photocatalytic ozonation process. Thus, the main features of this process are given below with emphasis on the appropriateness of this emerging AOP, the catalysts, the organics treated, the kinetics and the mechanism, etc.

2. The Solar Photocatalytic Ozonation Process

Photocatalytic ozonation with UVC or UVA lamps yields significant reductions of many organics and total organic carbon (TOC) of the contaminated water in a very reduced reaction time (minutes for organics and a few hours for TOC) if compared to ozone-free photocatalytic oxidation, but the process is expensive because both the generation of ozone and lamp functioning require electrical energy which can be significant in many cases. Conversely, solar photocatalytic ozonation only requires energy to produce ozone; moreover, this energy will eventually be able to be produced from the Sun with a suitable photovoltaic or similar system able to generate electrical energy from solar energy. Oyama et al. [40], for example, studied the removal of 2,4-dichlorophenoxyacetic acid, bisphenol A and two surfactant compounds with SPO, and the electric power for the reactor system, pumps, ozone generator, radiometer, and computer used was supplied entirely by a solar cell battery. The Sun can be a never-ending natural energy resource. The average incident solar energy on the Earth's surface is about 240 W m^{-2}, a part of which (about 5%) corresponds to UVA radiation with wavelengths from 290 to 400 nm, (5% UVB and 95% UVB) However, total solar energy (290–800 nm) would be approximately three thousand millions of MW, which is equivalent to about 15 million 2000 MW Nuclear Power Plants or about 600 million of 50 MW thermosolar or photovoltaic plants. Thus, solar energy could be the solution for environmental sustainable processes such as the photocatalytic degradation of water contaminants.

It might be said that works on SPO started in 2002, with an incipient study to remove *p*-nitrophenol and observe the effects of phosphates [41]. However, the first work in which SPO was applied as a main AOP was for the removal of 2,4-dichlorophenol with ozone or hydrogen peroxide, with Fe(III) as catalyst and different radiation sources from UVA blue fluorescent lamps emitting radiation between 300 and 420 nm as well as the Sun itself [42]. The main processes studied were solar photo-Fenton and SPO. In this paper, the greatest TOC removal achieved by SPO (approximately 90% with 20 µEinstein s^{-1}) was a clear indication of the high efficiency of this process compared to others, such as solar photo-Fenton, which only allowed 10% mineralization. A search on the Web of Science database from 2000 to 2017 produced 53 papers when the keywords solar photocatalytic ozonation and water were used. However, a closer look at these papers reveals that only 18 of them specifically study the SPO process. In addition, we found a further eight papers on SPO that did not appear in the search. Accordingly, it can be said that SPO is an emerging AOP since, so far, as many as five papers have been published in only one year. In Table 2, the main features of these papers regarding catalysts, organics treated, radiation sources, concentrations of ozone, types of AOPs, etc., are presented. In the following sections, explanations of the results obtained in these works are given.

Table 2. Research articles published so far on solar or UVA-Visible light photocatalytic ozonation of water contaminants [a].

Compounds Treated and Processes Applied	Data on Catalysts, Ozone, and Others	Radiation Source	Photoreactor Type	Observations	Ref.
Effluent from an urban wastewater lagoon. Total coliforms followed solar photolytic photocatalytic oxidation (SPO).	P25 TiO_2 alone: 2 g L^{-1}. Activated carbon (AC) alone: 0.3 g/L. Mixture TiO_2-AC: 2.3 g L^{-1}. Ozone dose or concentration: not given.	Solar light 10 to 14 h radiation.	Agitated or bubble vessels of 250 mL.	COD removal: 40% with SPO (best result) and AC-TiO_2. Total coliforms removal: >99.99%. Inorganic ions determined: NO_2^-, PO_4^{3-}. Fourier Transform FTR study.	[41]
2,4 dichlorophenol (DCP) at 100 mg L^{-1}. H_2O_2 concentration: 250 and 75 mg/L in UV/H_2O_2 and UV/Fe/H_2O_2, respectively. O_3 + Fe(II) + UV, photo-Fenton, UV + Fe(II), UV + H_2O_2, photocatalysis and photolysis.	Two catalysts: 1. Fe(II): 10–30 mg L^{-1} as $FeCl_3$. 2. TiO_2: 0.5 g L^{-1}. Ozone: 5.5 g h^{-1}.	1. Blue UVA lamp (350 nm); 2.5 µE s^{-1}. 2. Solar simulator: 6 µE s^{-1}. 3. The Sun: 236 µE s^{-1}. 4. Eight 15 W UVA blue fluorescent lamps (300–420 nm): 120 µE s^{-1} (µE s^{-1}; µEinstein s^{-1}).	1. UV, UV + H_2O_2, UV + Fe(II), Photo-Fenton runs in 12 × 13 cm photoreactor. 2. Solar simulator with parabolic mirrors and quartz tube reactor. 3. Three compound parabolic collectors CPC, each with eight tubes in series. 4. One 2 L gas-liquid contactor + tube photoreactor.	Scaling up factors for scaling up to pilot-plant size. Estimation of amount of waste water that could be treated. Analysis: DCP concentration and TOC. DCP degradation is plotted as a function of accumulated photons per liter entering the reactor. Apparent pseudo first order kinetics. Pseudo quantum yield as a function of incident energy.	[42]
2,4-dichlorophenoxyacetic acid, bisphenol A. Sodium butylnaphthalenesulfonate and benzyldodecyldimethyl-ammonium bromide surfactants. 2 mM concentration: 100, 175 and 250 mg L^{-1}. Processes: O_3, O_3/UV, O_2/UV/TiO_2, and O_3/UV/TiO_2.	P25 TiO_2 at 2 g L^{-1}.	One 75 W High Pressure Hg lamp (3 mW cm^{-2} at 360 nm) and the Sun.	Twenty Pyrex glass tubular reactors.	TOC measurement.	[40]
Phenol: 0.169 mg L^{-1}. Processes: O_3, O_3/Vis, O_2/Vis/cat, O_3/cat and O_3/Vis/cat. Ozone process: 0.45 g h^{-1}.	Commercial WO_3 powder and n-TiO_2: 0.2 g L^{-1}.	300 W Xe lamp with a cut-off filter (k >420 nm).	Pyrex inner-irradiation vessel placed in a water bath set at room temperature.	Catalyst characterization: SEM, XRD, SBET. AOP comparison results for compound concentration and TOC. Repetitive experiments for catalyst reuse.	[43]
Orange II dye: 10^{-4} M. Processes: O_3. 100 mL/min and 2 mg L^{-1} of ozone in influent gas. Processes: O_3 cat, O_2/UV/cat, O_3/UV/cat.	Bi_2O_3 and Au/Bi_2O_3 nanorods: 1 g L^{-1}.	55 W Xe lamp with 20,200 Lux and a 320 nm cut-off filter. Wavelength > 320 nm.	A 500-mL capacity borosilicate glass photoreactor with walls covered with aluminum foil to avoid the release of radiation.	Microwave and hydrothermal methods of catalyst synthesis. Catalyst characterization: DRX, SEM, TEM, XPS, EDX, DRUV-visible. Dye absorbance and concentration. Role of photosensitization. Reusability studies. Byproduct identification.	[44]
2-Chlorophenol, 2,4-dichlorophenoxyacetic acid, bisphenol A, sodium dodecylbenzenesulfonate, sodium butylnaphthalenesulfonate, and benzyldodecyldimethyl- ammonium bromide surfactants at 0.5 and 0.1 mM in simulated wastewater. Processes: UV/O_3, O_2/TiO_2/sunlight, and O_3/TiO_2/sunlight. 3 mg L^{-1} dissolved ozone.	Dispersed TiO_2 P25 (0–2 g L^{-1}) and TiO_2-coated glass.	Solar light.	Pilot plant: three modules of 40 Pyrex glass tubes (inner diameter, 1.76 cm; length: 145 cm), each connected in series. Photoreactor volume: 42.3 L and total solar light harvest area of the three modules: 3.06 m^2. Bubble ozone column connected in series with photoreactor modules. Solar cell for electric power to the photoreactor and ozonator.	Contaminants, chloride ion concentrations, and TOC followed vs. accumulated sunlight energy incident on the photoreactor per liter of the solution.	[45]

Table 2. *Cont.*

Compounds Treated and Processes Applied	Data on Catalysts, Ozone, and Others	Radiation Source	Photoreactor Type	Observations	Ref.
Dyes: Rhodamine B for UVA radiation and Methylene Blue for Visible radiation. At 10 mg L⁻¹.	TiO_2 and M-TiO_2 catalysts. $M = Ag^+$, Cu^{2+}, Mn^{2+}, Ce^{3+}, Fe^{3+} and Zr^{4+} ions. Main catalyst used: Mn-TiO_2 0.25 g L⁻¹ (0–7% Mn).	Lamp not given; full solar radiation wavelength (300 and above) and only visible light (λ > 400 nm).	Cylindrical Pyrex vessel surrounded by a cooling water jacket in a solar simulating box.	Catalyst characterization: DRX, SBET, DRUV-Vis. Absorbance of dyes solutions followed. Correlation between SBET and dye removal percentage.	[46]
Metoprolol (MTP) at 10 to 50 mg L⁻¹. Processes: O_3/Light, O_3/Cat/Light, and all combinations.	1. Fe_3O_4/TiO_2/AC 331 m² g⁻¹, 68% Anatase (0.38 g L⁻¹) 2. P25 TiO_2 (0.25 g L⁻¹). Ozone gas inlet: 20 L h⁻¹ and 6 mg L⁻¹.	1500 W Xe lamp with limited radiation above 300 nm with filters. 550 W/m².	Glass-made agitated tank provided with gas inlet, gas outlet, and liquid sampling ports.	Catalyst characterization: nitrogen adsorption, XRD, FTIR, SEM, EDX, SQUID magnetometer. Reusability and activity in five cycles. MTP Concentration. TOC followed.	[47]
Dichloroacetonitrile at 1 ppm. Ozone: 1–1.4 g L⁻¹ h⁻¹. Processes: UVsolar/TiO_2, O_3, O_3/TiO_2, UVsolar/O_3, and UVsolar/TiO_2/O_3.	P25 TiO_2 ozono dosage: 1–1.38 g L⁻¹ h⁻¹.	Three halide lamps (100, 250, and 400 W) and the Sun with similar light spectra from 300 to 800 nm.	1. Bench system: three halide lamps (100, 250, and 400 W), 60 cm at the top of three quartz tubes (40 cm × 2.7 cm). 2. CPC reactor (38° tilted) and cylindrical ozone reactor (1.1 m high, 10 cm internal diameter) in series. Turbulent regime.	Influence of different AOPs, pH (3, 6.5, and 10), W (4.6 to 33.8 W m⁻²), catalyst dosage 0.2–2.5 g L⁻¹. Temperature: 10 to 40 °C.	[48]
Bisphenol A (BPA) and oxalic acid, 10 mg L⁻¹. Ozone: 1 mL min⁻¹ and 500 mgh⁻¹. Processes: O_3, UV-Vis/cat/O_2, O_3/UV-Vis/cat.	Graphitic carbon nitride (g-C_3N_4). Composed of numerous interconnected nanosheets, 0.5 g L⁻¹.	High-pressure Xe long-arc lamp jacketed by a quartz thimble (GX2500 W). Filter with Na_2NO_2 to cut λ < 400 nm.	One 1 L glass tubular photoreactor (8.5 × 40 cm).	Synthesis (from urea) and characterization TEM, FTIR, BET (67 m² g⁻¹), XRD, UV-Vis (2.7 eV, 450 nm max.). Rates for oxalic acid and BPA higher than sum of rates of single processes. Tert-butanol (TBA), triethanolamine (TEOA), and benzoquinone (BQ) act as the hydroxyl radical, hole and O_2^- scavenger.	[49]
Metoprolol, 50 ppm. Processes: O_3/Light, O_3/Cat/Light, and all combinations.	1. Fe_3O_4/TiO_2/AC 331 m² g⁻¹, 68%. Anatase (0.38 g L⁻¹) 2. P25 TiO_2 (0.25 g L⁻¹). Ozone gas inlet: 20 L h⁻¹ and 6 mg L⁻¹.	1500 W Xe lamp with limited radiation above 300 nm with filters. 550 Wm⁻², 300–800 nm, 320–800 nm, and 390–800 nm.	Glass-made agitated tank provided with gas inlet, gas outlet, and liquid sampling ports.	Catalyst preparation and characterization (nitrogen adsorption, XRD, SEM, EDX, XPS and SQUID magnetometer). MTP concentration and TOC, dissolved O_3, H_2O_2, acid intermediates.	[50]
Atenolol, Hydrochlorothiazide, Ofloxacin, and Trimethoprim in ultrapure water (10 mg L⁻¹ doping) and WWSE (0.5 mg L⁻¹ doping). Processes: O_3, UVA-Vis, O_3/cat, O_3/UVA-Vis, O_3/cat/UVA-Vis, O_3/Cat/UVA-Vis.	P25 TiO_2 45 L h⁻¹ and 20 mg L⁻¹ for ozone in the inlet gas.	Solar light (visible + UVA) with: 35 ± 5 W m⁻².	CPC: four tubes in series, 300–400 L h⁻¹. Water flow rate for recirculation T = 18–30 °C.	ECs concentration, TOC, ecotoxicity (*Daphnia magna*), Phenolic compounds formed, BOD/COD. Ozone consumption.	[51]
Caffeine, metoprolol, and ibuprofen: 2 mg L⁻¹ each in Municipal Wastewater MWW. Processes: O_3, UVA-Vis, O_3/cat, O_3/UVA-Vis, O_3/cat/UVA-Vis, O_3/Cat/UVA-Vis. 20 mg/L O_3 and 20 L h⁻¹.	WO_3/TiO_2 (from P25 and titanate nanotubes) 3.8% WO_3 and 0.5 g L⁻¹.	1500 W Xe lamp with limited radiation restricted to wavelengths over 320 nm because of the presence of quartz, glass, and polyester cut-off filters with 550 W m⁻².	0.5L semi-batch glass-made spherical reactor, provided with a gas inlet, a gas outlet, and a liquid sampling port in a commercial solar simulator chamber.	Catalyst characterization: ICP, N_2 adsorption–desorption isotherms (SBET), XRD, TEM, Raman, XPS, and DRUV-Vis spectroscopy. Contaminant concentration and TOC followed.	[52]

Table 2. *Cont.*

Compounds Treated and Processes Applied	Data on Catalysts, Ozone, and Others	Radiation Source	Photoreactor Type	Observations	Ref.
Oxalic acid: 0.01 M (TOC = 240 ppm).	TiO2 P25, Nb2O5, SnO2, WO3, Fe2O3, In2O3, and BiVO4: 2 g/L. O3: 14 mg L^{-1} and 0.45 g h^{-1}.	300 W Xe lamp with an IR cut-off filter. Incident light was ca. 200 mW in the range of 360 to 470 nm. For only visible irradiation, another filter was used: $\lambda > 410$ nm.	Pyrex inner irradiation vessel placed in a thermostatic water bath.	SBET: 1.7 to 54.1 m^2 g^{-1}. TOC. Visible active properties of semiconductors (only WO3, Fe2O3, In2O3, and BiVO4). WO3: Best material for photocatalytic ozonation under visible light irradiation.	[53]
Ibuprofen: 10 mg L^{-1}, and a mixture of acetaminophen, metoprolol, caffeine, hydrochlorothiazide, antipyrine, sulfamethoxazole, carbamazepine, ketorolac, diclofenac, and ibuprofen in an MWWT: 0.5 mg L^{-1} each. Processes: adsorption, photolysis, O3, UVA-Vis, O3/cat, O3/UVA-Vis, O2/cat/UVA-Vis, O3/cat/UVA-Vis. Ozone processes: 10 mg L^{-1} and 20 L h^{-1} gas.	WO3: 0.25 g L^{-1}.	1500 W air-cooled Xe arc lamp with emission restricted to visible light ($\lambda > 390$ nm) because of quartz, glass, and polyester cut-off filters. 550 Wm^{-2}.	0.5-L glass-made spherical reactor in the chamber of a box simulator.	Preparation conditions: Calcination temperature and time. Characterization: TGA-DTA, XRD, N2 adsorption-desorption isotherms, pH$_{PZC}$, XPS, and DRUV-Vis spectra. Contaminant concentrations and TOC followed. Mechanism, kinetic regime of ozonation.	[54]
N,N-diethyl-meta-toluamide DEET: 5 mg L^{-1}. Processes: adsorption, photolysis, O3, UVA-Vis, O3/cat, O3/UVA-Vis, O3/cat/UVA-Vis, O3/cat/UVA-Vis. Ozone processes: 10 mg L^{-1} and 15 L h^{-1} gas.	Commercial and homemade WO3 catalysts: 0.25 g L^{-1}. Calcination temperature: 500 to 700 °C.	1500 W air-cooled Xe arc lamp with emission restricted to visible light ($\lambda > 390$ nm) because of quartz, glass, and polyester cut-off filters. 550 Wm^{-2}.	0.5-L glass-made spherical Reactor in the chamber of a box simulator	Synthesis method. Catalyst characterization: XRD, Raman, N2 adsorption-desorption isotherms (SBET), SEM, XPS. Contaminant concentration and TOC followed. Oxalic, acetic, and formic acids were followed.	[55]
Acetaminophen, antipyrine, bisphenol A, caffeine, metoprolol, testosterone. Concentrations: 1.5 and 2.9 mg L^{-1} (10^{-5} M each). Processes: Sun, O3, Fe(III), or P25 TiO2 combinations. Also, H2O2 for photo-Fenton.	Fe(III) (homog.: 2.79 mg L^{-1}, pH 3), TiO2 (heterog.: 200 mg L^{-1}, pH 7). In some cases: Fe(III)/H2O2 = 6.09 mass ratio. 13 mg L^{-1} ozone in gas.	The Sun. Average solar radiation: 40 W m^{-2}.	Four borosilicate glass tube CPCs (29.4 × 75 cm). Collector surface 0.25 m^2. Illumination volume: 1.8 L. Tilted 45° to the south. Parabolic anodized aluminum reflectors. Turbulent regime. Semi-batch mode.	Concentrations and TOC removal. Energy and ozone demands. Kinetic regimes. Kinetics.	[56]
Acetaminophen, antipyrine, bisphenol A, caffeine, metoprolol, testosterone in secondary WWSE effluent (BOD = 10, COD = 58.6; TOC = 20 mg L^{-1}). Presence of anionic ions. Concentrations between 1.5 and 0.2 mg L^{-1}. Processes: Sun, O3, Fe(III), or P25 TiO2 and their combinations. Also, H2O2 in some cases.	Fe(III) (homog.: 2.8 mg L^{-1}, pH 3), TiO2 (heterog.: 200 mg L^{-1}, pH 7). 13 mg L^{-1} ozone in gas.	The Sun. Average solar radiation: 40 Wm^{-2}.	Four borosilicate glass tube CPCs (29.4 × 75 cm). Collector surface 0.25 m^2. Illumination volume: 1.8 L. Tilted 45° to the south. Parabolic anodized aluminum reflectors. Turbulent regime. Semi-batch mode. Reaction time: 5 h (+30 min, previous for adsorption).	Concentrations and TOC removal. Also, ions and total phenol concentration. *Daphnia magna* ecotoxicity measurements. Measurements of HO-radicals concentration with *p*-chlorombenzoic acid as a probe compound. Biodegradability as BOD/COD. Economic considerations.	[57]
Diuron, o-phenylphenol, 2-methyl-4-chlorophenoxyaceticacid (MCPA), and *tert*-buthylazine (5 mg L^{-1} each). Processes: O3, UVA-Vis, O3/cat, O3/UVA-Vis, O3/cat/UVA-Vis, O3/cat/UVA-Vis.	TiO2 and 0.5–0.8 wt. % B-TiO2: 0.33 g L^{-1}. Ozone in the inlet gas: 5 mg L^{-1} and 10 L h^{-1}.	1000 W Xe lamp. Incident radiation flux: 8.96 × 10^{-4} Einstein min^{-1}. Radiation intensity: 500 W m^{-2}.	Glass-made agitated tank provided with gas inlet, gas outlet, and liquid sampling ports in a solar simulator box.	Synthesis and characterization: ICP-OES, N2 adsorption–desorption, XRD, XPS, and DRUV-Vis spectroscopy (3.01 and 3.03 eV band gap). Compound concentration. TOC, dissolved ozone, and H2O2 concentrations. B leached.	[58]

Table 2. *Cont.*

Compounds Treated and Processes Applied	Data on Catalysts, Ozone, and Others	Radiation Source	Photoreactor Type	Observations	Ref.
Acetaminophen (ACM), antipyrine (ANT), caffeine (CAF), ketorolac (KET), metoprolol, sulfamethoxazole (SFX), carbamazepine (CARB), hydrochlorothiazide (HCT), and diclofenac (DIC). In WWPE doped: 200 µg L⁻¹ each. Processes: solar photocatalysis with O_3/TiO_2, solar photo-Fenton, or ozonation.	Three catalysts: 1. pH 3 with 2.8 mg L⁻¹ Fe(III). 2. 150 mg L⁻¹ Fe_3O_4. 3. Natural pH with 250 mg L⁻¹ P25 TiO_2 doped. Ozone in the inlet gas: 33.6 L h⁻¹ and 13 mg L⁻¹.	Radiation source: Sun.	Aerobic tank: HRT of 7 h, biomass sludge aged 5–6 days. MLVSS.MLSS⁻¹: 0.8. Oxygen: 2–4 mg L⁻¹. Reactor Volume: 5 L (1.8 L of irradiated volume) compound. Parabolic collector. Four borosilicate glass tube CPCs (29.4 × 75 cm).	Aerobic degradation followed by AOP; solar photocatalysis. ECs concentration, TOC, COD, ecotoxicity. Accumulated UV-vis energy calculated.	[59]
Bisphenol A and oxalic acid (OA), 10 mg L⁻¹ at 1 mL min⁻¹ and 500 mgh⁻¹.	Graphitic carbon nitride: g-C_3N_4 composed of numerous interconnected nanosheets. Concentration: 0.5 g L⁻¹.	High-pressure Xe long-arc lamp, jacketed by a quartz thimble (GXZ500 W). Filter with Na_2NO_2 to cut λ < 400 nm.	1 L glass tubular photoreactor (8.5 × 40 cm).	Synthesis (from urea) and characterization (TEM, FTIR, BET (67 m²/g), XRD, UV-Vis (2.7 eV, 450 nm max.), Rates for oxalic acid and BPA higher than sum of single processes: O_3 and UV-Vis/cat/O_2. Tert-butanol, (TBA), triethanolamine (TEOA), and benzoquinone (BQ) as HO·, hole and O_2^- scavengers.	[60]
Oxalic acid, 0.11 mM.	g-C_3N_4-reduced graphene oxide (rGOxide) 0.2 g/L. Ozone: 75 mgh⁻¹.	As in Reference [56].	As in Reference [56].	Catalyst characterization. 2% rGO leads to best results for OA removal. Catalyst activity and stability. Basic mechanism.	[61]
Phenol in water and urban wastewater: 50 mg/L. (DOC: 20 mg L⁻¹, COD: 40 mg L⁻¹, pH 7.4). Processes: Ozonation, photocatalysis, and photocatalytic ozonation.	Ag, Cu, Fe on TiO_2: 0.5 g L⁻¹. Ozone: 20.83 mg L⁻¹ min⁻¹.	Heraeus TQ 150 W immersion medium-pressure Hg lamp 70 mW cm⁻². The lamp emission spectrum has main peaks at 253.7, 313, and 366 nm in the UV range and 436, 546, and 578 nm in the visible range. For solar runs: 37.6 mW cm⁻².	For UV runs: cylindrical quartz photochemical reactor (0.7 L) wrapped in aluminum foil. For solar runs: glass tubular reactor (1 m in length and 0.04 m in diameter) and a parabolic solar collector.	Best catalyst: Fe-TiO_2 which presents the highest BET area and higher λ visible absorption (530 nm). Phenol concentration. COD, Langmuir kinetics applied simplified to pseudo first order kinetics. Synergic index, pseudo first rate constants calculated.	[62]
t-Butilazina: 5 mg L⁻¹. Processes: Adsorption onto AC and multi-walled carbon nanotube (MWCNT), UV photolysis, UV/H_2O_2, single ozonation, O_3/H_2O_2, catalytic ozonation (AC, MWCNT and TiO_2 as catalysts) and some solar driven processes such as photo ozonation, TiO_2-photocatalytic oxidation, and TiO_2-photocatalytic ozonation.	AC (DARCO®, 12–20 mesh) and MWCNTs (purity >95%) carbon, P25 TiO_2, and other TiO_2 prepared and TiO_2-MWCNT. Ozone: 10 mg L⁻¹, 20 L h⁻¹.	Low-pressure Hg lamp with emission at 254 nm (Heraeus, model TNN15/32). Average fluence rate: 0.6 W-1 (2.9 mW cm⁻²). For UVA-visible: 1000 W Xe lamp (300–800 nm). 581 W m⁻² (62 Wm⁻² UV-A irradiance.	For UVC photolysis and UV/H_2O_2: photoreactor provided with a central quartz well. For O_3 and O_3/cat runs: 400 mL semi-batch spherical Pyrex-made reactor provided with magnetic agitation. For photoprocesses: The same reactor as in ozone processes inside a Suntest solar simulator equipment	Characterization: XRD, SBET, TEM, FTIR. Isotherm and different AOPs and adsorption kinetics TBA concentrations and intermediates by HPLC-qTOF.	[63]

Table 2. *Cont.*

Compounds Treated and Processes Applied	Data on Catalysts, Ozone, and Others	Radiation Source	Photoreactor Type	Observations	Ref.
N,N-diethyl-meta-toluamide (DEET): 5 mg L⁻¹. Processes: Different AOP's and adsorption.	Two CeO_2 catalysts: nanorod and nanocubes. Hydrothermal method. Concentration: 0.25 g L⁻¹. Ozone: 10 mg L⁻¹, 15 L h⁻¹.	1500 W Xe lamp: 550 W m⁻² and 300 to 800 nm or 400 to 800 nm with filter.	Semi-batch borosilicate glass-made round flask in a Suntest CPS solar simulator.	Catalyst preparation and characterization: XRD, XPS, SBET, DR UV-Vis. DEET concentration, TOC, O_3 dis. H_2O_2, short chain carboxylic acids. Pseudo first order kinetics.	[64]
N,N-diethyl-meta-toluamide (DEET): 15 mg L⁻¹. Ozone processes: 10 mg L⁻¹, 15 L h⁻¹ gas flow rate.	Monoclinic WO_3 calcined at 600 °C (see Mena et al., 2015): 0.25 g L⁻¹.	1500 W air-cooled Xe arc lamp with emission restricted to visible light ($\lambda >$ 390 nm) because of quartz, glass, and polyester cut-off filters. 550 W m⁻².	0.5 L glass-made spherical reactor in the chamber of a box simulator.	HPLC-qTOF identification of intermediates. Mechanism and kinetics based on TOC removal. Scavengers used: t-butanol and oxalate. Arrhenius equation determined for DEET-O_3 reaction.	[65]

[a] COD: Chemical Oxygen Demand. AC: Activated Carbon. FTIR: Fourier Transform Infrared. SEM: Scanning electron microscopy. XRD: X-ray diffraction. SBET: Surface area from Brunauer-Emmer-Teller isotherm. TEM: Transmission electron microscopy. XPS: X-ray photoelectron spectroscopy. DRUV-Vis: Diffuse reflectance UV-Visible spectroscopy. EDX: Energy dispersive X-ray spectroscopy. SQUID: Superconducting quantum interference device, WWSE: Wastewater secondary effluent. BOD: Biological Oxygen Demand. ECs: Emerging contaminants. WWPM: Wastewater primary effluent. ICP-OES: Inductively coupled plasma-optical emission spectroscopy. HRT: Hydraulic residence time. HPLC-qTOF: High performance liquid chromatography coupled with quadrupole time-of-flight mass spectrometry. TGA-DTA: Thermogravimetric and differential thermal analysis.

2.1. Radiation Sources and Photoreactors Used

SPO processes are fed with solar radiation that covers the range between 290 to 800 nm, mainly UVA and visible zones of the solar spectrum. This means that radiation sources are limited to the Sun itself and lamps simulating this radiation wavelength range. The exciting gas of these lamps is in many cases Xe, but blue fluorescent lamps emitting from 300 to 420 nm have also been used [42], in this latter case to study part of the visible wavelength range. Also, Xe lamps emit over a wider range, but radiations below 300 nm are avoided with the use of filters. Also, in some studies, filters are used to allow specific radiation wavelength ranges, 320–800 nm or 390–800 nm, to be emitted from Xe lamps in order to reach the water to be treated. For example, Quiñones et al. [50] used a 1500 W air-cooled Xe arc lamp with the emission restricted to wavelengths over 300 nm because of the presence of quartz and glass cut-off filters. The irradiation intensity was kept at 550 Wm^{-2} and the temperature of the system was maintained between 25 and 40 °C throughout the experiments. To cut off all the wavelengths below 390 nm and 320 nm, films of flexible polyester from Edmund Optics and Unipapel, respectively, were used. In some cases, halide lamps were used. For experiments in the laboratory, Shin et al., [48], for example, used three Osram metal halide lamps of different powers (100, 250, and 400 W), which were placed 60 cm above a compound parabolic collector (CPC) reactor (see below) at the top of the chamber as an artificial solar light. According to the authors, the irradiation wavelengths of the metal halide lamps (300–800 nm) and solar light ($\lambda > 300$ nm) were similar in terms of the light spectrum. In another study, Mano et al. [53] used a 300 W Xe lamp with an IR cut-off filter; the incident light power inside the vessel, measured by a radiometer, was about 200 mW in the wavelength range from 360 to 470 nm. An additional cut-off filter ($\lambda > 410$ nm) was used for visible light irradiation. Also, Liao et al. [49] used a 500 W Xe lamp jacketed by a quartz thimble filled with flowing and thermostated aqueous $NaNO_2$ solution (1 M) between the lamp and the reaction chamber as a filter to block UV light ($\lambda < 400$ nm). High or medium-pressure Hg lamps have also been used. Mecha et al. [62], for instance, used a Heraeus TQ 150 W medium-pressure Hg lamp that was surrounded by a quartz cooling water jacket and 70 mW cm^{-2} of light intensity. The lamp emission spectrum had peaks at 253.7, 313 and 366 nm in the UV range and 436, 546 and 578 nm in the visible range. Results when this lamp was used could not exactly be classified as due to SPO because of the UVC 254 nm radiation peak; however, this work also used sunlight.

Regarding photoreactors, two main installations have been used, depending on the radiation source. Thus, when lamps, mainly Xe lamps emitting radiation with wavelengths higher than 300 nm, were used, photoreactors were made of glass in cylindrical or spherical shapes where the water containing the catalyst and the organics to be treated was continuously charged or recirculated as it was pumped from a reservoir or a tank. The photoreactor, if it is of semi-batch type, has inlets for ozone feeding and sample taking as well as an outlet for the gas to exit. In many cases, the photoreactor was inside a box or solar simulator also containing the lamp. In all cases, the experimental installations were supplied with connections to the ozone analyzer and ozone destruction units. The second important photoreactor, used to receive natural solar radiation, was in most cases the so-called CPC photoreactor or compound parabolic collector. The CPC photoreactor consists of a number of borosilicate glass tubes connected in series situated above an anodized aluminum parabolic platform or collector oriented to the south and tilted at an angle equal to the latitude of the place where the reactions are carried out. For example, Quiñones et al. [56] used a CPC photoreactor supplied with four borosilicate glass tubes (32 mm external diameter, 1.4 mm thickness, 750 mm length), anodized aluminum reflectors tilted at 45 degrees (the latitude of the place was in this case 38°52′) and inlets and outlets for the gases. The total collector surface was 0.25 m^2 and the illuminated volume was 1.8 L. In these reactors, a radiometer is usually included to measure the instantaneous and accumulated UV light absorbed. Quiñones et al. [56] used a broadband UV radiometer (290–370 nm) tilted at the same angle as the CPC. In order to introduce the oxidizing gas (ozone-oxygen or ozone-air), at the edge of some of the tubes there were porous plates connected to the gas circuit. The system is usually completed with a reservoir tank for the water that is recirculated through the CPC at a turbulent regime with the aid of a pump.

2.2. Catalysts Used

Catalysts or semiconductors used in SPO can be classified into three different types: metal oxides, metal-doped metal oxides, and composites of different materials. All these catalysts are charged to the reacting systems as solids, usually of nanometer size, so heterogeneous photocatalytic oxidation develops. Metal or metal oxide-supported catalysts on solid structures (glass film, Raschig rings, granular activated carbon, etc.) have not yet been used for solar photocatalytic ozonation systems with the exception of the work of Oyama et al. [45] in which, in addition to the conventional TiO_2 suspension form, the authors also used TiO_2 coated on glass. Additionally, in some cases, Fe(III) as a homogeneous catalyst was used. In this case, Fe(III) through the formation of aqua complexes, mainly $Fe(OH)^{2+}$, undergoes photolysis with radiation ($\lambda > 300$ nm) to give hydroxyl radicals and Fe(II) [56]. Due to the presence of ozone, hydrogen peroxide is also formed from ozone direct reactions [66] or simply by ozone decomposition [17,67]. Then, Fe(II) is reconverted to Fe(III) by reacting with hydrogen peroxide, which is the Fenton reaction [68]. Thus, the addition of Fe(III) as a catalyst in solar photocatalysis allows different ways of hydroxyl radical formation, including those of Fenton and photo-Fenton processes.

Apart from the use of Fe(III), the rest of the works already published (see Table 2) used a solid as a main catalyst or semiconductor. From the three different types of solid catalysts indicated above, metal oxides are the most frequently used, and of these TiO_2, due to its high catalytic capacity, is the main catalyst used since it first appeared in 1978 [22]. Although TiO_2 has been prepared and applied in different works, the commercial P25 TiO_2 from Degussa is actually the most frequently used catalyst. This is due to its low cost, stability (it does not leach into the water), and activity, especially with UV radiation. P25 TiO_2 contains around 75% anatase and 25% rutile crystalline phases which confer a highly active character [28]. Consequently, this catalyst is often used as a model to compare the activity of new catalysts. These comparisons of catalyst performance can be seen in many of the works quoted in Table 2. Other metal oxide catalysts used in SPO are Bi_2O_3, Nb_2O_5, SnO_2, WO_3, Fe_2O_3, In_2O_3, Fe_3O_4, and CeO_2 [44,53,55,59,64] which have a band gap energy lower than that of TiO_2 and hence, a priori, they can be excited with visible light ($\lambda > 400$ nm). In addition, the conduction band redox potentials of these catalysts are more negative than that of ozone, so ozone can trap electrons from their conduction bands, generate hydroxyl radicals, and avoid electron-hole recombination [36]. WO_3 and TiO_2 have also been applied as a double metal oxide catalyst, as is seen in the work of Rey et al. [52], in which P25 TiO_2 and TiO_2 nanotubes were coated with nanosized WO_3 particles. These catalysts present lower band gap energy due to the tungsten oxide and the charge transfer between photogenerated electrons from the conduction band of TiO_2 to the WO_3 conduction band, and holes transfer from the valence band of WO_3 to the TiO_2 valence band is favored. Another group of catalysts used in SPO are metal-doped TiO_2 materials. Of these, Cu, Ag, Fe, Mn, Zr, Ce, and B have so far been used. An Au-Bi_2O_3 catalyst was also prepared and applied to remove the Orange II dye in the work of Anandan et al. [44]. One of the possible ways TiO_2 can be active with visible light is by doping it with different metals. For instance, oxygen atoms in the TiO_2 lattice can be substituted by B atoms by mixing the p orbital of B with O_2 p orbitals, narrowing the band gap and thus shifting the optical response into the visible range [69]. Boron can also be located in interstitial positions of the TiO_2 lattice, leading to the partial reduction of Ti(IV) to Ti(III), which can act as an electron trap enhancing the photocatalytic activity of TiO_2 [69,70]. A third group of catalysts are composites formed by the combination of metal oxides (mainly TiO_2) and activated carbon, carbon nanotubes, and some other metal oxides. Examples of composites are Fe_2O_3/TiO_2/activated carbon (FeTiC), WO_3/TiO_2, and gC_3N_4 or gC_3N_4-rGO [49,50,52,60,61], where rGO stands for reduced graphene oxide. The main components are the magnetite crystalline phase that confers a magnetic moment causing the FeTiC catalyst to be easily separated from water, enhanced adsorption capacity, and the lower band gap of Ti-W composites and the narrower band gap energy of WO_3/TiO_2 and gC_3N_4 catalysts, especially when reduced graphene oxide is present.

2.3. Catalyst Synthesis and Characterization

Both the synthesis and characterization of catalysts are important parts of SPO research. Thus, the ways these catalysts have been prepared and characterized are presented in this section. Regarding the synthesis or preparation, the sol-gel method has so far been the most-used method in SPO catalysts. A clear example of this was the TiFeC magnetic composite prepared by Rey et al. [47]. The catalyst was prepared in three steps: first a meso-microporous activated carbon was impregnated with an ethanol ferric nitrate solution. Once iron nitrate was adsorbed, it was dried and subsequently impregnated with ethylene glycol. This was then heated in an oven, allowed to cool, and finally milled into power. Secondly, a sol-gel of titania was prepared from titanium (IV) butoxide diluted in isopropanol. Finally, magnetic activated carbon (FeC) was dispersed in the titania sol and subjected to ultrasonic, evaporation, washing, and drying procedures. A similar but more simple procedure was used to prepare metal doped-TiO_2. For example, Mecha et al. [62] prepared Cu-, Ag-, and Fe-doped TiO_2 via sol-gel methods from Ti_3Cl and metal nitrates, while Quiñones et al. [58] prepared a B-TiO_2 catalyst from a sol gel where the principal precursors were boric acid in anhydrous ethanol and titanium butoxide. In another study [63], TiO_2 was prepared from a sol formed from titanium (IV) butoxide, isopropanol, and ultrapure water acidified with HNO_3. In this work, composites of TiO_2 and multiwalled carbon nanotubes (MWCNT) were also synthesized from the TiO_2 sol dispersed with an amount of MWCNT under sonication. Finally, drying and washing procedures were carried out to obtain the TiO_2/MWCNT composite with TiO_2 percentages of about 70–80%. Graphitic carbon nitride (g-C_3N_4) has also been obtained [49] through a sol-gel method. In this case, the starting material was powdered urea that was heated, washed with ethanol and water, filtrated, and dried. Apart from the sol-gel method, the hydrothermal method was applied in some of the works depicted in Table 2. With this method, Rey et al. [52,55,65] prepared TiO_2 nanotubes, WO_3 catalysts, and nanocubes (NC) and nanorods (NR) of CeO_2. In this latter case, the procedure basically consisted of heating an alkaline aqueous $Ce(NO_3)_3$ solution in an autoclave for a given time at $100\,^\circ C$ for nanorods or at $180\,^\circ C$ for nanocubes. After the hydrothermal treatment, the autoclave was cooled down to room temperature and then the precipitates were separated by centrifugation, washed sequentially with water and ethanol, and dried overnight. A different procedure was used by Mano et al. [53], who carried out a solid-state reaction with $Bi(NO_3) \times 35\,H_2O$ and NH_4VO_3 to prepare their $BiVO_4$ catalyst (see also [71]). Then, the final product was dried and heated.

The characterization of synthesized catalysts is essential to understand the results of any catalytic process and/or confirm the presence of some metals in any metal-supported catalyst. Thus, works on SPO where new catalysts were prepared also show a section dedicated to catalyst characterization. The main methodologies applied for SPO catalysts are N_2 adsorption-desorption to establish the specific surface area of the catalyst (SBET) and pore volume (micro and mesoporosities); X-ray diffraction (XRD) to determine the crystalline phases present in the catalyst and the crystalline size; scanning and transmission electron microscopy (SEM, TEM) to obtain photographical information of catalyst samples below 10 or 1 nm diameter, respectively; associated technique energy dispersive X-ray spectroscopy (EDX) to verify the distribution of metal ions in the catalyst particles; X-ray photoelectron spectroscopy (XPS) to provide information about the chemical composition of the catalyst surface; and Fourier Transform-Infrared spectroscopy (FTIR), usually in the wavenumber range between 200 and 4000 cm^{-1}, to supply data about the catalyst surface, the presence of functional groups, and interactions between adsorbate and adsorbent. Raman spectroscopy, based on the Raman shift, is also useful for structural characterization, presence of defects, or different crystal size effects in the catalysts. In addition, other important techniques used are: diffuse reflectance UV-Vis spectroscopy (DR-UV-Vis) to determine absorption of radiation in the UV-visible wavelength range of the catalyst, and its band gap energy for potential activation of the catalyst with UVA and visible radiation; inductively coupled plasma to measure metal contents (inductively coupled plasma-mass spectroscopy (ICP-MS) or inductively coupled plasma-optical emission spectroscopy (ICP-OES)); and, for magnetic catalysts, the use of superconducting quantum interference device (SQUID) magnetometry to measure the magnetic moment. In Table 2, the different techniques used in SPO works are listed.

Combinations of these techniques are needed to justify the results obtained. For example, the metal (Fe, Ag or Cu)-doped TiO_2 catalysts prepared by Mecha et al. [62] were characterized by the combination of specific surface area and DR-UV-vis results to explain the best results obtained with the Fe-TiO_2 catalyst which presented the highest SBET and red shift of the band edge absorption, with significant tail absorption at 530 nm. In Rey et al. [47], XRD analysis showed the presence of anatase, magnetite, and maghemite crystalline phases in their FeTiC composite. The catalyst was easily separated from water with a magnet and its superparamagnetic behavior was confirmed through SQUID magnetometry. In this work, a very uniform distribution of Fe on the composite was also observed by SEM and EDX and pore volume and specific surface area were measured for the three solids used, activated carbon (AC), Fe-impregnated activated carbon (FeC), and the final composite, FeTiC. The results showed the decrease of micropore volume after impregnation (from AC to FeC) and catalyst preparation (from FeC to FeTiC), which justifies the loss of specific interfacial area from 640 to 552 and finally to 331 m^2 g^{-1}, respectively. FTIR, on the other hand, allowed the justification of some functional surface groups belonging to phenols or TiO_2-OH bonds, quinones and other carbonyl groups and aromatic structures, amongst others. Mano et al. [53] used different metal oxide catalysts of specific interfacial areas ranging between 2.1 and 54 m^2 g^{-1}. DRUV-vis showed visible light absorption properties for WO_3, Fe_2O_3, In_2O_3, Bi_2O_3, and $BiVO_4$ catalysts used, but not for Nb_2O_5 and SnO_2, although some activity of these latter catalysts was attributed to photosensitizing effects. In another work by Yin et al. [61], graphitic carbon nitride linked to a reduced graphene oxide (g-C_3N_4–rGO) composite was characterized. TEM analysis showed that the rGO sheet was sandwiched between g-C_3N_4 through the polymerization of melamine molecules. On the other hand, through DRUV-vis, Yin et al. [61] reported a red shift to a longer wavelength in the absorption band edge of g-C_3N_4–rGO composite, probably due to the presence of rGO in the g-C_3N_4–rGO composite, which exhibits a stronger broad background absorption in the visible-light region. This brings about band-gap narrowing and enhanced visible light utilization. While doping TiO_2 with boron, Quiñones et al. [58] noticed an SBET increase compared to the B-free TiO_2 catalyst (68 to 125 m^2 g^{-1}) effect attributed to the lower crystal size of the anatase phase in the B-TiO_2 catalysts. B/Ti (atomic ratio) proportion was measured through both XPS (surface data) and ICP (bulk data). The authors suggest that most of B is located on the surface of TiO_2 during the sol–gel synthesis. Sassolite boron structure (H_3BO_3) and anatase were crystalline forms identified by XRD, revealing that the crystal size decreases with the increasing B content, an effect attributed to the restrained TiO_2 crystal growing due to the existence of a large amount of boron. XPS confirmed the formation of Ti–O–B structures (interstitial B). XRD patterns of MWCNT, MWCNT-TiO_2, TiO_2-P25, and TiO_2 prepared samples were obtained in a study by Alvarez et al. [63]. According to the results shown, the pristine MWCNTs had a graphite-like structure and only anatase crystalline form was present in TiO_2 catalysts. Average crystallite sizes were found for TiO_2 particles in MWCNT-TiO_2, TiO_2-P25, and TiO_2 samples, respectively. TEM images show that MWCNTs tend to aggregate together as bundles with sorption sites including the inner and outer surface of individual tubes, interstitial channels between nanotubes, and external groove sites. Also, heterogeneous, non-uniform coating of MWCNTs by TiO_2 particles was observed showing both bare MWCNTs and random agglomeration of TiO_2 particles on MWCNTs surface. Finally, Mena et al. [64] showed TEM images of a size distribution between 25 and 100 nm for CeO_2 nanocubes (NC), and a thickness of approximately 7.2 nm and lengths between 40 and 200 nm for CeO_2 nanorods (NR). The application of XRD allowed pure CeO_2 cubic phase (fluorite structure) to be identified in both types of catalysts. The content of Ce(III) was higher in CeO_2-NR as observed by XPS and from DRUV-Vis band gap energies of 3.32 eV for CeO_2-NC and 3.07 eV for CeO_2-NR were measured, with the result that this latter catalyst was very efficient when visible radiation was used during photocatalytic ozonation.

2.4. Organics Treated

The finding of many organic compounds of a pharmaceutical origin in urban WWTPs has given rise to the application of SPO and other AOPs for the study of the removal of these from water. These compounds, also called emerging contaminants (ECs), so-called because they do not yet have maximum contaminant levels assigned, are the most studied ones in SPO works both in ultrapure or urban wastewater. ECs can lead to a number of risks, such as sterility, feminization of aquatic organisms, and bacterial resistance [72,73], and as such they have to be removed from waters. In most of the papers (see Table 2), concentration of these compounds is some orders of magnitude higher than those found in actual wastewater, ranging from some mg L^{-1} to μg L^{-1}, when in reality they usually are at ng L^{-1}. There is a double reason for this. First, analytical laboratory equipment to measure ng L^{-1} concentrations is very expensive, and also inefficient, given the high number of analyses required (for instance, to perform a kinetic study). Second, the SPO works in many cases aim to check the activity of new catalysts so, in these cases, the main objective is not really the removal of a given contaminant but the performance of a catalyst. In addition to pharmaceutical compounds, other important groups studied in SPO works are pesticides (see Table 2), which are also recalcitrant to biological and physicochemical processes applied in classical WWTPs. Water from agriculture, which is the industrial activity with the highest water consumption (about 70% of the world's accessible freshwater [74]) may contain a high number of organic pesticides and fertilizers which are eventually dispersed in aqueous environments by runoff or leaching. Finally, phenolic compounds, such as phenol itself or bisphenol A, a plasticizer compound that has been found in many wastewaters [75], and oxalic acid, because its oxidation directly leads to mineralization [76], are also model compounds in SPO works.

2.5. Influence of Main Variables

Variables studied in SPO works are the nature of the catalyst, including here the synthesis procedure for a given catalyst, the concentrations of organics and oxidants such as ozone itself, the pH of water, and the intensity and wavelength range of radiation applied. Although this review deals with solar radiation, the use of filters (for instance when Xe lamps are used), permit the study of different wavelength range effects from λ > 290 nm. However, the most studied variable in SPO works is the comparison with other AOPs which could be called blank AOPs, such as single ozonation and ozone-free solar photocatalytic oxidation or solar ozone photolysis. In fact, comparison of these AOPs with SPO results is a necessary step before proceeding to the influence of other variables. It is the first step to know whether the catalyst applied together with ozone and light is worth being investigated to remove water contaminants. Examples of this comparison can be seen in most of the works in Table 2. The normal way to make the comparison is by determining the concentration of the organics and total organic carbon with reaction time, the latter is known as mineralization. In many cases, there are few differences between the organic removal rates observed during the different ozone processes applied (ozonation, solar ozone photolysis, catalytic ozonation, and solar photocatalytic ozonation) because the organics studied react very fast with ozone through their direct reactions so that there is no need to combine ozone with any other agent (catalyst and/or light). A clear example of these results is given in the study of Márquez et al. [51] in which a mixture of ECs including atenolol, ofloxacin, hydrochlorotiazide, and trimetropim is treated. Similar results, that is, scarce differences between ozone processes for the removal of a mixture of six ECs were also found in the works of Quiñones et al. [56,57], both in ultrapure water and urban wastewater. However, the fact that differences between reaction rates of organics are independent of the ozone process applied is not only due to the fast ozone-organic compound direct reaction rates, as can be seen in another study by Quiñones et al. [58]. Here, a mixture of four pesticides with rate constant values of their direct reactions with ozone lower than 400 M^{-1} s^{-1} was treated. As in the preceding case, no differences were observed in the reaction rates for different ozone processes applied, but now the contribution of HO radicals cannot be disregarded, especially in the cases of diuron and t-buthylazine which react slowly with ozone (the rate constants of direct ozone reactions are 3.7 and 20 M^{-1} s^{-1} for diuron and, terbuthylazine, respectively, [58]). It is likely that direct reactions with some of the organics present in water that yield hydrogen peroxide will trigger the formation of hydroxyl radicals and facilitate the removal of other more recalcitrant compounds such as terbuthylazine,. In fact, when this herbicide has been treated alone [63], differences between ozonation alone and solar photocatalytic ozonation are important. Thus, the complete removal of terbuthylazine, when treated alone, was observed after 30 min in a solar photocatalytic (P25-TiO$_2$) ozonation run, while at this reaction time ozonation only yielded about 63% removal. With other compounds, for example, the insect repellent DEET [55,64,65] or oxalic acid [49], SPO gives much better results than ozonation alone, which is a clear consequence of the recalcitrant character of these compounds towards direct ozonation (very low ozone reaction direct rate constants: <10 M^{-1} s^{-1}), meaning that hydroxyl radicals are the only means of oxidation [65]. Another important fact to highlight when comparing AOP processes to remove organics is that, in all cases, SPO is a better option than ozone-free photocatalytic oxidation, since much more time is always needed in this latter process to remove the organics. To give an example of this, the removal of dichloroacetonitrile with SPO and ozone-free solar photocatalytic oxidation with P25 TiO$_2$ as a catalyst, after 4 h of reaction, was 90% and 20%, respectively [48]. Significant differences can also be observed with other compounds (see Table 2). Another common result of all SPO works is related to TOC removal. In this case, regardless of whether the compounds studied are alone or in a mixture with other compounds, TOC removal or mineralization is always much better with SPO than with any of the other AOPs examined, that is, blank AOPs such as ozonation alone or combined with the catalyst or UVA-visible or solar radiation. This can be observed in any of the studies listed in Table 2.

The effect of catalyst nature is another important variable tested in SPO works. In an attempt to benefit from the visible zone of the solar electromagnetic spectrum, many catalysts are prepared to be active with radiation wavelengths higher than 400 nm. The TiO$_2$ catalyst, due to its high band gap (3.2 eV), is only active with UV radiation, though the presence of defects in the crystalline structure shifts its absorption capacity to the visible light radiation wavelength in some cases. In order to activate this situation, metal doping of TiO$_2$, composites of TiO$_2$, or other visible active catalysts have been used. In these cases, SPO studies show results with Xe lamps and filters that cut wavelength radiation range. For instance, Quiñones et al. [50] studied the removal of metoprolol (MTP) with a FeTiC composite and a Xe lamp so that the use of filters allowed the passing of three radiation wavelength ranges: 300 to 800 nm, 320 to 800 nm, and 390 to 800 nm. Thus, during ozone photolysis, after 5 h of reaction, mineralization was about 60% with irradiated light between 300 to 800 nm but was reduced to 40% with radiation lights from the other two wavelength ranges. This means that the main photolysis of ozone to yield hydroxyl radicals happens between 300 to 320 nm, which is in accordance with ozone quantum yields [77]. In the presence of the best FeTiC composite catalyst, mineralization percentages were about 92%, 90%, and 60%, for radiation lights of 300 to 800 nm, 320 to 800 nm, and 390 to 800 nm, respectively, which confirm the 300 to 320 nm interval as the most efficient for TOC removal. In another paper [61] in which a g-C$_3$N$_4$–rGO composite was prepared and used as a catalyst with a NaNO$_2$ aqueous solution filter, the authors reported that the contribution of UV light ($\lambda < 400$ nm) and visible light ($\lambda > 400$ nm) to the degradation of oxalic acid by O$_3$/UV-Vis/g-C$_3$N$_4$–rGO system (see Table 2) was 33.1% and 66.9%, respectively. Other visible light active catalysts prepared were Bi$_2$O$_3$, WO$_3$, or metal doping TiO$_2$ (with metal = Cu, Mn, Fe, Zr, etc.) [44,46,54,55]. For instance, in the study by Feng et al. [46] the M-TiO$_2$ catalysts showed extended absorption spectra into the visible-light region. Ag-TiO$_2$, Cu-TiO$_2$, Ce-TiO$_2$, and Fe-TiO$_2$ showed a relatively small absorption region between 400 and 580 nm, while Mn-TiO$_2$ and Zr-TiO$_2$ exhibited substantial and broad absorption shoulders of up to 700 nm.

Another variable studied is the effect of changes in some steps of the synthesis procedure. For example, Quiñones et al. [58], while preparing a boron-doped TiO$_2$ catalyst, changed the amount of B containing precursor (boric acid) to have different mass B percentages in the final catalyst. They observed that the catalyst with the highest amount of B leads to the highest removal rate (the maximum

B percentage was 12%, though subsequently they observed that some B had leached). According to the authors, the probable reason is that B tends to lose its three valence electrons which are transferred to the 3d orbitals of lattice Ti ions yielding Ti(III), diminishing the electron-hole recombination. In the FeTiC/solar/O_3 process [50], the synthesis procedure included changes that affected the amount of Fe incorporated into the catalyst composite, which obviously produced changes in the saturation magnetization of catalysts and in TiO_2 content. The increasing presence of Fe affected the specific surface area that became lower but with more TiO_2. These two effects changed the amount of organic compound adsorbed and reacted on the catalyst surface. Thus, the catalyst with the lowest Fe content had the highest adsorption capacity, while the catalyst with the highest TiO_2 content presented the highest activity. Calcination time and/or temperature effects were studied for the preparation of WO_3, which led to different crystalline structure catalysts [54,55]. These works show that monoclinic or orthorhombic forms of WO_3 were more active for the photocatalytic ozonation process using visible light from $\lambda > 390$ nm or $\lambda > 300$ nm for the total solar spectrum.

Apart from catalyst and AOP comparison, the influence of other main variables that affect the oxidation rate, that is, pH, organics, ozone and catalyst concentrations, and intensity of radiation, has not been extensively examined in SPO works. In fact, only in some cases [48] has the influence of pH, ozone and catalyst (P25 TiO_2) concentration, and temperature been studied. In ozone processes, but also in adsorption processes, pH is a fundamental variable and hence its effect on SPO must be optimized. pH therefore affects the charge distribution of the catalyst surface, which will depend on the pH_{pzc} of the catalyst and dissociated species present in water. It also affects ozone decomposition in free radicals and even ozone direct reaction rates with dissociated species [78]. Shin et al. [48] studied the SPO of dichloroacetonitrile (DCA) at pH 3, 6.5 and 10. They observed that DCA self-decomposes at pH 10 while it remains unaltered at the other two pH values. In the SPO runs, they observed a positive increase in reaction rate with the increasing pH. At pH 10, the reaction rate was probably due to the action of hydroxyl radicals coming from ozone decomposition and photocatalysis. However, the reaction rate at pH 6.5 was also significant and in this case, given the pH_{pzc} of TiO_2 (6.2–6.6 [79,80]), pH 6.5 is also likely to be the best value for SPO application since, as the authors indicate, there is no need to add any basic or acid substance.

The effect of the concentration of organics on the removal rate during SPO runs has only been studied by Rey et al. [47]. They treated MTP at concentrations of 10 and 50 mg/L and they observed a similar behavior to the other ozone processes were studied, that is, for a given reaction time, an increase in the initial organic compound concentration leads to an increase in removal rate (both for the compound and TOC). The logical consequence of these results is that reaction rates are proportional to the concentration of the organics treated, which usually follow a Langmuir reaction rate equation. Obviously, the complete disappearance of MTP takes more time when the initial concentration is higher. A comparison was also made between the results obtained in solar photolytic ozonation and SPO at different concentrations. These authors noted that at 10 mg L^{-1}, concentration differences in the TOC removal rate were not significant, but at 50 mg L^{-1}, the removal rate through SPO was much higher than with solar photolytic ozonation. The effect of the initial concentration of organics has been more extensively treated in non-solar photocatalytic ozonation processes with similar results to those presented above [81,82].

With regard to the effect of the catalyst concentration, an increase in this variable leads to an increase in the reaction rate due to the increase in surface, that is, active centers. However, this is observed up to an optimum value above which the reaction rate diminishes. The reason is poor light transference or photon absorption rate through a higher concentrated catalyst suspension. Shin et al. [48] reported an optimum value of 1 g L^{-1} for the SPO removal of DCA, which is similar to others found in some solar photocatalytic oxidation processes [31].

The ozone concentration effect is also positive up to a given value, above which no influence on the reaction rate is observed. This happens in any type of ozone process and is related to the ozone saturation of water [8]. In the study by Shin et al. [48], the optimum ozone dose found was 1.13 g L^{-1} h^{-1}. This optimum value is also highly dependent on the nature of the organics in the water, pH, and catalyst concentration because all of these affect the ozone driving force in the water.

Temperature also has an optimum value because of two opposing effects. A temperature increase leads to an increase in reaction rate constants, but at the same time to a decrease in ozone solubility, which are opposing aspects. Thus, in SPO runs of Reference [48], the optimum temperature was 20 °C in the range from 10 to 40 °C.

Finally, the study by Shin et al. [48] is the only one that has analyzed the effect of light intensity on an SPO system. As a rule of thumb, in photocatalytic oxidation processes, reaction rate and light intensity are proportional up to a given value of the latter, above which the reaction rate becomes proportional to the square root of light intensity [31]. Shin et al. [48] varied the light intensity between 4.6 and 33.8 Wm^{-2}, and found that about 20 Wm^{-2} was needed to change the proportionality of reaction rate/light intensity, which is in accordance with what has been mentioned above.

In a practical situation, SPO will probably be applied to remove contaminants in a given wastewater, simply for depuration or for recycling. Thus, the effect of the water matrix in SPO also needs to be examined. Marquez et al. [51] studied the removal of ECs in a mixture dissolved in ultrapure water and wastewater from a secondary effluent. These authors observed a decrease in reaction rates in wastewater, but the effect of the water matrix could not be established because they used different EC concentrations in each water type. In fact, there are no studies on SPO where the effect of the water matrix on organic compound or TOC removal rate has been studied with equal organic concentrations. However, the presence of hydroxyl radical inhibitors in wastewater actually makes the SPO process slower than in ultrapure water.

Given that in a real situation, SPO will be an additional process of a biological step (for instance, a secondary treatment of urban wastewater), Gimeno et al. [59] studied the aerobic biological oxidation followed by SPO of a mixture of nine ECs (see Table 2) that were doped in an urban wastewater primary effluent. First, aerobic biological oxidation was applied with a food to a microbial ratio of 0.5, a hydraulic residence time of 7 h, and mixed liquor volatile suspended solids (MLVSS) of 1.5 g L^{-1} with WWTP conventional activated sludge as biomass. The initial concentration of each EC compound was 0.2 mg L^{-1} so that the TOC contribution of ECs was less than 5% of initial TOC of the wastewater primary effluent. The authors observed 50% and 80% COD removals in the presence and absence of ECs, respectively. This indicates that some sort of disturbance to microorganisms present on the activated sludge occurred when ECs were present. The authors highlighted that it could be due to the presence of sulfamethoxazole, an antibiotic which alters microorganism concentration and nature at concentrations higher than 50 μg L^{-1} [83]. After 7 h of biological oxidation, only three ECs (caffeine, acetaminophen, and metoprolol) out of nine present in wastewater underwent some significant reductions (between 40% and 61%), the other ECs only obtaining percentage removals of less than 5%. Gimeno et al. [59] in subsequent experiments applied some AOPs (SPO with TiO$_2$, solar photo-Fenton: Fe(III) and magnetite, and single ozonation) to the biologically treated wastewater. With a total accumulated solar energy of 30 kJ L^{-1}, more than 80% ECs removal was achieved with SPO TiO$_2$. Mineralization was 40%. In the absence of ozone, mineralization was less than 10%. Finally, process efficiency followed this order: SPO magnetite > SPO TiO$_2$ > SPO Fe(III) > ozonation >> solar photocatalytic oxidation.

2.6. Intermediates, Ozone Consumption, Biodegradability, and Toxicity

In some works, during the course of SPO runs, identification and/or concentration measurement of intermediates was also carried out for two reasons: to check the toxic character of these compounds and/or to establish the mechanism of reactions (see the next section). Because of the simplicity of the analytical procedures, the main intermediates detected were in some cases phenolics (such as total phenolic compounds), saturated carboxylic acids, and inorganic ions. In some other cases, High Performance Liquid Chromatograph-Mass Spectrometry HPLC-MS was used to detect the first intermediates of a molecular structure similar to the initial organic compound studied. Thus, in the work by Quiñones et al. [50], MTP and six intermediates formed were analyzed by HPLC-qTOF. They represented the results as the variation of the area with MTP conversion observing the existence of an MTP conversion (between 40% and 60%), for which the concentration (area) of intermediates reached a maximum value. Mena et al. [65], with similar analytical techniques, detected 22 intermediates of large molecular weight (MW > 160) and a series of low molecular weight saturated carboxylic acids as end products, the latter with ion chromatography, during DEET degradation. Phenolic compound concentration was measured in the works of Marquez et al. [51] and Quiñones et al. [57]. In these studies, SPO application leads to an initial increase in phenolic compound concentration with time to reach a maximum value followed. Then, at more advanced times a continuous decrease in phenolic concentration with time is observed, though this depends on the nature of the starting compounds to be degraded and the catalyst used in the SPO process. Quiñones et al. [57] observed a continuous decrease in phenolic concentration of up to 75% removal with 10 kJ L^{-1} of accumulated Sun energy with SPO TiO_2 applied to a mixture of seven ECs in a secondary effluent of an urban WWTP. They also observed that ozone-free solar photocatalytic (TiO_2) oxidation led to an increase of 50% phenolic concentration with 15 kJ L^{-1} of accumulated Sun energy which was reduced at a more advanced reaction time but remained higher than the initial one at the end of the reaction (after 40 kJ L^{-1} of accumulated sun energy). In SPO, the concentration of carboxylic acids was followed in several works, as shown in Table 2. Apart from phenolics, carboxylic acids, and TOC measurements as a way of determining the remaining amount of all intermediates and final products, in many works hydrogen peroxide and dissolved ozone concentration are followed. Hydrogen peroxide is a key compound in ozonation processes since its presence triggers different routes of hydroxyl radical formation (see the later mechanism section). For instance, Mena et al. [64] measured the concentration of hydrogen peroxide with time in experiments of ozonation, ozone photolysis, catalytic ozonation, and SPO (with simulated solar light and CeO_2 nanocubes) and observed a rapid formation of hydrogen peroxide, much greater in ozonation alone than in the catalytic processes. In fact, the concentration of hydrogen peroxide in SPO was very low, probably not only as a consequence of its reaction with ozone, but also because of its direct photolysis and participation in capturing electrons. However, in Mena et al. [64], using a WO_3 catalyst, the presence of H_2O_2 did not have a beneficial effect as an electron acceptor.

The concentration of dissolved ozone is also followed in SPO works to determine ozone consumption, that is, the amount of ozone needed to remove a given amount of organic carbon measured as TOC. Márquez et al. [51] reported that ozone consumption was much lower in SPO than in other ozone processes to treat four ECs. In ultrapure water, for a 40% TOC removal, they observed 35, 20, 40 and 18 $mgO_3/mgTOC$ in ozonation alone, catalytic ozonation, solar ozonation, and SPO, respectively, with P25 TiO_2 as a catalyst. In urban wastewater, also for 40% TOC removal, the figures were 61, 52 and 37 $mgO_3/mgTOC$ in ozonation alone, solar ozonation, and SPO, respectively. It can be seen then that SPO again leads to the lowest ozone consumption, though it is about twice the value observed in ultrapure water. In wastewater, however, the contribution to TOC of the four ECs studied was negligible.

Biodegradability has been studied as the ratio BOD/COD. As a general rule, it can be said that biodegradability in wastewaters increases with the application of ozone in any ozonation process, SPO included. In particular, Márquez et al. [51] observed that BOD/COD ozone-free photocatalytic ozonation of an urban secondary effluent wastewater remains constant with time but increases from 0.22 (raw wastewater) to 0.78 with SPO. This was the highest value obtained from the ozonation processes applied. Quiñones et al. [57] also reported a 2.5-fold increase in the BOD/COD initial ratio of another urban secondary effluent wastewater, with the homogeneous solar-Fenton photocatalytic ozonation process.

Finally, ecotoxicity is another typical determination in AOP and, in particular, in SPO processes. *Daphnia magna* survival is usually the test performed. Gimeno et al. [59] reported the absence of inhibition in immobilization tests after applying a SPO process to urban wastewater with TiO_2 or Fe(III) as catalysts. Quiñones et al. [57] studied the toxicity of urban wastewater doped with six

ECs (0.2 mg L^{-1} concentration each). The bioassays showed an increase in toxicity after the addition of ECs to the secondary effluent. Thus, the percentage of inhibition increased from 5% (without added ECs) to about 25%. Additionally, changes in sample toxicity were observed during the course of the photocatalytic experiments. As a rule, the toxicity of samples increased at the beginning of the photocatalytic treatment, probably as a consequence of the accumulation of phenolic and other toxic intermediates, and then decreased. This result is in agreement with that reported for the removal of other ECs by ozonation and solar photocatalytic oxidation with TiO$_2$ [51,84]. Toxicity removal below 25% inhibition was only observed when a high degree of TOC removal (i.e., mineralization) was achieved. This applies especially for the Fe(III)/O$_3$/solar light (pH 3) and Fe(III)/H$_2$O$_2$/O$_3$/solar light (pH 3) photocatalytic ozonation systems.

2.7. Mechanism and Kinetics: The Use of Scavengers

The SPO process goes through a complex mechanism of direct reactions between compounds present and oxidants (ozone, generated hydrogen peroxide, oxidizing holes), as well as free radical reactions. This complexity is the result of the combination of the ozonation and photocatalytic oxidation processes. The former, ozonation, implies direct ozone-organic reactions and ozone decomposition reactions due to the appearance of hydrogen peroxide, increase in pH, and ozone photolysis, at least with radiation wavelengths of up to 320 nm. Photocatalytic oxidation involves hydroxyl radical formation from holes in the catalyst valence band and probably from the superoxide ion radical formed from excited electrons of the catalyst conduction band [24–27]. Detailed explanations of these mechanisms are available elsewhere [56,85]. When ozone, an active semiconductor, and light are simultaneously present to remove some organics in water, some new reactions occur, the main one being the formation of the ozonide ion radical, O$_3\bullet^-$, from the ozone capture of electrons in the catalyst valence band:

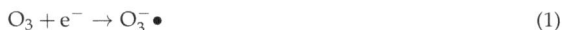

$$O_3 + e^- \rightarrow O_3^- \bullet \qquad (1)$$

which eventually leads to hydroxyl radicals:

$$O_3^- \bullet + H^+ \rightarrow HO_3 \bullet \rightarrow O_2 + HO \bullet \qquad (2)$$

In the SPO process, hydroxyl radicals can be formed from the individual mechanisms of ozonation and photocatalytic oxidation processes [17,85] but also from the synergism between these two processes, as it is observed from Reactions (1) and (2) [86]. Thus, in the SPO process, there are some possible oxidizing agents: ozone; hydrogen peroxide, formed in direct ozone reactions with some organics (aromatic compounds, olefines, etc.) [87]; oxidizing holes in the catalyst valence band; singlet oxygen, etc. [88].

The use of scavengers is an effective way to check the importance of the oxidation mechanism or the importance that some oxidants have in a given SPO process. These scavengers react exclusively with some oxidants like hydroxyl or superoxide ion radical or oxidant holes. Liao et al. [49], for example, used tert-butanol and triethanolamine as scavengers of hydroxyl and valence band holes, respectively. With these experiments, the authors observed that after 120 min, the oxalic acid removal efficiency reached 80% without scavengers. However, only 46.2% and 66.5% of oxalic acid was removed with the presence of tert-butanol and triethanolamine separately. This means that both hydroxyl radicals and holes contributed to the degradation of oxalic acid. However, the highest inhibition of removal rate was observed with the addition of tert-butanol, which suggests that HO· radicals play a dominant role in the SPO of oxalic acid. Furthermore, in another work [60], benzoquinone was used to scavenge any possible superoxide ion radical that could be formed from conduction band electrons trapped by oxygen:

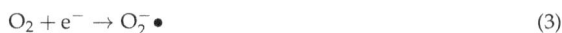

$$O_2 + e^- \rightarrow O_2^- \bullet \qquad (3)$$

Superoxide ion radicals can oxidize organic matter, though the rate constants of their reactions with organics are, in most cases, lower than those of hydroxyl radicals [89,90] and can also react with ozone to yield more ozonide ion radicals:

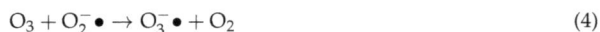

$$O_3 + O_2^- \bullet \rightarrow O_3^- \bullet + O_2 \tag{4}$$

In fact, Reactions (3) and (4) with Reactions (1) and (2) complete the synergism mechanism between ozonation and photocatalytic oxidation. In this respect, Liao et al. [60] observed a 95% and 37.8% removal of oxalic acid during SPO (with a g-C$_3$N$_4$ catalyst) in the absence and presence of benzoquinone. The contribution of superoxide ion radicals to remove oxalic acid was also confirmed.

The Langmuir equation, usually applied to follow the kinetics of ozone-free photocatalytic oxidation processes [85], is also used in SPO kinetics. The Langmuir equation is as follows:

$$-r_M = \frac{kKC_M}{1 + KC_M} \tag{5}$$

where k and K are the rate constant of the SPO process and the adsorption equilibrium constant of compound M, respectively, and r_M its removal rate. Mecha et al. [62] used simplified Langmuir kinetics reduced to a first order reaction rate because they observed negligible adsorption of the compound they studied, phenol, on the catalyst they prepared (see Table 2). With the obtained apparent pseudo first order rate constants, they calculated the synergy index of SPO, that is, the ratio between SPO rate and the sum of the ozonation and ozone-free photocatalytic oxidation rates:

$$SI = \frac{k_{SPO}}{k_{O3} + k_{SPoxidatio}} \tag{6}$$

The authors obtained values of 1.51, 1.03, 1.11 and 1.54 for the SPO processes conducted with TiO$_2$, Ag-TiO$_2$, Cu-TiO$_2$, and Fe-TiO$_2$ catalysts, respectively. Thus, bare TiO$_2$ and Fe-TiO$_2$ were the best catalysts investigated with the highest SPO removal rates.

As mentioned above, the contribution of hydroxyl radical reactions in SPO processes is one of the main ways of oxidation, and the concentration of these free radical species is necessary for kinetic studies. Quiñones et al. [57] applied the hydroxyl radical exposure concept to indirectly measure this concentration. This parameter is a measure of the concentration of hydroxyl radical during a given time in any AOP. It also allows for the determination of the R$_{CT}$, which is the ratio between the time integrating concentrations of hydroxyl radical and ozone in any water which is ozonated. It is based on the kinetics of any AOP process when an ozone non-reacting compound like *p*-chlorobenzoic acid (PCBA) is present in the problem water. PCBA does not react with ozone (the reaction rate constant is lower than 0.15 M^{-1} s^{-1} [91]) but it does react with hydroxyl radicals [7]. After taking into account the mass balance of PCBA in a semi-batch reactor where the SPO process was carried out, the hydroxyl radical exposure for a given reaction time can be obtained from Equation (7) [92]:

$$\int_0^t C_{HO}\, dt = \frac{ln\frac{C_{PCBA}}{C_{PCBA0}}}{k_{HO}} \tag{7}$$

where the left side of Equation (7) is the hydroxyl radical exposure; C$_{PCBA}$, C$_{PCBA0}$ are the concentrations of PCBA at time t and at the start of the SPO process, respectively; and k_{HO} the rate constant of the hydroxyl radical-PCBA reaction ($k_{OH} = 5 \times 10^9$ M^{-1} s^{-1} [7]). The authors observed that the hydroxyl radical exposure increased with the UV radiation dose for any of the systems tested (see Table 2). For a given radiation dose, the ozone/light systems (pH 3) led to higher HO· production than those systems tested in the absence of ozone. The Fe(III)/H$_2$O$_2$/O$_3$/light (pH 3) system can be highlighted as the one which produces the highest HO· exposure and, consequently, the highest TOC removal after 5 h of treatment (35%).

Additionally, process ozonation kinetics is studied by first determining the kinetic regime of ozone reactions. This can be established by calculating the Hatta number which, for second order irreversible reactions such as those ozone undergoes with organics in water [8], is defined as follows:

$$Hatta = \frac{\sqrt{k_{O3}D_{O3}C_M}}{k_L} \tag{8}$$

where k_{O3}, D_{O3}, and k_L are the ozone-organic M reaction rate constant, the ozone diffusivity in water, and the individual liquid side mass transfer coefficient, respectively. Quiñones et al. [56] studied the SPO kinetics of TOC removal from a mass balance in water in a semi-batch reacting system. In this balance, the different contributions to the TOC removal rates were considered as shown below. It should be noted that TOC is a parameter that depends on the type of water and nature of the compounds present and, hence, rate constants involved will also vary depending on the system. The authors studied two different SPO processes according to the catalyst used: Fe(III) and H_2O_2 at pH 3, and TiO_2 at pH 7. For the complex SPO process at pH 3: Fe(III)/O_3/H_2O_2/solar light, TOC mass balance applied was:

$$-\frac{dTOC}{dt} = \left(k_{T2}C_{Fe(II)}C_{H2O2} + k_{O3}C_{O3} + k_{T3}C_{O3}\right)TOC \tag{9}$$

In Equation (9) the terms of the right-hand side represent the contributions of the Fenton reaction and those of the TOC reactions with ozone (direct reactions) and with the fraction of hydroxyl radicals that comes from ozone decomposition initiation-promotion reactions. The apparent rate constant k_{T3} is the result of applying the R_{CT} concept, allowing the concentration of hydroxyl radicals to be expressed as a function of that of ozone [92]. In addition, the authors proposed the following TOC balance for the pH 7 SPO process: TiO_2/O_3/solar light:

$$-\frac{dTOC}{dt} = \left[k_{UV} + \frac{k_i C_{O3}}{1 + \Sigma K_i C_i} + k_{O3}C_{O3} + k_{HO}C_{HO}\right]TOC \tag{10}$$

where the four terms of the right-hand side are the contribution of direct photolysis, photocatalytic ozonation, direct ozone reactions, and reactions of hydroxyl radicals that come from non-photocatalytic ozonation reactions. As can be observed, Langmuir kinetics was considered for the contribution of photocatalysis due to oxidizing holes. Langmuir kinetics was reduced to a first order kinetics since most of the active sites were occupied with water molecules. Also, the direct photolysis of TOC was negligible. With these simplifications, Equation (10) eventually leads to a second order equation:

$$-\frac{dTOC}{dt} = k_T C_{O3}TOC \tag{11}$$

From experimental data, Quiñones et al. [56] determined all rate constant values of Equations (9) and (11). With the rate constants of ozone involving reactions, they calculated the Hatta number to obtain values of less than 0.3 which confirmed a slow kinetic regime and the validity of the mass balance equation applied [8]. They then determined the percentage contribution of the different pathways for the removal of TOC. For instance, for the Fe(III)/O_3/H_2O_2/Solar light/pH 3 system, they concluded that during the first seconds of reaction, the main contribution to the mineralization process corresponds to the photo-Fenton reaction. Minutes later, the percentage contribution of ozone-involving mechanisms increases due to the ozone accumulation and hydrogen peroxide partial consumption. Therefore, the contribution of ozone processes to mineralization increases with the increasing reaction time, accounting for 72% of mineralization after 5 min of reaction and for 98% after 1 h.

2.8. Stability and Activity of Catalysts

There are three important properties of any catalyst: selectivity, activity, and stability. In a few studies on SPO, both activity and stability have been analyzed. For example, regarding stability, Quiñones et al. [58] applied a B-doped TiO_2 catalyst to analyze the leaching phenomenon, after which catalysts were submitted to water washing at the same conditions as the reaction medium (catalyst concentration and pH). They observed that the boron concentration in the solution reached values as high as 5.5 mg L^{-1} in the case of the highest loading B-doped catalyst (12%), the loss of the total boron being from 46 to 70%. However, the authors observed no further loss of B after using the water-washed catalysts in three consecutive runs, which confirms a certain level of stability and activity. In another work, Rey et al. [47], with a magnetic FeTiC catalyst, observed Ti and Fe leaching at concentrations lower than 15 and 25 μg L^{-1}, respectively, in each 2 h SPO run they carried out. Then, after five consecutive cycles using the same catalyst, only 0.04% and 0.6% of the initial Ti and Fe, respectively, were leached into the water with a constant activity for TOC removal and good magnetic separability. Liao et al. [60] also reported high activity and stability of their gC_3N_4-rGO catalyst to remove oxalic acid. These authors found a slight decrease in oxalic acid removal percentage (96.7 to 93.8%) after five consecutive runs performed with the same amount of catalyst.

3. Conclusions and Future Steps

According to the few papers published on SPO compared to other similar technologies such as photocatalytic oxidation, the main conclusion that can be drawn is that SPO is an incipient but potentially attractive AOP since it projects the image of a green and environmentally sustainable process. Thus, solar energy can supply all the necessary energyneeded for an SPO process, that is, the energy for exciting catalysts (wavelength range depending on the nature of the catalyst), photolysing ozone (mainly in the range 290–320 nm), pump functioning, ozone production and even, in the case of cloudy days, accumulated energy could enable the use of new generation artificial lamps.

So far, SPO processes have been carried out at laboratory (solar simulator box with Xe lamps) and pilot plant scale (solar CPC photoreactors) with different types of catalysts: from the classical TiO_2 to new composite materials containing some metals, activated carbons, graphene, etc. The new synthesized catalysts that are now being investigated include, in many cases, TiO_2 as the main active material with other components such as metal or metal oxides (for example, Ag, Cu, Fe, Fe_2O_3, WO_3), or carbon materials (activated carbons, multiwalled carbon nanotubes, etc.) that provide energy states between the valence and conduction bands of TiO_2. This permits a decrease in TiO_2 band gap energy and allows it to be active under visible light.

Magnetic catalysts are also used in SPO. In this case, TiO_2, as the main active catalyst, and magnetite, for supplying magnetic properties, are supported on carbon materials such as activated carbons responsible for a wide surface area of adsorption. These magnetic catalysts are easily separated from water with the aid of a magnetic field.

Sol-gel and hydrothermal methods are still the most frequently used for the synthesis of catalysts for SPO. Condition changes of preparation steps lead to changes in the catalyst activity. Composition, crystalline structure, morphology, textural properties of SPO catalysts are obtained with classical techniques such as XRD, SEM, TEM, N_2 adsorption and desorption isotherms, etc.

At present, photoreactors are of a cylindrical type: tanks or tubes. Agitated tanks are used inside box simulators supplied, in general, with Xe lamps that emit in the solar radiation range ($\lambda > 290$ nm), that which reaches the Earth's surface. Different filters have been used to simulate radiation wavelength ranges between 290–320 nm, 290–400 nm, and above 400 nm (visible light). Several tubes are in many cases connected in series and placed above parabolic light reflecting structures oriented to the south as compound parabolic collectors (CPC).

A comparison of SPO with other AOPs that individually form the combination of ozone, UVA-visible light, and a catalyst is the first and principal study of papers. In all cases, TOC removal with SPO is much higher than that obtained from individual or blank processes (UVA-visible radiation, ozonation, catalytic ozonation, ozone photolysis, and ozone-free photocatalytic oxidation). The formation of hydrogen peroxide in the SPO process gives rise to an increase in TOC removal or mineralization by reacting with ozone to yield more hydroxyl radicals. However, the individual removal of many compounds is even faster with ozonation alone since in most of these cases direct ozone reactions are the main means of oxidation. This is particularly important in the removal of many pharmaceutical contaminants. The effects of other variables such as concentrations of ozone, organics, and catalysts are similar to those already reported for ozone-free photocatalytic oxidation and ozonation processes.

Ozone consumption results in much lower figures in SPO than in the rest of the ozone processes, and the water matrix strongly affects the organics and TOC removal rates, as has been shown from results in ultrapure and wastewater. Biodegradability, measured as BOD/COD, significantly increases with SPO application, rising by about 250% in an urban secondary wastewater effluent [59]. As a general rule, the toxicity of samples increases at the beginning of the photocatalytic treatment, probably as a consequence of the accumulation of phenolic and other toxic intermediates, and then at more advanced reaction times it decreases.

Reaction (1) or electron ozone capturing that generates the ozonide ion radical, eventually leading to hydroxyl radicals, is the main step of ozone-photocatalytic oxidation synergism. The participation of hydroxyl radicals has been confirmed in SPO works with the use of scavengers aiming at determining whether valence band holes and superoxide ion radicals could also intervene directly as other oxidizing species.

At the concentrations that the organics have in wastewaters (up to a few $\mu g\ L^{-1}$), the kinetic regime of ozone reactions is slow, which suggests AOPs and hence SPO as recommended processes to increase organic or TOC removal rates. This is because in slow kinetic regimes both ozone direct reactions and hydroxyl radical reactions compete, and the latter have much higher reaction rate constants. Thus, the application of AOPs that generate high concentration of hydroxyl radicals is recommended to increase the reaction rates.

It is foreseen that in the near future there will be an increase in studies on SPO due to its environmentally sustainable character, especially with new catalysts. The objectives of these studies will most likely be related to visible light activation and catalyst separation from water, two problems that so far have limited the practical application of photocatalytic processes in water treatment.

Acknowledgments: The authors thank the Spanish Ministry of Economy and Competitivity (MINECO) and the European Feder Funds for the economic support through Projects CTQ2015/64944-R and CTQ2015/73168/JIN(AEI/FEDER, UE).

Conflicts of Interest: The authors declare no conflict of interest.

References

1. Navarro, S.; Vela, N.; Gimenez, M.J.; Navarro, G. Persistence of four *s*-triazine herbicides in river, sea and groundwater samples exposed to sunlight and darkness under laboratory conditions. *Sci. Total Environ.* **2004**, *329*, 87–97. [CrossRef] [PubMed]
2. Stuart, M.; Lapworth, D.; Crane, E.; Hart, H. Review of risk from potential emerging contaminants in UK groundwater. *Sci. Total Environ.* **2011**, *416*, 1–21. [CrossRef] [PubMed]
3. Gracia-Lor, E.; Sancho, J.V.; Serrano, R.; Hernández, F. Occurrence and removal of pharmaceuticals in wastewater treatment plants at the Spanish Mediterranean area of Valencia. *Chemosphere* **2012**, *87*, 453–462. [CrossRef] [PubMed]
4. Bottoni, P.; Grenni, P.; Lucentini, L.; Barra Caracciolo, A. Terbuthylazine and other triazines in Italian water resources. *Microchem. J.* **2013**, *107*, 136–142. [CrossRef]
5. Barbosa, M.O.; Moreira, N.F.F.; Ribeiro, A.R.; Pereira, M.F.R.; Silva, A.M.T. Occurrence and removal of organic micropollutants: An overview of the watch list of EU Decision 2015/495. *Water Res.* **2016**, *94*, 257–279. [CrossRef] [PubMed]

6. Arnold, K.E.; Boxall, A.B.A.; Brown, A.R.; Cuthbert, R.J.; Gaw, S.; Hutchinson, T.H.; Jobling, S.; Madden, J.C.; Metcalfe, C.D.; Naidoo, V.; et al. Assessing the exposure risk and impacts of pharmaceuticals in the environment on individuals and ecosystems. *Biol. Lett.* **2013**, *9*. [CrossRef] [PubMed]

7. Buxton, G.V.; Greenstock, C.L.; Helman, W.P.; Ross, A.B. Critical review of data constants for reactions of hydrated electrons, hydrogen atoms and hydroxyl radicals ($\cdot$OH/$\cdot$O$^-$) in aqueous solution. *J. Phys. Chem. Ref. Data* **1988**, *17*, 513–886. [CrossRef]

8. Beltrán, F.J. *Ozone Reaction Kinetics for Water and Wastewater Systems*; CRC Press (Lewis Publishers): Boca Ratón, FL, USA, 2004.

9. Beltrán, F.J.; Aguinaco, A.; García-Araya, J.F. Mechanism and kinetics of sulfamethoxazole photocatalytic ozonation in water. *Water Res.* **2009**, *43*, 1359–1369. [CrossRef] [PubMed]

10. Rivas, F.J.; Sagasti, J.; Encinas, A.; Gimeno, O. Contaminants abatement by ozone in secondary effluents. Evaluation of second-order rate constants. *J. Chem. Technol. Biotechnol.* **2011**, *86*, 1058–1066. [CrossRef]

11. Huber, M.M.; Canonica, S.; Park, G.Y.; von Gunten, U. Oxidation of pharmaceuticals during ozonation and advanced oxidation processes. *Environ. Sci. Technol.* **2003**, *37*, 1016–1024. [CrossRef] [PubMed]

12. Dodd, M.C.; Buffle, M.O.; von Gunten, U. Oxidation of antibacterial molecules by aqueous ozone: moiety-specific reaction kinetics and application to ozone-based wastewater treatment. *Environ. Sci. Technol.* **2006**, *40*, 6519–6530. [CrossRef]

13. Jin, X.; Peldszus, S.; Huck, P.M. Reaction kinetics of selected micropollutants in ozonation and advanced oxidation processes. *Water Res.* **2012**, *46*, 6519–6530. [CrossRef] [PubMed]

14. Deborde, M.; Rabouan, S.; Duguet, J.P.; Legube, B. Kinetics of aqueous ozone-induced oxidation of some endocrine disruptors. *Environ. Sci. Technol.* **2005**, *39*, 6086–6092. [CrossRef] [PubMed]

15. Qiang, Z.; Adams, C.; Surampalli, R. Determination of ozonation rate constants for lincomycin and spectinomycin. *Ozone Sci. Eng.* **2004**, *26*, 526–537. [CrossRef]

16. Staehelin, S.; Hoigné, J. Decomposition of ozone in water: Rate of initiation by hydroxyde ions and hydrogen peroxide. *Environ. Sci. Technol.* **1982**, *16*, 666–681. [CrossRef]

17. Staehelin, S.; Hoigné, J. Decomposition of Ozone in Water the Presence of Organic Solutes Acting as Promoters and Inhibitors of Radical Chain Reactions. *Environ. Sci. Technol.* **1985**, *19*, 1206–1212. [CrossRef] [PubMed]

18. Kasprzyk-Hordern, B.; Ziolek, M.; Nawrocki, J. Catalytic ozonation and methods of enhancing molecular ozone reactions in water treatment. *Appl. Catal. B Environ.* **2003**, *46*, 639–669. [CrossRef]

19. Ikehata, K.; Naghashkar, N.J.; El-Din, M.G. Degradation of aqueous pharmaceuticals by ozonation and advanced oxidation processes: A review. *Ozone Sci. Eng.* **2006**, *28*, 353–414. [CrossRef]

20. Liu, Y.; He, H.P.; Wu, D.; Zhang, Y. Heterogeneous Catalytic Ozonation Reaction Mechanism. *Prog. Chem.* **2016**, *28*, 1112–1120.

21. Gmurek, M.; Olak-Kucharczyk, M.; Ledakowicz, S. Photochemical decomposition of endocrine disrupting compounds—A review. *Chem. Eng. J.* **2017**, *310*, 437–456. [CrossRef]

22. Noufi, R.N.; Kohl, P.A.; Frank, S.N.; Bard, A.J. Semiconductor electrodes XIV. Electrochemistry and electroluminescence at n-type TiO$_2$ in aqueous solutions. *J. Electrochem.* **1978**, *125*, 246–252. [CrossRef]

23. Fujishima, A.; Honda, K. Electrochemical photolysis of water at a semiconductor electrode. *Nature* **1972**, *238*, 37–38. [CrossRef] [PubMed]

24. *Photocatalysis and Water Purification*; Pichat, P., Ed.; Wiley-VCH: Weinheim, Germany, 2013.

25. *Photocatalysis: Fundamentals, Materials and Potential*; Pichat, P., Ed.; MDPI: Basel, Switzerland, 2016.

26. Schneider, J.; Bahnemann, D.; Ye, J.; Puma, L.G.; Dionysiou, D.D. (Eds.) *Photocatalysis, Vol. 1: Fundamentals and Perspectives*; The Royal Society of Chemistry (RSC): London, UK, 2016.

27. Pichat, P. Fundamentals of TiO$_2$ photocatalysis. Consequences for some environmental applications. In *Heterogeneous Photocatalysis*; Colmenares Quintero, J.C., Xu, Y.-J., Eds.; Springer: Basel, Switzerland, 2016; pp. 321–359.

28. Mills, A.; Davies, R.H.; Worsley, D. Water purification by semiconductor photocatalysis. *Chem. Soc. Rev.* **1993**, *22*, 417–425. [CrossRef]

29. Von Gunten, U. Ozonation of drinking water: Part I. Oxidation kinetics and product formation. *Water Res.* **2003**, *37*, 1443–1467. [CrossRef]

30. Von Gunten, U. Ozonation of drinking water: Part II. Disinfection and by-product formation in presence of bromide, iodide or chlorine. *Water Res.* **2003**, *37*, 1469–1487. [CrossRef]

31. Malato, S.; Fernandez-Ibañez, P.; Maldonado, M.; Blanco, J.; Gernjak, W. Decontamination and disinfection of water by solar photocatalysis: Recent overview and trends. *Catal. Today* **2009**, *147*, 1–59. [CrossRef]

32. Ahmad, R.; Ahmad, Z.; Khan, A.U.; Mastoi, N.R.; Aslam, M.; Kim, J. Photocatalytic systems as an advanced environmental remediation: Recent developments, limitations and new avenues for applications. *J. Environ. Chem. Eng.* **2016**, *4*, 4143–4164. [CrossRef]

33. Spasiano, D.; Marotta, R.; Malato, S.; Fernandez-Ibañez, P.; Di Somma, I. Solar photocatalysis: Materials, reactors, some commercial, and pre-industrialized applications. A comprehensive approach. *Appl. Catal. B Environ.* **2015**, *170*, 90–123. [CrossRef]

34. Hubner, U.; von Gunten, U.; Jekel, M. Evaluation of the persistence of transformation products from ozonation of trace organic compounds—A critical review. *Water Res.* **2015**, *68*, 150–170. [CrossRef] [PubMed]

35. Rakovsky, S.K.; Anachkov, M.P.; Iliev, V.I.; Elyias, A.E. Ozone Degradation of Alcohols, Ketones, Ethers and Hydroxybenzenes: Determination of Pathways and Kinetic Parameters. *J. Adv. Oxid. Technol.* **2013**, 31–51. [CrossRef]

36. Agustina, T.E.; Ang, H.M.; Vareek, V.K. A review of synergistic effect of photocatalysis and ozonation on wastewater treatment. *J. Photochem. Photobiol. C Photochem. Rev.* **2005**, *6*, 264–273. [CrossRef]

37. Mehrjouei, M.; Muller, S.; Moller, D. A review on photocatalytic ozonation used for the treatment of water and wastewater. *Chem. Eng. J.* **2015**, *263*, 209–219. [CrossRef]

38. Xiao, J.; Xie, Y.; Cao, H. Organic pollutants removal in wastewater by heterogeneous photocatalytic ozonation. *Chemosphere* **2015**, *121*, 1–17. [CrossRef] [PubMed]

39. Radwan, E.K.; Yu, L.L.; Achari, G.; Langford, C.H. Photocatalytic ozonation of pesticides in a fixed bed flow through UVA-LED photoreactor. *Environ. Sci. Pollut. Res.* **2016**, *23*, 21313–21318. [CrossRef] [PubMed]

40. Oyama, T.; Yanagisawa, I.; Takeuchi, M.; Koike, T.; Serpone, N.; Hidaka, H. Remediation of simulated aquatic sites contaminated with recalcitrant substrates by TiO_2/ozonation under natural sunlight. *Appl. Catal. B Environ.* **2009**, *91*, 242–246. [CrossRef]

41. Araña, J.; Herrer, J.A.; Doña, J.M.; González, O. TiO_2-photocatalysis as a tertiary treatment of naturally treated wastewater. *Catal. Today* **2002**, *76*, 279–289. [CrossRef]

42. Bayarri, B.; Gonzalez, O.; Maldonado, M.I.; Gimenez, J.; Esplugas, S. Comparative study of 2,4-dichlorophenol degradation with different advanced oxidation processes. *J. Sol. Energy Eng. Trans. ASME* **2007**, *129*, 60–67. [CrossRef]

43. Nishimoto, S.; Mano, T.; Kameshima, Y.; Miyake, M. Photocatalytic water treatment over WO_3 under visible light irradiation combined with ozonation. *Chem. Phys. Lett.* **2010**, *500*, 86–89. [CrossRef]

44. Anandan, S.; Lee, G.J.; Chen, P.K.; Fan, C.; Wu, J.J. Removal of Orange II Dye in Water by Visible Light Assisted Photocatalytic Ozonation Using Bi_2O_3 and Au/Bi_2O_3 Nanorods. *Ind. Eng. Chem. Res.* **2010**, *49*, 9729–9737. [CrossRef]

45. Oyama, T.; Otsu, T.; Hidano, Y.; Koike, T.; Serpone, N.; Hidaka, H. Enhanced remediation of simulated wastewaters contaminated with 2-chlorophenol and other aquatic pollutants by TiO_2-photoassisted ozonation in a sunlight-driven pilot-plant scale photoreactor. *Sol. Energy* **2011**, *85*, 938–944. [CrossRef]

46. Feng, H.; Zhang, M.H.; Yu, L.E. Hydrothermal synthesis and photocatalytic performance of metal-ions doped TiO2. *Appl. Catal. A Gen.* **2012**, *413–414*, 238–244. [CrossRef]

47. Rey, A.; Quiñones, D.H.; Alvarez, P.M.; Beltran, F.J.; Plucinski, P.K. Simulated solar-light assisted photocatalytic ozonation of metoprolol over titania-coated magnetic activated carbon. *Appl. Catal. B Environ.* **2012**, *111*, 246–253. [CrossRef]

48. Shin, D.; Jang, M.; Cui, M.; Na, S.; Khim, J. Enhanced removal of dichloroacetonitrile from drinking water by the combination of solar-photocatalysis and ozonation. *Chemosphere* **2013**, *93*, 2901–2908. [CrossRef] [PubMed]

49. Liao, G.Z.; Zhu, D.; Li, L.; Lan, B. Enhanced Photocatalytic Ozonation of Organics by g-C_3N_4 Under Visible Light Irradiation. *J. Hazard. Mater.* **2014**, *280*, 531–535. [CrossRef] [PubMed]

50. Quiñones, D.H.; Rey, A.; Alvarez, P.M.; Beltran, F.J.; Plucinski, P.K. Enhanced activity and reusability of TiO_2 loaded magnetic activated carbon for solar photocatalytic ozonation. *Appl. Catal. B Environ.* **2014**, *144*, 96–106. [CrossRef]

51. Marquez, G.; Rodriguez, E.M.; Beltran, F.J.; Alvarez, P.M. Solar photocatalytic ozonation of a mixture of pharmaceutical compounds in water. *Chemosphere* **2014**, *113*, 71–78. [CrossRef] [PubMed]

52. Rey, A.; García-Muñoz, P.; Hernández-Alonso, M.D.; Mena, E.; García-Rodríguez, S.; Beltrán, F.J. WO$_3$–TiO$_2$ based catalysts for the simulated solar radiation assisted photocatalytic ozonation of emerging contaminants in a municipal wastewater treatment plant effluent. *Appl. Catal. B Environ.* **2014**, *154–155*, 274–284. [CrossRef]

53. Mano, T.; Nishimoto, S.; Kameshima, Y.; Miyake, M. Water treatment efficacy of various metal oxide semiconductors for photocatalytic ozonation under UV and visible light irradiation. *Chem. Eng. J.* **2015**, *264*, 221–229. [CrossRef]

54. Rey, A.; Mena, E.; Chávez, A.M.; Beltrán, F.J.; Medina, F. Influence of structural properties on the activity of WO$_3$ catalysts for visible light photocatalytic ozonation. *Chem. Eng. Sci.* **2015**, *126*, 80–90. [CrossRef]

55. Mena, E.; Rey, A.; Contreras, S.; Beltrán, F.J. Visible light photocatalytic ozonation of DEET in the presence of different forms of WO$_3$. *Catal. Today* **2015**, *252*, 100–106. [CrossRef]

56. Quiñones, D.H.; Alvarez, P.M.; Rey, A.; Contreras, S.; Beltran, F.J. Application of solar photocatalytic ozonation for the degradation of emerging contaminants in water in a pilot plant. *Chem. Eng. J.* **2015**, *260*, 399–410. [CrossRef]

57. Quiñones, D.H.; Alvarez, P.M.; Rey, A.; Beltrán, F.J. Removal of emerging contaminants from municipal WWTP secondary effluents by solar photocatalytic ozonation. A pilot-scale study. *Separat. Purific. Technol.* **2015**, *149*, 132–139. [CrossRef]

58. Quiñones, D.H.; Rey, A.; Alvarez, P.M.; Beltran, F.J.; Puma, G.L. Boron doped TiO$_2$ catalysts for photocatalytic ozonation of aqueous mixtures of common pesticides: Diuron, *o*-phenylphenol, MCPA and terbuthylazine. *Appl. Catal. B Environ.* **2015**, *178*, 74–81. [CrossRef]

59. Gimeno, O.; Garcia-Araya, J.F.; Beltran, F.J.; Rivas, F.J.; Espejo, A. Removal of emerging contaminants from a primary effluent of municipal wastewater by means of sequential biological degradation-solar photocatalytic oxidation processes. *Chem. Eng. J.* **2016**, *290*, 12–20. [CrossRef]

60. Liao, G.; Zhu, D.; Li, C.; Lan, B.; Li, L. Degradation of Oxalic Acid and Bisphenol A by Photocatalytic Ozonation with g-C$_3$N$_4$ Nanosheet under Simulated Solar Irradiation. *Ozone Sci. Eng.* **2016**, *38*, 312–317. [CrossRef]

61. Yin, J.; Liao, G.; Zhu, D.; Lu, P.; Li, L. Photocatalytic ozonation of oxalic acid by g-C$_3$N$_4$/graphene composites under simulated solar irradiation. *J. Photochem. Photobiol. A Chem.* **2016**, *315*, 138–144. [CrossRef]

62. Mecha, A.C.; Onyango, M.S.; Ochieng, A.; Fourie, C.J.S.; Momba, M.N.B. Synergistic effect of UV-vis and solar photocatalytic ozonation on the degradation of phenol in municipal wastewater: A comparative study. *J. Catal.* **2016**, *341*, 116–125. [CrossRef]

63. Alvarez, P.M.; Quiñones, D.H.; Terrones, I.; Rey, A.; Beltran, F.J. Insights into the removal of terbuthylazine from aqueous solution by several treatment methods. *Water Res.* **2016**, *98*, 334–343. [CrossRef] [PubMed]

64. Mena, E.; Rey, A.; Rodriguez, E.M.; Beltran, F.J. Nanostructured CeO$_2$ as catalysts for different AOPs based in the application of ozone and simulated solar radiation. *Catal. Today* **2017**, *280*, 74–79. [CrossRef]

65. Mena, E.; Rey, A.; Rodríguez, E.M.; Beltrán, F.J. Reaction mechanism and kinetics of DEET visible light assisted photocatalytic ozonation with WO$_3$ catalyst. *Appl. Catal. B Environ.* **2017**, *202*, 460–472. [CrossRef]

66. Beltrán, F.J.; García-Araya, J.F.; Giráldez, I. Gallic acid water ozonation using activated carbon. *Appl. Catal. B Environ.* **2006**, *63*, 249–259. [CrossRef]

67. Tomiyasu, H.; Fukutomi, H.; Gordon, G. Kinetics and mechanism of ozone decomposition in basic aqueous solutions. *Inorg. Chem.* **1985**, *24*, 2962–2966. [CrossRef]

68. Bidga, R.J. Considering Fenton's chemistry for wastewater treatment. *Chem. Eng. Prog.* **1995**, *91*, 62–67.

69. Dozzi, M.V.; Selli, E. Doping TiO$_2$ with p-block elements: Effects on photocatalytic activity. *J. Photochem. Photobiol. C Photochem. Rev.* **2013**, *14*, 13–28. [CrossRef]

70. Begum, N.S.; Ahmed, H.M.F.; Hussain, O.M. Characterization and photocatalytic activity of boron-doped TiO$_2$ thin films prepared by liquid phase deposition technique. *Bull. Mater. Sci.* **2008**, *31*, 741–745. [CrossRef]

71. Yu, J.; Zhang, Y.; Kudo, A. Synthesis and photocatalytic performances of BiVO$_4$ by ammonia co-precipitation process. *J. Solid State Chem.* **2009**, *182*, 223–228. [CrossRef]

72. Kasprzyk-Hordern, B.; Dinsdale, R.M.; Guwy, A.J. The removal of pharmaceuticals, personal care products, endocrine disruptors and illicit drugs during wastewater treatment and its impact on the quality of receiving waters. *Water Res.* **2009**, *43*, 363–380. [CrossRef] [PubMed]

73. Halden, R.U. *Contaminants of Emerging Concern in the Environment: Ecological and Human Health Considerations*; ACS Symposium Series; American Chemical Society: Washington, DC, USA, 2010.

74. Calzadilla, A.; Rehdanz, K.; Tol, R. The economic impact of more sustainable water use in agriculture: A computable general equilibrium analysis. *J. Hydrol.* **2010**, *384*, 292–305. [CrossRef]

75. Fukazawa, H.; Hoshino, K.; Shiozawa, T.; Matsushita, H.; Terao, Y. Identification and quantification of chlorinated bisphenol-A in wastewater from wastepaper recycling plants. *Chemosphere* **2001**, *44*, 973–979. [CrossRef]

76. Beltrán, F.J.; Gimeno, O.; Rivas, F.J.; Carbajo, M. Photocatalytic Ozonation of Gallic Acid in Water. *J. Chem. Technol. Biotechnol.* **2006**, *81*, 1787–1796. [CrossRef]

77. Taube, H. Photochemical reactions of ozone in aqueous solution. *Trans. Faraday Soc.* **1957**, *53*, 656–665. [CrossRef]

78. Hoigné, J.; Bader, H. Rate constants of the reactions of ozone with organic and inorganic compounds. II. Dissociating organic compounds. *Water Res.* **1983**, *17*, 185–194. [CrossRef]

79. Wang, K.; Hsieh, Y.; Chou, M.; Chang, C. Photocatalytic degradation of 2-chloro and 2-nitrophenol by titanium dioxide suspensions in aqueous solution. *Appl. Catal. B Environ.* **1999**, *22*, 1–8. [CrossRef]

80. Doong, R.; Chen, C.; Maithreepala, R.; Chang, S. The influence of pH and cadmium sulfide on the photocatalytic degradation of 2-chlorophenol in titanium dioxide suspensions. *Water Res.* **2001**, *35*, 2880–2973. [CrossRef]

81. Delanghe, B.; Mekras, C.I.; Graham, N.J.D. Aqueous ozonation of surfactants—A review. *Ozone Sci. Eng.* **1991**, *13*, 639–673. [CrossRef]

82. Alvares, A.B.C.; Diaper, C.; Parsons, S.A. Partial oxidation by ozone to remove recalcitrance from wastewaters—A review. *Environ. Technol.* **2001**, *22*, 409–427. [CrossRef] [PubMed]

83. Collado, N.; Buttiglierib, G.; Martib, E.; Ferrando-Climent, L.; Rodriguez-Mozaz, S.; Barceló, D.; Comasa, J.; Rodriguez-Roda, L. Effects on activated sludge bacterial community exposed to sulfamethoxazole. *Chemosphere* **2013**, *93*, 99–106. [CrossRef] [PubMed]

84. Márquez, G.; Rodríguez, E.M.; Maldonado, M.I.; Álvarez, P.M. Integration of ozone and solar TiO$_2$-photocatalytic oxidation for the degradation of selected pharmaceutical compounds in water and wastewater. *Sep. Purif. Technol.* **2014**, *136*, 18–26. [CrossRef]

85. Turchi, S.C.; Ollis, D.F. Photocatalytic degradation of organic water contaminants: mechanisms involving hydroxyl radical attack. *J. Catal.* **1990**, *122*, 178–192. [CrossRef]

86. Jenks, W.S. Photocatalytic reactions pathways—Effects of molecular structure, catalyst and wavelength. In *Photocatalysis and Water Purification*; Pichat, P., Ed.; Wiley-VCH: Weinheim, Germany, 2013; pp. 25–51.

87. Beltrán, F.J.; Aguinaco, A.; García-Araya, J.F. Kinetic modelling of TOC removal in the photocatalytic ozonation of diclofenac aqueous solutions. *Appl. Catal. B Environ.* **2010**, *100*, 289–298. [CrossRef]

88. Pichat, P.; Disdier, J.; Hoang-Van, C.; Mas, D.; Goutailler, G.; Gaysse, C. Purification/deodorization of indoor air and gaseous effluents by TiO$_2$ photocatalysis. *Catal. Today* **2000**, *63*, 363–369. [CrossRef]

89. Ross, F.; Ross, A.B. *Selected Specific Rates of Reactions of Transients for Water in Aqueous Solution. III Hydroxyl Radical and Perhydroxyl Radical and Their Radical Ions*; US National Bureau of Standards: Washington, DC, USA, 1977.

90. Sawyer, D.T.; Valentine, J.S. How super is superoxide? *Acc. Chem. Res.* **1981**, *14*, 393–400. [CrossRef]

91. Yao, D.C.C.; Haag, W.R. Rate constants for direct reactions of ozone with several drinking water contaminants. *Water Res.* **1991**, *25*, 761–773.

92. Elovitz, M.S.; von Gunten, U. Hydroxyl radical/ozone ratios during ozonation processes. I. The R$_{ct}$ concept. *Ozone Sci. Eng.* **1999**, *21*, 239–260. [CrossRef]

molecules

MDPI

Article

Electrochemical Enhancement of Photocatalytic Disinfection on Aligned TiO$_2$ and Nitrogen Doped TiO$_2$ Nanotubes

Cristina Pablos [1,*], Javier Marugán [1], Rafael van Grieken [1], Patrick Stuart Morris Dunlop [2], Jeremy William John Hamilton [2], Dionysios D. Dionysiou [3] and John Anthony Byrne [2]

[1] Department of Chemical and Environmental Technology, ESCET, Universidad Rey Juan Carlos, c/Tulipán s/n, 28933 Móstoles, Madrid, Spain; javier.marugan@urjc.es (J.M.); rafael.vangrieken@urjc.es (R.v.G.)

[2] Nanotechnology and Integrated BioEngineering Centre (NIBEC), Ulster University, Newtownabbey BT37 0QB, Northern Ireland, UK; psm.dunlop@ulster.ac.uk (P.S.M.D.); jwj.hamilton@ulster.ac.uk (J.W.J.H.); j.byrne@ulster.ac.uk (J.A.B.)

[3] Environmental Engineering and Science program, University of Cincinnati, 705 Engineering Research Center, Cincinnati, OH 45221, USA; dionysios.d.dionysiou@uc.edu

* Correspondence: cristina.pablos@urjc.es; Tel.: +34-914884619

Academic Editor: Pierre Pichat
Received: 17 February 2017; Accepted: 26 April 2017; Published: 28 April 2017

Abstract: TiO$_2$ photocatalysis is considered as an alternative to conventional disinfection processes for the inactivation of waterborne microorganisms. The efficiency of photocatalysis is limited by charge carrier recombination rates. When the photocatalyst is immobilized on an electrically conducting support, one may assist charge separation by the application of an external electrical bias. The aim of this work was to study electrochemically assisted photocatalysis with nitrogen doped titania photoanodes under visible and UV-visible irradiation for the inactivation of *Escherichia coli*. Aligned TiO$_2$ nanotubes were synthesized (TiO$_2$-NT) by anodizing Ti foil. Nanoparticulate titania films were made on Ti foil by electrophoretic coating (P25 TiO$_2$). N-doped titania nanotubes and N,F co-doped titania films were also prepared with the aim of extending the active spectrum into the visible. Electrochemically assisted photocatalysis gave higher disinfection efficiency in comparison to photocatalysis (electrode at open circuit) for all materials tested. It is proposed that electrostatic attraction of negatively charged bacteria to the positively biased photoanodes leads to the enhancement observed. The N-doped TiO$_2$ nanotube electrode gave the most efficient electrochemically assisted photocatalytic inactivation of bacteria under UV-Vis irradiation but no inactivation of bacteria was observed under visible only irradiation. The visible light photocurrent was only a fraction (2%) of the UV response.

Keywords: titania nanotubes; nitrogen-doped nanotubes; photoelectrocatalysis; *E. coli*; visible light

1. Introduction

Chlorination is an effective approach for the disinfection of water; however, it can lead to the formation of disinfection by-products e.g., trihalomethane, which can be mutagenic and carcinogenic. Furthermore, some species of pathogenic microorganisms are resistant to chlorination and ozonation [1]. As a consequence, new technologies have been developed to overcome the drawbacks of current disinfection processes. Heterogeneous photocatalysis is an advanced oxidation process (AOP) which can operate under ambient temperature and pressure, and oxygen from the air can be utilized as the oxidant without the addition of consumable chemicals. If one can use solar energy to drive the photocatalytic process then it becomes a truly clean technology. Since the early work of

Matsunaga et al. [2], TiO_2 photocatalysis has been reported by many research groups to be effective for the inactivation of a wide range of microorganisms in water [3,4].

The use of TiO_2 suspensions for photocatalysis involves an additional post treatment step to separate the particles from the treated water, which may increase the complexity and cost of treatment. Another disadvantage of suspension systems is low quantum yield for hydroxyl radical generation due to the recombination of charge carriers. Electrochemically enhanced photocatalysis using TiO_2 photoanodes is a potential solution to improve charge carrier separation and this approach utilizes immobilized photocatalyst without the need for post-treatment separation. The application of an external electrical potential to the TiO_2 anode can improve charge carrier separation and thus reduce the rate of recombination. Under irradiation, photogenerated valence band holes migrate to the semiconductor surface where the water oxidation occurs, producing hydroxyl radicals ($\cdot OH$), and the photogenerated conduction band electrons migrate or diffuse to the supporting electrode, from where they are driven to the counter electrode and passed on to molecular oxygen creating superoxide radical anion ($O_2{}^{\cdot -}$). Subsequent reduction reactions yield hydrogen peroxide (H_2O_2) and $\cdot OH$. The use of electrochemically assisted photocatalysis for the disinfection of water and the degradation of organic pollutants has been previously reported [5–7]. One of the important advantages of the application of electrochemically assisted photocatalysis with respect to the inactivation of microorganisms is that mass transport may be enhanced through electromigration of negatively charged bacteria to a positively biased photoanode [8–10].

Nanoengineering of the titania (TiO_2) electrodes is an interesting approach to improve the efficiency for electrochemically assisted photocatalytic disinfection. The interest in the potential use of titanium dioxide nanotubes as photoanodes has been increasing. They may exhibit better photoelectrolytic properties compared with nanoparticle films, due to the short diffusion path for photogenerated holes and a direct path for photogenerated electrons to the supporting electrodes [11–13]. In 1999, anodically grown self-organized TiO_2 nanotubes were reported by Zwilling et al. [14]. These nanotubes can be prepared by anodic oxidation of a Ti substrate in fluoride containing electrolytes. TiO_2 nanotubes remain attached on the substrate and therefore can be used directly as photoanodes. It has been reported that photoelectrolytic degradation of chemical contaminants in water is more effective with the TiO_2-NT electrodes as compared to TiO_2 nanoparticulate electrodes [12,15,16]. However, few studies have been conducted to date on the use of nanotubular TiO_2/Ti for electrochemically assisted photocatalysis in water disinfection [9,17–20].

A major drawback of TiO_2 for solar applications is that it has a wide band gap (anatase = 3.2 eV) which means that it absorbs in the UV domain and only 4% of solar photons are of sufficient energy to excite TiO_2. Many efforts have been made to modify titania (particles, films, and nanotubes) in order to extend the action spectrum into the visible domain of the electromagnetic spectrum as around 45% of solar photons are in the visible wavelength range [13,21,22]. TiO_2 doping with non-metals, in particular with nitrogen, has attracted great interest since Asahi and co-workers [23] reported in 2001 that N-doped TiO_2 showed visible light photocatalytic activity. They reported that the mixing of N 2p with O 2p states in valence band results in the narrowing of the TiO_2 band-gap and shifting absorption onset of TiO_2 to lower energies. However, since heavy doping of the metal oxide semiconductor is required for narrowing the band gap [24,25], other mechanisms regarding the causes leading the absorption edge of TiO_2 to be shifted towards the visible region have been proposed. Several authors [26–28] have suggested that sub-band gap excitation is due to isolated N 2p states located above the valence band maximum as a result of N-doping. Other groups have pointed out an increase in the absorption of visible light due to the presence of oxygen vacancies, induced as a consequence of N doping, which give rise to the formation of Ti^{3+} defect states, also called color centers [28,29]. There is lack of consensus since these states have been reported to be lying just below the conduction band [27,29–32] but also just above the valence band [25,32–35]. Although some authors such as Yang et al. [36] and Nakamura et al. [37] have reported that these color centers may act as electron trapping sites, leading to new photo excitation processes, they have traditionally been reported as responsible for hole trapping, leading to charge carrier recombination [26,28,33,38]. It has been reported that while Ti^{3+} color centers can give rise to visible light absorption, they may not contribute to visible photocatalytic activity [38]. Furthermore, co-doping with N and F has been reported to yield visible light activity. The presence of fluorine as a dopant induces formation of shallow Ti^{3+} donor levels a few tenths of electron volts below the conduction band [39]. Calculations suggest that the co-doping N-TiO_2 with fluorine reduces the number of oxygen vacancies as compared to nitrogen doping alone [40].

It is worth noting that several reports have been published concerning the potential use of N-TiO_2 photocatalysts for disinfection applications [41]. However, no papers concerning electrochemically assisted photocatalytic disinfection involving N-doped titania nanotubes have been found. Therefore, the aim of this work is to study the effect of the application of an electric potential bias on the photocatalytic inactivation efficiency of *E. coli* under simulated solar irradiation by using TiO_2 particulate films, sol gel films, and N-doped titania nanotubes as photoanodes. In this work, N,F-TiO_2 has been used as a reference for a reported visible light active photocatalyst material [42].

2. Results and Discussion

2.1. Electrode Characterization

SEM analysis of TiO_2 nanostructures obtained by anodization of Ti foil were carried out for both, undoped and doped titania nanotube samples. The SEM image for the N-doped titania NT electrode is shown in Figure 1 as an example. There was no discernible difference observed by SEM between the titania nanotubes annealed in air (TiO_2-NT) and those annealed in NH_3 (N-TiO_2-NT).

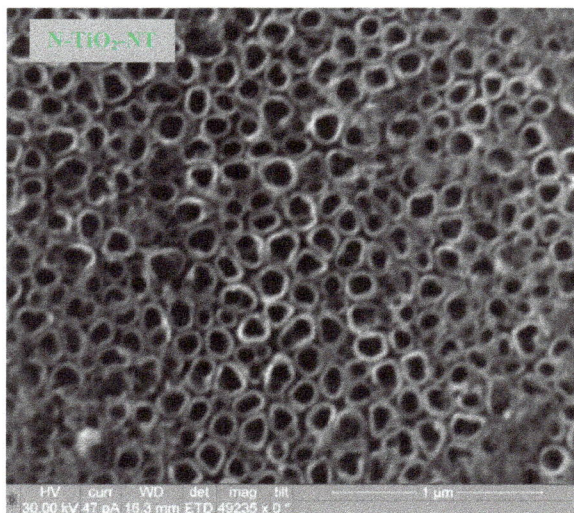

Figure 1. SEM image of titania nanotubes (N-TiO_2-NT) grown by anodization of Ti foil, followed by annealing in ammonia atmosphere.

The nanotubes are observed as discrete, hollow, and cylindrical tube-like features (both undoped and N-doped). They are uniformly distributed throughout the sample and densely populated. It must be noted that the undoped and doped nanotubular layers resist heat treatment. They show a shape similar to stacked rings, which has already been observed in the case of tubes grown in aqueous electrolytes containing fluoride ions [43,44]. For both samples, not only the diameter of nanotubes (ca. 100 nm) but also the wall thickness (ca. 8 nm) was similar, which is in agreement with the dimensions reported by other researchers using anodizing potentials between +20 to +25 V [12,43–46].

Previous XRD analysis of TiO_2-NTs showed that annealing at 450 °C in air gives the anatase crystal phase [47]. Annealing the NTs in NH_3 at 450 °C does not alter the crystal phase. SEM analysis of the electrophoretically deposited P25 electrode (P25 TiO_2) has already been reported [48]. A highly porous fractured appearance was obtained. The P25 TiO_2 is an 80:20 ratio of anatase to rutile and this does not change following annealing at 450 °C. The N,F-TiO_2 used in this work has been previously characterized by XRD and has been found to consist predominantly of the anatase phase [42]. The overall surface morphology of the N,F-TiO_2 synthesized film was previously reported and a rough but fairly uniform grain size distribution was observed [42].

Reported BET analysis of titania NTs has suggested that the surface area is between 12 [12] and 38 $m^2 g^{-1}$ [49]. Given that the NTs by mass are much less than 1.0 mg cm^{-2}, it is not then possible to measure the specific surface area on these samples, but it will be less than 0.038 m^2. The P25 TiO_2 has a specific surface area of 50 $m^2 g^{-1}$ as the free powder. Necking of the particles occurs through partial sintering at 450 °C, so the surface area will be somewhat less that of the free particles; however, the loading is 1.0 mg cm^{-2} of P25 (<0.05 m^2). The N,F-TiO_2 as free powder has a surface area of 136 $m^2 g^{-1}$, but the film thickness is around 1.5 μm, with 160 μg cm^{-2} (<0.02 m^2).

XPS analysis was used to confirm nitrogen incorporation in the nanotube electrodes (Figure 2).

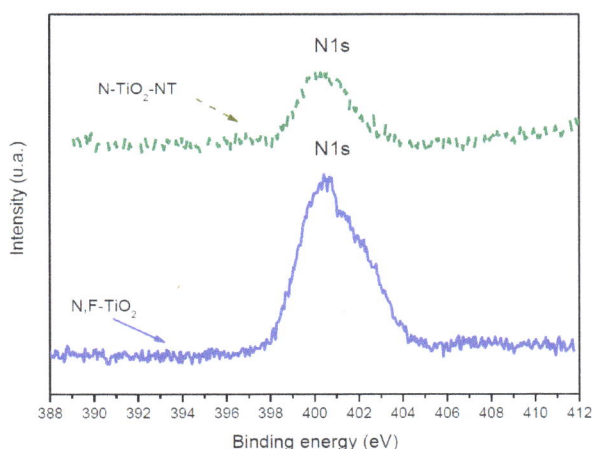

Figure 2. XPS spectra of the N 1s core level obtained for N,F-TiO_2 and N-TiO_2-NT.

Analysis of the N 1s region from the XPS spectra (Figure 2) shows a wide peak centered with binding energy at 399.5 eV indicating that nitrogen is incorporated at interstitial sites [42]. The presence of fluoride in the N,F-TiO_2, using this preparation method, was previously confirmed by a peak centered at 688 eV [42]; however, there was no significant peak in the analysis of the materials fabricated in this work. The concentration of nitrogen in the N,F-TiO_2 was determined to be 1.5 atom%. The concentration of nitrogen in the N-TiO_2-NT was 0.5 atom%. Asahi et al. reported an optimal N concentration of 0.25 atom% with samples prepared by sputter deposition [23]. Delegan et al. reported an optimal N doping at 0.4 atom% for band gap narrowing with no further effect at higher loadings [50]. Therefore, the N doping level of 0.5 atom% in the N-TiO_2-NTs is similar to that reported for optimal doping. According to the literature, peaks in the range of 400–402 eV may be due to molecularly chemisorbed N_2 [38,45] whilst NO_2 and NO_3 have been reported to occur at higher energies 403–408 eV [16,51,52]. Other groups have observed the N 1s peak alone at 399–400 eV and assigned it as NO [38]. However, the absence of peaks beyond 407 eV indicates the absence of chemisorbed NO or NO_2 species [53]. In general, the N 1s peak at ca. 400 eV is typically assigned to the interstitial nitrogen dopant while the peak at ca. 396 eV is associated to the substitutional nitrogen dopant. However, there is no consensus about which kind of doping is more efficient for giving rise to visible light activity [34].

To study the photoelectrochemical properties, linear sweep voltammetry (LSV) was carried out under chopped UV-Vis irradiation. The supporting electrolyte was a $\frac{1}{4}$ strength Ringers solution. The current-potential behavior of the different materials is shown in Figure 3.

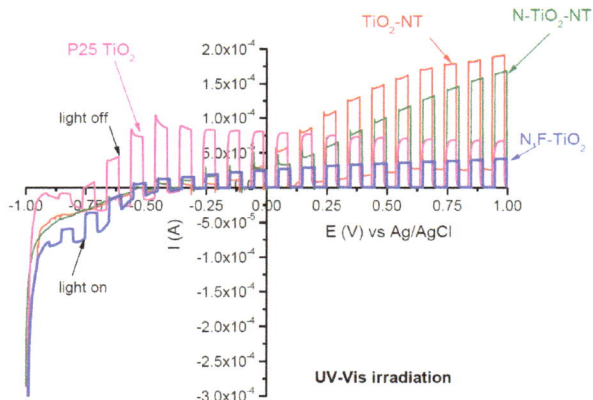

Figure 3. Linear sweep voltammograms for nanotube (TiO$_2$-NT and N-TiO$_2$-NT) and nanoparticle (P25 and N,F-TiO$_2$) titania electrodes in $\frac{1}{4}$ strength solution under chopped (10 s light on/ off) UV-Vis irradiation. The potential was swept from negative to positive with a sweep rate of 5 mV s^{-1}.

All samples showed negligible dark anodic current under UV-Vis irradiation, which is typical of n-type semiconductor electrodes in contact with an electrolyte. It has widely been reported that electron transport, in particulate electrodes, occurs by diffusion since the small size of the particles does not allow the formation of the depletion layer across the particle [54–56]. Moreover, as a consequence of inter-particle boundaries due to their porous structure, many localized states may exist leading to charge carrier recombination. Therefore, the photocurrent is not expected to be dependent on the applied potential. However, the Fermi level of the supporting electrode is potential dependent [57,58] and the application of a positive potential leads to a decrease in the Fermi level of the supporting electrode and consequently, to the increase in the efficiency of electron drift from the irradiated nanoparticulate film [15].

It must be noted that both nanotube electrodes (undoped and N-doped) show the highest values of photocurrent under UV-Vis illumination in comparison with both nanoparticle electrodes (P25 and N-doped sol-gel) at the same anodic potential. This is in agreement with other workers [11,12,15] who reported a higher photocurrent of titania nanotubes as compared to P25; and others who observed a higher photocurrent of titania nanotubes compared to titania sol-gels or films [46,59–61]. However, it should be noted that the onset potential for anodic photocurrent is more positive with the NT electrodes as compared to the P25 or N,F-TiO$_2$ electrodes.

Several authors have pointed out that despite the nanotubular morphology, these nanostructured materials give a lower surface area compared to that of mesoporous nanoparticle thick films (e.g., P25), yet the NT electrodes give a higher photocurrent response [11,12,15]. In nanoparticulate films the particles are partially sintered and electron diffusion to the supporting electrode follows a convoluted pathway via particle to particle across grain boundaries. Also, the electrolyte penetrates into the mesoporous film and conduction band electrons may be lost to surface recombination reactions resulting in relatively low photocurrent, as compared to compact oxide electrodes. The aligned nanotube morphology provides a more direct pathway for the electrons to transfer to the supporting electrode through the continuous TiO$_2$ tube wall. Also, the NTs are grown from the titanium substrate which favors good electron transfer from the NTs to the metal support [59]. It is also worth noting that there was only a small difference in photocurrent response for the doped and undoped nanotube electrodes.

The current-time behavior of all the electrodes at a fixed potential of +1.0 V is given in Figure 4. The same trends in photocurrent intensity were observed at fixed potential as were observed in the LSV. Both the P25 and N,F-TiO$_2$ samples gave a similar photocurrent response. The NT samples gave the highest photocurrent values. The third cycle of amperometry is represented for each electrode. No decrease in photocurrent is observed between the first (data not shown) and the third cycle in any sample, therefore there is no observed decrease in the photocurrent activity in repeat runs.

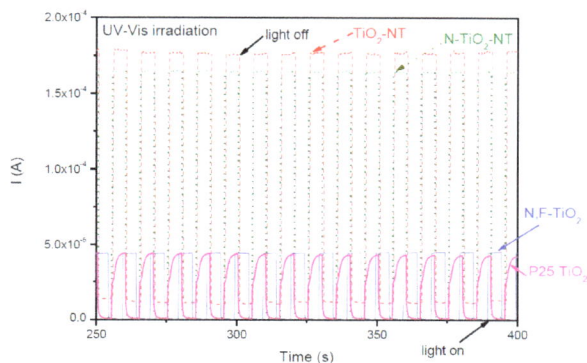

Figure 4. Current-time behavior of nanotube (TiO$_2$-NT and N-TiO$_2$-NT) and nanoparticle (P25 and N,F-TiO$_2$) electrodes in $\frac{1}{4}$ strength solution under chopped UV-Vis irradiation. UV-Vis exposure time: 5 s. Potential bias: +1.0 V. 3rd cycle of current measurement for each electrode.

Linear sweep voltammetry was also carried out under visible only irradiation (Figure 5). The photocurrent response under only visible light is two orders of magnitude lower than that observed under UV-Vis irradiation (Figure 4) which agrees with previous results [62]. This would indicate that the visible light activity for water oxidation is two orders of magnitude lower than that under band gap excitation. Previously, it has been reported that the hydroxyl radical is the main species responsible for *E. coli* inactivation with UV excited nonmetal doped TiO$_2$ [63]. If the anodic photocurrent is a measure of water oxidation, and formation of the hydroxyl radical, therefore, one might expect a rate of two orders of magnitude lower under visible as compared to UV (not considering direct photolytic inactivation) and as such, no inactivation would be observed under visible irradiation in the timescale of the experiments. The nitrogen doped titania nanotube sample (N-TiO$_2$-NT) showed the highest photocurrent (I_{ph}) under visible irradiation. The N,F-TiO$_2$ electrode also showed a small visible photocurrent response. Interestingly, the P25 shows a small visible photocurrent response but it is noted that P25 is an 80:20 mixture of anatase and rutile, with the rutile band gap at 3.0 eV, just into the visible region (413 nm).

The N-TiO$_2$-NT photoresponse is also higher than that observed for the other doped titania sol gel sample (N,F-TiO$_2$). Again, it suggests that nanotubular structure offers a less resistant path for the electrons to reach the supporting electrode, and consequently, it leads to a more effective separation of the electron—hole pairs. In addition, if both undoped and doped nanoparticulate samples are compared, their photocurrent response is similar.

It must be highlighted that the photocurrent response under visible irradiation is only a fraction of the band gap response observed under UV-Vis irradiation for the N-TiO$_2$-NT electrode (and all electrodes). This correlates with previous work investigating the photoelectrochemical response of N,F-TiO$_2$ [62] where it was found that the visible light photocurrent was only a fraction of the band gap response and the photocurrent action spectrum did not correlate to the optical absorbance spectrum for the material; however, the open circuit photopotential gave a better correlation to the optical spectra.

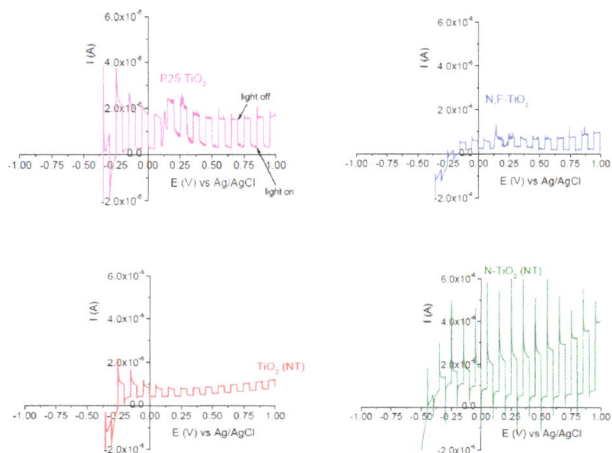

Figure 5. Linear sweep voltammograms for nanoparticle (P25 and N,F-TiO$_2$) and nanotube (TiO$_2$-NT and N-TiO$_2$-NT) electrodes in $\frac{1}{4}$ strength solution under chopped visible irradiation (10 s light on/off). The potential was swept positive at a sweep rate of 5 mV s^{-1}.

The current-time behavior of all the electrodes, shown in Figure 6, is typical of an n-type semiconductor under chopped illumination. An initial anodic photocurrent spike ($I_{\text{ph in}}$) is instantly observed under exposure to illumination which corresponds to separation of photogenerated e$^-$-h$^+$ pairs. Then, photocurrent decay is observed until a steady-state photocurrent ($I_{\text{ph st}}$) is reached, due to surface charge carrier recombination and/or accumulation of holes at the surface. A cathodic photocurrent spike (i^-_{in}) is observed when the light is turned off, representing back reaction of conduction band electrons with the hole trap sites at the surface [33,64]. In this case, a higher recombination might be suggested by the observed decay in the initial I_{ph} produced in the instant of illumination for N-TiO$_2$-NT under visible irradiation as compared to UV-Vis irradiation (Figure 4).

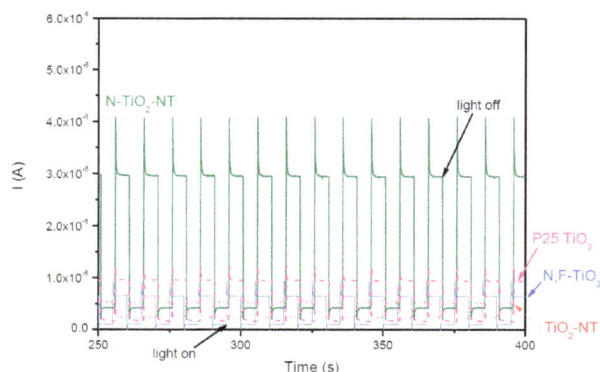

Figure 6. Current-time behavior of nanotube (TiO$_2$-NT and N-TiO$_2$-NT) and nanoparticle (P25 and N,F-TiO$_2$) electrodes in $\frac{1}{4}$ strength solution under chopped visible irradiation (light on/off 10 s) at fixed potential (+1.0 V). Data shown is from the third cycle of current measurement for each electrode.

2.2. Electrochemically Assisted Photocatalytic Disinfection

Figure 7 shows *E. coli* inactivation obtained with the different photoelectrodes: P25 TiO$_2$, N,F-TiO$_2$, TiO$_2$-NT, and N-TiO$_2$-NT. Firstly, different control experiments were carried out. The chemical composition of the supporting electrolyte, electrochemical oxidation under an electric potential bias of +1.0 V, and photolysis did not lead to any significant decrease in the viable concentration of *E. coli* after 4 hours of treatment for any of the electrodes analyzed.

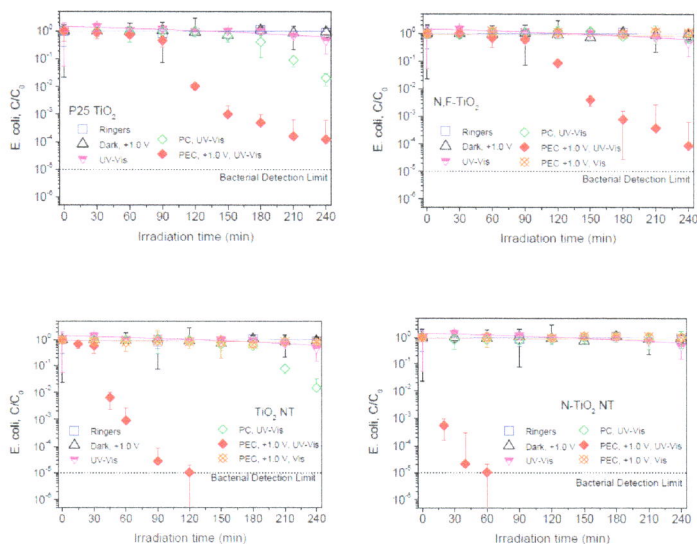

Figure 7. Photocatalytic and electrochemically assisted photocatalytic inactivation of *E. coli* for the different electrodes in $\frac{1}{4}$ strength solution under UV-Vis irradiation, and visible only irradiation for N,F-TiO$_2$, TiO$_2$-NT, and N-TiO$_2$-NT electrodes. The applied potential was +1.0 V for electrochemically assisted photocatalysis. PC = Photocatalysis, Ringers = dark control, Dark, +1.0 V = electrochemical control, UV-Vis = light control. All the experiments have been performed in triplicate using the same electrode.

The extent of bacterial inactivation for photocatalysis (electrodes at open circuit) was small over the timescale of the experiments. This is to be expected as the irradiated surface area to volume ratio (Incident Illuminated Density, ICD) for this reactor system is 0.066 cm^2 cm^{-3}. In a previous study, complete inactivation was observed at around 100 min for *E. coli* with photocatalysis on P25 TiO$_2$ films in a stirred tank reactor with an ICD of 0.28 cm^2 cm^{-3} [65]. However, the photocatalytic bacterial inactivation is notably improved by the application of a positive electrical potential for all the TiO$_2$ electrodes under UV-Vis irradiation. Dunlop et al. [6] reported that the rate of inactivation of *E. coli* was increased using an applied potential of +1.0 V (SCE) as compared to photocatalysis alone for particulate electrodes (Degussa and Aldrich (anatase)) as well as an increase in bacterial inactivation with application of a positive bias. Also, Dunlop et al. [66] reported that the application of an external electrical bias significantly increased the rate of photocatalytic disinfection of *C. perfringens* spores on particulate electrodes (P25). Thus, it suggests that the increase in inactivation efficiency for nanoparticulate electrodes is probably due an improvement of mass transport of the negatively charged bacteria to the positively charged photoanode by electromigration [8–10].

Comparing the photocurrent at fixed potential (+1.0 V) under UV-Vis irradiation (Figure 4) it is clear that the NT electrodes give a much better response i.e., 175 µA for the TiO$_2$-NT and 165 µA for the N-TiO$_2$-NT electrodes, as compared to less than 50 µA for the NF-TiO$_2$ and P25 TiO$_2$ electrodes. Comparing the photocurrent at fixed potential (+1.0 V) under visible only irradiation (Figure 6), the N-TiO$_2$-NT electrode gives the highest photocurrent (3 µA) as compared to P25 TiO$_2$ (1 µA), N,F-TiO$_2$ (0.6 µA) and TiO$_2$-NT (0.5 µA). The order of photocurrent magnitude correlates with the observed electrochemically assisted photocatalytic disinfection rates under UV-Vis irradiation (Figure 7) in that both NT electrodes reach a 5 log inactivation within the timescale of the experiments (120 min for the TiO$_2$-NT and 60 min for the N-TiO$_2$-NT), but neither the N,F-TiO$_2$, nor the P25 TiO$_2$ electrodes achieve a 5 log kill within 240 min. None of the electrodes gave any significant inactivation of bacteria under only visible irradiation. It is noted that the visible light photocurrent is two orders of magnitude lower than the photocurrent observed under UV-Vis irradiation. The correlation with photocurrent and disinfection rate fails when comparing the N-TiO$_2$-NT and the TiO$_2$-NT under UV-Vis irradiation. The time taken to reach a 5 log inactivation for the N-TiO$_2$-NT is only 30 min while it takes 120 min for the TiO$_2$-NT under UV-Vis irradiation. Both NT electrodes give similar UV-Vis photocurrent at +1.0 V, but the N-TiO$_2$-NT electrode gives a greater visible only photocurrent, yet this is only 2% of the UV-Vis photocurrent. If one considers that the main species responsible for photocatalytic inactivation of bacteria under UV irradiation on TiO$_2$ is the hydroxyl radical [63], then one would expect the time taken for inactivation on both NT electrodes to be similar, based on photocurrent. It is not clear why the N-TiO$_2$-NT electrode should give a much faster rate of inactivation, but this must be related to the mechanism of disinfection. Differences in the attachment/association of bacteria to the surface of the catalyst should not be excluded as the materials may have different properties concerning surface interactions. Additionally, Hamilton et al. [62] previously reported that the mid-gap state introduced by N-doping of titania could yield some visible photocatalytic activity by the visible light excitation of electrons from the mid-gap state to the conduction band leading to the reduction of molecular oxygen to superoxide and hydrogen peroxide, and/or the oxidation of superoxide by visible light generated holes in the mid-gap state to yield singlet oxygen. The additional ROS generated under visible excitation, in combination with the hydroxyl radicals generated by UV excitation, yields a much faster inactivation rate as compared to the UV photocatalytic mechanism alone. Previous work reported on the activity of N-TiO$_2$ thin-films produced using atmospheric-pressure chemical vapor deposition (APCVD) with tert-butylamine as the nitrogen source [67]. They attributed the enhancement of the UV activity of the N-TiO$_2$ to surface N species which, were not stable, leading to a decrease in activity over time. However, in this work the disinfection experiments were undertaken in triplicate with no decrease in disinfection efficiency observed between the experiments. Furthermore, no decrease in the photocurrent response was observed in repeat measurements, suggesting that these materials are relatively stable.

Further research is required to elucidate the difference in the electrochemically assisted photocatalytic disinfection mechanism on N-TiO$_2$-NT and TiO$_2$-NT. Nevertheless, the N-TiO$_2$-NT electrode gives a much faster rate of disinfection under UV-Vis irradiation than any of the other materials studied.

3. Experimental Section

3.1. TiO$_2$ Electrode Preparation

Ti foil (Sigma-Aldrich, Irvine, UK, 0.127 mm, 99.7%) was cut into 2.5 × 2.0 cm^2 pieces and cleaned in an ultrasonic bath in methanol for 15 min. An area of ca. 1 cm^2 was exposed for coating or iodization.

Two particulate photoelectrodes were prepared by: (i) electrophoretic coating of a TiO$_2$ suspension. A Ti foil was submerged in a 1% P25 TiO$_2$:CH$_3$OH suspension of P25 TiO$_2$ (P25 Evonik industries, Rhine-Main Germany, CH$_3$OH, 99.9% Sigma Aldrich, Scotland, UK) and an electric potential bias of +20 V vs. a Pt paddle (Windsor scientific, Berkshire, UK) used as the anode was applied for 15 s [15]. The samples were annealed at 450 °C for 1 h in air with a heating rate of 2 °C min^{-1}; (ii) dip-coating of N and F co-doped TiO$_2$ suspensions (N,F-TiO$_2$) onto a Ti foil. This catalyst sol was prepared according to Pelaez et al. [42], using titanium (IV) isopropoxide (TTIP, 97%, Aldrich, Irvine, UK) as the titania precursor and anhydrous ethylenediamine (EDA, Fisher, Loughborough, UK) as the nitrogen source. The chosen concentration of the non-ionic fluorosurfactant in molar ratio corresponded to five. The sol was deposited onto the Ti substrate at a withdrawal speed rate of 4.9 mm s^{-1}. The samples were annealed at 400 °C for 30 min in air. The temperature was increased at a ramp rate of 1 °C min^{-1} and cooled down at 4 °C min^{-1}.

Titania nanotubes were grown by the electrochemical oxidation of Ti in a fluoride-based electrolyte according to Dale et al. [15]. Ti foil pieces were anodized in a custom designed two electrode electrochemical cell made of *Perspex* with a 100 mL capacity. Each Ti foil piece was fixed in the cell with copper back-plate electrical contact. An O-ring was used to seal the foil in the cell, leaving an area of ca. 1 cm^2 exposed to the electrolyte (1 M Na_2SO_4 + 0.12 M NaF in aqueous solution). A platinum electrode served as the counter electrode. The distance between the two electrodes was 3 cm. Conducting wires soldered to the copper back-plates allowed connection to a power supply. The Ti foil was treated at constant potential of +25 V for 4 h using the power supply. After anodization, the Ti foil was ultrasonicated in methanol for 5 min to wash off any remnant salts from electrolyte. Undoped titania samples were annealed in air at 450 °C at the ramp rate of 2 C min^{-1} for 1 h (TiO_2-NT) and those doped with nitrogen (N-TiO_2-NT) were annealed at 450 °C at the same ramp rate but in NH_3 (BOC gas and gear, BOC, Belfast, UK). All the samples were made into electrodes by cleaning part of the Ti foil to attach a copper wire using silver loaded epoxy. Afterwards, the contact, wire and titanium were masked with negative photoresist (KPR, Cassio chemicals, Hertfordshire, UK) which was cured under UV-A exposure for 10 min. The electrodes were finally sealed with epoxy resin (Araldite) leaving an area of TiO_2 of 1 cm^2 exposed.

3.2. Materials Characterisation

Scanning electron microscopy (SEM, FEI Quanta 200, FEI Eindhoven, The Netherlands) was used to confirm the formation of nanotubes for the TiO_2-NT and N-TiO_2-NT electrodes. Chemical composition analysis was undertaken by X-ray photoelectron spectroscopy (XPS) with a Kratos Axis Ultra employing an Al Kα source (Kratos Analytical, Eppstein, Germany). The binding energies of the samples were calibrated relative to the C 1 s peak at 284 eV.

3.3. Photoelectrochemical Analysis

Linear sweep voltammetry (LSV) was used to determine the current-potential characteristics of the electrodes under chopped irradiation. Photocurrent measurements were recorded at a scan rate of 5 mV·s^{-1} sweeping from −1.0 to +1.0 V under irradiation. Photocurrent-time measurements at a fixed potential (+1.0 V) were also carried out. A shutter (Unblitz, WMM-T1, Vincent associates, Rochester, NY, USA) was also used to chop the light. The titania samples were used as the working electrodes (WE), a Pt mesh paddle (5.9 cm^2) was used as the counter electrode (CE) and an Ag/AgCl electrode was used as the reference electrode (RE) (Figure 8). An electrochemical workstation with PC control (Autolab PGStat 30, Metrohm UK, Runcorn, UK) provided potentiostatic control. Quarter strength Ringers solution (Oxoid, Fisher, Hampshire, UK) was used as electrolyte, consisting of 2.25×10^3 mg dm^{-3} NaCl, 1.05×10^2 mg dm^{-3} KCl, 1.2×10^2 mg dm^{-3} $CaCl_2$, and 50 mg dm^{-3} $NaHCO_3$ in deionized water [6].

3.4. Photoreactor Configuration

The photoreactor used was a 35 cm^3 Pyrex water-jacketed reactor (Figure 8). The working volume used corresponded to 15 mL. Air was bubbled in through a Pasteur pipette suspended in the working solution. Distilled water was pumped from a thermostated water tank and circulated round the reactor water jacket. The suspension was magnetically stirred to ensure effective mixing and the temperature was maintained at 25 °C throughout the experiments. The suspension was irradiated with UV-Vis using a 450 W xenon lamp (Horiba Jobin Yvon, Horiba UK LTD, Northampton, UK) full spectrum as solar simulated irradiation placed 4 cm away from the reactor. The incident light intensity reaching the suspension was determined to be 197 W m^{-2} (measured between 200–800 nm using a spectral radiometer, Gemini 180, Horiba Jobin Yvon, Horiba UK LTD, Northampton, UK). For visible light only, a UV filter ($\lambda > 420$ nm) (UQG Optics Ltd., Cambridge, UK) was placed between the reactor and the irradiation source. The incident light intensity corresponded to 150 W m^{-2}.

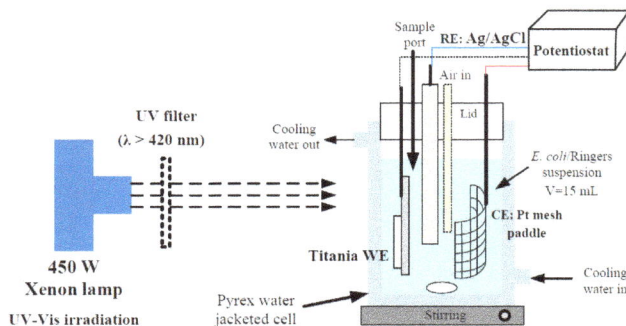

Figure 8. Photoelectrochemical cell.

The same reactor was used for the electrochemically assisted photocatalytic experiments with the photoanode under a fixed potential of +1.0 V vs. Ag/AgCl. The photocurrent was recorded against time for each experiment. Quarter strength Ringers solution was used as the working suspension.

3.5. Bacterial Growth and Detection

Escherichia coli K12 was used as the model microorganism for the inactivation experiments. Fresh liquid cultures were prepared by inoculation in a Luria-Bertani nutrient medium (Sigma-Aldrich, Irvine, UK) and incubation at 37 °C for 24 h under constant stirring on a rotary shaker. The cell density of the culture was checked prior to the experiment by measuring its absorbance at 520 nm as prediction of the initial bacterial concentration. Cells were harvested by centrifugation (at 4000 rpm for 10 min) washing twice the bacteria with sterile $\frac{1}{4}$ strength Ringers solution before resuspension in the same solution. Finally, a volume of the bacterial suspension corresponding to 0.15 mL was diluted in 15 mL of sterile $\frac{1}{4}$ strength Ringers solution to start the experiments at an initial concentration approximately 10^6 CFU mL^{-1}. The bacterial inactivation was followed by evaluating the concentration of viable bacteria in the samples taken throughout the reaction. The quantification was carried out following a standard serial dilution procedure. Serial dilutions in sterile $\frac{1}{4}$ strength Ringers solution were carried out prior to spotting 10 μL drops of each decimal dilution 4 times on LB nutrient agar plates. At longer irradiation times (low bacterial concentrations), 2 drops of 100 μL were obtained from the working suspension of the undiluted suspension were plated directly onto LB agar to reduce the limit of bacterial detection to 10 CFU mL^{-1}. All LB agar plates were incubated at 37 °C for 24 h, and the number of colonies were manually counted. Key experiments were also repeated three times to test the reproducibility of the disinfection results. Full details of components of the LB medium are available elsewhere [6].

4. Conclusions

Aligned self-organized titania nanotubes were grown on titanium foil by anodic oxidation in fluoride containing electrolyte (TiO$_2$-NT). The nanotube samples were doped with nitrogen by annealing in ammonia. Nanoparticulate films of N,F-TiO$_2$ and P25 were prepared for comparison purposes.

This study demonstrates that electrochemically assisted photocatalysis is much more effective than photocatalysis alone (electrodes at open circuit) for the inactivation of *E. coli* under solar simulated

UV-Vis irradiation for all of the different electrodes tested. No significant inactivation of bacteria was observed under only visible light irradiation of any of the materials tested, with or without applied potential. The photocurrent response of the electrodes at fixed potential under UV-Vis irradiation correlated well with the disinfection rates observed under the same conditions, in that the NT electrodes gave a much higher photocurrent response than the nanoparticulate electrodes and a much better disinfection efficiency. The most interesting finding of this work was that the N-TiO$_2$-NT electrode gave a much better disinfection rate at +1.0 V under UV-Vis irradiation as compared to any of the other materials studied. The time taken for a 5 log kill on the N-TiO$_2$-NT was half that observed for the TiO$_2$-NT electrode. However, the UV-Vis photocurrent was slightly higher for the TiO$_2$-NT electrode at a fixed potential. This suggests that the N doping generates mid gap states which yield visible light activity through the formation of superoxide and singlet oxygen. Although, these ROS are less effective for the inactivation of bacteria than hydroxyl radical (produced under UV excitation), their presence should not be dismissed. Also, the visible light excitation of mid gap excitation does not generate significant photocurrent (only 2% of the UV-Vis response). The higher UV-Vis disinfection efficiency of the N-TiO$_2$-NT electrodes must be due to differences in the disinfection mechanism and further research is required to elucidate these differences. Nevertheless, N doping of titania nanotubes gives electrodes with excellent UV-Vis activity for the electrochemically assisted disinfection of water.

Acknowledgments: The authors gratefully acknowledge the financial support of the Spanish Ministry of Economy and Competitiveness (MINECO) through the project WATER4FOOD (CTQ2014-54563-C3-1-R) and Comunidad de Madrid through the program REMTAVARES (S2013/MAE-2716). Also the Department of Employment and Learning Northern Ireland and the National Science Foundation USA for funding under the US-Ireland collaborative partnership initiative (NSF-CBET-1033317). D.D.D. also acknowledges support from the University of Cincinnati through a UNESCO co-Chair Professor position on "Water Access and Sustainability".

Author Contributions: C. Pablos and J.A. Byrne designed the experiments and C. Pablos carried out the experiments; C. Pablos and J.A. Byrne were the lead authors of the paper; J.W.J Hamilton, J. Marugán, and R. van Grieken revised and wrote a part of the manuscript; J.W.J Hamilton undertook characterization of the electrodes; P.S.M. Dunlop designed the disinfection experiments; D.D. Dionysiou was responsible for providing the protocol to prepare N,F-TiO$_2$ photoanodes and wrote part of the manuscript. All authors read, revised and approved the final manuscript.

Conflicts of Interest: The authors declare no conflict of interest.

References

1. Richardson, S.D. Disinfection By-Products and other emerging contaminants in drinking water. *Trends Anal. Chem.* **2003**, *20*, 666–684. [CrossRef]

2. Matsunaga, T.; Tomoda, R.; Nakajima, T.; Wake, H. Photoelectrochemical sterilization of microbial cells by semiconductor powders. *FEMS Microbiol. Lett.* **1985**, *29*, 211–214. [CrossRef]

3. Malato, S.; Fernández-Ibáñez, P.; Maldonado, M.I.; Blanco, J.; Gernjak, W. Decontamination and disinfection of water by solar photocatalysis: Recent overview and trends. *Catal. Today* **2009**, *147*, 1–60. [CrossRef]

4. Byrne, J.A.; Dunlop, P.S.M.; Hamilton, J.W.J.; Fernández-Ibáñez, P.; Polo-López, I.; Sharma, P.K.; Vennard, A.S.M. A Review of Heterogeneous Photocatalysis for Water and Surface Disinfection. *Molecules* **2015**, *20*, 5574–5615. [CrossRef] [PubMed]

5. Vinodgopal, K.; Stafford, U.; Gray, K.A.; Kamat, P.V. Electrochemically Assisted Photolysis. II. The Role of Oxygen and Reaction Intermediates in the Degradation of 4-Chlorophenol on Immobilized TiO$_2$ Particulate Films. *J. Phys. Chem.* **1994**, *98*, 6797–6803. [CrossRef]

6. Dunlop, P.S.M.; Byrne, J.A.; Manga, N.; Eggins, B.R. The photocatalytic removal of bacterial pollutants from drinking water. *J. Photochem. Photobiol. A Chem.* **2002**, *148*, 355–563. [CrossRef]

7. Christensen, P.A.; Curtis, T.P.; Egerton, T.A.; Kosa, S.A.M.; Tinlin, J.R. Photoelectrocatalytic and photocatalytic disinfection of *E. coli* suspensions by titanium dioxide. *Appl. Catal. B Environ.* **2003**, *41*, 371–386. [CrossRef]

8. Butterfield, I.M.; Christensen, P.A.; Curtis, T.P.; Gunlazuardi, J. Water disinfection using an immobilised titanium dioxide film in a photochemical reactor with electric field enhancement. *Water Res.* **1997**, *31*, 675–677. [CrossRef]

9. Baram, N.; Starosvetsky, D.; Starosvetsky, J.; Epshtein, M.; Armon, R.; Ein-Eli, Y. Enhanced inactivation of *E. coli* bacteria using immobilized porous TiO_2 photoelectrocatalysis. *Electrochim. Acta* **2009**, *54*, 3381–3386. [CrossRef]

10. Cho, M.; Cates, E.L.; Kim, J.H. Inactivation and surface interactions of MS-2 bacteriophage in a TiO_2 photoelectrocatalytic reactor. *Water Res.* **2011**, *45*, 2104–2110. [CrossRef] [PubMed]

11. Macak, J.M.; Zlamal, M.; Krysa, J.; Schmuki, P. Self-organized TiO_2 nanotube layers as highly efficient photocatalysts. *Small* **2007**, *3*, 300–304. [CrossRef] [PubMed]

12. Zlamal, M.; Macak, J.M.; Schmuki, P.; Krýsa, J. Electrochemically assisted photocatalysis on self-organized TiO_2 nanotubes. *Electrochem. Commun.* **2007**, *9*, 2822–2826.

13. Yuan, B.; Wang, Y.; Bian, H.; Shen, T.; Wu, Y.; Chen, Z. Nitrogen doped TiO_2 nanotube arrays with high photoelectrochemical activity for photocatalytic applications. *Appl. Surf. Sci.* **2013**, *280*, 523–529. [CrossRef]

14. Zwilling, V.; Aucouturier, M.; Darque-Ceretti, E. Anodic oxidation of titanium and TA6V alloy in chromic media. *Electrochim. Acta* **1999**, *45*, 921–929. [CrossRef]

15. Dale, G.R.; Hamilton, J.W.J.; Dunlop, P.S.M.; Byrne, J.A. Electrochemically assisted photocatalysis on anodic titania nanotubes. *Curr. Top. Electrochem.* **2009**, *14*, 89–97.

16. Mazierski, P.; Nischk, M.; Gołkowska, M.; Lisowski, W.; Gazda, M.; Winiarski, M.J.; Klimczuk, T.; Zaleska-Medynska, A. Photocatalytic activity of nitrogen doped TiO_2 nanotubes prepared by anodic oxidation: The effect of applied voltage, anodization time and amount of nitrogen dopant. *Appl. Catal. B Environ.* **2016**, *196*, 77–88. [CrossRef]

17. Kang, Q.; Lu, Q.Z.; Liu, S.H.; Yang, L.X.; Wen, L.F.; Luo, S.L.; Cai, Q.Y. A ternary hybrid CdS/Pt–TiO_2 nanotube structure for photoelectrocatalytic bactericidal effects on *Escherichia coli*. *Biomaterials* **2010**, *31*, 3317–3326. [CrossRef] [PubMed]

18. Hayden, S.C.; Allam, N.K.; El-Sayed, M.A. TiO_2 nanotube/CdS hybrid electrodes: Extraordinary enhancement in the inactivation of Escherichia coli. *J. Am. Chem. Soc.* **2010**, *132*, 14406–14408. [CrossRef] [PubMed]

19. Zhang, F.-J.; Chen, M.-L.; Oh, W.-C. Photoelectrocatalytic properties and bactericidal activities of silver-treated carbon nanotube/titania composites. *Compos. Sci. Technol.* **2011**, *71*, 658–665.

20. Brugnera, M.F.; Miyata, M.; Zocolo, G.J.; Queico Fujimura Leite, C.; Valnice Boldrin Zanoni, M. Inactivation and disposal of By-Products from Mycobacterium smegmatis by photoelectrocatalytic oxidation using Ti/TiO_2-Ag nanotube electrodes. *Electrochim. Acta* **2012**, *85*, 33–41. [CrossRef]

21. Liu, Y.; Li, J.; Qiu, X.; Burda, C. Bactericidal activity of nitrogen-doped metal oxide nanocatalysts and the influence of bacterial Extracellular Polymeric Substances (EPS). *J. Photochem. Photobiol. A Chem.* **2007**, *190*, 94–100. [CrossRef]

22. Cheng, Y.H.; Subramaniam, V.P.; Gong, D.; Tang, Y.; Highfield, J.; Pehkonen, S.O.; Pichat, P.; Schreyer, M.K.; Chen, Z. Nitrogen-sensitized dual phase titanate/titania for visible-light driven phenol degradation. *J. Solid State Chem.* **2012**, *196*, 518–527. [CrossRef]

23. Asahi, R.; Morikawa, T.; Ohwaki, T.; Aoki, K.; Taga, Y. Visible-light photocatalysis in nitrogen doped titanium oxides. *Science* **2001**, *293*, 269–271. [CrossRef] [PubMed]

24. Serpone, N. Is the band gap of pristine TiO_2 Narrowed by Anion- and Cation-Doping of Titanium Dioxide in Second-Generation Photocatalysts? *J. Phys. Chem. B* **2006**, *110*, 24287–24293. [CrossRef] [PubMed]

25. Wu, H.-C.; Lin, Y.-S.; Lin, S.-W. Mechanisms of visible light photocatalysis in N-doped anatase TiO_2 with oxygen vacancies from GGA + U calculations. *Int. J. Photoenergy* **2013**, *2013*, 1–7.

26. Irie, H.; Watanabe, Y.; Hashimoto, K. Nitrogen-Concentration Dependence on Photocatalytic Activity of $TiO_{2-x}N_x$ Powders. *J. Phys. Chem. B* **2003**, *107*, 5483–5486. [CrossRef]

27. Nakamura, R.; Tanaka, T.; Nakato, Y. Mechanism for Visible Light Responses in Anodic Photocurrents at N-doped TiO_2 Film Electrodes. *J. Phys. Chem. B* **2004**, *108*, 10617–10620. [CrossRef]

28. Batzill, M.; Morales, E.H.; Diebold, U. Influence of nitrogen doping on the defect formation and surface properties of TiO_2 rutile and anatase. *Phys. Rev. Lett.* **2006**, *96*, 1–4. [CrossRef] [PubMed]

29. Emeline, A.V.; Kuznetsov, V.N.; Rybchuk, V.K.; Serpone, N. The case of N-doped TiO_2s—Properties and some fundamental issues. *Int. J. Photoenergy* **2008**, *2008*, 1–19. [CrossRef]

30. Wang, J.; Tafen, D.N.; Lewis, J.P.; Hong, Z.; Manivannan, A.; Zhi, M.; Li, M.; Wu, N. Origin of photocatalytic activity of nitrogen-doped TiO_2 nanobelts. *J. Am. Chem. Soc.* **2009**, *131*, 12290–12297. [CrossRef] [PubMed]

31. Etacheri, V.M.; Seery, K.; Hinder, S.J.; Pillai, S.C. Highly visible light active $TiO_{2-x}N_x$ heterojunction photocatalysts. *Chem. Mater.* **2010**, *22*, 3843–3853. [CrossRef]

32. Tafalla, D.; Salvador, P.; Benito, R.M. Kinetic Approach to the Photocurrent Transients in Water Photoelectrolysis at n-TiO_2 Electrodes II. Analysis of the Photocurrent-Time Dependence. *J. Electrochem. Soc.* **1990**, *137*, 1810–1815. [CrossRef]

33. Diwald, O.; Thompson, T.L.; Zubkov, T.; Walck, S.D.; Yates, J.T. Photochemical activity of nitrogen-doped rutile TiO_2(110) in visible light. *J. Phys. Chem. B* **2004**, *108*, 6004–6008. [CrossRef]

34. Rumaiz, A.K.; Woicik, J.C.; Cockayne, E.; Lin, H.Y.; Jaffari, G.H.; Shah, S.I. Oxygen vacancies in N doped anatase TiO_2: Experiment and first-principles calculations. *Appl. Phys. Lett.* **2009**, *95*, 1–3. [CrossRef]

35. Hu, L.; Wang, J.; Zhang, J.; Zhang, Q.; Liu, Z. An N-Doped anatase/rutile TiO_2 hybrid from low-temperature direct nitridization: Enhanced photoactivity under UV-/Visible-light. *RSC Adv.* **2014**, *4*, 420–427. [CrossRef]

36. Yang, G.; Wang, T.; Yang, B.; Yan, Z.; Ding, S.; Xiao, T. Enhanced visible-light activity of F-N Co-doped TiO_2 nanocrystals via nonmetal impurity, Ti^{3+} ions and oxygen vacancies. *Appl. Surf. Sci.* **2013**, *287*, 135–142. [CrossRef]

37. Nakamura, I.; Negishi, N.; Kutsuna, S.; Ihara, T.; Sugihara, S.; Takeuchi, K. Role of oxygen vacancy in the plasma-treated TiO_2 photocatalyst with visible light Activity for NO removal. *J. Mol. Catal. A Chem.* **2000**, *161*, 205–212. [CrossRef]

38. Wang, Y.; Feng, C.; Zhang, M.; Yang, J.; Zhang, Z. Enhanced visible light photocatalytic activity of N-doped TiO_2 in relation to single-electron-trapped oxygen vacancy and doped-nitrogen. *Appl. Catal. B Environ.* **2010**, *100*, 84–90. [CrossRef]

39. Di Valentin, C.; Pacchioni, G.; Selloni, A. Reduced and n-Type Doped TiO_2: Nature of Ti^{3+} Species. *J. Phys. Chem. C* **2009**, *113*, 20543–20552. [CrossRef]

40. Di Valentin, C.; Finazzi, E.; Pacchioni, G.; Selloni, A.; Livraghi, S.; Czoska, A.M.; Paganini, M.C.; Giamello, E. Density Functional Theory and Electron Paramagnetic Resonance Study on the Effect of N-F Codoping of TiO_2. *Chem. Mater.* **2008**, *20*, 3706–3714. [CrossRef]

41. Liou, J.-W.; Chang, H.-H. Bactericidal effects and mechanisms of visible light-responsive titanium dioxide photocatalysts on pathogenic bacteria. *Arch. Immunol. Ther. Exp.* **2012**, *60*, 267–275. [CrossRef] [PubMed]

42. Pelaez, M.; Falaras, P.; Likodimos, V.; Kontos, A.G.; de la Cruz, A.A.; O'shea, K.; Dionysiou, D. Synthesis, structural characterization and evaluation of sol-gel-based NF-TiO_2 films with visible light-photoactivation for the removal of microcystin-LR. *Appl. Catal. B Environ.* **2010**, *99*, 378–387. [CrossRef]

43. Mor, G.K.; Varghese, O.K.; Paulose, M.; Shankar, K.; Grimes, C.A. A Review on highly ordered, vertically oriented TiO_2 nanotube arrays: Fabrication, material properties, and solar energy applications. *Sol. Energ. Mater. Sol. C.* **2006**, *90*, 2011–2075. [CrossRef]

44. Jaroenworaluck, A.; Regonini, D.; Bowen, C.R.; Stevens, R.; Allsopp, D. Macro, micro and nanostructure of TiO_2 anodised films prepared in a fluorine-containing electrolyte. *J. Mater. Sci.* **2007**, *42*, 6729–6734. [CrossRef]

45. Macak, J.M.; Ghicov, A.; Hahn, R.; Tsuchiya, H.; Schmuki, P. Photoelectrochemical properties of N-doped self-organized titania nanotube layers with different thicknesses. *J. Mater. Res.* **2006**, *21*, 2824–2828. [CrossRef]

46. Wang, N.; Li, X.; Wang, Y.; Quan, X.; Chen, G. Evaluation of bias potential enhanced photocatalytic degradation of 4-chlorophenol with TiO_2 nanotube fabricated by anodic oxidation method. *Chem. Eng. J.* **2009**, *146*, 30–35. [CrossRef]

47. Dale, G.R. Electrochemical Growth and Characterization of Self-Organized Titania Nanotubes. Ph.D. Thesis, Ulster University, Newtownabbey, UK, 2009.

48. Byrne, J.A.; Eggins, B.R.; Brown, N.M.D.; McKinney, B.; Rouse, M. Immobilization of titanium dioxide for the treatment of polluted water. *Appl. Catal. B Environ.* **1998**, *17*, 25–36. [CrossRef]

49. Paulose, M.; Prakasam, H.E.; Varghese, O.K.; Peng, L.; Popat, K.C.; Mor, G.K.; Desai, T.A.; Grimes, C. TiO_2 Nanotube Arrays of 1000 μm Length by Anodization of Titanium Foil: Phenol Red Diffusion. *J. Phys. Chem. C* **2007**, *111*, 14992–14997. [CrossRef]

50. Delegan, N.; Daghrir, R.; Drogui, P.; El Khakani, M.A. Bandgap tailoring of in-situ nitrogen doped TiO_2 sputtered films intended for electrophotocatalytic applications under solar light. *J. Appl. Phys.* **2014**, *116*, 153510. [CrossRef]

51. Huang, L.H.; Sun, C.; Liu, Y.L. Pt/N-codoped TiO_2 nanotubes and its photocatalytic activity under visible light. *Appl. Surf. Sci.* **2007**, *253*, 7029–7035. [CrossRef]

52. Rodriguez, J.A.; Jirsak, T.; Liu, G.; Hrbek, J.; Dvorak, J.; Maiti, A. Chemistry of NO$_2$ on oxide surfaces: formation of NO$_3$ on TiO$_2$(110) and NO$_2 \leftrightarrow$ O vacancy interactions. *J. Am. Chem. Soc.* **2001**, *123*, 9597–9605. [CrossRef] [PubMed]

53. Navio, J.A.; Cerrillos, C. Photo-induced transformation, upon UV illumination in air, of hyponitrite species N$_2$O$_2^{2-}$ preadsorbed on TiO$_2$ surface. *Real Surf. Interface Anal.* **1996**, *24*, 355–359. [CrossRef]

54. Cao, F.; Oskam, G.; Meyer, G.J.; Searson, P.C. Electron Transport in Porous Nanocrystalline TiO$_2$ Photoelectrochemical Cells. *J. Phys. Chem.* **1996**, *100*, 17021–17027. [CrossRef]

55. Waldner, G.; Krýsa, J.; Jirkovský, J.; Grabner, G. Photoelectrochemical properties of sol-gel and particulate TiO$_2$ layers. *Int. J. Photoenergy* **2003**, *5*, 115–122. [CrossRef]

56. Marugán, J.; Christensen, P.; Egerton, T.; Purnama, H. Synthesis, characterization and activity of photocatalytic sol–gel TiO$_2$ powders and electrodes. *Appl. Catal. B Environ.* **2009**, *89*, 273–283. [CrossRef]

57. Vinodgopal, K.; Kamat, P.V. Electrochemically assisted photocatalysis using nanocrystalline semiconductor thin films. *Sol. Energ. Mat. Sol. C* **1995**, *38*, 401–410. [CrossRef]

58. Byrne, J.A.; Davidson, A.; Dunlop, P.S.M.; Eggins, B.R. Water treatment using nano-crystalline TiO$_2$ electrodes. *J. Photochem. Photobiol. A Chem.* **2002**, *148*, 365–374. [CrossRef]

59. Xie, Y. Photoelectrochemical application of nanotubular titania photoanode. *Electrochim. Acta* **2006**, *51*, 3399–3406. [CrossRef]

60. Liang, H.-C.; Li, X.-Z. Effects of structure of anodic TiO$_2$ nanotube arrays on photocatalytic activity for the degradation of 2,3-dichlorophenol in aqueous solution. *J. Hazard. Mater.* **2009**, *162*, 1415–1422. [CrossRef] [PubMed]

61. Zhang, Y.; Wang, D.; Pang, S.; Lin, Y.; Jiang, T.; Xie, T. A study on photo-generated charges property in highly ordered TiO$_2$ nanotube arrays. *Appl. Surf. Sci.* **2010**, *256*, 7217–7221. [CrossRef]

62. Hamilton, J.W.J.; Byrne, J.A.; Dunlop, P.S.M.; Dionysiou, D.D.; Pelaez, M.; O'Shea, K.; Synnott, D.; Pillai, S.C. Evaluating the Mechanism of Visible Light Activity for N,F-TiO$_2$ Using Photoelectrochemistry. *J. Phys. Chem. C* **2014**, *118*, 12206–12215. [CrossRef]

63. Rengifo-Herrera, J.A.; Pulgarin, C. Photocatalytic activity of N, S co-doped and N-doped commercial anatase TiO$_2$ powders towards phenol oxidation and *E. coli* inactivation under simulated solar light irradiation. *Sol. Energy* **2010**, *84*, 37–43. [CrossRef]

64. Byrne, J.A.; Eggins, B.R. Photoelectrochemistry of oxalate on particulate TiO$_2$ electrodes. *J. Electroanal. Chem.* **1998**, *457*, 61–72. [CrossRef]

65. Alrousan, D.M.A.; Dunlop, P.S.M.; McMurray, T.A.; Byrne, J.A. Photocatalytic inactivation of *E. coli* in surface water using immobilized nanoparticle TiO$_2$ films. *Water Res.* **2009**, *43*, 47–54. [CrossRef] [PubMed]

66. Dunlop, P.S.M.; McMurray, T.A.; Hamilton, J.W.J.; Byrne, J.A. Photocatalytic inactivation of Clostridium perfringens spores on TiO$_2$ electrodes. *J. Photochem. Photobiol. A Chem.* **2008**, *196*, 113–119. [CrossRef]

67. Quesada-Cabrera, R.; Sotelo-Vazquez, C.; Darr, J.A.; Parkin, I.P. Critical influence of surface nitrogen species on the activity of N-doped TiO$_2$ thin-films during photodegradation of stearic acid under UV light irradiation. *Appl. Catal. B Environ.* **2014**, *160–161*, 582–588. [CrossRef]

Sample Availability: Samples of the compounds TiO$_2$-NT, N-TiO$_2$-NT, N,F-TiO$_2$, and P25 TiO$_2$ are not available.

molecules   MDPI

Article

Transparent Nanotubular TiO$_2$ Photoanodes Grown Directly on FTO Substrates

Šárka Paušová [1,*], Štěpán Kment [2], Martin Zlámal [1], Michal Baudys [1], Zdeněk Hubička [2] and Josef Krýsa [1,*]

[1] University of Chemistry and Technology Prague, Technická 5, 166 28 Prague 6, Czech Republic; zlamalm@vscht.cz (M.Z.); baudysm@vscht.cz (M.B.)
[2] Palacký University, RCPTM, Joint Laboratory of Optics, 17. Listopadu 12, 771 46 Olomouc, Czech Republic; stepan.kment@upol.cz (Š.K.); hubicka@fzu.cz (Z.H.)
* Correspondence: sarka.pausova@vscht.cz (Š.P.); josef.krysa@vscht.cz (J.K.); Tel.: +420-220-444-112; Fax: +420-220-444-420

Academic Editor: Pierre Pichat
Received: 17 March 2017; Accepted: 4 May 2017; Published: 10 May 2017

Abstract: This work describes the preparation of transparent TiO$_2$ nanotube (TNT) arrays on fluorine-doped tin oxide (FTO) substrates. An optimized electrolyte composition (0.2 mol dm^{-3} NH$_4$F and 4 mol dm^{-3} H$_2$O in ethylene glycol) was used for the anodization of Ti films with different thicknesses (from 100 to 1300 nm) sputtered on the FTO glass substrates. For Ti thicknesses 600 nm and higher, anodization resulted in the formation of TNT arrays with an outer nanotube diameter around 180 nm and a wall thickness around 45 nm, while for anodized Ti thicknesses of 100 nm, the produced nanotubes were not well defined. The transmittance in the visible region (λ = 500 nm) varied from 90% for the thinnest TNT array to 65% for the thickest TNT array. For the fabrication of transparent TNT arrays by anodization, the optimal Ti thickness on FTO was around 1000 nm. Such fabricated TNT arrays with a length of 2500 nm exhibit stable photocurrent densities in aqueous electrolytes (~300 µA cm^{-2} at potential 0.5 V vs. Ag/AgCl). The stability of the photocurrent response and a sufficient transparency ($\geq$65%) enables the use of transparent TNT arrays in photoelectrochemical applications when the illumination from the support/semiconductor interface is a necessary condition and the transmitted light can be used for another purpose (photocathode or photochemical reaction in the electrolyte).

Keywords: nanotubular; transparent; TiO$_2$; sputtered Ti; FTO; anodization; photocurrent

---

1. Introduction

With an increasing demand on energy supply, the necessity of using an efficient renewable source of energy is growing. One of the possibilities is the conversion of solar light to electricity or fuel. Since photocatalytic water splitting on TiO$_2$ was discovered by Fujishima and Honda [1], great effort has been devoted to the application of TiO$_2$ in energy conversion. Although TiO$_2$ is a suitable candidate for a photoanode, some major drawbacks need to be overcome, such as the high electron–hole recombination. The use of 1-D nanostructures (nanotubes, nanorods, or nanowires) might be a way of improving the efficiency of electron transport through TiO$_2$ film. Nanotubular structures have a mechanical stability that is superior to that of nanorod or nanowire structures. Nanotubular structures of TiO$_2$ can be prepared by various methods such as the anodization of Ti substrates [2–5] or of Ti thin films [6–9], template synthesis [10], and hydrothermal synthesis [5].

Because of their ability to utilize backside illumination, transparent films of TiO$_2$ on a conductive substrate are considered to be very promising photoanodes for either DSSCs [11–13] or photoelectrochemical water splitting [14,15]. So far, most transparent TiO$_2$ nanotubular arrays have consisted of nanotubes prepared on bulk material and then transferred and attached on a transparent substrate [16,17]. However, the fully transparent nanotubular TiO_2 arrays prepared directly by the complete anodization of a metallic Ti thin layer deposited on a fluorine-doped tin oxide (FTO) glass substrate would represent the most straightforward solution [6–9,18–24]. Nanotubes prepared on FTO glass are mainly tested as photoanodes in DSSCs [7,21,23,24] and not in PEC (photoelectrochemical) water splitting [9,22], so there is not much work dealing with photoelectrochemical measurements in an aqueous phase for TNTs prepared directly on FTO glass. We recently showed that the magnetron sputtering of Ti on an FTO substrate appears to be the most promising technique for the anodic fabrication of transparent TNT arrays [25]. This work was thus devoted to the preparation of TNT arrays via anodic oxidation of sputtered Ti films of different thicknesses and to a comparison of their photo(electro)chemical properties with TNT arrays prepared via anodic oxidation of Ti foil.

2. Results and Discussion

2.1. Optimizing Electrolyte Composition

Water content in electrolytes has been shown to be a very important parameter [26]. Therefore, this parameter has been studied in detail. Figure 1 shows the surface morphology of TNTs prepared by the anodization of Ti foil in electrolytes with different water concentrations influenced by annealing at 500 °C. Other conditions such as NH_4F concentration, temperature, applied voltage, and anodization time were kept constant.

For TNTs prepared with a water content of 1 mol dm^{-3} and lower, cracks appeared even before annealing (Figure 1a). The increase in water concentration to 2 mol dm^{-3} led to a significant decrease in the number of cracks in the TNT array; however, after annealing cracks appeared again in the TNT film (Figure 1c). The optimal amount of water was determined as 4 mol dm^{-3}, all films prepared using this electrolyte composition were free of cracks, even after annealing. These results agree well with findings of Tsui et al. [27], who report that a water content around 10% (5.5 mol dm^{-3}) results in a well-formed nanotube film with the highest photocurrent density. Additionally, the influence of different NH_4F concentrations in the 0.1–0.2 mol dm^{-3} range was studied. Resulting layers were similar and no significant influence on their structure was observed, which is in agreement with published results by Acevedo-Pena et al. [28]. Concentrations of 0.2 mol dm^{-3} NH_4F and 4 mol dm^{-3} H_2O were thus chosen for further study.

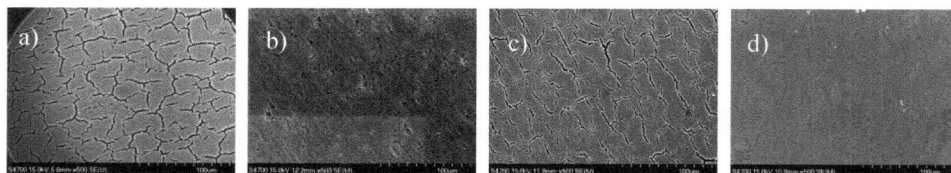

Figure 1. SEM images of TiO_2 nanotubes (TNTs) prepared by anodization of Ti foil in electrolytes with different water content. Non-annealed: (**a**) 1 mol dm^{-3} H_2O; (**b**) 2 mol dm^{-3} H_2O. Annealed at 500 °C; (**c**) 2 mol dm^{-3} H_2O; (**d**) 4 mol dm^{-3} H_2O.

2.2. The Growth of TNT Array and Optical Properties

Figure 2 shows the current density transient during anodization of a sputtered Ti metal layer of different thicknesses (100, 600, 1000, and 1300 nm) on FTO glass. All current transients show the typical trend of TNT formation. A fast decrease of current corresponds to the formation of compact TiO_2 followed by a slow increase of current and its stabilization typical of TNT formation. The anodization of Ti films on FTO was terminated when the transparent film was formed. The passed charge depends almost linearly on the Ti film thickness (see the insert in Figure 2).

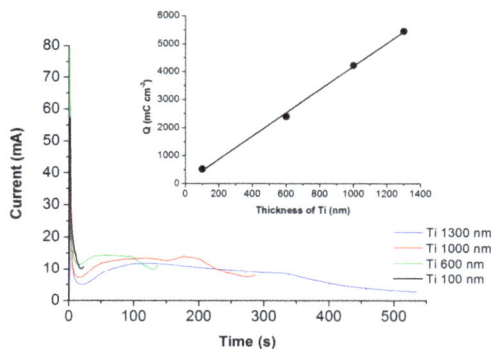

Figure 2. Current density transients during anodization of sputtered Ti films of four different thicknesses to transparent TiO$_2$ nanotubes on fluorine-doped tin oxide (FTO) glass. Inset: Passed charge as a function of Ti layer thickness.

The UV-Vis spectra of anodized Ti films of four different thicknesses were corrected on the transmittance of the FTO substrate and are shown in Figure 3a. The transmittance of the TNT film in the whole range of spectra decreases as layer thickness increases—however, the difference between anodized Ti films of thicknesses 1000 and 1300 nm is very small.

The inserts in Figure 3a shows that all TNT arrays on FTO were transparent. The next step is the evaluation of the transparency in the visible region of solar light and the amount of absorbed light in the UV region. The transparency of the fabricated TNT array is expressed as the transmittance (T) at 500 nm in Figure 3b. This transmission is significantly higher than that in the UV region (90% and 35% for the anodized 100 nm and 1000 nm Ti film, respectively). This means that, even for the thickest TNT; more than 60% of the light in the visible part of the solar spectra is transmitted.

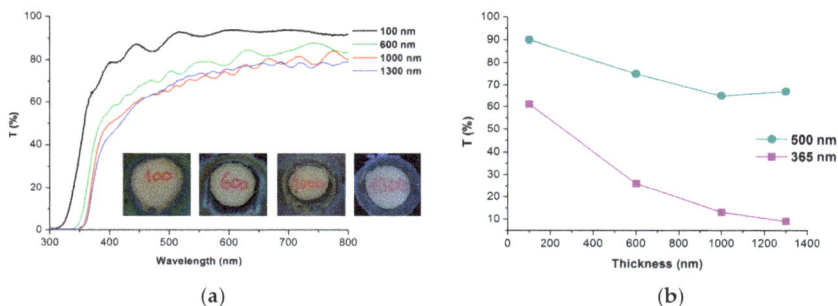

(a)   (b)

Figure 3. (a) UV-Vis spectra of transparent TNT films prepared via the anodization of Ti films of various thicknesses on FTO glass and annealing at 500 °C. Inset: Optical images of anodized Ti films on FTO glass; (b) The dependence of T at 365 nm and T at 500 nm (corresponding to annealed TNT films) on the thickness of Ti films on FTO.

Figure 3b shows the values of T for 365 nm. It is known that the amount of reflection component of the light is not negligible; therefore, the absorbed amount of light cannot be simply obtained as (1 − T). Therefore, for the calculation of absorbed light, we also have to measure the total reflectance (diffuse plus specular) and correct the measured transmittance. This task is possible in the case of single dense layers on a transparent support, as we show for the case of TiO$_2$ films of various thicknesses on quartz glass in our previous work [29]. However, in the case of a two-layer structure (TNT arrays on the FTO layer) on a glass support, due to the numerous layer interfaces and the nanostructure character of TNT arrays, the exact subtraction of the reflectance is almost impossible due to light interference. The experimentally obtained reflectance of TNT arrays on FTO was thus only used to estimate the amount of reflected light. For all fabricated TNT arrays with thicknesses of 180–3500 nm, the average value of R in the range from 350 to 800 nm lies in the range 15 ± 5%. This value has been taken for the estimation of absorbed UV light at 365 nm. As expected, the amount of absorbed UV light increases with TNT length (from 25 ± 5% for anodized 100 nm Ti film to 70 ± 5% and 75 ± 5% for anodized 1000 and 1300 nm Ti film).

2.3. Crystallinity and Morphology of TNT Arrays

Figure 4 shows the diffractograms of the TNT arrays grown on FTO glass substrates. For comparison, the diffractogram for TNTs grown on Ti foil under the same anodization conditions with a passed charge similar to anodization of the 1000 nm Ti film is also shown. After anodization and annealing in an air atmosphere at 500 °C, the TNT arrays on FTO glass and on Ti foil were well crystallized, and the presence of anatase phase was confirmed via the observation of the most intense plane orientations (**110**), (**200**), and (**105**) (Figure 4). There is almost negligible signal corresponding to Ti for anodized Ti films on FTO, implying that Ti was completely oxidized during anodization.

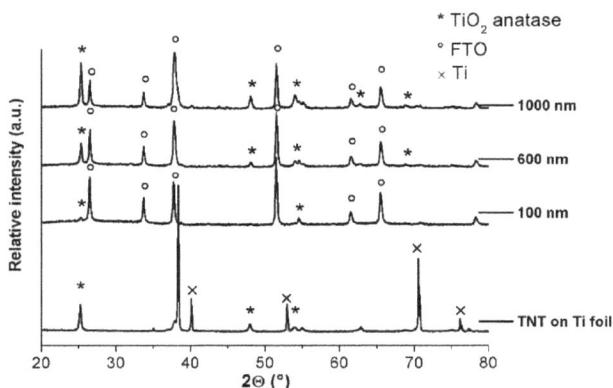

Figure 4. XRD patterns of TNT arrays prepared on Ti foil and on FTO glass and annealed at 500 °C.

The surface morphology of anodized Ti films was followed by top-view SEM images; the thickness and structure of anodized Ti films was determined from cross-section SEM images (see Figure 5). The structure of the thinnest transparent film was not completely nanotubular—the beginning of the formation of the nanotubular structure is shown in Figure 5a—but in the cross section, the nanotubular structure was not observed. This is due to the very short (25 s) anodization period, which is not sufficient for the development of a proper nanotubular structure. The anodization of thicker Ti films resulted in the fabrication of well-formed nanotubes; however, they were covered with a thin and irregular nanoporous structure. The presence of this top layer has been reported elsewhere and can be, for some purposes, removed by sonication in dilute HCl or by Ar ion sputtering [30].

The anodization of the 100 nm Ti film resulted in the formation of a transparent nanoporous film with a thickness of 180 ± 20 nm. The anodization of the 600 nm and 1000 nm Ti films resulted in the formation of a transparent TNT array with a thickness of 1000 ± 50 nm and 2450 ± 50 nm, respectively. The expansion factor F_{ex} for the anodization of the 100 nm and 600 nm Ti films was in the 1.6–1.8 range. For the 1000 nm Ti film, F_{ex} increased to 2.5. The anodization of the 1300 nm Ti film resulted in the formation of a TNT array with a thickness of 3500 ± 50 nm (not shown in Figure 5) and an expansion factor F_{ex} very similar (2.6) to that for the anodized 1000 nm Ti film.

F_{ex} is highly dependent on electrolyte composition (mainly water content) and applied potential [8]. For the present anodization conditions (a water content of 7 wt %, a potential of 60 V), Albu et al. [8] reported for an evaporated 1000 nm Ti layer on FTO a much lower F_{ex} value (1.7). For a lower water content (3 wt %), Albu et al. [8] reported a value of F_{ex} around 2.6, which was also observed in our recent work [25] but for the anodization of a sputtered Ti film on FTO.

This suggests that, for sputtered Ti films on FTO, the water content is not as critical as it is for evaporated films, and the starting thickness of a sputtered Ti film is more significant. The reason for the increase in the expansion factor with thickness of sputtered Ti film on FTO is not completely clear at the moment. In a detailed previous study, Albu et al. [8] concludes that the variation in the expansion factor can be due to the variation in parameters such as efficiency of oxide growth, chemical composition, density, and porosity of the TiO$_2$ nanotubular array. We also think that the influence of the chemical composition, density, and the porosity of Ti film is crucial. In our work, Ti films were grown by magnetron sputtering with emphasis on the adhesion of Ti to FTO. Therefore, the composition and porosity of Ti films close to FTO glass can be different from that of the bulk of Ti films, and this influences the expansion factors of TNTs obtained by anodization of Ti with less thickness, where the different interfacial parts of the Ti films close to FTO represent significant parts of the entire Ti film.

The outer diameter of TNTs for samples prepared from 600 nm, 1000 nm, and 1300 nm Ti films was between 160 and 190 nm, with a wall thickness between 40 and 50 nm. These characteristics differ for samples 100 nm thick; the nanotubes are not that well defined, the wall thickness varies between 40 and 70 nm and the inner diameter is about 50 nm.

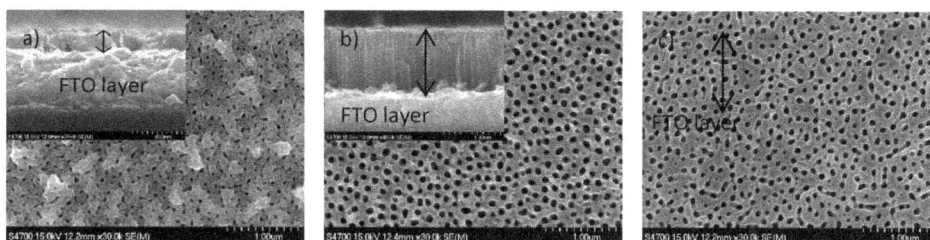

Figure 5. SEM top view and cross-sectional images of TNTs on FTO prepared by the anodization of Ti films of thickness: (**a**) 100 nm; (**b**) 600 nm; and (**c**) 1000 nm and annealed at 500 °C. The thickness of anodized Ti films is marked with a double arrow.

2.4. Photoelectrochemical Activity

The photoelectrochemical behavior of anodized films on FTO glass has been compared in aqueous electrolytes with the aim of evaluating the efficiency of PEC water splitting. Figure 6 shows the comparison of chopped light polarization curves under simulated solar light for TNTs prepared from Ti films on FTO glass annealed at 500 °C. The photocurrent onset is around -0.4 V (Ag/AgCl), and at 0 V the photocurrent plateau starts to develop. The photocurrent density in plateau increases as the anodized Ti thicknesses increase from 100 to 1000 nm, but for Ti thicknesses 1000 and 1300 nm, it is almost identical. This is in agreement with the fact that for thicknesses of 1000–1300 nm, the amount of absorbed UV light is very similar (see Figure 3b). The chronoamperometry measurements enabled the evaluation of the photocurrent stability as a function of irradiation time. This was performed on all TNT films on FTO, and the photocurrent density in plateau (at 1 V (Ag/AgCl)) was stable for at least 5 min. Values are shown in Table 1. At least two samples of Ti films with identical thicknesses on FTO were anodized to obtain transparent TNT films. The deviation in the measured photocurrents for each TNT film was lower than 10%. The photocurrent density of the TNT film with a thickness of 2500 nm (prepared from the 1000 nm Ti film) reached 300 μA cm^{-2} at 0.5 V (Ag/AgCl), which is 25 times higher than the value published by Szkoda et al. [9] (although they prepared transparent TNT films from Ti films with a greater thickness (2000 nm). Bai et al. [22] reported a substantially high value of photocurrent density (750 $\mu A\ cm^{-2}$ at 0.2 V (Ag/AgCl)) for transparent TNT films. However, the measurement was carried out in 1 mol dm^{-3} Na$_2$S electrolytes, which can act as a hole scavenger [31] and thus increase the PEC performance of a TNT layer.

A long test was performed for the anodized Ti film with a thickness of 1000 nm on FTO. The photocurrent density at the beginning of the test was 296 $\mu A/cm^{-2}$ and after 3 h it decreased to 282 $\mu A\ cm^{-2}$ (about 95% of the initial value). The highest decrease was observed after the first 30 min; afterwards, the value of the photocurrent remained almost stable at 282 $\pm$ 2 $\mu A\ cm^{-2}$. The summarized values of the photocurrents are shown in Table 1. The TNT film on FTO prepared from 1000-nm-thick Ti was compared with TNT films on Ti foil (with a comparable charge passed during anodization). The photocurrent densities obtained by both types of TNT films were similar (about 300 $\mu A\ cm^{-2}$ at 1 V against Ag/AgCl). This means that such fabricated transparent TNT arrays can be successfully used in photoelectrochemical applications that require illumination from the support/semiconductor interface.

Table 1. Main characteristics of all prepared samples.

Name of Sample	Ti Film Thickness (nm)	Anodization Time (s)	Passed Charge (mC cm^{-2})	TNT Thickness (nm)	Photocurrent Density * ($\mu A\ cm^{-2}$)
100 nm	100	25	524	180 $\pm$ 20	43 $\pm$ 2
600 nm	600	138	2404	1000 $\pm$ 50	196 $\pm$ 5
1000 nm	1000	290	4226	2450 $\pm$ 50	300 $\pm$ 5
1300 nm	1300	538	5460	3500 $\pm$ 50	294 $\pm$ 5
Ti foil	-	887	4155	-	315 $\pm$ 5

* Results obtained by chronoamperometry measurement at 1 V vs. Ag/AgCl in Na$_2$SO$_4$ electrolytes.

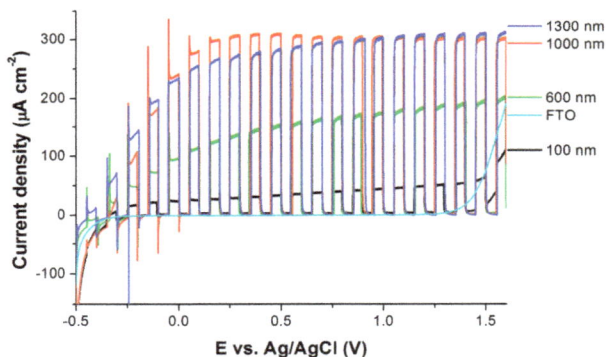

Figure 6. Chopped light polarization curve for TNTs of different anodized Ti thicknesses on FTO glass annealed at 500 °C, electrolytes of 0.1 mol dm^{-3} Na$_2$SO$_4$.

2.5. Photocatalytic Activity

For each TNT array sample annealed at 500 °C, the photoactivity test using Resazurin indicator ink was performed. Resazurin ink works via a photo-reductive mechanism and photocatalytic activity is determined by its color change from blue to pink. The rate of the color change of the ink, measured using digital photography, provides a direct measurement of the photocatalytic activity.

The plot of the R vs. the time of UV exposure for TNTs of various thicknesses is shown in Figure 7. R(t)90 (90% of overall bleaching) was used to calculate the time to bleach 90% of the color, ttb(90) [21] and thus to compare the photocatalytic activity of prepared TNT films. ttb(90) values are marked by gray circles in Figure 6 and shown together with the reciprocal values 1/ttb(90) (proportional to photoactivity) in Table 2. All annealed TNT arrays were photoactive. The photocatalytic activity of

TNT arrays on FTO glass significantly increased with an increase in anodized Ti thickness from 100 to 600 nm; however, further increases in the thickness of anodized Ti did not result in an additional photoactivity increase. The ttb90 value for non-anodized Ti foil and non-annealed TNTs was for both samples higher than 420 s, which implies that those samples exhibit about two orders of magnitude lower photoactivity. In the case of Ti foil, the ttb90 value corresponds to the very thin (several nm) film of native TiO_2 oxide on Ti foil; in the case of non-annealed TNTs, it corresponds to the amorphous TiO_2.

Table 2. ttb(90) values of Resazurin reduction for various anodized Ti thickness.

Thickness of Ti on FTO	TNT Thickness	ttb(90) (s)	1/ttb (90) (s^{-1})
100 nm	180 nm	50	0.020
600 nm	1000 nm	10	0.100
1000 nm	2500 nm	9	0.111
Ti foil	-	$\geq$420	$\leq$0.0024
non-annealed TNT	-	$\geq$420	$\leq$0.0024

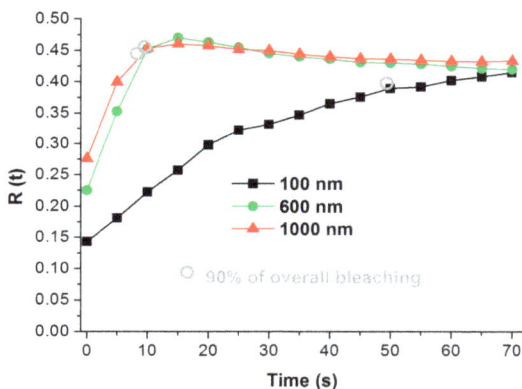

Figure 7. Dependence of R(t) corresponding to Resazurin ink on UV exposition time for TNT arrays of different anodized Ti thicknesses on FTO glass, annealed at 500 °C.

The fact that the above Ti thickness of 600 nm, the photocatalytic activity of fabricated TNT arrays does not increase, while the photoelectrochemical activity increases to a Ti thickness of 1000 nm, which is rather surprising. There is one possible explanation. The deposited ink film is in contact only with the upper part of the TNT array. The photoactivity of TNTs with thicknesses of 180 and 1000 nm differ by a factor of 5, so photoactivity is proportional to the TNT length and roughly to the exposed surface area of TNT arrays. An increase in TNT length from 1000 to 2500 nm then results in a photoactivity increase of about 10%, which does not correspond to the increase in TNT length. This suggests that the bottom part of the TNTs of a higher thickness (than approximately 1000 nm) is not in contact with ink film and thus cannot be utilized for photocatalytic reactions.

3. Experimental Part

3.1. TNT Preparation and Characterization

Self-organized TiO_2 nanotubes (TNTs) were grown via the electrochemical anodization of a Ti foil (purity 99.6%, Advent Research Materials Ltd., Oxford, UK) or of titanium film deposited on FTO.

Titanium films were deposited on FTO substrates (FTO-TCO22-7, Solaronix, Aubonne, Switzerland) via pulsed magnetron sputtering using a pure titanium target. An operating pressure of 0.2 Pa was kept constant during the deposition. The duty cycle of the pulse was 90% and the frequency was 50 kHz. In order to improve the adhesion of titanium, the FTO glass substrate was treated by radio-frequency (RF) plasma before the deposition. The substrate holder worked as an RF electrode connected to the RF power supply working at 13.6 MHz. A gas mixture of Ar–O_2 was used for this purpose, with a pressure of 10 Pa in the reactor chamber. The deposition process followed immediately after RF plasma treatment without the interruption of a vacuum in the reactor chamber. The deposition time differed according to the prepared Ti film thickness: 100 nm—3.5 min, 600 nm—21 min, 1000 nm—35 min, 1300 nm—45.5 min; deposition speed was the same for all samples.

The Ti foil was chemically polished in a mixture of HF and HNO_3 and then washed in ethanol and acetone in an ultrasonic bath. The Ti films on FTO glass were just washed with ethanol. TNTs were grown at 60 V using a power source (STATRON 3253.3, Statron AG, Mägenwil, Switzerland) in a two-electrode configuration with a counter electrode made of platinum in electrolytes containing 0.1–0.2 mol dm^{-3} NH_4F + 0.5–8 mol dm^{-3} H_2O in ethylene glycol. After the anodization process, the samples were washed in ethanol and then dried in a nitrogen stream. To transfer amorphous TiO_2 nanotubes to a crystalline anatase structure, anodized samples were annealed at 500 °C for 1 h in air using a cylindrical furnace (Clasic CLARE 4.0, CLASIC CZ Ltd., Řevnice, Czech Republic) with a temperature increase of 5 °C min^{-1}.

3.2. Characterization Methods

The titania nanotubes produced were characterized by XRD (PANanalytical X´Pert PRO, Cu tube, 1D XCelerator detector, PANanalytical B.V., Almelo, Netherlands), SEM (Hitachi SEM S-4700, Hitachi, Tokyo, Japan), and UV-Vis spectroscopy (Varian CARY 100 with integrating sphere DRA-CA-30I and 8 reflectance geometry, Varian, Palo Alto, CA, USA).

3.3. Photocatalytic Activity

Photocatalytic activity was determined using degradation of Resazurin in model ink. Photocatalytic activity indicator inks work on the basis that (i) the ink film is deposited onto the surface of the photocatalytic film under test; (ii) upon ultra-band gap illumination of the underlying photocatalyst, the photogenerated holes oxidize a sacrificial electron donor (such as glycerol), which is present in the ink. The photogenerated electrons are then free to reduce the redox dye present in the ink and in doing so change the color of the ink. The redox dye is chosen so that it is readily and irreversibly reduced by the photogenerated electrons and that the associated color change is striking. A typical ink consisted of 10 g of 1.5 wt % of hydroxyethyl cellulose, 1 g of glycerol, 20 mg of polysorbate 20 surfactant, and 10 mg of the Resazurin dye [32]. The amount of dye in the ink was chosen based on previous experience with an aim to achieve decolorization times in the range from 1–90 min [33,34]. Ink was coated on a photocatalytically active surface using a paintbrush. After application, the ink-coated sample was dried for 1 h in the dark and then exposed to UVA light from a black light (BL) lamp (λ_{max}(emission) = 352 nm; irradiance: 2 mW/cm^2). Throughout the irradiation, digital images of the ink-coated samples were periodically recorded using a handheld document scanner (CopyCat). The central part of the ink-coated digital image (with minimal dpi 300) of each sample was then analyzed in terms of its *RGB* values ($RGB(red)_t$, $RGB(green)_t$, and $RGB(blue)_t$) using a free graphical software package, ImageJ. The normalized color component $R(t)$ value for red was calculated according to Equation (1):

$$R(t) = \frac{RGB(red)_t}{RGB(red)_t + RGB(green)_t + RGB(bleu)_t}. \tag{1}$$

The present method of ink deposition (paintbrush) was different from that in our previous work [29,32], where a 24 micron K-bar was used to obtain the same thickness and homogeneous distribution of the ink film on smooth photocatalytic surfaces. The amount of deposited ink could be monitored by the measurement of the initial normalized color component $R(t = 0)$ value. This value was for all tested TNT samples in the 0.14–0.27 range, which was found to be sufficient for the achievement of the reproduction (20% deviation) of the amount of deposited ink on the TNT arrays of various lengths.

3.4. Photoelectrochemical Activity

A 150 W Xe arc lamp (Newport) with an AM1.5G filter (100 mW cm^{-2}) was used for the photoelectrochemical characterization of TiO$_2$ nanotubes. Photocurrents were acquired using a photoelectrochemical cell in aqueous 0.1 mol/dm^{-3} Na$_2$SO$_4$ electrolytes using a conventional three-electrode configuration (a Pt counter electrode, an Ag/AgCl reference electrode, and the TNT array as the working electrode). The photocurrent was recorded versus the applied potential using a potentiostat interfaced to a computer. The linear voltammetry of prepared layers was measured with a sweep rate of 5 mV/s^{-1} while being periodically illuminated (5 s light/5 s dark). Chronoamperometry measurement was performed at 0.5 V against an Ag/AgCl reference electrode (1.45 V vs. RHE) under continuous illumination.

4. Conclusions

The insufficient water amount in the electrolytes led to a high amount of cracks in the TNT layer, so the electrolyte composition had to be optimized. The anodization of Ti films with thicknesses of 100–1300 nm on FTO resulted in transparent TNT arrays; the observed expansion factor depended on the Ti thickness and varied from 1.8 for Ti thicknesses lower than ≈600 nm to 2.5 for Ti thicknesses higher than ≈1000 nm. The transmittance in the visible field (500 nm) varied from 90% for the thinnest TNT array to 65% for the thickest TNT array. The photocatalytic activity was successfully demonstrated by the decolorization of Rz ink. Fabricated TNT arrays on FTO glass showed an increase in photocurrent density as TNT length increased, but for TNT lengths higher than 2500 nm, the photocurrent was fairly constant. For the fabrication of transparent TNT arrays, the optimal Ti thickness is around 1000 nm. This is due to the absorption of a significant amount of UV light (≥70%) and a sufficient transparency (≥65%), which enables their use in photoelectrochemical applications when the illumination from the support/semiconductor interface is a necessary condition.

Acknowledgments: The authors acknowledge the financial support from Grant Agency of Czech Republic (15-19705S).

Author Contributions: Šárka Paušová was involved in writing the manuscript and performed anodic oxidation of Ti (foils and thin films) and photoelectrochemical measurements. Martin Zlámal was involved in data collection and the characterization of samples. Michal Baudys measured the photocatalytic activity of the prepared samples. Zdeněk Hubička and Štěpán Kment worked on plasmatic film deposition. Josef Krýsa managed the literature search and was involved in writing the manuscript. All authors read and approved the final manuscript.

Conflicts of Interest: The authors declare no conflict of interest.

References

1. Fujishima, A.; Honda, K. Electrochemical photolysis of water at a semiconductor electrode. *Nature* **1972**, *238*, 37–38. [CrossRef] [PubMed]
2. Roy, P.; Berger, S.; Schmuki, P. TiO$_2$ nanotubes: Synthesis and applications. *Angew. Chem. Int. Ed.* **2011**, *50*, 2904–2939. [CrossRef] [PubMed]
3. Macak, J.M.; Tsuchiya, H.; Ghicov, A.; Yasuda, K.; Hahn, R.; Bauer, S.; Schmuki, P. TiO$_2$ nanotubes: Self-organized electrochemical formation, properties and applications. *Curr. Opin. Solid State Mater. Sci.* **2007**, *11*, 3–18. [CrossRef]
4. Grimes, C.A.; Mor, G.K. *TiO$_2$ Nanotube Arrays: Synthesis, Properties, and Applications*; Springer: New York, NY, USA, 2009.
5. Kasuga, T.; Hiramatsu, M.; Hoson, A.; Sekino, T.; Niihara, K. Formation of titanium oxide nanotube. *Langmuir* **1998**, *14*, 3160–3163. [CrossRef]
6. Mor, G.K.; Varghese, O.K.; Paulose, M.; Grimes, C.A. Transparent highly ordered TiO$_2$ nanotube arrays via anodization of titanium thin films. *Adv. Funct. Mater.* **2005**, *15*, 1291–1296. [CrossRef]

7. Kathirvel, S.; Su, C.; Yang, C.-Y.; Shiao, Y.-J.; Chen, B.-R.; Li, W.-R. The growth of TiO$_2$ nanotubes from sputter-deposited ti film on transparent conducting glass for photovoltaic applications. *Vacuum* **2015**, *118*, 17–25. [CrossRef]

8. Albu, S.P.; Schmuki, P. Influence of anodization parameters on the expansion factor of TiO$_2$ nanotubes. *Electrochim. Acta* **2013**, *91*, 90–95. [CrossRef]

9. Szkoda, M.; Lisowska-Oleksiak, A.; Grochowska, K.; Skowroński, Ł.; Karczewski, J.; Siuzdak, K. Semi-transparent ordered TiO$_2$ nanostructures prepared by anodization of titanium thin films deposited onto the fto substrate. *Appl. Surf. Sci.* **2016**, *381*, 36–41. [CrossRef]

10. Hoyer, P. Formation of a titanium dioxide nanotube array. *Langmuir* **1996**, *12*, 1411–1413. [CrossRef]

11. Jinsoo, K.; Jonghyun, K.; Myeongkyu, L. Laser welding of nanoparticulate TiO$_2$ and transparent conducting oxide electrodes for highly efficient dye-sensitized solar cell. *Nanotechnology* **2010**, *21*, 345203.

12. Lin, C.-J.; Yu, W.-Y.; Chien, S.-H. Transparent electrodes of ordered opened-end TiO$_2$-nanotube arrays for highly efficient dye-sensitized solar cells. *J. Mater. Chem.* **2010**, *20*, 1073–1077. [CrossRef]

13. Kim, J.Y.; Noh, J.H.; Zhu, K.; Halverson, A.F.; Neale, N.R.; Park, S.; Hong, K.S.; Frank, A.J. General strategy for fabricating transparent TiO$_2$ nanotube arrays for dye-sensitized photoelectrodes: Illumination geometry and transport properties. *ACS Nano* **2011**, *5*, 2647–2656. [CrossRef] [PubMed]

14. Krysa, J.; Zlamal, M.; Kment, S.; Brunclikova, M.; Hubicka, Z. TiO$_2$ and Fe$_2$O$_3$ films for photoelectrochemical water splitting. *Molecules* **2015**, *20*, 1046–1058. [CrossRef] [PubMed]

15. Freitas, R.G.; Santanna, M.A.; Pereira, E.C. Dependence of TiO$_2$ nanotube microstructural and electronic properties on water splitting. *J. Power Source* **2014**, *251*, 178–186. [CrossRef]

16. Cho, I.S.; Chen, Z.; Forman, A.J.; Kim, D.R.; Rao, P.M.; Jaramillo, T.F.; Zheng, X. Branched TiO$_2$ nanorods for photoelectrochemical hydrogen production. *Nano Lett.* **2011**, *11*, 4978–4984. [CrossRef] [PubMed]

17. Lei, B.-X.; Liao, J.-Y.; Zhang, R.; Wang, J.; Su, C.-Y.; Kuang, D.-B. Ordered crystalline TiO$_2$ nanotube arrays on transparent fto glass for efficient dye-sensitized solar cells. *J. Phys Chem C* **2010**, *114*, 15228–15233. [CrossRef]

18. Abdi, F.F.; Firet, N.; Dabirian, A.; van de Krol, R. Spray-deposited co-pi catalyzed bivo4: A low-cost route towards highly efficient photoanodes. *MRS Proc.* **2012**, *1446*. [CrossRef]

19. Chappanda, K.; Smith, Y.; Mohanty, S.; Rieth, L.; Tathireddy, P.; Misra, M. Growth and characterization of tio2 nanotubes from sputtered ti film on si substrate. *Nanoscale Res. Lett.* **2012**, *7*, 1–8. [CrossRef] [PubMed]

20. Tang, Y.; Tao, J.; Zhang, Y.; Wu, T.; Tao, H.; Bao, Z. Preparation and characterization of TiO$_2$ nanotube arrays via anodization of titanium films deposited on fto conducting glass at room temperature. *Acta Phys.-Chim. Sin.* **2008**, *24*, 2191–2197. [CrossRef]

21. Pugliese, D.; Lamberti, A.; Bella, F.; Sacco, A.; Bianco, S.; Tresso, E. TiO$_2$ nanotubes as flexible photoanode for back-illuminated dye-sensitized solar cells with hemi-squaraine organic dye and iodine-free transparent electrolyte. *Org. Electron.* **2014**, *15*, 3715–3722. [CrossRef]

22. Bai, J.; Li, J.; Liu, Y.; Zhou, B.; Cai, W. A new glass substrate photoelectrocatalytic electrode for efficient visible-light hydrogen production: Cds sensitized TiO$_2$ nanotube arrays. *Appl. Catal. B Environ.* **2010**, *95*, 408–413. [CrossRef]

23. Krumpmann, A.; Dervaux, J.; Derue, L.; Douhéret, O.; Lazzaroni, R.; Snyders, R.; Decroly, A. Influence of a sputtered compact TiO$_2$ layer on the properties of TiO$_2$ nanotube photoanodes for solid-state dsscs. *Mate. Des.* **2017**, *120*, 298–306. [CrossRef]

24. Lim, S.L.; Liu, Y.; Li, J.; Kang, E.-T.; Ong, C.K. Transparent titania nanotubes of micrometer length prepared by anodization of titanium thin film deposited on ito. *Appl. Surf. Sci.* **2011**, *257*, 6612–6617. [CrossRef]

25. Krysa, J.; Lee, K.; Pausova, S.; Kment, S.; Hubicka, Z.; Ctvrtlik, R.; Schmuki, P. Self-organized transparent 1d TiO$_2$ nanotubular photoelectrodes grown by anodization of sputtered and evaporated ti layers: A comparative photoelectrochemical study. *Chem. Eng. J.* **2017**, *308*, 745–753. [CrossRef]

26. Ratnawati; Gunlazuardi, J.; Slamet. Development of titania nanotube arrays: The roles of water content and annealing atmosphere. *Mater. Chem. Phys.* **2015**, *160*, 111–118.

27. Tsui, L.-K.; Homma, T.; Zangari, G. Photocurrent conversion in anodized TiO$_2$ nanotube arrays: Effect of the water content in anodizing solutions. *J. Phys. Chem. C* **2013**, *117*, 6979–6989. [CrossRef]

28. Acevedo-Peña, P.; Lartundo-Rojas, L.; González, I. Effect of water and fluoride content on morphology and barrier layer properties of TiO$_2$ nanotubes grown in ethylene glycol-based electrolytes. *J. Solid State Electrochem.* **2013**, *17*, 2939–2947. [CrossRef]

29. Krýsa, J.; Baudys, M.; Mills, A. Quantum yield measurements for the photocatalytic oxidation of acid orange 7 (ao7) and reduction of 2,6-dichlorindophenol (dcip) on transparent TiO_2 films of various thickness. *Catal. Today* **2015**, *240*, 132–137. [CrossRef]

30. Kuzmych, O.; Nonomura, K.; Johansson, E.M.J.; Nyberg, T.; Hagfeldt, A.; Skompska, M. Defect minimization and morphology optimization in TiO_2 nanotube thin films, grown on transparent conducting substrate, for dye synthesized solar cell application. *Thin Solid Films* **2012**, *522*, 71–78. [CrossRef]

31. Kang, Q.; Liu, S.; Yang, L.; Cai, Q.; Grimes, C.A. Fabrication of pbs nanoparticle-sensitized TiO_2 nanotube arrays and their photoelectrochemical properties. *ACS Appl. Mater. Interfaces* **2011**, *3*, 746–749. [CrossRef] [PubMed]

32. Mills, A.; Hepburn, J.; Hazafy, D.; O'Rourke, C.; Krysa, J.; Baudys, M.; Zlamal, M.; Bartkova, H.; Hill, C.E.; Winn, K.R.; et al. A simple, inexpensive method for the rapid testing of the photocatalytic activity of self-cleaning surfaces. *J. Photochem. Photobiol. A Chem.* **2013**, *272*, 18–20. [CrossRef]

33. Mills, A.; Hepburn, J.; Hazafy, D.; O'Rourke, C.; Wells, N.; Krysa, J.; Baudys, M.; Zlamal, M.; Bartkova, H.; Hill, C.E.; et al. Photocatalytic activity indicator inks for probing a wide range of surfaces. *J. Photochem. Photobiol. A Chem.* **2014**, *290*, 63–71.

34. Baudys, M.; Krýsa, J.; Mills, A. Smart inks as photocatalytic activity indicators of self-cleaning paints. *Catal. Today* **2017**, *280*, 8–13. [CrossRef]

Sample Availability: Samples of the compounds are not available from the authors.

Section 4:
Photocatalysts: Modeling; Efficacy Effects of Composition, Characteristics, Supports and Modifications

molecules   MDPI

Review

The Role of Molecular Modeling in TiO$_2$ Photocatalysis

Zekiye Cinar

Department of Chemistry, Yildiz Technical University, 34220 Istanbul, Turkey; cinarz@yildiz.edu.tr;
Tel.: +90-212-383-4179

Academic Editor: Pierre Pichat
Received: 7 January 2017; Accepted: 27 March 2017; Published: 30 March 2017

Abstract: Molecular Modeling methods play a very important role in TiO$_2$ photocatalysis. Recent advances in TiO$_2$ photocatalysis have produced a number of interesting surface phenomena, reaction products, and various novel visible light active photocatalysts with improved properties. Quantum mechanical calculations appear promising as a means of describing the mechanisms and the product distributions of the photocatalytic degradation reactions of organic pollutants in both gas and aqueous phases. Since quantum mechanical methods utilize the principles of particle physics, their use may be extended to the design of new photocatalysts. This review introduces molecular modeling methods briefly and emphasizes the use of these methods in TiO$_2$ photocatalysis. The methods used for obtaining information about the degradabilities of the pollutant molecules, predicting reaction mechanisms, and evaluating the roles of the dopants and surface modifiers are explained.

Keywords: TiO$_2$; photocatalysis; DFT; quantum mechanics; molecular modeling

---

1. Introduction

Molecular modeling is the art of representing molecular structures mathematically and simulating their behavior with quantum mechanical methods. Quantum mechanical methods allow us to study chemical phenomena by running calculations on computers rather than carrying out experiments. Geometries and properties of the transition states, excited states or other short-lived species can only be calculated by using quantum mechanical methods. They utilize the principles of particle physics and examine structure as a function of electron distribution. Therefore, they can be used to design new photocatalysts, to even analyze reactions which have not yet been carried out. Quantum mechanical methods generate data on geometries (bond lengths, bond angles, and torsion angles), energies (total energies, heats of formation, activation energies, and thermodynamic properties), electronic (frontier orbital energies, charge distributions, and dipole moments) and spectroscopic properties (absorption thresholds, vibrational frequencies, and chemical shifts).

TiO$_2$ photocatalysis is an advanced oxidation process to destroy hazardous compounds in water or air [1–4]. The process is non-energy intensive, operates at ambient conditions and able to mineralize organic pollutants using only atmospheric oxygen as the additional chemical species. Because of its low cost, long-time stability, chemical inertness and high activity, and TiO$_2$ has been proven to be the most effective photocatalyst for this process [5–8]. The photocatalytic reactions on TiO$_2$ are initiated by band-gap excitation and subsequent generation of electron–hole (e$^-$/h$^+$) pairs that can initiate redox reactions on the surface. Electrons are trapped at surface defect sites (Ti^{3+}) and removed by reactions with adsorbed molecular O$_2$ to produce superoxide anion radical O$_2\bullet^-$, while holes react with adsorbed water molecules or OH$^-$ ions to produce $\bullet$OH radicals. $\bullet$OH radicals are considered to be the principal reactive species responsible for the degradation reactions. However, TiO$_2$ has a wide band-gap (~3.2 eV) and is only excited by UV-light; it is inactive under visible light irradiation.

This feature of TiO_2 inhibits the utilization of solar energy as a sustainable energy source for its excitation because only 5% of the incoming solar energy on the earth's surface is in the UV range. Besides, electron–hole recombination speed is too fast to allow any chemical reaction, due to short charge separation distances within the particle. The majority of the e^-/h^+ pairs generated upon band-gap excitation are lost through recombination instead of being involved in redox processes at the surface. The e^-/h^+ recombination process not only decreases the quantum yield but also decreases the oxidation capability of TiO_2 [8,9]. Therefore, in recent years, in order to utilize sunlight instead of UV irradiation, studies have begun to develop the next generation of TiO_2, well-tailored photocatalysts with high photocatalytic activities under visible light irradiation. One way to achieve this is doping of impurities into the TiO_2 matrix in order to reduce the band-gap. The methods used are transition metal doping [10–16], metal-ion implanting, surface modification, non-metal doping [17–29] and cooping [30–38]. However, each method has shown both positive and negative effects.

The common positive effect of doping and surface modification of TiO_2 is that the absorption edge shifts to the red region of the spectrum and the photocatalytic activity increases, but, the key question to be answered is how doping or surface modification achieves this. On the other hand, photocatalytic degradation reactions of organic contaminants may take place through the formation of harmful intermediates that are more toxic than the original compound. In order to eliminate certain reaction paths yielding such hazardous compounds, the mechanisms and the nature of the reactions should be known. As for the mixtures of different pollutants, the reactivities of the individual molecules are also needed. All of these problems can be solved by computational techniques based on the principles of quantum mechanics. The aim of this review is to introduce molecular modeling methods briefly and emphasize the use of these methods in TiO_2 photocatalysis. The methods used for obtaining information about the degradabilities of the pollutant molecules, predicting reaction mechanisms and evaluating the roles of the dopants and surface modifiers are explained.

2. Molecular Modeling Methods

Molecular modeling studies start with generating a model of the molecule under investigation. Models are generated in the computer by defining the relative positions of the atoms in space by a set of Cartesian coordinates. A reasonable and reliable starting geometry essentially determines the quality of the calculations. There are mainly four classes of molecular modeling methods; molecular mechanics (MM), electronic structure, post-ab initio and molecular dynamics methods [39–41]. All molecular modeling methods compute the energy and related properties of a particular molecular structure. However, the generated model of a given molecule does not have ideal geometry; therefore, a geometry optimization must be performed subsequently. Geometry optimizations locate the lowest energy molecular structure in close proximity to the specified starting structure. Geometry optimizations depend primarily on the gradient of the energy which is the first derivative of the energy with respect to atomic positions. Gradient is the force acting on the structure. The lowest energy structure obtained through energy-minimization techniques corresponds to the geometry with zero gradient. Vibrational frequencies of molecules resulting from interatomic motions within the molecule may also be calculated by molecular modeling methods. Frequencies depend upon the second derivative of the energy with respect to atomic structure and they can be used to predict thermodynamic properties of the molecule.

2.1. Molecular Mechanics Methods

Molecular mechanics (MM) methods use the laws of classical physics and a set of experimental data for atom types. MM methods do not treat electrons; instead, they perform computations based upon the interactions among the nuclei. This approximation makes MM calculations quite inexpensive so they can be used for very large systems.

2.2. Electronic Structure Methods

Electronic structure methods use the laws of quantum mechanics rather than classical physics. The fundamental equation of quantum mechanics is the Schrödinger equation;

$$H\Psi = E\Psi \tag{1}$$

where E is the energy of the system, Ψ is the wavefunction which defines Cartesian and spin coordinates of atoms, and H is the Hamiltonian operator including kinetic and potential energy terms. However, exact solutions to the Schrödinger equation are not computationally practical. Approximations must be introduced in order to apply the method to multi-electronic and polyatomic systems. Electronic structure methods are characterized by their various mathematical approximations to the solution of the Schrödinger equation.

"*Semi-empirical methods*" use experimental data to simplify the computation and solve an approximate form of the Schrödinger equation ignoring complex differentials. They reduce the computational cost by reducing the number of complex integrals. "*Ab initio*" methods use no experimental parameters. Instead, they use the well-known physical constants, and solve the equation directly using only the principles of quantum mechanics.

Semi-empirical methods are relatively inexpensive and provide reasonable qualitative descriptions of molecular systems. In contrast, ab initio methods provide accurate quantitative predictions of energies and structures. Both semi-empirical and ab initio methods depend upon Hartree–Fock (HF) theory. In HF theory, single-electron wavefunctions are used. The one-electron wavefunctions are molecular orbitals which are given as a product of a spatial orbital times a spin function. The use of these functions implies that electron correlation is neglected. The electron–electron repulsion is only included as an average effect.

2.3. Post-Ab Initio Methods

The third class of electronic structure methods, such as the Density Functional Theory (DFT), configuration interaction (CI) and Moller–Plesset Perturbation Theory (MP) methods, known as "*post-ab initio methods*" has recently been developed. These methods are attractive because they take the electron correlation into account. The DFT methods are the most widely used post-ab initio methods because they are thought as being the least expensive. The DFT methods use the electron density instead of the wavefunction. While the complexity of a wavefunction increases with the number of electrons, the electron density has the same number of variables. The DFT methods design functionals connecting the electron density with the energy.

2.4. Molecular Dynamics Method

The molecular dynamics methods combine energy calculations with the laws of classical mechanics. The simulation is performed numerically integrating Newton's equation of motion over small time steps. Once the velocities are computed, new atom locations and the temperature of the assembly can be calculated.

2.5. Solvent Effect

All of the molecular modeling methods, molecular mechanics, semiempirical, ab initio, post-ab initio and molecular dynamics methods assume that the molecules are isolated from each other as in the gas phase. However, solvation plays a decisive role in determining the energetics of the reactions in aqueous media. Molecular modeling allows us also to compute the properties of the systems in solution.

In solutions, solvent molecules affect the properties of the solute molecules and the kinetics of the reactions. In aqueous media, solvent water affects the energetics of the solute species and also induces geometry relaxation for systems containing hydrogen-bonded complexes. However, previous results indicate that geometry changes have a negligible effect on the energy of the solute in water for both open and closed shell structures [42,43].

Solvation effects are generally modeled by polarizable continuum models (PCMs). In these methods, solvent is treated as a polarizable continuum rather than separate, individual molecules in order to reduce the computational cost. The solute is placed into a cavity within the solvent. The construction of the cavity is different in different PCM methods. In most cases, it is constructed as an assembly of atom-centered spheres, while the cavity surface is approximated by segments.

Among PCMs, the so-called conductor-like screening model COSMO is popular to use for aqueous media [42]. In COSMO, the solvent is treated as a dielectric continuum surrounding the solute molecule. Solute molecule forms a cavity with a similar shape. The charge distribution on the solute polarizes the dielectric medium. This effect causes the generation of screening charges on the cavity surface. Therefore, solvent is described by apparent polarization charges included in the solute Hamiltonian, so that it is possible to perform iterative procedures leading to self-consistence between the solute wavefunction and the solvent polarization. The COSMO method describes the dielectric continuum by means of apparent polarization charges distributed on the cavity surface, which are determined by imposing that the total electrostatic potential cancels out on the surface. This condition describes the solvation in polar liquids. Hence, COSMO can be accepted as a suitable method to be used in TiO_2 photocatalysis.

3. Degradation Reaction Model

3.1. Active Species

In TiO_2 photocatalysis studies, the reaction model used so far has been the reaction between the pollutant molecule and the •OH radical [44,45]. However, the governing role of active species leading to initial photocatalytic process is still a matter of controversy. The interfacial transfer of conduction band electrons to the adsorbed oxygen acting as primary electron acceptor has been accepted as the rate-determining step of the whole photocatalytic reaction. The most important primary chemical process is the formation of •OH radicals from adsorbed OH groups. These radicals either diffuse in solution or migrate on the surface. During migration, other species like H_2O_2 or peroxyl radicals are formed. However, •OH radicals are considered to be the principle reactive species responsible for the photocatalytic reaction.

In order to compare the behavior of the two species, the frontier orbitals of the free and the adsorbed •OH radicals on TiO_2 have been calculated by DFT/B3LYP/6-31G* method [46]. The results obtained indicate that the •OH radical is strongly bound to the TiO_2 surface, the distance between the oxygen atom of the •OH radical and the surface titanium cation has been calculated to be 1.822 A. Moreover, it has been also found that the frontier-orbital energy of the •OH radical does not change much, the SOMO (singly occupied molecular orbital) of the adsorbed •OH radical has been calculated to be -8.871 eV, slightly higher than the energy of the free •OH, -8.953 eV. The frontier orbitals of 4-chlorophenol (4-CP) and the adsorbed •OH radical are presented in Figure 1. Therefore, it may be concluded that the photocatalytic degradation reactions of compounds in the presence of TiO_2 may be based on hydroxyl radical chemistry. The most plausible reaction pathway for hydroxyl radical having a strong electrophilic character is a direct attack on one of the atoms of the pollutant molecule, generally the one with the highest electron density.

3.2. Reaction Center

Most of the organic pollutant molecules are aromatic in nature. •OH radicals react with aromatic molecules through addition to yield hydroxycyclohexadienyl type radicals which then form intermediates. Due to its highly electrophilic character, •OH radical has a very low-lying SOMO (singly occupied molecular orbital) the energy of which is -8.871 eV, while the energies of the HOMO (highest occupied molecular orbital) and LUMO (lowest unoccupied molecular orbital) of the aromatic molecules are around -5.4 and -2.7 eV respectively. Therefore, in the hydroxylation of the aromatic molecules, the SOMO of the •OH radical interacts with the HOMOs of the molecules, as displayed in Figure 1 for 4-chlorophenol. It may be concluded that the photocatalytic degradation reactions on TiO_2 are orbital-controlled reactions.

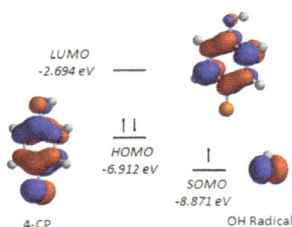

LUMO
-2.694 eV

HOMO
-6.912 eV

SOMO
-8.871 eV

4-CP   OH Radical

Figure 1. The frontier orbitals of 4-chlorophenol and the adsorbed •OH radical.

It has been experimentally proven that •OH radicals attack aromatic molecules at the ring positions. Therefore, the intermediates and products of the degradation reactions of aromatic molecules depend upon the position of attack of the •OH radicals. There are several theoretical shortcut methods in the literature for the determination of the position of attack of the •OH radical in such reactions [47]. One of the most successful ones is the "frontier orbital theory (FMO)" which states that in electrophilic reactions, the point of attack is at the position of the greatest electron density in the HOMO of the aromatic molecule. The HOMO coefficients for 4-nitrophenol (4-NP) have been calculated by using a semiempirical PM3 (Parametric Model number 3) method to be 0.4 for the two -*ortho* positions and 0.1 for the two-*meta* positions with respect to the functional –OH group. These values indicate a high preference for the -*ortho* carbons and the most probable primary intermediate that forms in the photocatalytic degradation of 4-NP has been predicted to be 1,2-dihydroxy-4-nitro-cyclohexadienyl radical which then forms 4-nitrocatechol [48].

Another theory for the determination of the position of attack of the •OH radical to the aromatic molecule is the localization approach of Wheland [47]. According to this theory, the position of attack is determined by the energy of the intermediate. The most probable intermediate is the one with the lowest energy. The calculated total energies and the heats of formation of the hydroxylated radicals forming in the photocatalytic degradation reaction of 4-NP indicate that the attack is at the -*ortho* position in agreement with the FMO Theory [48].

4. Prediction of Degradability

Quantum mechanical methods are very useful tools for obtaining information about the degradabilities of the pollutant molecules. In fact, the degradation rate depends upon the electronic structure of the compound. However, there are relatively few correlations between the degradability and the molecular structure, reported in the literature [49–51]. In all of these equations, empirical parameters, such as 1-octanol/water partition coefficient, the Hammett constant, molecular refractivity and Brown constant have been used as the molecular properties. Therefore, the correlations obtained contain either experimental or approximate structural parameters. The use of molecular modeling techniques gives better relationships.

4.1. Electronic Molecular Properties

The energies of the frontier molecular orbitals are important quantum mechanical indices, the application of which in quantitative structure activity relations (QSAR) has gained considerable attention in the last decades. The energy of HOMO (E_{HOMO}) is a measure of the ease of oxidation of the compound. On the other hand, the energy of the lowest unoccupied orbital (E_{LUMO}) generally shows the reduction potency of the compound. The excitation energy, which is defined as the difference in energies of the HOMO and LUMO, reflects the electronic stability of the pollutant molecule.

In order to obtain a relationship predicting the degradabilities of monosubstituted anilines, San and Cinar [52] have used the semi-empirical PM3 method to calculate electronic properties of the compounds and correlated the logarithm of the experimental degradation rate constant (k) with the energies of the HOMOs, (E_{HOMO}) and the sum of the electron densities (q) of the substituents. E_{HOMO} is a measure of the ease of oxidation of the compound while 1-octanol/water partition coefficient (K_{OW}) is a measure of the distribution of the compound between TiO_2 particles and the solution. The sum of the electron densities has been used in order to take into account the electrophilic character of the •OH radical. As a result, a linear relationship of the form

$$\log k = Aq + BE_{HOMO} + C \log K_{OW} + D \tag{2}$$

has been obtained by using multiple regression. In this equation, A, B, C and D are the constants obtained by regression. The correlation coefficient r has been calculated to be 0.9937 with A = 0.02, B = 0.22, C = −0.27 and D = −0.32.

4.2. DFT Reactivity Descriptors

The best descriptors showing degradability can be obtained by the application of the Conceptual Density Functional Theory (DFT). According to the Conceptual DFT, the reactivity of a molecule depends upon its response to the perturbations caused by the attacking chemical species, in TiO_2 photocatalysis •OH radicals. The typical perturbations for a chemical reaction are changes in external potential and the number of electrons N. The Conceptual DFT discusses reactions in terms of these changing properties. This approach leads to a series of reactivity descriptors, such as; the electronic chemical potential, hardness, softness and Fukui function. These descriptors then may be connected to different reactivity principals to be used in the determination of the regioselectivity and the reactivity of the compound under investigation.

There are mainly two classes of DFT descriptors; global and local descriptors. Perturbations due to changes in the number of electrons are defined as global descriptors and are related to overall molecular stability. Perturbations due to changes in external potential are called local descriptors and determine the site selectivity of a molecule for a specific reaction type.

Global hardness η is defined as the second derivative of the energy E with respect to the number of electrons N. It is equal to the reciprocal of global softness (S) [53,54]. Using the finite

$$\eta = \frac{1}{2}\left(\frac{\partial^2 E}{\partial N^2}\right)_{v(r)} = \frac{1}{2S} \tag{3}$$

difference approach together with Koopman's theorem, hardness can be written in terms of the first ionization potential (I) and the electron affinity (A) of the molecule, whereas, in the frozen-core approximation, global hardness equals the gap between the frontier orbitals;

$$\eta = \frac{I - A}{2} = \frac{E_{LUMO} - E_{HOMO}}{2} \tag{4}$$

Global hardness is a measure of the stability of the molecule. It is also a measure of the resistance of a chemical species to change its electronic configuration. Therefore, stable molecules are likely to be harder than less stable molecules and thus they have low reactivities. On the other hand; global softness is related with the polarizability of the molecule. Soft molecules have a high polarizability, which can allow a large deformation of the electron cloud. Soft molecules are more reactive, thus their degradation rates are faster than the hard molecules.

Generally, local properties are used in the determination of the reactivities of different sites of a molecule. Fukui function $f(r)$ is the most important local DFT descriptor. It is defined as the mixed second derivative of the energy of the molecule with respect to the number of electrons and the external potential:

$$f(r) = \left(\frac{\partial^2 E}{\partial N.\partial v(r)}\right) = \left[\frac{\partial \mu}{\partial v(r)}\right]_N = \left[\frac{\partial \rho(r)}{\partial N}\right]_{v(r)} \tag{5}$$

Fukui function reflects the reactivity of a certain site of the molecule and it is the change in the electron density driven by a change in the number of electrons. The larger the value of the Fukui function, the higher the reactivity of that site. The fundamental equations defining the Fukui functions per atom i in a molecule are;

$$f_i^- = [q_i(N) - q_i(N-1)] \tag{6}$$

$$f_i^o = [f_i^+ + f_i^-]/2 \tag{7}$$

where q_i is the electron population of atom i in the molecule. $f^-(r)$ is used when the system undergoes an electrophilic attack, whereas f_i^o governs radical attack. The local softness $s(r)$ provides intermolecular reactivity information about regioselectivity. It is related to the Fukui function through $s(r) = Sf(r)$. This equation indicates that $f(r)$ redistributes the global softness among different parts of the molecule.

San et al. [55] have calculated global and local DFT descriptors for monosubstituted phenols and examined correlations between the experimental degradation rate constants and the calculated DFT descriptors. However, the results indicate that the global DFT descriptors do not well describe the degradabilities of the phenol derivatives due to their different adsorptive capacities and characteristics of the substituents. The calculated local descriptors give better results. This finding indicates that the photocatalytic degradation reactions of phenol derivatives are orbital-controlled reactions rather than radical attack.

The use of *"softness-matching principle"* which is based on the HSAB principle has given the best descriptor for phenol derivatives studied. According to the HSAB principle, soft atoms react preferentially with other soft atoms and hard atoms with other hard atoms. However, the principle also indicates that the interaction between two chemical species will not necessarily occur through their softest atoms but through those whose softnesses are approximately equal. The reactions investigated are orbital-controlled reactions in which soft–soft interactions dominate. For these, reactions, Δs between the reacting atoms must be as small as possible. Therefore, $\Delta s = s^+(O) - s^-(C)$ the difference between the local softnesses of the oxygen atom of the attacking •OH radical and of the carbon atom of the aromatic ring that undergoes the electrophilic attack has been calculated for each of the phenol molecules.

Consequently, a simple linear equation giving log k in terms of Δs and E_{HOMO} with a regression coefficient $r = 0.9816$ has been obtained. The experimental and calculated k values are presented in Figure 2. Therefore, it may be concluded that local DFT descriptors describe the reactivities of the phenol molecules in their photocatalytic degradation reactions better than the global ones.

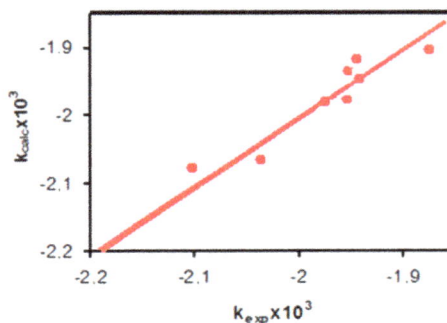

Figure 2. Observed vs. calculated rate constants for the photocatalytic degradation reactions of phenols.

5. Reaction Mechanisms

Molecular modeling methods can also be used to determine degradation reaction mechanisms. One way is to apply a short-cut method based on the use of DFT reactivity descriptors. However, the most reliable method is to apply Transition State Theory through quantum mechanical calculations.

5.1. Short-Cut Method

Short-cut method based on the use of DFT reactivity descriptors is computationally less demanding than transition state calculations. This method describes the preferred reaction energetics and thus kinetics in terms of the properties of the reactants in the ground state and is a successful tool to gain insight into how photocatalytic degradation reactions occur. The calculations based on reactivity indices are computationally less intensive but also less detailed because all information is obtained through study of the reactants only. Consequently, only information about the onset of the chemical reaction should be expected.

In an earlier study [56], the degradation mechanism of cefazolin has been determined by using the DFT descriptors. Cefazolin is a semi-synthetic antibiotic. It belongs to the first generation of cephalosporines that are the most widely-used group of antibiotics and inhibits cell-wall biosynthesis in a manner similar to that of penicillins. As seen in Figure 3, cefazolin with molecular formula $C_{14}H_{14}N_8O_4S_3$, consists of a fused β-lactam-Δ^3-dihydrothiazine two-ring system.

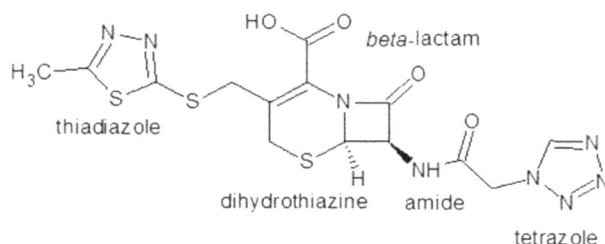

Figure 3. Cefazolin molecule.

The reaction model used is the reaction between the cefazolin molecule and the photogenerated •OH radicals. Therefore, all the calculations have been based on hydroxyl radical chemistry. First the geometries of the reactants have been optimized and their electronic properties have been calculated. Then, the calculations have been repeated by adding and subtracting an electron from the reactant structures. Geometry optimizations of the reactants have been performed with the DFT method. The DFT calculations have been carried out using the hybrid B3LYP functional, which combines HF and Becke exchange terms with the Lee-Yang-Parr correlation functional in order to obtain an exact solution. 6-31G* has been used as the basis set.

Using the electron densities of the atoms in the optimized reactant structures Fukui functions have been calculated for each of the atoms in the molecule. Then, by using the highest f values local softnesses have been calculated. In order to determine the reaction centers and the reaction intermediates accordingly, softness-matching principle has been used. Therefore, for each of the atoms of the molecule $\Delta s^o = s^o(O) - s^o(X)$, the difference between local softnesses of the oxygen atom of the attacking •OH radical and the atom (X) of the cefazolin molecule that undergoes the radical attack has been calculated.

Then, the softnesses of the atoms have been compared with that of the •OH radical and the ones with softnesses being close to that of the •OH radical have been chosen. As a result, three main competing reaction pathways have been determined. The local softness s^o and softness difference Δs^o values indicated that sulfur atoms on the thiadiazole ring are the prime targets of hydroxyl radical attack.

In *Pathway I*, upon the attack of the •OH radical, S–C bond cleavage occurs leading to the formation of 5-methyl-1,3,4-thiadiazole-2-thiol as seen in Figure 4. Oxidation of the other sulfur by •OH radicals precedes ring opening. Hydrogens on the dihydrothiazine are the two possible positions of •OH attack. Therefore, in the first step of *Pathway II*, •OH radical bonds to the carbon atom of the radical formed through H-abstraction. In the second step, lactonization occurs between hydroxyl and carboxy groups to yield a lactone as a by-product. The reactivity of the carbon on β-lactam ring is high according to its local softness value, in the third step, upon the attack of the •OH radical, β-lactam ring opens to give a β-lactam ring-opened lactone. The local softness and softness difference values of the atoms indicated that sulfur atom on the dihydrothiazine is another reactive site for •OH attack and it has been predicted that dihydrothiazine ring disappears when β-lactam ring is opened. In *Pathway III*, N–C bond cleavage occurs in the tetrazole part of the molecule. Successive •OH attacks results in ring opening and mineralization due to comparable reactivities of the nitrogen atoms on the tetrazole ring. All the predicted intermediates have been confirmed through FTIR, HPLC and UV-vis analyses.

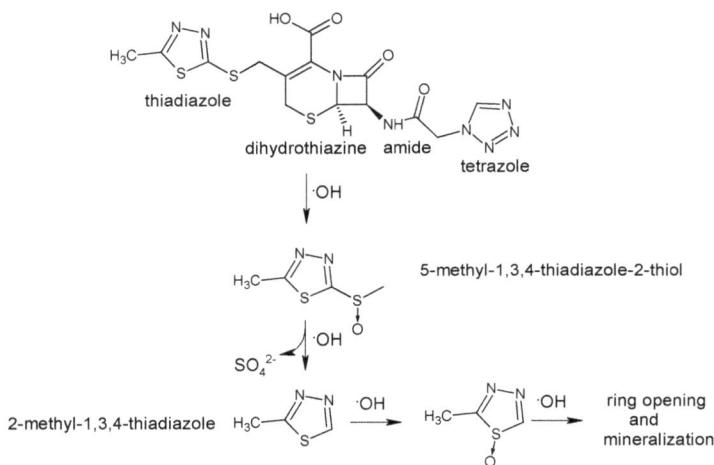

Figure 4. *Pathway I* for the photocatalytic degradation mechanism of cefazolin.

5.2. Transition State Calculations

In order to determine the most plausible reaction path and the product distributions for the photocatalytic degradation reactions of organic pollutants the best way is to use Transition State Theory. For transition state calculations, the first thing to do is to carry out a conformer search for the reactants and the products to determine the most stable structures. Then, geometry optimizations of the reactants and the transition state complexes are performed by quantum mechanical methods. Zero-point corrections are made and the thermodynamic properties of all the species involved in the reactions are calculated.

Vibrational frequencies are calculated for the determination of the reactant and product structures as stationary points and true minima on the potential energy surfaces. All possible stationary geometries located as minima are generated by free rotation around single bonds. The forming C–O bonds in the OH-addition paths and the H–O bonds in the H-abstraction paths are chosen as the reaction coordinates in the determination of the transition states. Each transition state is characterized with only one negative eigenvalue in its force constant matrix.

Product distributions are determined by calculating the rate constant for each of the possible reaction paths by using the Transition State Theory. The classical rate constant k in the Transition State Theory is given by Equation (8);

$$k = \frac{k_B T}{h} \frac{q_{TS}}{q_R \cdot q_{OH}} e^{-E_a/RT} \qquad (8)$$

where k_B is Boltzmann's factor, T is temperature, h is Planck's constant and q's are molecular partition functions for the transition state complex (TS) and the reactant species, (R) and •OH and E_a is the activation energy. Each of the molecular partition functions is the product of the translational, rotational, vibrational and electronic partition functions of the corresponding species. The equations used for computing partition functions are those given in standard texts on thermodynamics. In the computation of the translational partition function, the molecule is treated as a particle with mass equal to the molecular mass of the molecule confined in a three-dimensional box. Vibrational partition function is composed of the sum of the contributions from each vibrational mode. The bottom of the potential well of the molecule is used as the zero of energy. Rotational partition function depends upon the geometry of the molecule, which is determined in the optimization step of the calculations, while electronic partition function is composed of the sum of the contributions from each electronic energy level. Thus, it depends upon the energies of the orbitals and their degeneracies. The first and higher excited states are assumed to be inaccessible at any temperature since the first excitation energy is much greater than $k_B T$.

In an earlier study [57], in order to determine the product distributions and also to develop a short-cut model for the photocatalytic degradation reactions of phenol derivatives, their reactions with the photogenerated •OH radicals have been modeled. Forty-three different reaction paths for the reactions of 11 phenol derivatives with the •OH radical have been determined by nature of the carbon atoms of the aromatic ring, the substituent and the functional –OH group. •OH radical additions to the aromatic rings yielding dihydroxycyclohexadienyl type radicals and direct H-atom abstraction from the phenolic functionalities have been modeled. For all the possible reaction routes, calculations of the geometric parameters, the electronic and thermodynamic properties of the reactants, the product radicals and the transition state complexes have been performed with the semi-empirical PM3 and DFT-COSMO methods successively. The molecular orbital calculations have been carried out by a self-consistent field SCF method using the restricted RHF or unrestricted UHF Hartree–Fock formalisms depending upon the multiplicity of each species. Single point energies were then refined by DFT calculations. Based on the results of the calculations, the rate constant k for each reaction path has been calculated by using the transition state theory for 300 K. The branching ratios and the product distributions of all the possible reaction paths have been calculated by dividing the corresponding rate constant of each reaction path by overall k taking the number of similar addition centers into account.

The results obtained have been compared with the available experimental data in order to assess the reliability of the proposed model. The values indicate that DFT/B3LYP/6-31G*//PM3 calculations underestimate the rate constants. The reason may be attributed to the use of the PM3 method which is wavefunction based. The use of higher basis sets could modify the results, but they are not affordable in terms of computational time and resources. However, the primary intermediates determined in this study are in perfect agreement with the experimental ones. Therefore, it may be concluded that the proposed theoretical model, even by the PM3 method can be used for the estimation of the rates and the product distributions for the photocatalytic degradation reactions of phenol derivatives or similar pollutant molecules in gas and aqueous media [48,58–60].

With the aim of describing the mechanism of the photocatalytic degradation reaction of 4-chlorophenol (4-CP) in detail, the reaction between •OH radical and 4-CP has been modeled by means of the DFT method for geometry optimizations in order to determine the identities and the relative concentrations of the primary intermediates [46]. The DFT calculations have been performed by the hybrid B3LYP functional by using 6-31G* basis set. As seen in Figure 5, four different possible reaction paths for the reaction of 4-CP with the •OH radical have been determined by nature of the carbon atoms of the aromatic ring and the functional –OH group. The first three of the reaction paths, *ortho*-addition, *meta*-addition and *ipso*-addition are OH-addition reactions, which yield dihydroxy-chlorocyclohexadienyl type radicals. The fourth reaction path, H-abstraction is hydrogen abstraction from the functional –OH group producing 4-chlorophenoxyl radical and a water molecule. The unexpected result obtained in this study is that the energies of the transition states, optimized with the DFT/B3LYP/6-31G* method are lower than the energies of the reactants, 4-CP and the •OH radicals. This finding indicates that in all the reaction paths, pre-reactive complexes which lower the energy barriers are formed before the formation of the transition state complexes. Therefore, pre-reactive complexes (PCs) have been located on the potential energy surfaces. Similar results have been obtained for the degradation reactions of nitrobenzene and toluene [61,62].

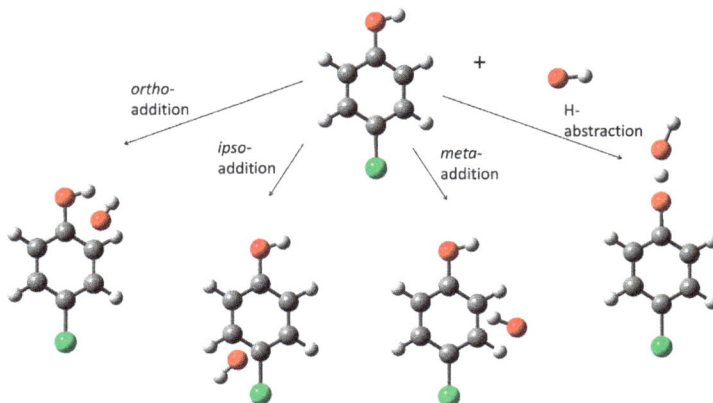

Figure 5. Possible reaction paths for 4-CP + •OH reaction.

The product distribution obtained shows that the major primary intermediate that is formed in the photocatalytic degradation of 4-CP is 1,2-dihydroxy-4-chlorocyclohexadienyl radical which then forms 4-chlorocatechol. The reaction also yields 1,4-dihydroxy-4-chlorohexadienyl radical through *ipso*-addition to the aromatic ring. As the chlorine substituent is then released due to the presence of water molecules and steric effects, this radical is converted to hydroquinone. The results obtained indicate that the major intermediates of the 4-CP + •OH reaction are 4-chlorocatechol (4-CC) and hydroquinone (HQ) with [4-CC] > [HQ], confirming the experimental findings of earlier studies reported in the literature [39].

In an earlier study, a combination of experimental and quantum mechanical methods has been used for the reaction of toluene with •OH radical in both gas and aqueous media in order to determine the most probable reaction path and the product distribution [62]. The obtained experimental and theoretical results are in perfect agreement. Both of them indicate that the dominant reaction path is the *ortho*-addition yielding *ortho*-cresol. Quantum mechanical results show that the primary intermediate is 1-hydroxy-2-methylcyclohexadienyl radical which then forms *ortho*-cresol through abstraction of the redundant ring hydrogen by molecular oxygen.

6. TiO$_2$ Surfaces

Molecular modeling techniques can also be used to examine TiO$_2$ surfaces either modified or doped. Localized and delocalized modeling methods are the two main modeling techniques to be used in the quantum mechanical studies of crystalline solids and surfaces. In delocalized modeling technique, it is assumed that the presence of translational symmetry in crystalline solids leads to periodic functions as solutions of the Schrödinger equation. The simplest periodic functions are known as plane waves. Periodic models are advantageous, because they have no surfaces, but they need exceptionally large repeating units that increase the computational cost. In localized modeling technique, finite clusters are modeled by using molecular orbitals instead of plane waves. In contrast to periodic models, cluster models use small representative portions of the crystal that can be treated with molecular quantum mechanical methods. Plane waves are delocalized and do not refer to a particular site in the crystal lattice. However, surface modifiers, dopants and defects are localized, thus their description by cluster models is favorable over periodic models. There is one problem in the use of cluster models: they have more surface area than the real crystal. Free clusters have borders such as corners, surfaces and edges that are not present in the bulk of the crystal. Therefore, the surface area should be reduced by saturating the unsaturated atoms at the surface by other type of atoms or groups that are similar to the ones in the environment of the crystal.

6.1. Doped TiO₂ Surfaces

The anatase phase is the most used TiO_2 photocatalyst. Among the low-index planes, (101), (100) and (001) lattice planes present on the surface of anatase, (001) surface is the most stable surface with a high photocatalytic activity [63,64]. Therefore, to determine the electronic properties of the doped or surface modified TiO_2 surfaces, the non-defective undoped anatase (001) surface is modeled first with finite, neutral and stoichiometric cluster models. The reason for using neutral clusters is to avoid associating formal charges with the cluster. The bare TiO_2 cluster models can be constructed by using the structure of the anatase unit cell. The unit cell for anatase has a tetragonal structure with the bulk lattice constants $a = b = 3.78$ A and $c = 9.51$ A [65]. The building stone is a slightly distorted TiO_6 octahedron, with the oxygens at the corners. Each octahedron is in contact with eight neighbors, four sharing an edge and four a corner. Ti^{4+} cations are coordinated to six O^{2-} anions and the oxygen atoms are coordinated to three titanium atoms. The small cluster models can be enlarged by extending the lattice vectors to construct larger models, supercells. However, the anatase surface is Lewis acidic due to the presence of adsorbed water molecules [66]. Water adsorption on anatase occurs mostly by dissociative adsorption. Therefore, in the cluster models, to saturate the free valence at the surface and also to keep the coordination of the surface atoms the same as that in the bulk, the unsaturated oxygens are terminated with hydrogens and titaniums with OH groups.

In an earlier study, the bare (001) surface of anatase has been modeled by two different sized cluster models and the electronic properties of the cluster surfaces have been calculated by quantum chemical methods [63]. All the calculations have been performed using the DFT/B3LYP method within the GAUSSIAN 03 package [67] because it takes electron correlation into account. The double-zeta LanL2DZ basis set has been used in order to take relativistic effects into account. The cluster geometry has been kept frozen throughout all the calculations, but the terminal hydrogens and the OH groups are relaxed. Electronic structure calculations indicate that the upper states of the valence band (VB) are dominated by O 2p orbitals, while the bottom of the conduction band (CB) is mainly due to Ti 3d orbitals. The DFT/B3LYP calculations for such clusters generally underestimate band-gap energies, due to the well-known shortcoming of the exchange–correlation potential used within the framework of DFT [68,69]. However, calculated band-gap energies for small clusters do not reflect this effect, because the energy of the first excited state increases as the model size decreases. Band-gap energies for large clusters are corrected by using a scissors operator that displaces the empty and occupied bands relative to each other by a rigid shift to bring the minimum band-gap in line with experimental band-gap of anatase.

In order to determine the location and the bonding status of the dopants, doped anatase clusters are modeled. The doped models can be constructed by locating the dopant either substitutionally or interstitially depending upon the radius of the dopant into the bare anatase cluster. Then, site preference of the dopant on the surface is determined by changing the position of the dopant and calculating total energies of the doped clusters. The geometric parameters, the band edges, band-gap energies and Mulliken charge distributions of the surface atoms can be calculated for the doped model with the optimum dopant position.

Gurkan et al. [63] have examined selenium (IV)-doped TiO_2 experimentally and theoretically. Only substitutional sites for Se (IV) ion have been analyzed. The doped models have been constructed by replacing one titanium atom by one selenium atom as seen in Figure 6. DFT/B3LYP/LanL2DZ calculations indicate that four-fold coordinated Ti site substitution is favored over five-fold coordinated Ti site substitution. Moreover, the results show that doping with Se (IV) does not cause a significant change in the positions of the band edges; instead, it introduces three empty mid-gap levels into the band-gap. The calculated coefficients of the wavefunctions indicate that they are mainly originating from the Se 3p orbitals. These energy levels are not donor states because they are not populated by electrons, but they are allowed energy states. These levels separate the band-gap of the Se(IV)-doped TiO_2 into two parts; a wider lower gap and a significantly narrower upper gap. These intermediate energy levels offer additional steps for the absorption of low energy photons through the excitation of VB electrons to these intermediate energy levels, from where they can be excited again to the CB. The lower gap has been calculated to be 2.85 eV corresponding to a 435 nm photon. The result is consistent with the UV-DRS spectrum of the sample which shows two absorption thresholds. Therefore, it may be stated that the lower gap is responsible for the absorption in the first region of the spectrum between 420 and 580 nm, while the second region between 580 and 650 nm corresponds to the excitation of electrons from mid-gap levels to the CB.

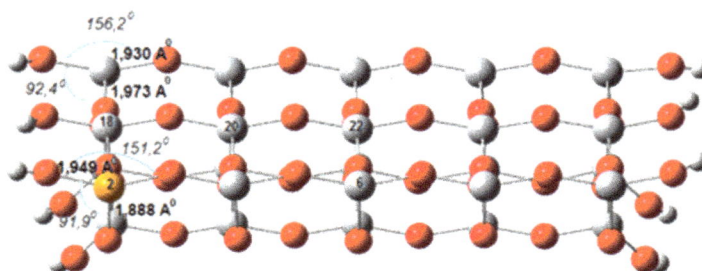

Figure 6. Optimized structure of the Se(IV)-doped TiO_2 cluster model (grey, titanium; red, oxygen; orange, selenium; white, hydrogen).

6.2. Surface Modifiers

Surface modifiers enhance the surface coverage of pollutant molecules on TiO_2, inhibit e^-/h^+ recombination process by separating the charge pairs and expand the wavelength response range. Benzoic acid, salicylic acid and ascorbic acid have been used as surface modifiers to increase the photocatalytic activity of TiO_2 [70,71]. The experimental results indicate that these compounds modify the surface of the particles through the formation of π–π donor–acceptor complexes [72]. Salicylic acid SA has been used as a modifier for TiO_2 and it has been reported that the onset of absorption of the SA-modified TiO_2 shifts to the red and the threshold of absorption is at 420 nm compared with 380 nm for the unmodified TiO_2 [70]. Although it has been known that surface modification by aromatic carboxylic acids increases the photocatalytic activity of TiO_2, the key question to be answered is the electronic nature of the surface and the role of the surface complexes that lead the absorption edge to be red-shifted.

In order to investigate the effect of surface modification with salicylic acid SA, the non-defective anatase (001) surface modified with SA has been modeled with a finite, neutral, stoichiometric cluster model SA-$Ti_3O_{11}H_8$ [73]. Almost 40% of the TiO_2 surface consists of Ti atoms whose coordination is incomplete [74]. These atoms are four-fold coordinated to oxygen and have two unfilled orbitals. Therefore, they can accept two lone electron pairs from electron donors to complete the octahedral coordination [71]. In the SA-modified TiO_2, the oxygen atoms of SA bond to these Ti cations, resulting in a chelate structure as seen in Figure 7. Both carboxyl and hydroxyl groups of SA are involved in binding. A bidentate binding of SA occurs through the two oxygen atoms yielding a six-membered ring structure which is the favorable configuration of the surface Ti atoms. As a result of surface modification, titanium(IV) salicylate charge-transfer complex is formed on TiO_2 particles through the charge transfer reaction from the modifier (SA) to the conduction band of TiO_2.

Electronic structure calculations indicate that conduction band minimum (CBM) shifts to a lower energy level in the SA-modified TiO_2 with respect to the bare TiO_2 due to the electron transfer from SA to the conduction band (CB) of TiO_2. The formation of the bidentate charge transfer complex on TiO_2 introduces additional electronic states into the band gap. The results indicate that the formation of the SA-bidentate complex on TiO_2 particles causes an important reduction in the band gap of the photocatalyst. As a result of the study, 396 nm for the absorption threshold of the SA-TiO_2 has been obtained which is in agreement with the experimental result reported in the literature [70].

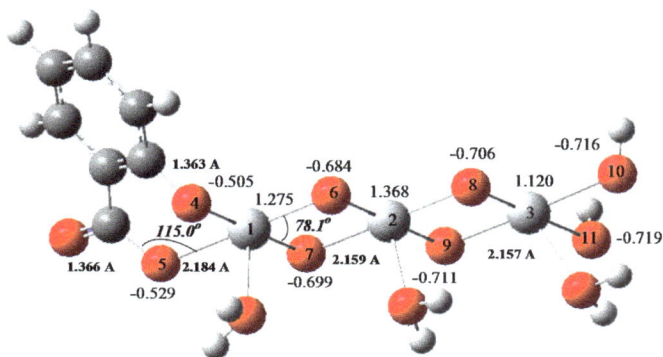

Figure 7. SA-modifed TiO_2 cluster model (Grey, titanium; red, oxygen; white, hydrogen; black, carbon).

The surface modification does not only reduce the band gap, but it also causes a change in the local charge distribution of the atoms on TiO_2 surface. The Mulliken charge distribution of the atoms on the surface of the SA-TiO_2 indicates that electron transfer occurs from SA to the conduction band of TiO_2, causing the separation of the charge pairs on the surface into two phases. The holes on the complex can readily oxidize the adsorbed OH^- ions or H_2O molecules and form •OH radicals. As the holes are on the bidentate complex and the electrons are on the TiO_2 cluster, the electron–hole recombination process will be greatly inhibited, resulting in an increase in the photocatalytic degradation rates of organic contaminants. Similar results have been obtained for ascorbic acid modified TiO_2 samples through quantum mechanical calculations [71].

7. Summary and Outlook

This review highlights the most important molecular modeling methods and applications in TiO_2 photocatalysis. The relative advances in the synthesis and preparation of TiO_2 photocatalysts have led to a huge range of novel materials with improved properties and surface phenomena. The photocatalytic degradation reactions of pollutants have gained considerable attention in the last decade due to an increased number of hazardous synthetic pollutants that are being consumed. Selected examples of the applications of molecular modeling methods to the problems of TiO_2 photocatalysis have been explained.

Molecular modeling techniques are very important tools for TiO_2 photocatalysis. They use quantum mechanical methods that allow us to study chemical phenomena by running calculations on computers rather than carrying out experiments. Quantum mechanical calculations appear promising as a means of describing the mechanisms and the product distributions of the photocatalytic degradation reactions of organic pollutants in both gas and aqueous phases. Since quantum mechanical methods utilize the principles of particle physics, their use may be extended to the design of new photocatalysts.

In TiO$_2$ photocatalysis the primary process is the generation of •OH radicals from the adsorbed OH groups. •OH radicals are considered the principle reactive species responsible for the photocatalytic reaction. Therefore, photocatalytic degradation reactions of aromatic pollutants on TiO$_2$ may be based on hydroxyl radical chemistry. Local DFT descriptors describe the reactivities of phenol derivatives and indicate that the photocatalytic degradation reactions are orbital-controlled reactions in which soft–soft interactions dominate. Product distributions may be predicted using even semiempirical quantum mechanical methods. However, DFT methods describe the reaction mechanisms better than the semiempirical ones. COSMO method is a suitable solvation model to be used to describe the reaction kinetics in aqueous media. Surface modification by aromatic carboxylic acids reduces the band gap of TiO$_2$ and inhibits electron–hole recombination process by causing the separation of charges on the surface. Dopants reduce the band-gap either by introducing additional electronic energy levels into the band gap or contributing electrons to the VB of TiO$_2$.

Novel TiO$_2$ photocatalysts are visible-light active and they have higher photocatalytic activities than the standard anatase phase. The key question to be answered is how doping, codoping or surface modification achieve these phenomena. Molecular modeling methods can be used to obtain information about the degradabilities of the pollutants, to predict photocatalytic reaction mechanisms and evaluate the roles of impurities used to develop new photocatalysts. Many of these phenomena or the improved properties of the photocatalysts have not yet been explained in detail at the molecular level. I hope that the brief introduction of molecular modeling methods explained in this review, their applications in TiO$_2$ photocatalysis and selected examples can provide readers necessary background and ideas to broaden the area of application of TiO$_2$. There is no doubt that cooperative studies of computational chemists and experimentalists will lead to the fabrication of novel TiO$_2$ photocatalysts and to the success of removing hazardous pollutants from air and water resources in the future.

Conflicts of Interest: The author declares no conflict of interest.

References

1. Helz, G.R.; Zepp, R.G.; Crosby, D.G. *Aquatic and Surface Photochemistry*; Lewis Publishers: Baco Raton, FL, USA, 1994; p. 261.
2. Pelaez, M.; Nolan, N.T.; Pillai, S.C.; Seery, M.K.; Falaras, P.; Kontos, A.G.; Dunlop, P.S.M.; Hamilton, J.W.J.; Byrne, J.A.; O'Shea, K.; et al. A review on the Visible Light Active Titanium Dioxide Photocatalysts for Environmental Applications. *Appl. Catal. B* **2012**, *125*, 331–349. [CrossRef]
3. Pichat, P. (Ed.) *Photocatalysis and Water Purification*; Wiley-VCH: Weinheim, Germany, 2013.
4. Schneider, J.; Bahnemann, D.; Ye, J.; Puma, L.G.; Dionysios, D.D. (Eds.) *Photocatalysis: Fundamentals and Perspectives*; Royal Society of Chemistry: London, UK, 2016.
5. Bahnemann, D.; Bockelmann, D.; Goslich, R. Mechanistic Studies of Water Detoxification in Illuminated TiO$_2$ Suspensions. *Sol. Energy Mater.* **1991**, *24*, 564–583. [CrossRef]
6. Ollis, D.F.; Pelizzetti, E.; Serpone, N. Photocatalyzed Destruction of Water Contaminants. *Environ. Sci. Technol.* **1991**, *25*, 1522–1529. [CrossRef]
7. Nosaka, Y.; Nosaka, A. *Introduction to Photocatalysis: From Basic Science to Applications*; Royal Society of Chemistry: London, UK, 2016.
8. Suib, S.L. (Ed.) *New and Future Developments in Catalysis: Solar Photocatalysis*; Elsevier: Amsterdam, The Netherlands, 2013; Volume 7.
9. Pichat, P. *Handbook of Heterogeneous Photo-Catalysis*; Ertl, G., Knozinger, H., Weitkamp, J., Eds.; Wiley-VCH: Weinheim, Germany, 1997; Volume 4, p. 2111.
10. Choi, W.; Termin, A.; Hoffmann, M.R. The Role of Metal Ion Dopants in Quantum-Sized TiO$_2$: Correlation Between Photoreactivity and Charge Carrier Recombination Dynamics. *J. Phys. Chem.* **1994**, *98*, 13669–13679. [CrossRef]
11. Di Paola, A.; Marcì, G.; Palmisano, L.; Schiavello, M.; Uosaki, K.; Ikeda, S.; Ohtani, B. Preparation of Polycrystalline TiO$_2$ Photocatalysts Impregnated with Various Transition Metal Ions: Characterization and Photocatalytic Activity for the Degradation of 4-nitrophenol. *Phys. Chem. B* **2002**, *106*, 637–645. [CrossRef]

12. Karakitsou, K.E.; Verykios, X.E. Effects of Altervalent Cation Doping of Titania on its Performance as a Photocatalyst for Water Cleavage. *J. Phys. Chem.* **1993**, *97*, 1184–1189. [CrossRef]

13. Mu, W.; Herrmann, J.-M.; Pichat, P. Room Temperature Photocatalytic Oxidation of Liquid Cyclohexane into Cyclohexanone over Neat and Modified TiO_2. *Catal. Lett.* **1989**, *3*, 73–84. [CrossRef]

14. Nagaveni, K.; Hegde, M.S.; Madras, G. Structure and Photocatalytic Activity of $Ti_{1-x}M_xO_{2\pm\delta}$ (M = W, V, Ce, Zr, Fe, and Cu) Synthesized by Solution Combustion Method. *Phys. Chem. B* **2004**, *108*, 20204–20212. [CrossRef]

15. Yalcin, Y.; Kilic, M.; Cinar, Z. Fe^{+3}-doped TiO_2: A Combined Experimental and Computational Approach to the Evaluation of Visible Light Activity. *Appl. Catal. B* **2010**, *99*, 469–477. [CrossRef]

16. Zhu, J.; Chen, F.; Zhang, J.; Chen, H.; Anpo, M. Fe^{3+}-TiO_2 Photocatalysts Pprepared by Combining Sol–Gel Method with Hydrothermal Treatment and their Characterization. *J. Photochem. Photobiol. A* **2006**, *180*, 196–204. [CrossRef]

17. Asahi, R.; Morikawa, T.; Ohwaki, T.; Aoki, K.; Taga, Y. Visible-Light Photocatalysis in Nitrogen-Doped Titanium Oxides. *Science* **2001**, *293*, 269–271. [CrossRef] [PubMed]

18. Choi, H.; Antoniou, M.G.; Pelaez, M.; de la Cruz, A.A.; Shoemaker, J.A.; Dionysiou, D.D. Mesoporous Nitrogen-Doped TiO_2 for the Photocatalytic Destruction of the Cyanobacterial Toxin Microcystin-lr under VisibleLight Irradiation. *Environ. Sci. Technol.* **2007**, *41*, 7530–7535. [CrossRef] [PubMed]

19. Di Valentin, C.; Pacchioni, G.; Selloni, A.; Livraghi, S.; Giamello, E. Characterization of Paramagnetic Species in N-doped TiO_2 Powders by EPR Sspectroscopy and DFT Calculations. *J. Phys. Chem. B* **2005**, *109*, 11414–11419. [CrossRef] [PubMed]

20. Emeline, A.V.; Kuzmin, G.N.; Serpone, N. Wavelength-Dependent Potostimulated Adsorption of Molecular O_2 and H_2 on Second Generation Titania Photocatalysts: The Case of the Visible-Light-Active N-doped TiO_2 System. *Chem. Phys. Lett.* **2008**, *454*, 279–283. [CrossRef]

21. Jagadale, T.C.; Takale, S.P.; Sonawane, R.S.; Joshi, H.M.; Patil, S.I.; Kale, B.B.; Ogale, S.B. N-doped TiO_2 Nanoparticle Based Visible Light Photocatalyst by Modified Peroxide Sol–Gel Method. *J. Phys. Chem. C* **2008**, *112*, 14595–14602. [CrossRef]

22. Lu, N.; Zhao, H.; Li, J.; Quan, X.; Chen, S. Characterization of Boron-Doped TiO_2 Nanotube Arrays Prepared by Electrochemical Method and its Visible Light Activity. *Sep. Purif. Technol.* **2008**, *62*, 668–673. [CrossRef]

23. Ohno, T.; Akiyoshi, M.; Umebayashi, T.; Asai, K.; Mitsui, T.; Matsumura, M. Preparation of S-Doped TiO_2 Photocatalysts and their Photocatalytic Activities under Visible Light. *Appl. Catal. A* **2004**, *265*, 115–121. [CrossRef]

24. Sakthivel, S.; Janczarek, M.; Kisch, H. Visible Light Activity and Photoelectrochemical Properties of Nitrogen-Doped TiO_2. *J. Phys. Chem. B* **2004**, *108*, 19384–19387. [CrossRef]

25. Sakthivel, S.; Kisch, H. Daylight Photocatalysis by Carbon-Modified Titanium Dioxide. *Angew. Chem. Int. Ed.* **2003**, *42*, 4908–4911. [CrossRef] [PubMed]

26. Sathish, M.; Viswanathan, B.; Viswanath, R.P.; Gopinath, C.S. Synthesis, Characterization, Electronic Structure, and Photocatalytic Activity of Nitrogen-Doped TiO_2 Nanocatalyst. *Chem. Mater.* **2005**, *17*, 6349–6353. [CrossRef]

27. Sato, S.; Nakamura, R.; Abe, S. Visible-Light Sensitization of TiO_2 Photocatalysts by Wet-Method N Doping. *Appl. Catal. A* **2005**, *284*, 131–137. [CrossRef]

28. Yalcin, Y.; Kilic, M.; Cinar, Z. The Role of Non-Metal Doping in TiO_2 Photocatalysis. *J. Adv. Oxid. Technol.* **2010**, *13*, 281–296.

29. Zheng, R.; Lin, L.; Xie, J.; Zhu, Y.; Xie, Y. State of Doped Phosphorus and its Influence on the Physicochemical and Photocatalytic Properties of P-Doped Titania. *J. Phys. Chem. C* **2008**, *112*, 15502–15509. [CrossRef]

30. Jaiswal, R.; Patel, N.; Kothari, D.C.; Miotello, A. Improved Visible Light Photocatalytic Activity of TiO_2 Co-Doped with Vanadium and Nitrogen. *Appl. Catal. B* **2012**, *126*, 47–54. [CrossRef]

31. Katsanaki, A.V.; Kontos, A.G.; Maggos, T.; Pelaez, M.; Likodimos, V.; Pavlatou, E.A.; Dionysiou, D.D.; Falaras, P. Photocatalytic Oxidation of Nitrogen Oxides on N-F-Doped Titania Thin Films. *Appl. Catal. B* **2013**, *140–141*, 619–625. [CrossRef]

32. Li, D.; Ohashi, N.; Hishita, S.; Kolodiazhnyi, T.; Haneda, H. Origin of Visible-Light-Driven Photocatalysis: A Comparative Study on N/F-Doped and N-F-Codoped TiO_2 Powders by means of Experimental Characterizations and Theoretical Calculations. *J. Solid State Chem.* **2005**, *178*, 3293–3302. [CrossRef]

33. Ling, Q.; Sun, J.; Zhou, Q. Preparation and Characterization of Visible-Light-Driven Titania Photocatalyst Co-Doped with Boron and Nitrogen. *Appl. Surf. Sci.* **2008**, *254*, 3236–3241. [CrossRef]

34. Márquez, A.M.; Plata, J.J.; Ortega, Y.; Sanz, J.F.; Colón, G.; Kubacka, A.; Fernández-García, M. Making Photo-Selective TiO$_2$ Materials by Cation–Anion Codoping: From Structure and Electronic Properties to Photoactivity. *J. Phys. Chem. C* **2012**, *116*, 18759–18767. [CrossRef]

35. Sun, H.; Bai, Y.; Cheng, Y.; Jin, W.; Xu, N. Preparation and Characterization of Visible-Light-Driven Carbon–Sulfur-CoDoped TiO$_2$ Photocatalysts. *Ind. Eng. Chem. Res.* **2006**, *45*, 4971–4976. [CrossRef]

36. Sun, H.; Zhou, G.; Liu, S.; Ang, H.M.; Tadé, M.O.; Wang, S. Visible Light Responsive Titania Photocatalysts Codoped by Nitrogen and Metal (Fe, Ni, Ag, or Pt) for Remediation of Aqueous Pollutants. *Chem. Eng. J.* **2013**, *231*, 18–25. [CrossRef]

37. Wang, X.; Lim, T.-T. Solvothermal Synthesis of C–N Codoped TiO$_2$ and Photocatalytic Evaluation for Bisphenol a Degradation using a Visible-Light Irradiated LED Photoreactor. *Appl. Catal. B* **2010**, *100*, 355–364. [CrossRef]

38. Yu, J.; Zhou, M.; Cheng, B.; Zhao, X. Preparation, Characterization and Photocatalytic Activity of in situ N,S-Codoped TiO$_2$ Powders. *J. Mol. Catal. A Chem.* **2006**, *246*, 176–184. [CrossRef]

39. Foresman, J.B.; Frisch, A. *Exploring Chemistry with Electronic Structure Methods*; Gaussian Inc.: Pittsburg, PA, USA, 1995.

40. Holtje, H.-D.; Sippl, W.; Rognan, D.; Folkers, G. *Molecular Modeling*; Wiley VCH: Weinheim, Germany, 2003.

41. Jensen, F. *Introduction to Computational Chemistry*; Wiley: New York, NY, USA, 1999.

42. Barone, V.; Cossi, M. Quantum Calculation of Molecular Energies and Energy Gradients in Solution by a Conductor Solvent Model. *J. Phys. Chem. A* **1998**, *102*, 1995–2001. [CrossRef]

43. Andzelm, J.; Kölmel, C.; Klamt, A. Incorporation of Solvent Effects into Density Functional Calculations of Molecular Energies and Geometries. *J. Chem. Phys.* **1995**, *103*, 9312–9320. [CrossRef]

44. Jenks, W.S. Photocatalytic Reactions Pathways: Effect of Molecular Structure, Catalyst and Wavelength. In *Photocatalysis and Water Purification*; Pichat, P., Ed.; Wiley-VCH: Weinheim, Germany, 2013; pp. 25–51.

45. Turchi, C.S.; Ollis, D.F. Photocatalytic Degradation of Organic Water Contaminants: Mechanisms Involving Hydroxyl Radical Attack. *J. Catal.* **1990**, *122*, 178–192. [CrossRef]

46. Kilic, M.; Cinar, Z. Hydroxyl Radical Reactions with 4-chlorophenol as a Model for Heterogeneous Photocatalysis. *J. Mol. Struct. THEOCHEM* **2008**, *851*, 263–270. [CrossRef]

47. Eberhardt, M.K.; Yoshida, M. Radiation-Induced Homolytic Aromatic Substitution. I. Hydroxylation of Nitrobenzene, Chlorobenzene, and Toluene. *J. Phys. Chem.* **1973**, *77*, 589–597. [CrossRef]

48. San, N.; Hatipoglu, A.; Kocturk, G.; Cinar, Z. Photocatalytic Degradation of 4-nitrophenol in Aqueous TiO$_2$ Suspensions: Theoretical Prediction of the Intermediates. *J. Photochem. Photobiol. A* **2002**, *146*, 189–197. [CrossRef]

49. Amalric, L.; Guillard, C.; Blanc-Brude, E.; Pichat, P. Correlation Between the Photocatalytic Degradability over TiO$_2$ in Water of meta and para Substituted Methoxybenzenes and their Electron Density, Hydrophobicity and Polarizability Properties. *Water Res.* **1996**, *30*, 1137–1142. [CrossRef]

50. D'Oliveira, J.-C.; Minero, C.; Pelizzetti, E.; Pichat, P. Photodegradation of dichlorophenols and trichlorophenols in TiO$_2$ Aqueous Suspensions: Kinetic Effects of the Positions of the Cl Atoms and Identification of the intermediates. *J. Photochem. Photobiol. A* **1993**, *72*, 261–267. [CrossRef]

51. Parra, S.; Olivero, J.; Pacheco, L.; Pulgarin, C. Structural Properties and Photoreactivity Relationships of Substituted Phenols in TiO$_2$ Suspensions. *Appl. Catal. B* **2003**, *43*, 293–301. [CrossRef]

52. San, N.; Cinar, Z. Structure-Activity Relations for the Photodegradation Reactions of Monosubstituted anilines in TiO$_2$ Suspensions. *J. Adv. Oxid. Technol.* **2002**, *5*, 85–92. [CrossRef]

53. Vos, A.M.; Nulens, K.H.L.; de Proft, F.; Schoonheydt, R.A.; Geerlings, P. Reactivity Descriptors and Rate Constants for Electrophilic Aromatic Substitution: Acid Zeolite Catalyzed Methylation of Benzene and Toluene. *J. Phys. Chem. B* **2002**, *106*, 2026–2034. [CrossRef]

54. Torrent-Sucarrat, M.; de Proft, F.; Geerlings, P. Stiffness and Raman Intensity: A Conceptual and Computational DFT Study. *J. Phys. Chem. A* **2005**, *109*, 6071–6076. [CrossRef] [PubMed]

55. San, N.; Kilic, M.; Cinar, Z. Reactivity Indices for ortho/para Monosubstituted Phenols. *J. Adv. Oxid. Technol.* **2007**, *10*, 51–59. [CrossRef]

56. Gurkan, Y.Y.; Turkten, N.; Hatipoglu, A.; Cinar, Z. Photocatalytic Degradation of Cefazolin over N-doped TiO$_2$ under UV and Sunlight Irradiation: Prediction of the Reaction Paths via Conceptual DFT. *Chem. Eng. J.* **2012**, *184*, 113–124. [CrossRef]

57. Kilic, M.; Kocturk, G.; San, N.; Cinar, Z. A Model for Prediction of Product Distributions for the Reactions of Phenol Derivatives with Hydroxyl Radicals. *Chemosphere* **2007**, *69*, 1396–1408. [CrossRef] [PubMed]

58. San, N.; Kilic, M.; Tuiebakhova, Z.; Cinar, Z. Enhancement and Modeling of the Photocatalytic Degradation of Benzoic Acid. *J. Adv. Oxid. Technol.* **2007**, *10*, 43–50. [CrossRef]

59. Hatipoglu, A.; San, N.; Cinar, Z. An Experimental and Theoretical Investigation of the Photocatalytic Degradation of *meta*-cresol in TiO₂ Suspensions: A Model for the Product Distribution. *J. Photochem. Photobiol. A* **2004**, *165*, 119–129. [CrossRef]

60. San, N.; Hatipoglu, A.; Kocturk, G.; Cinar, Z. Prediction of Primary Intermediates and the Photodegradation Kinetics of 3-aminophenol in Aqueous TiO₂ Suspensions. *J. Photochem. Photobiol. A* **2001**, *139*, 225–232. [CrossRef]

61. Vione, D.; de Laurentiis, E.; Berto, S.; Minero, C.; Hatipoglu, A.; Cinar, Z. Modeling the Photochemical Transformation of Nitrobenzene under Conditions Relevant to Sunlit Surface Waters: Reaction Pathways and Formation of Intermediates. *Chemosphere* **2016**, *145*, 277–283. [CrossRef] [PubMed]

62. Hatipoglu, A.; Vione, D.; Yalcin, Y.; Minero, C.; Cinar, Z. Photo-Oxidative Degradation of Toluene in Aqueous Media by Hydroxyl Radicals. *J. Photochem. Photobiol. A* **2010**, *215*, 59–68. [CrossRef]

63. Gurkan, Y.Y.; Kasapbasi, E.; Cinar, Z. Enhanced Solar Photocatalytic Activity of TiO₂ by Selenium(iv) Ion-Doping: Characterization and DFT Modeling of the Surface. *Chem. Eng. J.* **2013**, *214*, 34–44. [CrossRef]

64. Homann, T.; Bredow, T.; Jug, K. Adsorption of Small Molecules on the Anatase(1 0 0) Surface. *Surf. Sci.* **2004**, *555*, 135–144. [CrossRef]

65. Sekiya, T.; Igarashi, M.; Kurita, S.; Takekawa, S.; Fujisawa, M. Structure Dependence of Reflection Spectra of TiO₂ Single Crystals. *J. Electron. Spectrosc. Relat. Phenom.* **1998**, *92*, 247–250. [CrossRef]

66. Onal, I.; Soyer, S.; Senkan, S. Adsorption of Water and Ammonia on TiO₂-Anatase Cluster Models. *Surf. Sci.* **2006**, *600*, 2457–2469. [CrossRef]

67. Frisch, M.J.; Trucks, G.W.; Schlegel, H.B.; Scuseria, G.E.; Robb, M.A.; Cheeseman, J.R.; Montgomery, A.M., Jr.; Vreven, T.; Kudin, K.N.; Burant, J.C.; et al. *Gaussian 03*; B01; Gaussian Inc.: Pittsburgh, PA, USA, 2003.

68. Di Valentin, C.; Finazzi, E.; Pacchioni, G.; Selloni, A.; Livraghi, S.; Paganini, M.C.; Giamello, E. N-doped TiO₂:Theory and Experiment. *Chem. Phys.* **2007**, *339*, 44–56. [CrossRef]

69. Bredow, T.; Pacchioni, G. A Quantum-Chemical Study of Pd Atoms and Dimers Supported on TiO₂ (110) and their Interaction with CO. *Surf. Sci.* **1999**, *426*, 106–122. [CrossRef]

70. Li, S.X.; Zheng, F.Y.; Cai, W.L.; Han, A.-Q.; Xie, Y.K. Surface Modification of Nanometer Size TiO₂ with Salicylic Acid for Photocatalytic Degradation of 4-nitrophenol. *J. Hazard. Mater.* **2006**, *135*, 431–436. [PubMed]

71. Mert, E.H.; Yalcin, Y.; Kilic, M.; San, N.; Cinar, Z. Surface Modification of TiO₂ with Ascorbic Acid for Heterogeneous Photocatalysis: Theory and Experiment. *J. Adv. Oxid. Technol.* **2008**, *11*, 199–207.

72. Rajh, T.; Nedeljkovic, J.M.; Chen, L.X.; Poluektov, O.; Thurnauer, M.C. Improving Optical and Charge Separation Properties of Nanocrystalline TiO₂ by Surface Modification with Vitamin C. *J. Phys. Chem. B* **1999**, *103*, 3515–3519. [CrossRef]

73. Kilic, M.; Cinar, Z. A Quantum Mechanical Approach to TiO₂ Photocatalysis. *J. Adv. Oxid. Technol.* **2009**, *12*, 37–46.

74. Xagas, A.P.; Bernard, M.C.; Hugot-Le Goff, A.; Spyrellis, N.; Loizos, Z.; Falaras, P. Surface Modification and Photosensitisation of TiO₂ Nanocrystalline Films with Ascorbic Acid. *J. Photochem. Photobiol. A* **2000**, *132*, 115–120. [CrossRef]

molecules

MDPI

Article

Influence of Se/N Codoping on the Structural, Optical, Electronic and Photocatalytic Properties of TiO$_2$

Yelda Y. Gurkan [1], Esra Kasapbasi [2], Nazli Turkten [3] and Zekiye Cinar [3,*]

[1] Department of Chemistry, Namik Kemal University, 59030 Tekirdag, Turkey; yyalcin@nku.edu.tr
[2] Department of Molecular Biology and Genetics, Halic University, 34220 Istanbul, Turkey; esrakasapbasi@halic.edu.tr
[3] Department of Chemistry, Yildiz Technical University, 34220 Istanbul, Turkey; cinarz@yildiz.edu.tr
* Correspondence: cinarz@yildiz.edu.tr; Tel.: +90-212-383-4179

Academic Editor: Pierre Pichat
Received: 9 January 2017; Accepted: 27 February 2017; Published: 7 March 2017

Abstract: Se^{4+} and N^{3-} ions were used as codopants to enhance the photocatalytic activity of TiO$_2$ under sunlight irradiation. The Se/N codoped photocatalysts were prepared through a simple wet-impregnation method followed by heat treatment using SeCl$_4$ and urea as the dopant sources. The prepared photocatalysts were well characterized by X-ray diffraction (XRD), X-ray photoelectron spectroscopy (XPS), UV-diffuse reflectance spectroscopy (UV-DRS), scanning electron microscopy (SEM) and Raman spectroscopy. The codoped samples showed photoabsorption in the visible light range from 430 nm extending up to 580 nm. The photocatalytic activity of the Se/N codoped photocatalysts was evaluated by degradation of 4-nitrophenol (4-NP). The degradation of 4-NP was highly increased for the Se/N codoped samples compared to the undoped and single doped samples under both UV-A and sunlight irradiation. Aiming to determine the electronic structure and dopant locations, quantum chemical modeling of the undoped and Se/N codoped anatase clusters was performed using Density Functional Theory (DFT) calculations with the hybrid functional (B3LYP) and double-zeta (LanL2DZ) basis set. The results revealed that Se/N codoping of TiO$_2$ reduces the band gap due to mixing of N2p with O2p orbitals in the valence band and also introduces additional electronic states originating from Se3p orbitals in the band gap.

Keywords: TiO$_2$; DFT calculations; Se/N-codoping; sunlight; heterogeneous photocatalysis

1. Introduction

In the last few decades, TiO$_2$ has gained an enormous interest due to its potential application in photocatalysis, solar cells and waste remediation. TiO$_2$-mediated photocatalysis is an efficient and economic method to eliminate recalcitrant contaminants from water or air, because it is non-energy intensive, operates at ambient conditions and able to mineralize organic pollutants using only atmospheric oxygen as the additional chemical species [1–4]. Owing to its high chemical and photo stability, environmental friendliness, water insolubility, low-cost, non-toxicity and high oxidative power, TiO$_2$ has been proven to be the most efficient photocatalyst for this process [5–8].

The photocatalytic reactions on TiO$_2$ are initiated by band-gap excitation and subsequent generation of electron/hole (e$^-$/h$^+$) pairs that can initiate redox reactions on the surface. Electrons are trapped at surface defect sites (Ti^{3+}) and removed by reactions with adsorbed molecular O$_2$ to produce superoxide anion radical O$_2$$^{\bullet-}$, while holes react with adsorbed water molecules or OH$^-$ ions to produce $^\bullet$OH radicals. $^\bullet$OH radicals are considered to be the principal reactive species responsible for the degradation reactions. However, the wide band-gap of TiO$_2$ (~3.2 eV) requires an excitation wavelength that falls in the UV region. This disadvantage of TiO$_2$ limits the utilization of solar energy as a sustainable energy source for its excitation because only 5% of the incoming solar energy on the earth's surface is in the UV-range. In order to utilize natural solar light in TiO_2 photocatalysis, the band gap of TiO_2 must be reduced to be active under visible light irradiation. Recombination of photogenerated charge carriers is another major drawback associated with TiO_2. The majority of the e^-/h^+ pairs generated upon band gap excitation are lost through recombination instead of being involved in redox processes at the surface. The e^-/h^+ recombination process not only decreases the quantum yield but also decreases the oxidation capability of TiO_2 [8,9]. Therefore, in recent years research on TiO_2 has been focused on extending its optical absorption to the visible region of the spectrum in order to substitute UV-light by sunlight and also to increase its photocatalytic activity by decreasing the recombination rate of the charge carriers.

In the past decades, considerable efforts have been devoted to modify the electronic structure of TiO_2. The most common method is doping in which impurities are introduced into the TiO_2 matrix in order to reduce the band gap. The dopants develop electronic energy levels within the band gap for absorption of photons or contribute electrons to the valence band (VB). The dopants also behave as trapping sites for electrons and holes to significantly reduce the recombination processes thus prolonging the lifetime of the charge carriers. Metal ion doped TiO_2 photocatalysts have been extensively studied and found to enhance photocatalytic activity in the visible range [10]. However, some investigators have reported that doping with metal ions enhances the photocatalytic activity while some research groups have found that the presence of cations in TiO_2 is detrimental for the photocatalytic degradation reactions of organics in aqueous systems [11–16]. Moreover, thermal instability and increase in the charge carrier recombination centers have caused metal ion dopants to be unfavorable. Therefore, non-metal doping of TiO_2 has gained considerable attention as an approach to overcome the drawbacks of metal doping.

Non-metal C, N, S, F, B-doped TiO_2 photocatalysts have been found to show a relatively high level of activity under visible-light irradiation [17–22]. These anion dopants either reduce the band gap of TiO_2 through mixing their p orbitals with O2p orbitals or introduce additional energy levels into the band gap. Nitrogen seems to be more attractive than all the other anionic dopants because of its comparable atomic size with oxygen, small ionization energy and stability. After Asahi et al. [23] have reported that nitrogen doping of TiO_2 extends its light absorption to visible light range, nitrogen-doped (N-doped) TiO_2 has been extensively studied [24–29]. However, at high dopant concentrations, the impurity levels in the non-metal doped TiO_2 act as charge recombination centers and reduce photoactivity.

Recently, it has been reported that the photocatalytic activity of TiO_2 doped with non-metals can be further increased by the presence of a non-metal ion as a codopant [30]. Codoping of TiO_2 has exhibited significant improvement in photocatalytic activity as compared to single doping due to synergistic effects of two different non-metals. Yu et al. [31] have investigated N/S-codoped TiO_2 and obtained a high day-light induced photocatalytic activity. Wang et al. [32] have synthesized N/C-codoped TiO_2 by a solvothermal method and reported that the surface of TiO_2 was modified by both C and N via formation of Ti-C bonds, carbonate species and oxynitrides. They have evaluated the photocatalytic activity of their samples by investigating the degradation reaction of bisphenol. Li et al. [33] have obtained high visible light activity for N/F-codoped TiO_2. In another study, visible light activated TiO_2 with N and F codopants have been prepared by the surfactant assisted sol-gel method and immobilized on glass substrates [34]. The prepared films have been examined for the oxidation of NO and the modified catalysts have exhibited significant photocatalytic activity under daylight illumination. B and N codoped titania photocatalyst has been synthesized by Ling et al. [35]. The results of their study have shown that the codoping of B and N played an important role in the band gap decrease, which led to the rise of the photocatalytic activity.

Although codoping with two non-metals has been believed to be superior to single doping, codoping at two anionic sites induces significant crystal distortion and charge unbalance resulting in a high recombination rate of the charge carriers [36]. Therefore, more recently, most researchers have concentrated on codoping with metal and non-metal combinations. In this case, the former contributes to the VB while the later forms additional levels in the band gap. N/V-codoped TiO$_2$ has been investigated and found that codoping with V and N induces isolated energy levels near the conduction band (CB) and VB causing an effective narrowing of the band gap [37]. Kubacka et al. [38] have synthesized micro-crystalline W/N-codoped TiO$_2$ that showed high activity under sunlight. The structural and electronic properties of the codoped photocatalyst have been explored by combining spectroscopic data with Density Functional Theory (DFT) calculations. Effect of metal ions (Fe, Ni, Ag, Pt) on the physicochemical properties of N-doped TiO$_2$ has been investigated experimentally [36]. A negative effect of Fe and Ni was observed while Ag and Pt codopants have positive effects.

In our previous study [39], we doped TiO$_2$ with Se^{4+} ions. Characterization techniques showed that Se^{4+} is in O–Se–O linkages in the crystal lattice. The absorption threshold of the Se^{4+}-doped photocatalyst shifted to the visible region of the spectrum. We obtained a higher photocatalytic activity for the degradation of 4-nitrophenol (4-NP) for the Se^{4+}-doped TiO$_2$ compared to the undoped TiO$_2$. However, we did not observe any direct correlation between the visible light activity and the photocatalytic activity of the doped samples. Our DFT calculations indicated that Se^{4+}-doping of TiO$_2$ does not cause a significant change in the positions of the band edges; in contrast, it produces additional electronic states originating from the Se 3p orbitals in the band-gap. The visible-light photocatalytic activity of the Se^{4+}-doped TiO$_2$ is due to these localized mid-gap levels. In one of our earlier studies [40], we doped TiO$_2$ with N^{3-} ions. Characterization techniques showed that nitrogen anions are in O–Ti–N linkages and the dopant nitrogen led to an important reduction in the band-gap through substitutional N-doping. We obtained a higher photocatalytic activity for the degradation of 4-NP. Our DFT calculations indicated that band gap reduction arises from the contribution of N 2p to the O 2p and Ti 3d states in the VB of TiO$_2$.

Based on these results, we attempted to dope TiO$_2$ with Se^{4+} and N^{3-} ions simultaneously to obtain a more active, visible-light driven photocatalyst. This paper has the purpose of determining the electronic structure, optical and photocatalytic properties of Se/N-codoped TiO$_2$, to elucidate the chemical nature, the position and the synergistic effect of the dopants on the activity of the photocatalyst. For this purpose, a combination of experimental and quantum mechanical methods were used. In the experimental part of the study, a series of Se/N-codoped TiO$_2$ photocatalysts were prepared by means of a simple wet impregnation method and characterized by structural techniques. The photocatalytic activity of the Se/N-codoped TiO$_2$ was also determined by investigating the kinetics of the photocatalytic degradation of 4-NP in the presence of the undoped and Se/N-codoped TiO$_2$. Modeling of the undoped and Se/N-codoped clusters was performed using DFT calculations to provide a framework for the interpretation of the experimental data and to elucidate the structural and electronic properties of the Se/N-codoped titania.

2. Experimental and Computational Details

2.1. Materials

TiO$_2$ Evonik P-25 grade (Degussa Limited Company, Istanbul, Turkey) with a particle size of about 21 nm and a surface area of 50 m$^2 \cdot$g^{-1} was used as the photocatalyst without further treatment. Evonik P-25 powder, which is a mixture of anatase and rutile phases (80% anatase, 20% rutile) was chosen as the precursor for Se/N-codoping, in order to compare the results with the previous ones. Moreover, Evonik P-25 is the standard photocatalyst with high activity and has a well-known structure and photocatalytic data. SeCl$_4$, urea and 4-NP were purchased from Merck (Istanbul, Turkey). All the chemicals that were used in the experiments were of laboratory reagent grade and used as received without further purification. The solutions were prepared with doubly distilled water.

2.2. Preparation of Se/N-Codoped TiO$_2$

Doping was performed by an incipient wet impregnation method in order to prevent penetration of the dopant ions into the bulk of TiO$_2$, since bulk doping increases the recombination rate of charge carriers resulting in a decrease in photocatalytic activity. $SeCl_4$ was used as the Se-source and urea as the N-source. 10 g TiO_2 Evonik P-25 was mixed with 10 mL of aqueous solutions of $SeCl_4$ and urea and stirred at room temperature for 1 h. During this period, the mixture changed color into a pinkish-beige depending upon the dopant concentration. Five different Se/N-codoped photocatalysts containing (wt. %) 0.1 N–0.25 Se, 0.25 N–0.1 Se, 0.5 N–0.5 Se, 0.25 N–0.25 Se, 0.1 N–0.1 Se were prepared. Then, the prepared photocatalysts were washed with water and centrifugally separated three times, heat-treated at 378 K for 24 h to eliminate water, calcined at 623 K for 3 h, ground and sieved. Three different temperatures (623, 723 and 823 K) and three different times (1, 3 and 5 h) were applied to the sample containing 0.5% N–0.5% Se in order to determine the effects of calcination temperature and period on the structure of the photocatalyst.

2.3. Characterization Techniques

In order to determine the effect of Se/N-codoping on the crystal structure of TiO_2, X-ray Diffraction (XRD) patterns were obtained. XRD measurements were carried out at room temperature by using a Philips Panalytical X'Pert Pro X-ray (Philips, Eindhoven, The Netherlands) powder diffraction spectroscope with Cu Kα radiation (λ = 1.5418 A). The accelerating voltage and emission current were 45 kV and 40 mA respectively. The scan ranged from 20 to 70 (2 theta degree) with a scan rate of $3°·min^{-1}$. Crystallite size was determined using the Scherrer equation:

$$d = \frac{(0.9\lambda 180)}{(\pi FWHM_{hkl} \cos\theta)}$$ (1)

where $FWHM_{hkl}$ is the full width at half-maximum of an *hkl* peak at θ value. The crystal structure was further analyzed by Raman spectroscopy. Raman spectra were acquired by a PerkinElmer 400F dispersive Raman spectrometer (Perkin Elmer, Waltham, MA, USA) equipped with dielectric edge filters and a cooled CCD detector. Samples were excited using a near infrared 765 nm laser pulse. To examine the morphological structure of the Se/N-codoped TiO_2 photocatalysts, scanning electron microscopy (SEM) was performed on gold-coated samples by using a SEM apparatus (JEOL JSM 5410 LV, Peabody, MA, USA) operated at an accelerating voltage of 10 kV. The UV-visible diffuse reflectance spectra (UV-DRS) were recorded on a Perkin Elmer Lambda 35 spectrometer equipped with an integrating sphere assembly using $BaSO_4$ as the reference material. The analysis range was from 200 to 800 nm. Surface properties of the codoped samples were examined by X-ray photoelectron spectroscopy (XPS). XPS measurements were performed on a SPECS ESCA (Berlin, Germany) system with MgKα source (hv = 1253.6 eV) at 10.0 kV and 20.0 mA respectively. All the binding energies were referenced to the C 1s peak at 284.5 eV. Gaussian/Lorentzian peak shapes were utilized for curve fitting.

2.4. Photocatalytic Experiments

The performance of the Se/N-codoped TiO_2 was assessed on 4-NP by carrying out the photocatalytic degradation reactions under both UV-A and sunlight irradiation. The photocatalytic activity experiments were carried out in a Pyrex double-jacket photoreactor. A water bath connected to a pump was used to maintain the reaction temperature constant. 5 × 8 W blacklight fluorescent lamps emitting light between 300 and 400 nm with a maximum at 365 nm were used as the light source for UV-A irradiation. Total photonic fluence was determined by potassium ferrioxalate actinometer [41] as 3.1×10^{-7} Einstein·s^{-1}. The experiments under solar light were performed in the second week of May (the outside temperature was 29 °C) in Istanbul (41°02′ latitude, 28°97′ longitude). The daily average solar light intensity was 650 W/m^2.

In the experiments, a stock solution of 4-NP at a concentration of 1.0×10^{-2} mol·L^{-1} was used. The suspension was prepared by mixing specific volumes of this solution containing the desired amount of 4-NP with TiO_2 Evonik P-25 and the Se/N-codoped TiO_2. The suspension was agitated in an ultrasonic bath for 15 min in the dark before introducing it into the photoreactor, to ensure adsorption equilibrium between the photocatalyst and 4-NP. The concentration of 4-NP was constant before irradiation. The volume of the suspension was 600 mL. The amount of the photocatalyst used was 0.2 g/100 mL, which was determined as the corresponding optimum photocatalyst concentration. The suspension was stirred mechanically throughout the reaction period in order to prevent TiO_2 particles from settling. The temperature of the reaction solution was 23 ± 2 °C. Under these conditions, the initial pH was at the natural pH of 4-NP, 5.8 ± 0.1 as measured by a pH-meter (Metrohm 632, Istanbul, Turkey). Duplicate experiments were performed unless otherwise stated.

All the samples, each 10 mL in volume were taken intermittently for analysis. The samples were then filtered through 0.45 μm cellulose acetate filters (Millipore HA, Istanbul, Turkey). The concentration of 4-NP was measured by a UV-Visible spectrophotometer (Agilent 8453, Santa Clara, CA, USA) at 318 nm which was the wavelength of maximum absorption of 4-NP. The calibration curves were prepared for a concentration range of $(1.0–10.0) \times 10^{-5}$ mol·L^{-1} and the detection limit for 4-NP was calculated to be 3.79×10^{-6} mol·L^{-1}. In the experiments, the pH of the reaction solution decreased slightly. For 120 min of degradation the change in the pH was ± 0.1, which did not affect the wavelength of maximum absorption in the UV-spectrum of 4-NP.

2.5. Computational Models and Methodology

Quantum mechanical modeling techniques were employed in order to determine the effect of the codopants Se^{4+} and N^{3-} on the electronic and optical properties of TiO_2. Of the two theoretical modeling techniques used for crystalline solids and surfaces, localized modeling technique was used in this study, since dopant ions in crystals are localized. This technique describes small representative portions of the crystal by molecular orbitals.

The anatase phase is the most abundant phase of Evonik P-25 powder. (001) surface is known to have the highest stability and photocatalytic activity among the low index planes of anatase [42]. Therefore, in order to determine the location and the bonding status of the dopant ions, the non-defective anatase (001) surface was modeled with saturated, finite, neutral, and stoichiometric cluster models, cut from the anatase bulk structure. For the free cluster models "water saturation technique" was used in order to avoid spin localization and boundary effects [43].

Two different sized cluster models were considered. The primitive cluster $Ti_7O_{18}H_8$ was constructed by using the structure of the anatase unit cell [44]. The primitive cell was then enlarged by extending the lattice vectors resulting in a supercell $Ti_{25}O_{55}H_{10}$ with $4 \times 2 \times 1$ repetitive units respectively. The construction and the properties of the two undoped TiO_2 clusters have been reported and explained in detail previously [39].

In the Se/N codoped models, substitutional locations of Se^{4+} ion were analyzed. The structures of the codoped models were constructed by replacing one titanium atom by one selenium atom. For the codopant N, both substitutional and interstitial locations were analyzed. For substitutional models one oxygen was replaced by one nitrogen. In the interstitial model, one nitrogen was added and one OH group was removed. In order to keep the number of atoms the same as in the substitutional model, an oxygen vacancy was also created by using a dummy atom. The anatase surface is Lewis acidic due to the presence of adsorbed water molecules. Water adsorption on anatase surface occurs mostly by dissociative adsorption. Therefore, in the clusters developed, the unsaturated oxygen atoms were terminated with hydrogens and titanium atoms with OH groups, in order to saturate the free valence at the surface and also to keep the average coordination of the surface cluster atoms the same as that in the bulk.

All the calculations were carried out using the Density Functional Theory DFT method within the GAUSSIAN 09 package [45]. The DFT calculations were performed by the hybrid B3LYP functional which combines Hartree-Fock (HF) and Becke exchange terms with the Lee-Yang-Parr correlation functional. The double-zeta LanL2DZ basis set was used in order to take the relativistic effects into account. The dopant positions were optimized by changing their locations in the clusters to find the lowest energy configuration. Optimized geometries of the clusters were calculated to obtain the geometric parameters, the band edges and the band gap energies E_g of the undoped and Se/N-codoped photocatalysts.

3. Results and Discussion

3.1. Crystal Structure

Figure 1a shows XRD diffractograms of the undoped and Se/N-codoped TiO$_2$ samples containing 0.5% Se–0.5% N. The XRD diffractogram of the undoped TiO$_2$ (Evonik P-25) shows the presence of both anatase and rutile phases. XRD diffractograms of the Se/N-codoped TiO$_2$ have typical peaks of anatase and rutile without any detectable dopant-related peaks. This result reveals that neither Se^{4+} ions nor N^{3-} react with TiO$_2$ to form new crystalline phases, the dopants may have moved into the substitutional or interstitial sites of the TiO$_2$ crystal structure. The peaks for Se/N-codoped TiO$_2$ samples show peak broadening with the dopant-content, which indicates a reduction in the crystallite size and a higher disorder or defectiveness of the crystallites, since doping can lead to formation of new defects and disorder in the particles. The average crystallite sizes of the samples were estimated using the Scherrer equation and presented in Table 1.

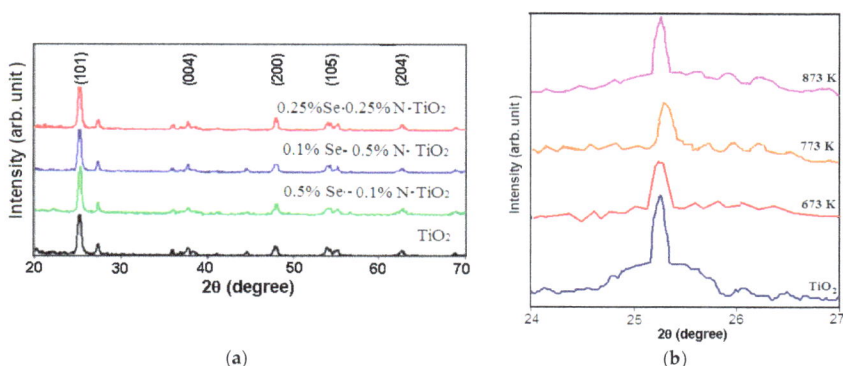

(a)   (b)

Figure 1. (a) X-ray Diffraction (XRD) diffractograms for undoped and 0.5% Se–0.5% N-codoped TiO$_2$; (b) XRD peaks for (101) planes of undoped and 0.5% Se–0.5% N-codoped TiO$_2$.

Table 1. Crystallite sizes, band gap energies E_g and absorption wavelengths λ for the undoped and Se/N-codoped TiO$_2$ samples.

Samples	Calcination Temperature (K) [1]	Crystallite Size (nm)	λ (nm)	Eg (eV)
TiO$_2$ Evonik P-25	623	22.3	411	3.01
0.25% Se–0.1% N	623	19.0	453	2.73
	723	19.3	442	2.80
	823	19.2	437	2.83
0.1% Se–0.25% N	623	17.9	460	2.69
	723	18.5	455	2.72
	823	19.0	451	2.74
0.5% Se–0.5% N	623	16.8	473	2.62
	723	17.4	467	2.65
	823	17.9	458	2.70
0.25% Se–0.25% N	623	17.3	482	2.57
	723	17.6	476	2.60
	823	17.9	469	2.64
0.1% Se–0.1% N	623	19.6	495	2.50
	723	20.1	488	2.54
	823	20.4	480	2.56

[1] All the values are for a calcination period of 3 h.

A slight shift in the peak position corresponding to (101) plane of anatase to a higher angle was observed as displayed in Figure 1b. This finding indicates that the crystal is distorted by the incorporation of the dopants. Due to a smaller ionic radius (64.0 pm) of Se^{4+} ion than Ti^{4+} ion (74.5 pm) and a higher ionic radius (14.6 pm) of N^{3-} ion than O^{2-} ion (14.0 pm), substitution of Se for Ti and N for O in TiO_2 crystal lattice resulted in a decrease in the interplanar distance. In addition, a smaller shift in the peak position corresponding to (004) plane of anatase was observed. This shift suggests a slight lattice variation in the vertical direction also. It can also be seen from Table 1 that crystallite size increases with the calcination temperature. The reason may be attributed to the fact that calcination at high temperatures or in long periods causes the doped ions to be desorbed.

Raman spectra of the undoped and Se/N codoped samples in Figure 2 support XRD results. Three well-resolved Raman peaks at 398 (B_{1g}), 516 (E_g) and 638 (E_g) cm^{-1} in the spectra of all the samples were obtained indicating that anatase nanoparticles are the predominant species. The weak peaks at 447, 612 and 826 cm^{-1} could be assigned to E_g, A_{1g} and B_{2g} modes in rutile phase respectively. No Raman lines due to other crystalline phases can be observed in the Se/N-codoped sample. Three anatase peaks shifted to lower values, confirming the presence of the dopant ions in the crystal lattice. Generally shifting in Raman spectra is caused by defect structures within the material or changes in grain size. For TiO_2, defect structures, mostly oxygen vacancies not grain size strongly affect the Raman spectrum by producing shifting [46]. Therefore, it may be concluded that Se/N-codoping increases oxygen vacancies in TiO_2 lattice.

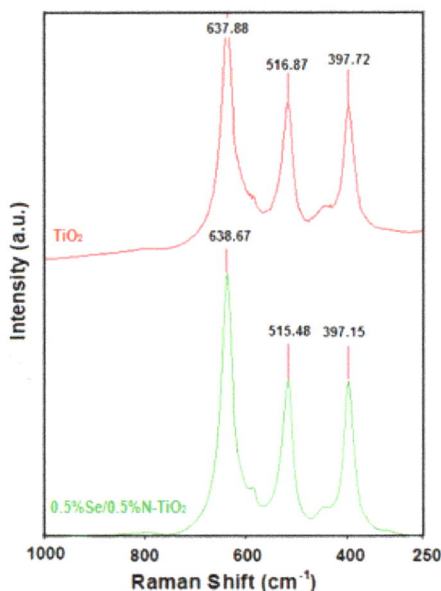

Figure 2. Raman spectra for the undoped and 0.5% Se–0.5% N-codoped TiO_2.

3.2. Morphological Structure

Figure 3a shows the SEM micrograph obtained for the Se/N-codoped TiO_2 (0.5% Se–0.5% N). As it can be seen, the sample consists of small, nearly spherical and some larger, elongated particles. SEM micrograph in Figure 3b shows that the undoped TiO_2 consists of uniform sized spherical particles of around 20–25 µm in diameter. In contrast, the Se/N-codoped TiO_2 consists of significantly larger particles with an average size of approximately 30–40 µm due to the fact that doping of TiO_2 causes agglomeration of the crystallites. The tendency of agglomeration may be attributed to the fact that impurity doping leads to the formation of new defects and dislocations in the crystal lattice. The sizes of these aggregates enlarge up to 50 μm.

(a)   (b)

Figure 3. Scanning electron (SEM) micrographs for (**a**) 0.5% Se–0.5% N-codoped TiO_2; (**b**) undoped TiO_2.

3.3. Optical Absorption and Band Gap Energies

UV-visible diffuse reflectance spectra for the undoped and Se/N-codoped TiO_2 are displayed in Figure 4. The spectrum for the undoped TiO_2 has a sharp absorption edge at around 380 nm, however the absorption threshold of the Se/N codoped TiO_2 shifted towards the visible region of the spectrum. In contrast to the undoped TiO_2, a high visible light absorption band from ca. 430 nm extending up to ca. 580 nm was obtained, which is consistent with the color of the samples.

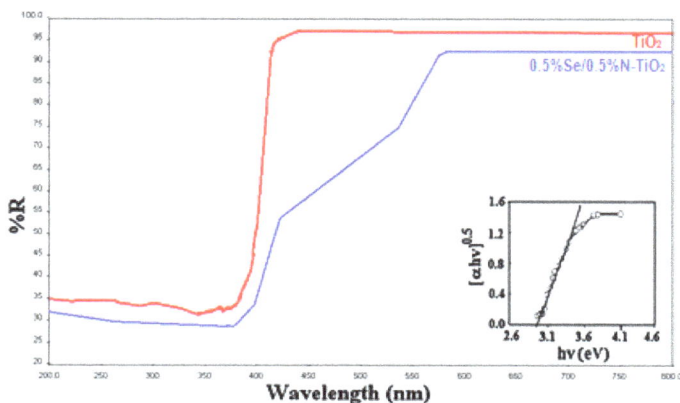

Figure 4. UV-diffuse reflectance (UV-DRS) spectra of the undoped and 0.5%Se-0.5% N-codoped TiO_2 samples (Red, TiO_2; blue, Se/N-codoped TiO_2).

In the UV-DRS spectrum of the Se/N-codoped TiO_2, two optical absorption thresholds were observed, one in the UV-region at around 430 nm, the other in the visible region at 550 nm. The first one is a rather sharp absorption edge indicating that the dopant ions are localized in the TiO_2 lattice, occupying Ti^{4+} and O^{2-} positions. It can be seen that the codoped sample presents a significant absorption in the visible region between 430–550 nm. In between 550–580 nm, there is a tailing which may be attributed to the presence of mid-gap levels in the band-gap of the codoped TiO_2.

The band gap energies of the codoped photocatalyst samples were calculated through the use of the Kubelka-Munk formula:

$$F(R) = \frac{(1-R)^2}{2R} \qquad (2)$$

where R is the reflectance read from the spectrum. Using the Tauc equation by plotting $[F(R).h\nu]^n$ vs. $h\nu$, where $h\nu$ is the photon energy and $n = 1/2$ [47], the band gap energies were deduced from the intersection of the Tauc's linear portion extrapolation with the photon energy axis as depicted in the insert in Figure 4. The calculated band gap energies and the corresponding wavelengths are presented in Table 1. The values indicate that the absorbance in the visible region of the Se/N-codoped samples increases with the concentration of the dopants in TiO_2. The presence of both ions caused an even more decrease in the band gap and an increase in the absorption in the visible region as compared to single Se-doped and N-doped TiO_2 [39,40].

3.4. XPS Analyses

X-ray photoelectron spectroscopy (XPS) was used to examine the bonding and status of the dopants in the Se/N-codoped TiO_2. Five areas of the XPS spectra, displayed in Figure 5 were examined, Ti 2p region near 460 eV, O 1s region near 530 eV, Se 3p region near 165 eV, Se 3d region near 55 eV and N 1s near 400 eV. In Figure 5a, the two peaks at ca. 460 and 465 eV correspond to the photo-splitting electrons Ti^{4+} $2p_{3/2}$ and Ti^{4+} $2p_{1/2}$ indicating that titanium in the sample is in the form of Ti^{4+}. In the XPS spectrum of the Se/N-codoped sample, Ti $2p_{3/2}$ peak appears at 461.1 eV higher than 459.9 eV for the undoped TiO_2 but lower than 461.3 eV for the Se-doped TiO_2. The higher binding energy confirms the presence of substitutional Se^{4+} cations in the crystal. Since the electronegativity of Se^{4+} is more than titanium, the electron density around titanium cations decreases causing an increase in the binding energy. On the other hand, the lower binding energy than that for Se-doped TiO_2 indicates the presence of substitutional and/or interstitial N anions in the same crystal. Since the tendency of nitrogen to attract the bonding electrons toward itself is lower than that of oxygen, the electron density around Ti atoms increases leading to a decrease in the binding energy. The broadness of Ti peaks for the codoped sample may be attributed to the presence of titanium atoms bonded to two different atoms, oxygen and nitrogen.

The O 1s binding energy of the codoped sample is located at 530.8 eV which is assigned to the metallic oxide (O^{2-}) in the TiO_2 lattice. There is a second shoulder peak at 529.9 which corresponds to surface hydroxyl groups. This implies that the oxygen environment is the same as in the undoped TiO_2 indicating the presence of substitutional Se and N atoms (Ti–O–Se, Ti–N–Ti) rather than interstitial ones (Ti–O–N) in the crystal lattice. The signals of the Se dopant were found to be weaker than Ti and O peaks, due to the low doping level. The peak at 165.6 eV corresponds to Se $3p_{3/2}$ electrons indicating that Se in the codoped sample is in the form of Se^{4+} [48]. The presence of the peak at 56.1 eV corresponding to Se $3d_{5/2}$ of Se^{4+} cation confirms this finding [49]. The characteristic $3d_{5/2}$ peaks at 55.5 eV [50] and 53.0–54.0 eV [51] corresponding to elemental Se and Se^{2-} were not observed. These observations reveal that selenium in the as-prepared sample is in the form of Se^{4+} that can penetrate into the TiO_2 lattice and substitute Ti^{4+} cations.

The N 1s spectrum in Figure 5e has two peaks at 397.8 and 402.3 eV. The first peak at 397.8 corresponds to anionic N substitutionally incorporated in TiO_2 in O–Ti–N linkages. The peak is 0.9 eV higher than the characteristic binding energy of 396.9 eV in TiN [52]. Therefore, it may be attributed to the 1s binding energy of the N atom in the environment O–Ti–N. This shift to a higher energy results from the fact that when N substitutes for O in TiO_2, O–Ti–N structures form, thus the electron density around N is less than that in TiN (N–Ti–N). On the other hand, the second peak at 402.3 eV may be assigned to oxidized N such as the ones in Ti–O–N species as in interstitial doped TiO_2 or adsorbed NO, NO_2 species on the surface, since the binding energy is higher than the typical binding energy of 396.9 eV in TiN indicating that the formal charge on the doped N is more positive than the one in TiN [26]. Even though the presence of interstitial N atoms in the prepared Se/N codoped TiO_2 cannot be ruled out, this peak is likely to result from the formation of nitrogen-containing species such as NO, NO_2, NO_2^-, NO_2^{2-} adsorbed on the surface.

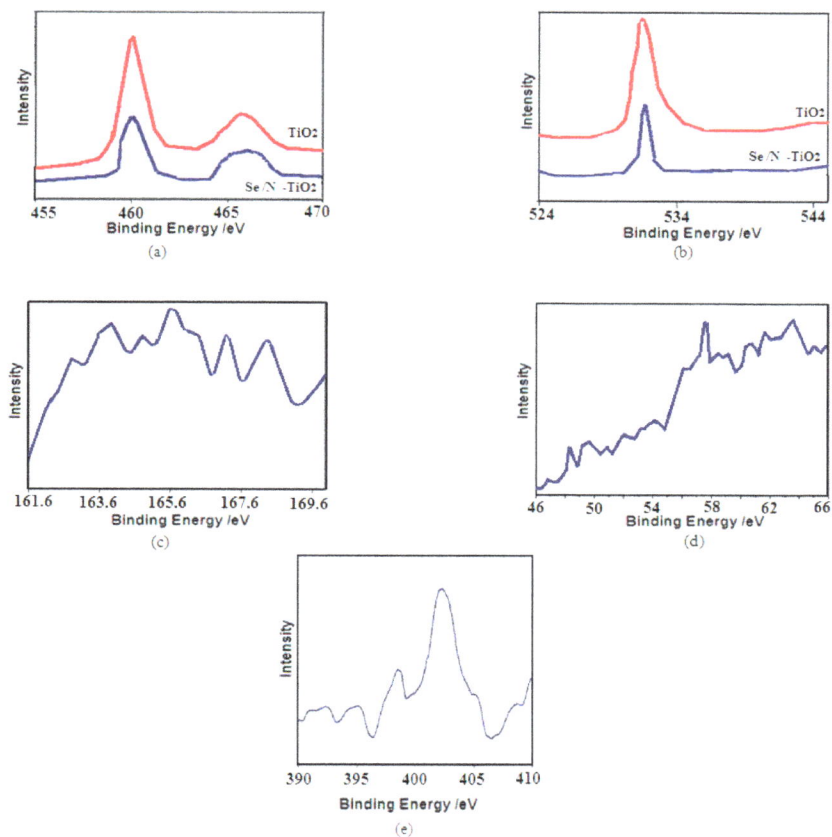

Figure 5. X-ray photoelectron (XPS) spectra of the undoped and 0.5% Se–0.5% N-codoped TiO_2 samples, (**a**) Ti 2p; (**b**) O 1s; (**c**) Se 3p; (**d**) Se 3d; (**e**) N 1s.

3.5. Photocatalytic Activity

To explore the photocatalytic activity of the Se/N-codoped TiO_2 samples, the degradation reaction of 4-NP was investigated in aqueous suspensions under both UV-A and natural solar light irradiation. Figure 6 shows the kinetics of disappearance of 4-NP from an initial concentration of 1.0×10^{-4} mol·L^{-1} which was determined as the optimum concentration under four conditions. In non-irradiated suspensions, there was a slight loss, ca. 4.3%, due to adsorption onto TiO_2 particles. As seen in Figure 6, there was no direct photolysis taking place. The degradation of 4-NP is due entirely to photocatalysis. In the presence of TiO_2, the concentration change amounts to 70% after irradiating for 120 min. The semi-logarithmic plots of concentration data gave a straight line. This finding indicates that the photocatalytic degradation of 4-NP in aqueous TiO_2 suspensions can be described by a pseudo-first order kinetic model, ln C = −kt + ln C_0, where C_0 is the initial concentration and C is the concentration of 4-NP at time t.

In the presence of Se/N-codoped TiO_2, the degradation rate of 4-NP increased, as expected. The concentration data gave a straight line, indicating that the kinetics of the degradation reaction of 4-NP in the presence of the Se/N-codoped TiO_2 also obeys the first-order kinetic model. The Se/N-codoped TiO_2 also exhibited substantial photocatalytic activity under direct sunlight irradiation, with 90% of 4-NP removed in 60 min as compared to 73% removal with the undoped TiO_2 and 88% removal with single Se-doped sample. The result is that the prepared codoped samples are photocatalytically active under solar light.

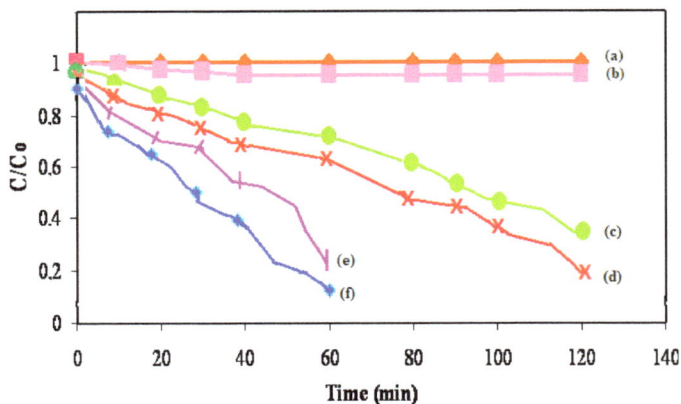

Figure 6. Kinetics of the photocatalytic disappearance of 4-NP on the undoped and Se/N-codoped TiO_2 (0.5% Se–0.5% N) (**a**) with light; (**b**) with TiO_2; (**c**) with TiO_2 + light; (**d**) with Se/N-codoped TiO_2 + light; (**e**) with TiO_2 + sunlight; (**f**) with 0.5% Se–0.5% N-codoped TiO_2 + sunlight.

The enhanced photocatalytic activity of the Se/N-codoped TiO_2 is due to several factors such as; synergistic effect of the dopants, formation of the oxygen vacancies, improved structures and the enhanced photo absorption. Due to their favorable energy levels (2.27 eV), Se^{4+} centers may act either as electron or hole traps so that charge carriers are temporarily separated. On the other hand, substitutional N^{-3} inhibits e^-/h^+ recombination due to charge compensation between N^{3-} and Ti^{4+}. Thus, the lifetime of the charge carriers increases leading to an enhancement of the photocatalytic activity. The role of the dopant nitrogen is not only to decrease e^-/h^+ recombination rate, but it also induces a substantial reduction of the formation energy of oxygen vacancy on TiO_2 [24]. This implies that N-doping causes oxygen vacancy formation on the surface of the particles in agreement with the Raman spectrum. The formation of the oxygen vacancies on the surface favors the adsorption of water molecules and thus increases the amount of hydroxyl radicals which are responsible of the degradation of 4-NP.

The high photocatalytic activity of the Se/N codoped TiO_2 is also due to the fact that it has smaller particle size thus higher adsorption area toward the organic pollutant. Moreover, the increase in the light absorbance extending up to visible light range with Se/N-codoping indicates that more electrons and holes are generated and participate in the surface redox reactions causing an increase in the amount of hydroxyl radicals which are responsible of the degradation of the pollutant molecule.

The results presented in Table 2 show the effect of Se and N concentrations of the codoped photocatalysts on the photocatalytic degradation of 4-NP. As it can be seen from the values, the photocatalytic degradation rate of 4-NP first increased and then decreased passing through the maximum degradation for the photocatalyst containing 0.5% Se and 0.5% N. There appears to be an optimal dopant concentration, 0.5%, above which the observed photoreactivity decreases. The reason may be attributed to the fact that at lower concentrations below the optimal value, photoreactivity increases with an increasing dopant concentration because there are available trapping sites. The dopants provide more trap sites for electrons and holes in addition to the surface trap sites, adsorbed O_2 and OH^-. However, at high dopant concentrations, the photocatalytic activity of the codoped samples decreased. This is because the recombination rate of the charge carriers increases exponentially with the dopant concentration. The average distance between trap sites decreases with increasing the number of dopants confined within a particle. Thus, it may be concluded that the number of trapped carriers is the highest in 0.5% Se–0.5% N codoped sample for which the highest photoreactivity was obtained.

Table 2. Apparent first order rate constants k for the photocatalytic degradation of 4-NP in the presence of the Se/N-codoped TiO_2 samples.

Photocatalyst	k $(10^{-3} \cdot min^{-1})$	r	% Degradation
TiO_2 Evonik P-25	9.21 ± 0.009	0.991	69.83
	14.15 ± 0.008 [1]	*0.996*	*73.15*
0.25% Se–0.1% N	14.85 ± 0.005	0.991	77.19
	17.21 ± 0.009	*0.998*	*79.83*
0.1% Se–0.25% N	17.52 ± 0.008	0.985	79.58
	18.89 ± 0.002	*0.982*	*81.93*
0.5% Se–0.5% N	20.21 ± 0.007	0.994	87.71
	23.37 ± 0.006	*0.990*	*89.25*
0.25% Se–0.25%N	18.99 ± 0.001	0.987	82.70
	20.78 ± 0.002	*0.983*	*85.82*
0.1% Se–0.1% N	14.97 ± 0.003	0.986	73.67
	16.88 ± 0.001	*0.995*	*75.17*

[1] Values in italics are the results of sunlight experiments.

In addition, the experiments demonstrated that there is no direct correlation between the visible light activity and the photocatalytic activity. The optimum dopant concentration was found to be 0.5% Se–0.5% N. However, the UV-DRS spectrum of this sample revealed intermediate values of the band-gap.

3.6. Electronic Structures

The structures obtained for the undoped and Se/N-codoped TiO_2 cluster models are presented in Figure 7. Electronic structure calculations of the models gave structures with deviations, which are not as symmetrical as that of the undoped TiO_2 model. The results indicate that the size and electronegativity difference between the two codopants induce structural changes.

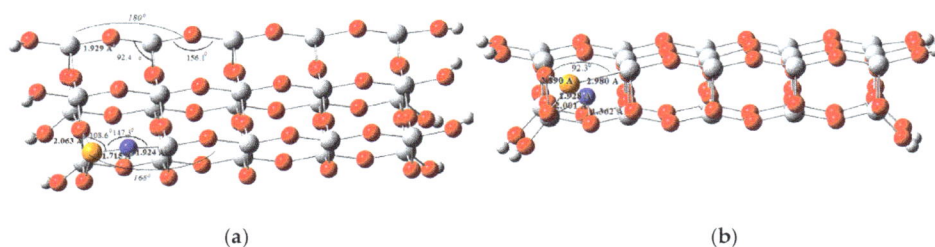

(a)   (b)

Figure 7. Optimized structures of Se/N-codoped TiO_2 clusters (**a**) substitutional Se/N-codoped model; (**b**) interstitial Se/N-codoped model (Grey, Ti; red, O; orange, Se; white, H; blue, N).

(001) surface of the undoped TiO_2 cluster contains the four- and five-fold-coordinated titanium atoms representing Lewis acid sites and the two- and three-fold-coordinated oxygen atoms which act as Lewis base sites. Site preferences of the dopants on (001) surface were determined by calculating

the total energies of the codoped clusters. The results indicate that for Se^{4+} four-fold-coordinated Ti site substitution is favored over five-fold-coordinated Ti site substitution by ~36 kcal·mole^{-1}. For substitutional nitrogen, two-fold-coordinated O site is favored over three-fold-coordinated O site, and nitrogen prefers to be at the position closest to Se^{4+}. In the interstitial model, the optimum position for the vacancy was found to be the one near Se^{4+} dopant.

The visible light activity of a photocatalyst depends upon the magnitude of the band-gap and the presence or absence of any intermediate electronic states within the band-gap. On the other hand, the photocatalytic activity of TiO_2 is governed by the positions of the band edges. A schematic diagram of the electronic energy levels for the undoped and Se/N-codoped anatase models obtained from electronic structure calculations are presented in Figure 8. For the clusters developed in this study, the energies of the highest occupied HOMO and the lowest unoccupied molecular orbitals LUMO were used to represent the VB and CB edges, while the occupied and unoccupied molecular orbitals correspond to the electronic states in the VB and CB respectively. An examination of the calculated band-gap energies of the undoped and codoped clusters in Figure 8 shows that the DFT/B3LYP method underestimates the band-gap energy due to the well-known shortcoming of the exchange-correlation potential used within the framework of DFT. The experimental band-gap energy of the undoped TiO_2 (3.2 eV) was adopted as the benchmark to correct the calculated values. The calculated band-gap was corrected using a scissors operator that displaces the empty and occupied bands relative to each other by a rigid shift of 0.40 eV to bring the minimum band-gap in line with experiment for the band-gap of anatase.

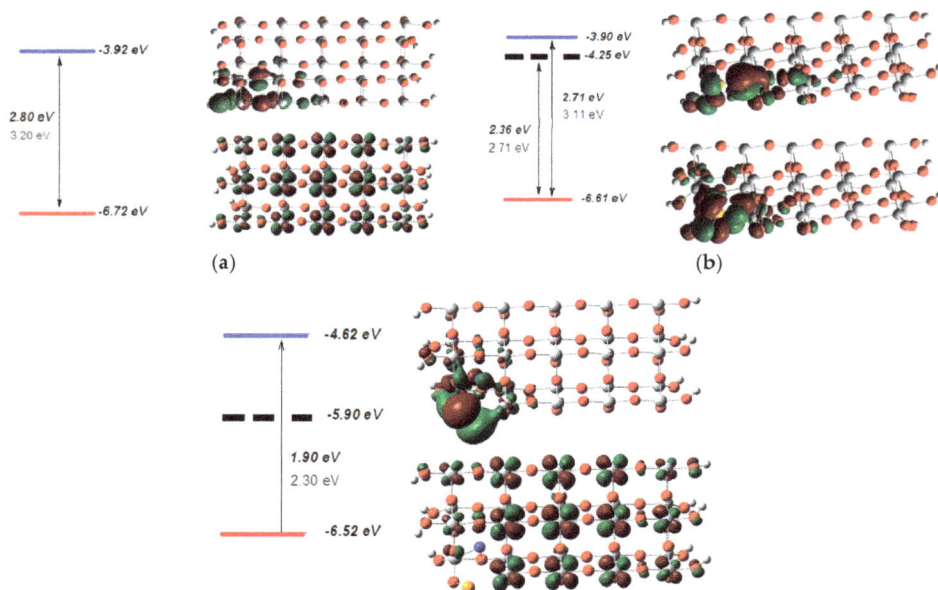

Figure 8. Energy level diagrams and the frontier orbitals of the (**a**) undoped TiO_2; (**b**) substitutional Se/N-codoped TiO_2; (**c**) interstitial Se/N-codoped TiO_2 clusters computed with DFT/B3LYP method. (Grey, Ti; red, O; orange, Se; white, H; blue, N) (Values in italics are the DFT results).

The computational results show that codoping with Se^{4+} and substitutional nitrogen causes a significant change in the position of the valence band edge. The reason is that N 2p states mix with O 2p states and reduce the band gap. For the substitutional model, the calculations indicated the presence of three empty mid-gap levels in the band-gap as shown in Figure 8. These intermediate electronic states were determined to be mainly originating from the Se3p states hybridized with the O 2p states by examining the calculated coefficients of the orbital wave functions. These energy levels are not populated by electrons. They are not donor states but allowed energy states. Thus, they induce a decrease in the band gap as the dopant concentration increases as obtained by UV-DRS analysis. The increase in the concentration of the dopant Se^{4+} introduces more electronic states into the band gap, thus enhances the density of the electronic states in the gap. The presence of these intermediate levels separates the band-gap of the Se/N-codoped TiO_2 into two parts; a wider lower gap and a significantly narrower upper gap. These intermediate energy levels offer additional steps for the absorption of low energy photons through the excitation of VB electrons to these intermediate energy levels, from where they can be excited again to the CB. The experimentally observed absorptions in the range 430–550 nm and 550–580 nm and the rather diffused character of the UV-DRS spectrum of the Se/N-codoped TiO_2 samples may be attributed to the excitation of electrons to or from these additional electronic levels. The lower gap was calculated to be 2.71 eV corresponding to a 458 nm photon which is in agreement with the experimental results obtained from the UV-DRS spectra of the codoped samples. Therefore, it may be stated that the lower gap is responsible for the absorption in the first region of the spectrum between 430–550 nm, while the second region between 550–580 nm corresponds to the excitation of electrons from mid-gap levels to the CB.

On the other hand, in the interstitial model, N 2p states mix with Se 3p orbitals and thus form a mid-gap level between the VB and CB of TiO_2. The contribution of Se 3p orbitals to the lowest unoccupied orbital was found to be less than the one in substitutional model. Although we may not rule out the presence of interstitial nitrogens, the codopant N is in O–Ti–N structures while Se ion substitutes for Ti in our samples. Moreover, comparison of the energies of the two models indicated that substitutional model is more stable than interstitial model.

4. Conclusions

Codoping of TiO_2 with Se^{4+} and N^{3-} ions was performed through a simple wet-impregnation method using $SeCl_4$ and urea as the dopant sources. The characterization results reveal that Se^{4+} is in O–Se–O while N^{3-} is in O–Ti–N linkages in the crystal lattice. The Se/N codoped samples showed photoabsorption in the visible light range from 430 nm extending up to 580 nm. The degradation of 4-NP was highly increased for the Se/N codoped samples compared to the undoped and single doped samples under both UV-A and sunlight irradiation. The enhanced photocatalytic activity of the codoped samples may be attributed to the increase in the number of trap sites for electrons and holes, increase in the photoabsorption, smaller particle size and the formation of oxygen vacancies on the surface. The experiments demonstrated that there is no direct correlation between the visible light activity and the photocatalytic activity. 623 K, 3 h and 0.5% Se–0.5% N were determined to be the most suitable calcination temperature, calcination period and the codopant concentration to prepare the photocatalyst with the highest photocatalytic activity. Eventually, on the basis of experimental results combined with DFT calculations, it may be concluded that Se/N-codoping of TiO_2 reduces the band gap due to mixing of N 2p with O 2p orbitals in the VB and also introduces additional electronic states originating from the Se 3p orbitals in the band gap.

Acknowledgments: The authors express their thanks to Yildiz Technical University Research Foundation for financial support (Project No. 29-01-02-KAP01), to Degussa Limited Company in Turkey for the generous gift of TiO_2 and to the National Center for High Performance Computing of Turkey (UYBHM) Grant No. 1001162011.

Author Contributions: Zekiye Cinar contributed to the design of the study. Yelda Y. Gurkan performed the laboratory experiments. Esra Kasapbasi performed the quantum mechanical computations. Zekiye Cinar contributed to the construction of the models, evaluated the results and wrote the manuscript. Nazli Turkten helped with the interpretation of the spectra and writing the manuscript.

Conflicts of Interest: The authors declare no conflict of interest.

References

1. Bahnemann, D.; Cunningham, J.; Fox, M.A.; Pelizzetti, E.; Pichat, P.; Serpone, N. *Aquatic and Surface Photochemistry*; Lewis Publishers: Baca Raton, FL, USA, 1994; p. 261.
2. Pelaez, M.; Nolan, N.T.; Pillai, S.C.; Seery, M.K.; Falaras, P.; Kontos, A.G.; Dunlop, P.S.M.; Hamilton, J.W.J.; Byrne, J.A.; O'Shea, K.; et al. A review on the Visible Light Active Titanium Dioxide Photocatalysts for Environmental Applications. *Appl. Catal. B* **2012**, *125*, 331–349. [CrossRef]
3. Pichat, P. (Ed.) *Photocatalysis and Water Purification*; Wiley-VCH: Weinheim, Germany, 2013.
4. Schneider, J.; Bahnemann, D.; Ye, J.; Puma, L.G.; Dionysios, D.D. (Eds.) *Photocatalysis: Fundamentals and Perspectives*; Royal Society of Chemistry: London, UK, 2016.
5. Ollis, D.F.; Pelizzetti, E.; Serpone, N. Photocatalyzed Destruction of Water Contaminants. *Environ. Sci. Technol.* **1991**, *25*, 1522–1529. [CrossRef]
6. Bahnemann, D.; Bockelmann, D.; Goslich, R. Mechanistic Studies of Water Detoxification in Illuminated TiO_2 Suspensions. *Sol. Energy Mater.* **1991**, *24*, 564–583. [CrossRef]
7. Nosaka, Y.; Nosaka, A. *Introduction to Photocatalysis: From Basic Science to Applications*; Royal Society of Chemistry: London, UK, 2016.
8. Suib, S.L. (Ed.) *New and Future Developments in Catalysis: Solar Photocatalysis*; Elsevier: Amsterdam, The Netherlands, 2013; Volume 7.
9. Kilic, M.; Cinar, Z. A Quantum Mechanical Approach to TiO_2 Photocatalysis. *J. Adv. Oxid. Technol.* **2009**, *12*, 37–46.
10. Zhu, J.; Chen, F.; Zhang, J.; Chen, H.; Anpo, M. Fe^{3+}-TiO_2 Photocatalysts prepared by Combining Sol–Gel Method with Hydrothermal Treatment and their Characterization. *J. Photochem. Photobiol. A* **2006**, *180*, 196–204. [CrossRef]
11. Yalcin, Y.; Kilic, M.; Cinar, Z. Fe^{+3}-doped TiO_2: A Combined Experimental and Computational Approach to the Evaluation of Visible Light Activity. *Appl. Catal. B* **2010**, *99*, 469–477. [CrossRef]
12. Choi, W.; Termin, A.; Hoffmann, M.R. The Role of Metal Ion Dopants in Quantum-Sized TiO_2: Correlation between Photoreactivity and Charge Carrier Recombination Dynamics. *J. Phys. Chem.* **1994**, *98*, 13669–13679. [CrossRef]
13. Nagaveni, K.; Hegde, M.S.; Madras, G. Structure and Photocatalytic Activity of $Ti_{1-x}M_xO_{2\pm\delta}$ (M = W, V, Ce, Zr, Fe, and Cu) Synthesized by Solution Combustion Method. *Phys. Chem. B* **2004**, *108*, 20204–20212. [CrossRef]
14. Di Paola, A.; Marcì, G.; Palmisano, L.; Schiavello, M.; Uosaki, K.; Ikeda, S.; Ohtani, B. Preparation of Polycrystalline TiO_2 Photocatalysts Impregnated with Various Transition Metal Ions: Characterization and Photocatalytic Activity for the Degradation of 4-nitrophenol. *Phys. Chem. B* **2002**, *106*, 637–645. [CrossRef]
15. Mu, W.; Herrmann, J.-M.; Pichat, P. Room Temperature Photocatalytic Oxidation of Liquid Cyclohexane into Cyclohexanone over Neat and Modified TiO_2. *Catal. Lett.* **1989**, *3*, 73–84. [CrossRef]
16. Karakitsou, K.E.; Verykios, X.E. Effects of Altervalent Cation Doping of Titania on its Performance as a Photocatalyst for Water Cleavage. *J. Phys. Chem.* **1993**, *97*, 1184–1189. [CrossRef]
17. Jagadale, T.C.; Takale, S.P.; Sonawane, R.S.; Joshi, H.M.; Patil, S.I.; Kale, B.B.; Ogale, S.B. N-doped TiO_2 Nanoparticle Based Visible Light Photocatalyst by Modified Peroxide Sol–Gel Method. *J. Phys. Chem. C* **2008**, *112*, 14595–14602. [CrossRef]
18. Sakthivel, S.; Kisch, H. Daylight Photocatalysis by Carbon-Modified Titanium Dioxide. *Angew. Chem. Int. Ed.* **2003**, *42*, 4908–4911. [CrossRef] [PubMed]
19. Ohno, T.; Akiyoshi, M.; Umebayashi, T.; Asai, K.; Mitsui, T.; Matsumura, M. Preparation of S-Doped TiO_2 Photocatalysts and their Photocatalytic Activities under Visible Light. *Appl. Catal. A* **2004**, *265*, 115–121. [CrossRef]
20. Zheng, R.; Lin, L.; Xie, J.; Zhu, Y.; Xie, Y. State of Doped Phosphorus and its Influence on the Physicochemical and Photocatalytic Properties of P-Doped Titania. *J. Phys. Chem. C* **2008**, *112*, 15502–15509. [CrossRef]
21. Lu, N.; Zhao, H.; Li, J.; Quan, X.; Chen, S. Characterization of Boron-Doped TiO_2 Nanotube Arrays Prepared by Electrochemical Method and its Visible Light Activity. *Sep. Purif. Technol.* **2008**, *62*, 668–673. [CrossRef]
22. Yalcin, Y.; Kilic, M.; Cinar, Z. The Role of Non-Metal Doping in TiO_2 Photocatalysis. *J. Adv. Oxid. Technol.* **2010**, *13*, 281–296.

23. Asahi, R.; Morikawa, T.; Ohwaki, T.; Aoki, K.; Taga, Y. Visible-Light Photocatalysis in Nitrogen-Doped Titanium Oxides. *Science* **2001**, *293*, 269–271. [CrossRef] [PubMed]

24. Di Valentin, C.; Pacchioni, G.; Selloni, A.; Livraghi, S.; Giamello, E. Characterization of Paramagnetic Species in N-doped TiO_2 Powders by EPR Spectroscopy and DFT Calculations. *J. Phys. Chem. B* **2005**, *109*, 11414–11419. [CrossRef] [PubMed]

25. Sakthivel, S.; Janczarek, M.; Kisch, H. Visible Light Activity and Photoelectrochemical Properties of Nitrogen-Doped TiO_2. *J. Phys. Chem. B* **2004**, *108*, 19384–19387. [CrossRef]

26. Sato, S.; Nakamura, R.; Abe, S. Visible-Light Sensitization of TiO_2 Photocatalysts by Wet-Method N Doping. *Appl. Catal. A* **2005**, *284*, 131–137. [CrossRef]

27. Sathish, M.; Viswanathan, B.; Viswanath, R.P.; Gopinath, C.S. Synthesis, Characterization, Electronic Structure, and Photocatalytic Activity of Nitrogen-Doped TiO_2 Nanocatalyst. *Chem. Mater.* **2005**, *17*, 6349–6353. [CrossRef]

28. Emeline, A.V.; Kuzmin, G.N.; Serpone, N. Wavelength-Dependent Photostimulated Adsorption of Molecular O_2 and H_2 on Second Generation Titania Photocatalysts: The Case of the Visible-Light-Active N-doped TiO_2 System. *Chem. Phys. Lett.* **2008**, *454*, 279–283. [CrossRef]

29. Choi, H.; Antoniou, M.G.; Pelaez, M.; de la Cruz, A.A.; Shoemaker, J.A.; Dionysiou, D.D. Mesoporous Nitrogen-Doped TiO_2 for the Photocatalytic Destruction of the Cyanobacterial Toxin Microcystin-lr under Visible Light Irradiation. *Environ. Sci. Technol.* **2007**, *41*, 7530–7535. [CrossRef] [PubMed]

30. Sun, H.; Bai, Y.; Cheng, Y.; Jin, W.; Xu, N. Preparation and Characterization of Visible-Light-Driven Carbon-Sulfur-Codoped TiO_2 Photocatalysts. *Ind. Eng. Chem. Res.* **2006**, *45*, 4971–4976. [CrossRef]

31. Yu, J.; Zhou, M.; Cheng, B.; Zhao, X. Preparation, Characterization and Photocatalytic Activity of in situ N,S-Codoped TiO_2 Powders. *J. Mol. Catal. A Chem.* **2006**, *246*, 176–184. [CrossRef]

32. Wang, X.; Lim, T.-T. Solvothermal Synthesis of C–N Codoped TiO_2 and Photocatalytic Evaluation for Bisphenol a Degradation using a Visible-Light Irradiated LED Photoreactor. *Appl. Catal. B* **2010**, *100*, 355–364. [CrossRef]

33. Li, D.; Ohashi, N.; Hishita, S.; Kolodiazhnyi, T.; Haneda, H. Origin of Visible-Light-Driven Photocatalysis: A Comparative Study on N/F-Doped and N–F-Codoped TiO_2 Powders by means of Experimental Characterizations and Theoretical Calculations. *J. Solid State Chem.* **2005**, *178*, 3293–3302. [CrossRef]

34. Katsanaki, A.V.; Kontos, A.G.; Maggos, T.; Pelaez, M.; Likodimos, V.; Pavlatou, E.A.; Dionysiou, D.D.; Falaras, P. Photocatalytic Oxidation of Nitrogen Oxides on N-F-Doped Titania Thin Films. *Appl. Catal. B* **2013**, *140–141*, 619–625. [CrossRef]

35. Ling, Q.; Sun, J.; Zhou, Q. Preparation and Characterization of Visible-Light-Driven Titania Photocatalyst Co-Doped with Boron and Nitrogen. *Appl. Surf. Sci.* **2008**, *254*, 3236–3241. [CrossRef]

36. Sun, H.; Zhou, G.; Liu, S.; Ang, H.M.; Tadé, M.O.; Wang, S. Visible Light Responsive Titania Photocatalysts Codoped by Nitrogen and Metal (Fe, Ni, Ag, or Pt) for Remediation of Aqueous Pollutants. *Chem. Eng. J.* **2013**, *231*, 18–25. [CrossRef]

37. Jaiswal, R.; Patel, N.; Kothari, D.C.; Miotello, A. Improved Visible Light Photocatalytic Activity of TiO_2 Co-Doped with Vanadium and Nitrogen. *Appl. Catal. B* **2012**, *126*, 47–54. [CrossRef]

38. Márquez, A.M.; Plata, J.J.; Ortega, Y.; Sanz, J.F.; Colón, G.; Kubacka, A.; Fernández-García, M. Making Photo-Selective TiO_2 Materials by Cation-Anion Codoping: From Structure and Electronic Properties to Photoactivity. *J. Phys. Chem. C* **2012**, *116*, 18759–18767. [CrossRef]

39. Gurkan, Y.Y.; Kasapbasi, E.; Cinar, Z. Enhanced Solar Photocatalytic Activity of TiO_2 by Selenium(IV) Ion-Doping: Characterization and DFT Modeling of the Surface. *Chem. Eng. J.* **2013**, *214*, 34–44. [CrossRef]

40. Gurkan, Y.Y.; Turkten, N.; Hatipoglu, A.; Cinar, Z. Photocatalytic Degradation of Cefazolin over N-doped TiO_2 under UV and Sunlight Irradiation: Prediction of the Reaction Paths via Conceptual DFT. *Chem. Eng. J.* **2012**, *184*, 113–124. [CrossRef]

41. Calvert, J.G.; Pitts, J.N. *Photochemistry*; Wiley: New York, NY, USA, 1966; pp. 783–786.

42. Homann, T.; Bredow, T.; Jug, K. Adsorption of Small Molecules on the Anatase (1 0 0) Surface. *Surf. Sci.* **2004**, *555*, 135–144. [CrossRef]

43. Wahab, H.S.; Bredow, T.; Aliwi, S.M. MSINDO Quantum Chemical Modeling Study of Water Molecule Adsorption at Nano-Sized Anatase TiO_2 Surfaces. *Chem. Phys.* **2008**, *354*, 50–57. [CrossRef]

44. Sekiya, T.; Igarashi, M.; Kurita, S.; Takekawa, S.; Fujisawa, M. Structure Dependence of Reflection Spectra of TiO$_2$ Single Crystals. *J. Electron. Spectrosc. Relat. Phenom.* **1998**, *92*, 247–250. [CrossRef]

45. Frisch, M.J.; Trucks, G.W.; Schlegel, H.B.; Scuseria, G.E.; Robb, M.A.; Cheeseman, J.R.; Scalmani, G.; Barone, V.; Mennucci, B.; Petersson, G.A.; et al. *Gaussian 09, Revision D.01*; Gaussian, Inc.: Wallingford, CT, USA, 2009.

46. Parker, J.C.; Siegel, R.W. Calibration of the Raman Spectrum to the Oxygen Stoichiometry of Nanophase TiO$_2$. *Appl. Phys. Let.* **1990**, *57*, 943. [CrossRef]

47. Kuvarega, A.T.; Krause, R.W.M.; Mamba, B.B. Nitrogen/Palladium-Codoped TiO$_2$ for Efficient Visible Light Photocatalytic Dye Degradation. *J. Phys. Chem C* **2011**, *115*, 22110–22120. [CrossRef]

48. Badrinayaran, S.; Mandale, A.B.; Gunjikar, V.G.; Sinha, A.P.B. Mechanism of High Temperature Oxidation of Tin Selenide. *J. Mater. Sci.* **1986**, *21*, 3333–3338. [CrossRef]

49. Shenasa, M.; Sainkar, S.; Lichtman, D. XPS Study of Some Selected Selenium Compounds. *J. Electron. Spectrosc. Relat. Phenom.* **1986**, *40*, 329–337. [CrossRef]

50. Cahen, D.; Ireland, P.J.; Kazmerski, L.L.; Thiel, F.A. X-ray Photoelectron and Auger Electron Spectroscopic Analysis of Surface Treatments and Electrochemical Decomposition of CuInSe$_2$ Photoelectrodes. *J. Appl. Phys.* **1985**, *57*, 4761–4772. [CrossRef]

51. Song, L.; Chen, C.; Zhang, S.; Wei, Q. Synthesis of Se-Doped InOOH as Efficient Visible-Light-Active Photocatalysts. *Catal. Commun.* **2011**, *12*, 1051–1054. [CrossRef]

52. Emeline, A.V.; Kuznetsov, V.N.; Rybchuk, V.K.; Serpone, N. Visible-Light-Active Titania Photocatalysts: The Case of N-Doped TiO$_2$s Properties and Some Fundamental Issues. *Int. J. Photoenergy* **2008**, 258394. [CrossRef]

Sample Availability: Samples of the compounds are not available from the authors.

molecules

MDPI

Article

Growth, Structure, and Photocatalytic Properties of Hierarchical V₂O₅-TiO₂ Nanotube Arrays Obtained from the One-step Anodic Oxidation of Ti-V Alloys

María C. Nevárez-Martínez [1,2], Paweł Mazierski [3,*], Marek P. Kobylański [3],
Grażyna Szczepańska [4], Grzegorz Trykowski [4], Anna Malankowska [3], Magda Kozak [3],
Patricio J. Espinoza-Montero [2] and Adriana Zaleska-Medynska [3,*]

[1]   Facultad de Ingeniería Química y Agroindustria, Escuela Politécnica Nacional, Ladrón de Guevara E11-253,
      P.O. Box 17-01-2759, Quito 170525, Ecuador; ma.cristina.nevarez@gmail.com
[2]   Centro de Investigación y Control Ambiental "CICAM", Departamento de Ingeniería Civil y Ambiental,
      Facultad de Ingeniería Civil y Ambiental, Escuela Politécnica Nacional, Ladrón de Guevara E11-253,
      P.O. Box 17-01-2759, Quito 170525, Ecuador; patricio.espinoza@epn.edu.ec
[3]   Department of Environmental Technology, Faculty of Chemistry, University of Gdansk,
      Gdansk 80-308, Poland; marek.kobylanski@phdstud.ug.edu.pl (M.P.K.);
      anna.malankowska@ug.edu.pl (A.M.); magda.kozak@ug.edu.pl (M.K.)
[4]   Faculty of Chemistry, Nicolaus Copernicus University, Torun 87-100, Poland; gina@chem.umk.pl (G.S.);
      grazyna.szczepanska@umk.pl (G.T.)
*   Correspondence: pawel.mazierski@phdstud.ug.edu.pl (P.M.); adriana.zaleska@ug.edu.pl (A.Z.-M.);
      Tel.: +48-58-523-52-29 (P.M.); +48-58-523-52-20 (A.Z.-M.)

Academic Editor: Pierre Pichat
Received: 29 January 2017; Accepted: 1 April 2017; Published: 5 April 2017

Abstract: V₂O₅-TiO₂ mixed oxide nanotube (NT) layers were successfully prepared via the one-step anodization of Ti-V alloys. The obtained samples were characterized by scanning electron microscopy (SEM), UV-Vis absorption, photoluminescence spectroscopy, energy-dispersive X-ray spectroscopy (EDX), X-ray diffraction (DRX), and micro-Raman spectroscopy. The effect of the applied voltage (30–50 V), vanadium content (5–15 wt %) in the alloy, and water content (2–10 vol %) in an ethylene glycol-based electrolyte was studied systematically to determine their influence on the morphology, and for the first-time, on the photocatalytic properties of these nanomaterials. The morphology of the samples varied from sponge-like to highly-organized nanotubular structures. The vanadium content in the alloy was found to have the highest influence on the morphology and the sample with the lowest vanadium content (5 wt %) exhibited the best auto-alignment and self-organization (length = 1 μm, diameter = 86 nm and wall thickness = 11 nm). Additionally, a probable growth mechanism of V₂O₅-TiO₂ nanotubes (NTs) over the Ti-V alloys was presented. Toluene, in the gas phase, was effectively removed through photodegradation under visible light (LEDs, λ_{max} = 465 nm) in the presence of the modified TiO₂ nanostructures. The highest degradation value was 35% after 60 min of irradiation. V₂O₅ species were ascribed as the main structures responsible for the generation of photoactive e⁻ and h⁺ under Vis light and a possible excitation mechanism was proposed.

Keywords: V₂O₅-TiO₂ nanotubes; visible-light-driven photocatalysis; alloys; toluene degradation; air treatment

1. Introduction

Over the past few decades, photocatalytic processes on the surface of TiO₂ have been intensively studied due to a wide range of industrially oriented applications based on the conversion of sunlight into usable chemical energy [1–6]. Being non-toxic, abundant, chemically and physically stable, and photostable [7,8], TiO_2 is a semiconductor material of great interest for environmental remediation [9,10], hydrogen evolution from water splitting [11,12], dye-sensitized solar cells [12,13], CO_2 reduction [12,14], and self-cleaning surfaces [15,16]. However, the usage of TiO_2 is limited not only by its wide bandgap (3.0–3.2 eV), which allows the absorption of only UV light corresponding to 4% of the incident solar energy [17], but also by the fast recombination rate of charge carriers [18,19]. In order to harvest sunlight, many TiO_2 modification approaches have been developed [20,21], such as metal [22], nonmetal [23–25], or rare earth element doping [26], dye sensitization with organic and inorganic dyes [27], and the formation of photocatalytic heterostructures (coupling) with other semiconductors [28] or noble metals [29–31]. In particular, tuning TiO_2 with V_2O_5 is an efficient way of improving TiO_2 performance [32]. V_2O_5 is a small-bandgap semiconductor (~2.3 eV) which can extend the light absorption to the visible range [33]. Furthermore, photogenerated electrons and holes can be efficiently separated, and the surface charge carrier transfer rate is enhanced [34,35]. V_2O_5 itself has been used as a photocatalyst under UV light [36–39], visible light [40], and sunlight [41]. Xie, et al. [42] obtained photoactive V_2O_5-TiO_2 nanocomposites for the oxidation of As(III). They stated that under visible light irradiation, h^+ and O_2^- are the main active species responsible for the photoreaction. Choi, et al. [43] synthesized V_2O_5-TiO_2 nanocomposite powder by DC arc plasma. They found that, in the presence of the nanocomposite, Rhodamine B was decomposed under visible light, while it was not decomposed in the presence of TiO_2 nanopowder. They also reported visible photoactivation and an enhanced charge separation in the case of toluene removal in a dielectric barrier discharge reactor. These aspects make the V_2O_5-TiO_2 system an attractive material for visible-light-driven photocatalytic applications.

Moreover, TiO_2 performance also critically depends on mass transfer, charge transfer, and charge/ion transport on its surface and bulk [7,44]. These processes are mainly controlled by morphology, which can be 0D (nanoparticles), 1D (nanowires, rods and tubes), 2D (layers and sheets), or 3D (spheres) [7]. Among 1D structures, TiO_2 nanotubes (NTs) have become an interesting material because of their high electron mobility, excellent electron hole separation ability, long-distance transport capability, high specific surface area, mechanical strength, and extremely high aspect ratio [45,46]; however, no major improvement was reported for photocatalytic air purification with respect to nanoparticles under similar conditions [47].

Electrochemical anodization under specific conditions appears to be the simplest, least expensive, and most straightforward technique to obtain self-organized, auto-aligned NT arrays [48,49] over the surface of various metals, e.g., Ti [45,50,51], Zr [52], Hf [53], or alloys, e.g., TiNb, TiZr, TiTa [54], TiV [55,56], TiW [57], TiMn [58], TiMoNi [59], Ti_6Al_4V [60], and TiAg [61]. Anatase TiO_2 nanotube array films with exposed {001} nanofacets, obtained by a low temperature hydrothermal method, exhibited enhanced UV activity, which was attributed to the enhanced charge separation derived from the synergy between {001} and {101} facets [62]. However, an electrochemical method is the most efficient for preparing mixed oxide nanotubes from a Ti suitable alloy. V_2O_5-TiO_2 NTs have been successfully fabricated by electro-synthesis using Ti-V alloys as a substrate by the Schmuki research group [55] and Yang, Kim, Yang and Schmuki [56]. These mixed oxide NTs showed, respectively, improved electrochromic and capacitive properties compared with those of pure TiO_2 NTs. Nevertheless, despite the proven visible light absorption of V_2O_5-TiO_2 nanotubes, there is still a lack of data regarding the photoactivity of the V_2O_5-TiO_2 NTs obtained from the anodization of Ti-V alloys. In our previous work [63], self-organized TiO_2-MnO_2 NTs were successfully obtained by the one-step anodization of Ti-Mn alloys in a fluoride-containing ethylene glycol (EG)-based electrolyte. The as-prepared layers were highly organized and showed visible-light photoactivity towards the degradation of toluene in the gas phase. It was demonstrated that a Vis-excited composite of wide and narrow bandgap oxides could be obtained by the anodization of Ti/V alloys, and that the preparation parameters (e.g., applied voltage, content of the MnO_2 in nanocomposite) affected both the morphology and photoactivity of the TiO_2/MnO_2 NTs.

In view of this, this work focuses on the synthesis of visible-light photoactive V_2O_5-TiO_2 NTs through the one-step anodic oxidation of Ti-V alloys in an ethylene glycol-based electrolyte, and their application in the photocatalytic degradation of toluene. The effect of the vanadium content in the alloy, applied voltage, and electrolyte composition (water content) was systematically studied to determine the influence of these parameters on the morphology and gas phase photoactivity, evaluated for the first time, of the obtained nanomaterials. The as-prepared NTs were characterized by using scanning electron microscopy (SEM), X-ray diffraction (XRD), energy-dispersive X-ray spectroscopy (EDX), micro-Raman spectroscopy, UV-Vis absorption, and photoluminescence spectroscopy. A possible mechanism of toluene degradation at the surface of V_2O_5-TiO_2 NTs under the influence of visible light was also proposed.

2. Results and Discussion

2.1. Morphology and Growth Mechanism

Ti foils and Ti-V alloys of technical grade were anodically oxidized for 60 min, under the specific parameters summarized in Table 1. The effect of the applied potential (30, 40, and 50 V), vanadium content in the alloy (5, 10, and 15 wt %), and water content in the electrolyte (2, 5, and 10 vol %) on the morphology of the as-prepared samples were studied by scanning electron microscopy. The top-view and cross-sectional scanning electron microscopy (SEM) images are presented in Figure 1. The anodization of Ti sheets led to the formation of uniform and self-organized NTs with an open tube top and smooth walls, and the tube diameter and length ranged from 81 to 120 nm and from 1.5 to 16.2 μm, respectively (Ti_30V, Ti_50V, respectively). The samples anodized from the Ti-V alloys presented a different morphology, depending on the preparation parameters. The series of samples synthesized from alloys with a 10 wt % vanadium content generally exhibited a sponge-like structure integrated by overlapped layers with a tubular appearance. The registered diameters of these structures varied from 61 to 101 nm and the average thickness of the mixed oxide layers was 0.3–0.8 μm. The samples prepared from alloys with 15 wt % of vanadium and using electrolytes with different water contents showed different morphologies. The $Ti_{85}V_{15}$_40V_2% and $Ti_{85}V_{15}$_40V_10% samples presented a sponge-like structure made up of interconnected disordered bundles. Conversely, the $Ti_{85}V_{15}$_40V_5% sample had a tubular structure with ripples on the tube wall, although the nanotubular layer was not highly organized. NTs presented a diameter (103 nm) similar to that of pristine TiO_2 NTs (100 nm) obtained at the same voltage (40 V), while the length (0.9 μm) was smaller than that of the analogous pristine sample (5 μm). The highest level of self-organization was achieved with the sample obtained from the anodization of the alloy with a 5 wt % of vanadium content ($Ti_{95}V_5$_40V), for which the synthesized NTs appeared to be composed of interconnected rings with a diameter of 86 nm and a length of 1 μm. As can be seen, the vanadium content in the alloy has a strong influence on the morphology of the samples. According to Yang, Kim, and Schmuki [55], the absence of a self-organized nanotube layer can be attributed to the low stability of the vanadium oxide, and therefore, the sample ($Ti_{95}V_5$_40V) synthesized from the alloy with the lowest vanadium content exhibited the best auto-alignment and self-organization. The influence of the other parameters, applied potential and water content, on the morphology of the samples was not clear due to the strong influence of the vanadium content in the alloy.

Table 1. Sample labels, preparation parameters, characterization, and photocatalytic activity of V_2O_5–TiO_2 nanotubes under Vis irradiation.

| Sample Label | Preparation Parameters | | External Diameter (nm) | Tube Length (µm) | Wall Thickness (nm) | Average Crystallite Size (nm) | EDX Analysis | | | | Photoactivity Vis Light (λ_{max} = 465 nm) | |
	Electrolyte	Applied Potential (V)					Ti (wt %)	V (wt %)	C (wt %)	O (wt %)	Initial Reaction Rate $\times 10^2$ (μmol·dm^{-3}·min^{-1})	Reaction Rate Constant $\times 10^3$ (min^{-1})
Ti$_{90}$V$_{10}$_30V	EG 98% (v/v), H$_2$O 2% (v/v), NH$_4$F 0.09 M	30	61	0.8	13	31	74.61	7.61	0.01	17.78	5.34	5.98
Ti$_{90}$V$_{10}$_40V	EG 98% (v/v), H$_2$O 2% (v/v), NH$_4$F 0.09 M	40	91	0.3	19	32	73.84	7.39	0.01	18.78	6.76	7.57
Ti$_{90}$V$_{10}$_50V	EG 98% (v/v), H$_2$O 2% (v/v), NH$_4$F 0.09 M	50	101	0.4	30	30	69.34	6.85	0.01	23.79	6.56	7.35
Ti$_{85}$V$_{15}$_40V_2%	EG 98% (v/v), H$_2$O 2% (v/v), NH$_4$F 0.09 M	40	Sponge-like structure			31	73.09	12.14	0.01	14.78	6.62	7.41
Ti$_{85}$V$_{15}$_40V_5%	EG 95% (v/v), H$_2$O 5% (v/v), NH$_4$F 0.09 M	40	103	0.9	20	36	67.06	9.08	1.10	22.77	7.08	7.92
Ti$_{85}$V$_{15}$_40V_10%	EG 90% (v/v), H$_2$O 10% (v/v), NH$_4$F 0.09 M	40	Sponge-like structure			32	66.10	8.91	1.14	23.86	4.50	5.04
Ti$_{85}$V$_5$_40V	EG 98% (v/v), H$_2$O 2% (v/v), NH$_4$F 0.09 M	40	86	1.0	11	33	72.59	3.25	0.02	24.14	5.39	6.04
Ti_30V	EG 98% (v/v), H$_2$O 2% (v/v), NH$_4$F 0.09 M	30	81	1.5	10	33	71.47	0.00	0.19	28.34	0.37	0.42
Ti_40V	EG 98% (v/v), H$_2$O 2% (v/v), NH$_4$F 0.09 M	40	100	5.0	13	34	66.73	0.00	0.03	33.24	0.43	0.49
Ti_50V	EG 98% (v/v), H$_2$O 2% (v/v), NH$_4$F 0.09 M	50	120	16.2	18	38	67.69	0.00	0.03	32.28	0.64	0.72

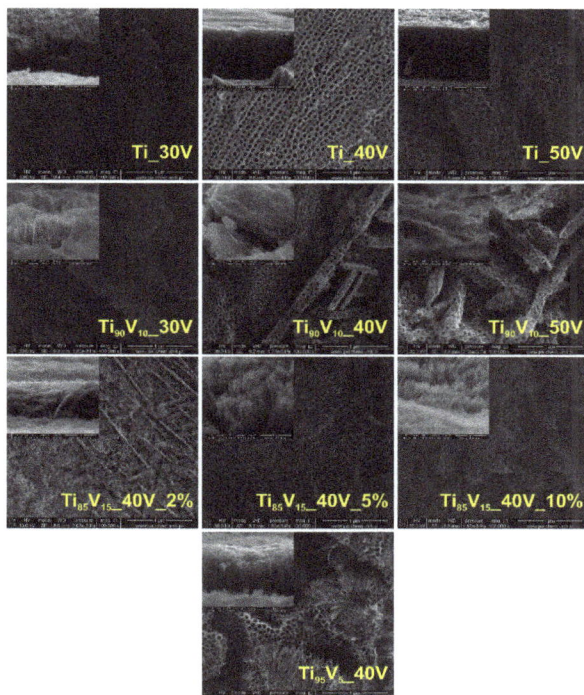

Figure 1. Top-view and cross-sectional scanning electron microscopy (SEM) images of pristine TiO$_2$ nanotubes (NTs) and Ti-V anodized alloys.

Considering these results, the SEM images of the Ti90V10_40V sample anodized during 4, 15, and 60 min (Figure 2d–f), together with literature data, a probable growth mechanism of V$_2$O$_5$-TiO$_2$ NTs has been described. As can be seen in Figure 2a–c, the shape of the current density-time curves recorded for the V$_2$O$_5$-TiO$_2$ samples were very similar to those of pristine TiO$_2$ NTs. During the first stage, the formation of the V$_2$O$_5$-TiO$_2$ oxide layer induced an exponential decrease in the current density, because of the reaction of Ti and V with the O$_2^-$ and OH$^-$ ions from the water. The presence of this mixed oxide layer can be observed in Figure 2d, corresponding to the Ti90V10_40V sample after 4 min of anodization. Then, the current density progressively increased throughout the second stage due to the dissolution of the oxide layer, which led to an increase in the surface area of the electrode with the initiation of pore growth [64]. These soluble species correspond to the fluoride complexes, [TiF$_6$]$^{2-}$ and [VF$_6$]$^-$ [65,66]. Figure 2e shows the initial pores in the sample after 15 min of anodic oxidation. Finally, a regular and self-ordered NT layer, which can be appreciated in Figure 2f, is formed under a quasi-steady state, which is stablished due to the equilibrium between the formation and dissolution of the oxide layer. During this stage, pores equally share the available current [45].

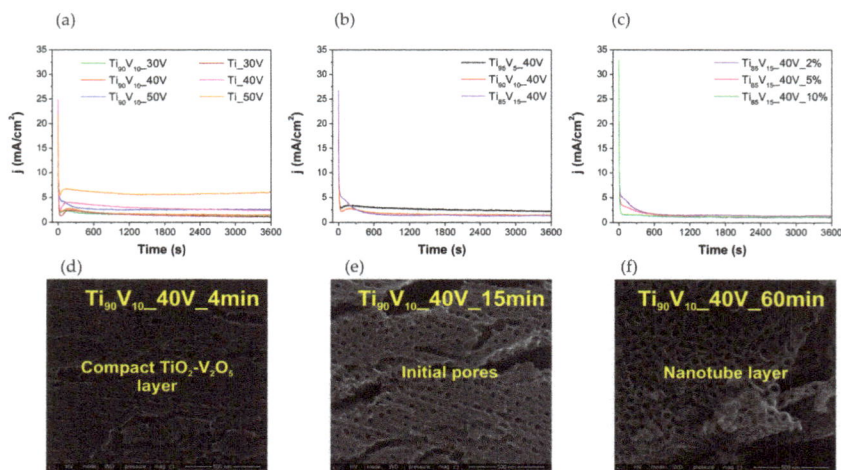

Figure 2. Current density-time curves recorded for the anodization of technical grade Ti foil and Ti-V alloys for the study of (**a**) applied voltage; (**b**) vanadium content in the alloy; and (**c**) water content in the electrolyte. SEM images of $Ti_{90}V_{10}$_40V sample anodized during (**d**) 4 min; (**e**) 15 min; and (**f**) 60 min.

The elemental composition of the obtained samples was analyzed through energy-dispersive X-ray spectroscopy (EDX) and the results presented in Table 1 show that the mass ratios between Ti and V in the V_2O_5-TiO_2 mixed oxides nanostructures (NS) agree well with the nominal content of the alloy. In addition, no trace of elements other than Ti, V, C, and O, was observed. These findings confirm the chemical homogeneity of the nanotube layer. Furthermore, from the EDX mapping presented in Figure 3, it can be concluded that the aggregation of Ti and V was not observed.

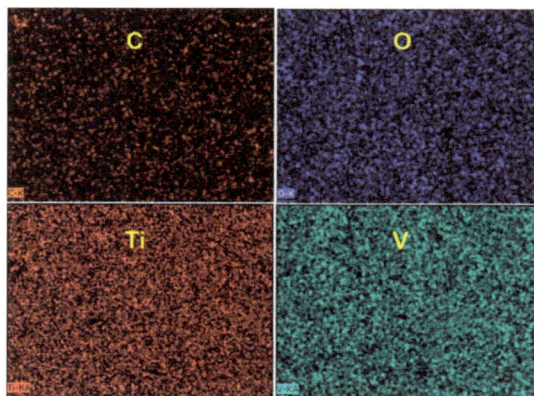

Figure 3. Energy-dispersive X-ray spectroscopy (EDX) mapping of the $Ti_{85}V_{15}$_40V_5% sample.

2.2. Optical Properties

The UV-Vis spectra of the obtained samples were compared with those of pristine TiO_2 NTs. Figure 4a clearly shows that the samples prepared from $Ti_{90}V_{10}$ alloys exhibited a stronger absorbance in the broad visible range of 400–750 nm than TiO_2 NTs. The spectra of the series with different vanadium contents, displayed in Figure 4b, indicated that an increase in the vanadium content in the alloy led to an increase in the absorbance intensity in the visible range, together with a red-shift. In particular, the spectrum of the sample $Ti_{85}V_{15}\_40V$ presented a peak of maximum absorbance near 500 nm, which, according to literature data, corresponds to V_2O_5 [42,67]. The spectra of the series of samples prepared in an electrolyte with different water contents and plotted in Figure 4c are consistent with the previous statements and no clear effect of the water content on the UV-Vis properties was found. All of the spectra for this series of samples showed a peak of absorption in the Vis range near 500 nm, and the spectrum of the $Ti_{85}V_{15}\_40V\_5\%$ sample showed the highest absorbance intensity peak. It can be concluded that the presence of the V_2O_5 in V_2O_5-TiO_2 matrix enhanced the light absorption in the range of 400–750 nm.

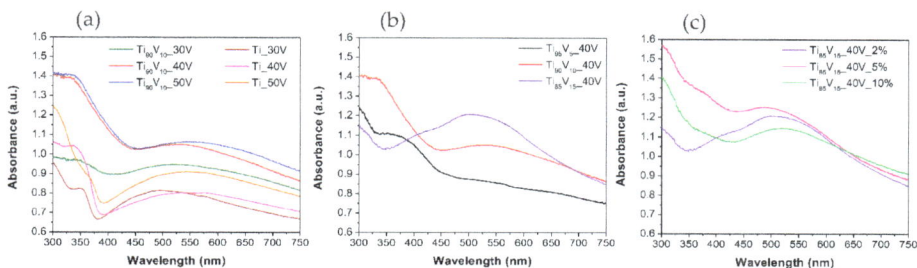

Figure 4. UV-Vis spectra of pristine TiO_2 NTs and V_2O_5-TiO_2 nanostructures (NS). Effect of (**a**) anodization potential; (**b**) vanadium content in the alloy; and (**c**) water content in the electrolyte.

It is known that photoluminescence (PL) spectroscopy is a powerful tool for determining the presence of surface defects, trap states, and sub-band states in the mid-gap level of photocatalysts [68]. The PL spectra of the obtained photocatalysts are presented in Figure 5. It should be noted that the same emission and position peaks were observed among all series. Notably, the emission peak at approximately at 420 nm can be ascribed to the existence of self-trapped excitons from the TiO_6^{8-} octahedron, while the two emission peaks at 450 and 485 nm could be assigned to the presence of surface defects, in the form of oxygen vacancies, which can create intermediate energy states located below the conduction band and which are able to trap electrons. The last peak at approximately 525 nm can be associated with the radiative recombination of the charge carriers [69,70].

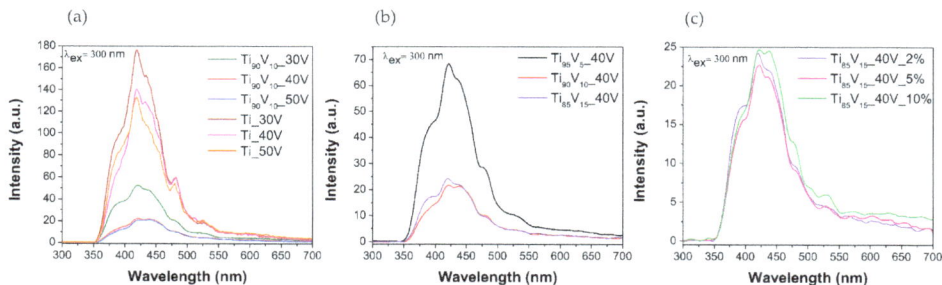

Figure 5. Photoluminescence spectra of pristine TiO_2 NTs and V_2O_5-TiO_2 NS. Effect of (**a**) anodization potential; (**b**) vanadium content in the alloy; and (**c**) water content in the electrolyte.

The results mentioned above confirm the presence of surface/structural defects, which can play a role in the photocatalytic degradation of pollutants.

2.3. Structural Properties

XRD patters of the obtained photocatalysts are presented in Figure 6. The calculated average crystallite size for pristine and modified TiO_2 NTs are gathered in Table 1. The average crystallite size was calculated using the Scherrer equation, based on the (101) diffraction peak. In the registered region, peaks at 2θ values of 25.67°, 37.97°, 48.31°, 54.16°, and 55.30° can be ascribed to (101), (004), (200), (105), and (211) planes, respectively, which are characteristic of the anatase phase (JCPDS card). The other peaks at 2θ = 35.4°, 38.70°, 40.77°, and 53.31° can be ascribed to planes of metallic Ti substrate. As was mentioned above, the diffraction peaks corresponding to the pure anatase TiO_2 phase were found, but other phases assigned to V_2O_5 were not observed. There are three possible explanations for this. Firstly, it could be because V_2O_5 diffraction peaks exist; however, the intensity of peaks is too low for this to be true. The absence of peaks corresponding to V_2O_5 in the XRD patters may be due to the low content and amorphous character of V_2O_5 or the short-range crystalline. Eventually, the vanadium species are incorporated into the TiO_2 lattice. On the other hand, in modified samples, the bands assigned to the anatase phase had a smaller and wider intensity. In particular, the intensity of the pick ascribed to the characteristic (101) plane of anatase decreased with the increase in the vanadium content in the alloy. This is related to the smaller crystallite size of V_2O_5-TiO_2 NS than that of pristine TiO_2 NTs [71].

Furthermore, it can be seen that the intensity of the anatase reflexes increased, while those of the substrate decreased, with the increase of the anodizing voltage. This is caused by the increasing thickness of the nanotube layer.

The average crystallite size varied from 30 to 36 nm among Ti-V series, and from 33 to 38 nm for pristine TiO_2 NTs. The smallest crystallite size was found for the $Ti_{90}V_{10}\_50V$ sample, which reached 30 nm. A clear correlation between the crystallite size and (i) anodization potential; (ii) vanadium content in the alloy; and (iii) water content in the electrolyte, was not observed.

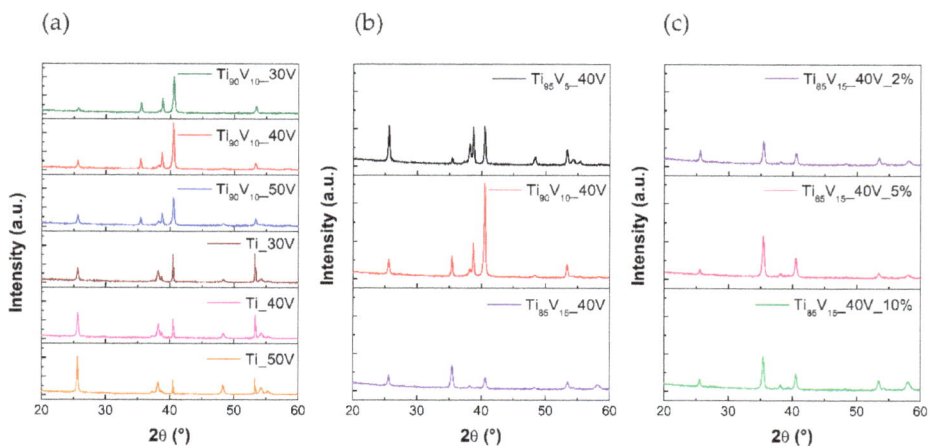

Figure 6. X-ray diffraction (XRD) spectra of pristine TiO_2 NTs and V_2O_5-TiO_2 NS. Effect of (**a**) anodization potential; (**b**) vanadium content in the alloy; and (**c**) water content in the electrolyte.

Micro-Raman spectroscopy was performed to determine the microstructure of the prepared samples. A 532 nm laser was used for the excitation. Figure 7 shows the Raman spectra of pristine TiO_2 and V_2O_5-TiO_2 NTs. The observed peaks at approximately 150, 396, 515, and 636 cm^{-1} are ascribed to the E_g, B_{1g}, A_{1g} + B_{1g}, and E_g modes of the anatase phase, respectively, in agreement with previous reports [42,72–74]. The E_g modes are assigned to TiO_2 symmetry, B_{1g} to O-Ti-O bending, and A_{1g} + B_{1g} to Ti-O stretching [75]. All of the spectra also registered a weak combination band at ca. 800 cm^{-1}, which is characteristic of the Raman signature of anatase [76]. No distinguishable crystalline V_2O_5 Raman bands were present at 703 and 997 cm^{-1} in any spectra, probably due to the low content of vanadium in the alloy precursors or to the highly dispersed state of V_2O_5 in V_2O_5-TiO_2 NS. This was also reported by former publications for composites with the V_2O_5-TiO_2 system [32,75,77].

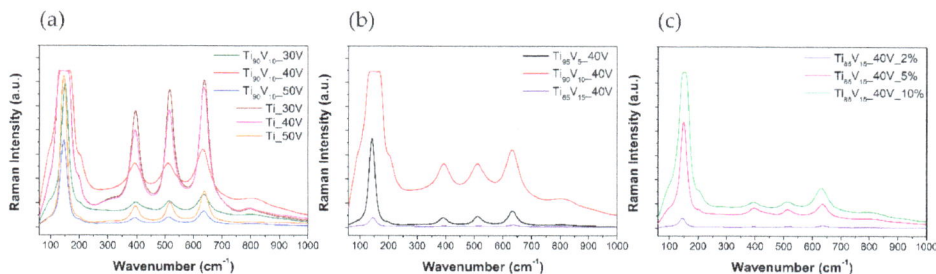

Figure 7. Raman spectra of pristine TiO_2 NTs and V_2O_5-TiO_2 NS. Effect of (**a**) anodization potential; (**b**) vanadium content in the alloy; and (**c**) water content in the electrolyte.

2.4. Photocatalytic Performance

The effect of the anodization voltage, vanadium content in the alloy, and water content in the electrolyte on the photocatalytic activity was evaluated through the degradation of toluene from an air mixture (200 ppmv of toluene) under Vis irradiation (LEDs array, λ_{max} = 465 nm). Figure 8 presents the degradation curves for the above-mentioned series and their comparison with the photoactivity of reference pristine TiO_2 NTs. These plots show that V_2O_5-TiO_2 samples from all series were active in the photodegradation reaction, in contrast with pristine TiO_2 NTs which exhibited negligible toluene removal (ca. 5%). The highest degradation of toluene in the presence of samples prepared from the $Ti_{90}V_{10}$ alloys (see Figure 8a), after 60 min of irradiation, was observed for the sample anodized under 40 V (34%). The toluene removal reached by samples anodized under 30 V and 50 V were not that different from the best one (27% and 33%, respectively). In view of this, 40 V was selected as the potential for further synthesis, to determine the vanadium content in the alloy and the composition of the electrolyte solution, which are favorable for the photodegradation reaction. Figure 8b presents similar results, for the samples obtained from alloys with different vanadium contents. It can be observed that the vanadium content in the alloy slightly affected the photoactivity of the samples. The maximum toluene removal was found to be achieved for the sample with 10 wt % of vanadium in the alloy ($Ti_{90}V_{10}$_40V, 34% of degradation). The analysis of the effect of water content in the electrolyte was carried out with NS obtained from $Ti_{85}V_{15}$ alloys. As can be seen in Figure 8c, there is a slight difference in the photocatalytic performance between these samples, among this series. The highest degradation of toluene was exhibited by the sample anodized in the electrolyte containing 5% of water and it corresponded to 35% of toluene removal ($Ti_{85}V_{15}$_40V_5%). For a more detailed comparison of the obtained results, the initial reaction rate and reaction rate constants were calculated and presented in Table 1. The highest value for the initial reaction rate, among all series, was achieved in the presence of the $Ti_{85}V_{15}$_40V_5% sample (7.08 × 10^{-2} µmol·dm^{-3}·min^{-1}), which also exhibited the highest absorbance intensity peak near 500 nm and consisted of a NT layer which was not highly organized. This suggests that this NT composite effectively enhanced visible light harvesting and the consequent photocatalytic reaction, owing to the presence of V_2O_5 [35,43]. Furthermore, no correlation between the morphology and the photocatalytic performance of the samples was observed.

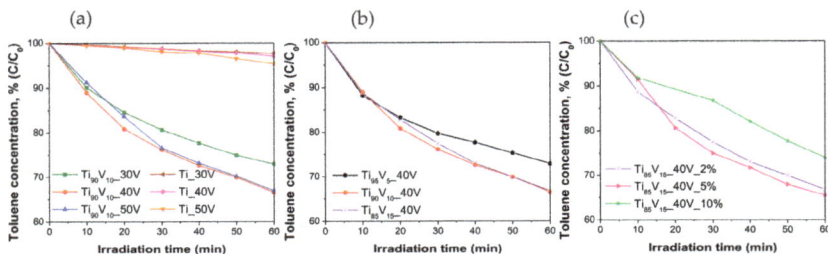

Figure 8. Photoactivity of pristine TiO_2 NTs and V_2O_5-TiO_2 NS in gas phase degradation of toluene under Vis-light irradiation (λ_{max} = 465 nm). Effect of (**a**) applied voltage; (**b**) vanadium content in the alloy; and (**c**) water content in the electrolyte.

In conclusion, the highest photoactivity under visible light (465 nm) was observed in the presence of the $Ti_{85}V_{15}$_40V_5% sample. This sample not only exhibited the highest absorbance intensity at a wavelength of about 500 nm, but also reported the highest diameter (103 nm), the second longest NTs (0.9 µm), and the largest crystallite size (36 nm), from the modified samples. Its vanadium content, based on EDX analysis, was 9.08 wt %. On the other hand, the $Ti_{85}V_{15}$_40V_10% sample showed the lowest photoactivity. It had a sponge-like morphology with a vanadium content of 8.91%, based on EDX analysis, which is lower than the content of the sample with the highest photoactivity, considering that both were prepared from $Ti_{85}V_{15}$ alloys. Its crystallite size was 32 nm, smaller than that of the $Ti_{85}V_{15}$_40V_5% sample. The initial reaction rate achieved in the presence of this sample was 4.50×10^{-2} µmol·dm^{-3}·min^{-1}, which is 1.6 times lower than that reported for the most photoactive one (7.08×10^{-2} µmol·dm^{-3}·min^{-1}).

To further analyze the photocatalytic properties of the synthesized composites, the effect of different irradiation wavelengths was studied using the most photoactive sample ($Ti_{85}V_{15}$_40V_5%). The gas phase degradation of toluene was tested under 375, 415, and 465 nm and the obtained results are displayed in Figure 9. It can be observed that the highest degradation (52%) after 60 min of irradiation was achieved under UV light (375 nm). This can be explained by the presence of TiO_2 in the NT matrix, which is the main active species under UV light irradiation. On the other hand, under the influence of visible light irradiation, 415 and 465 nm, the photocatalytic degradation reached almost the same level, in both cases, with values of 34% and 35%, respectively. This indicates that under Vis light irradiation, V_2O_5 are the main species responsible for the generation of e$^-$ and h$^+$ (as presented in Figure 10, excitation mechanism) over the surface of NTs, which led to the photodegradation of toluene, and this is supported by the negligible degradation reported for pristine TiO_2 NTs under Vis light.

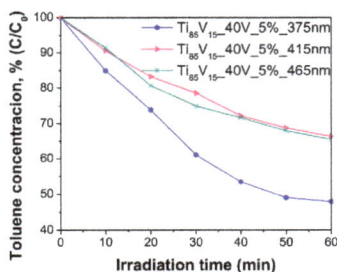

Figure 9. Photoactivity of $Ti_{85}V_{15}$_40V_5% sample in gas phase degradation of toluene under different wavelengths of irradiation (λ_{max} = 375, 415, 465 nm).

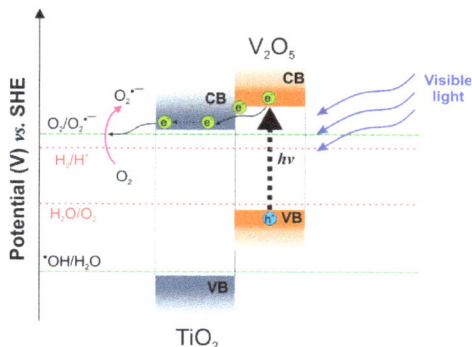

Figure 10. Excitation mechanism of V_2O_5-TiO_2 samples under visible light irradiation.

3. Materials and Methods

3.1. Materials

Acetone, isopropanol, and methanol were purchased from P.P.H. "STANLAB" Sp. J. (Lublin, Poland), ethylene glycol (EG) was acquired from CHEMPUR (Piekary Śląskie, Poland), and ammonium fluoride was bought from ACROS ORGANICS (Geel, Belgium). Technical grade Ti foils and Ti-V alloys with 5, 10, and 15 wt % of vanadium content were provided by HMW-Hauner Metallische Werkstoffe (Röttenbach, Germany). Deionized (DI) water with a conductivity of 0.05 μS was used to prepare all of the aqueous solutions.

3.2. Synthesis of Pristine TiO₂ and V₂O₅-TiO₂ Nanotubes

Ti foils and Ti-V alloys were ultrasonically cleaned in acetone, isopropanol, methanol, and deionized water for 10 min. Then, the foils were dried in an air stream. The anodization processes were carried out at room temperature, in an electrochemical cell consisting of a platinum mesh as the counter electrode, and the Ti-V alloy (2.5 cm × 2.5 cm) as the working electrode. A reference electrode of Ag/AgCl connected to a digital multimeter (BRYMEN BM857a, New Taipei City, Taiwan) was used to control and record information about the actual potential and current on the alloy. The anodization was conducted in an electrolyte composed of EG, water, and NH₄F 0.09 M, during 60 min, with a voltage in the range of 30–50 V which was applied with a programmable DC power supply (MANSON SDP 2603, Hong Kong, China). Three electrolyte solutions with different water contents were used (volume ratios of EG:water of 98:2, 95:5, and 90:10). The obtained samples were rinsed with deionized water, sonicated in deionized water (1 min), dried in air (80 °C for 24 h), and calcined (450 °C, heating rate 2 °C/min) for 1 h.

3.3. Characterization of Pristine TiO₂ and V₂O₅-TiO₂ Nanotubes

The morphology of the synthesized pristine TiO₂ and V₂O₅-TiO₂ nanotubes was determined by using scanning electron microscopy (SEM, FEI QUANTA 3D FEG, FEI Company, Brno, Czech Republic). Energy-dispersive X-ray spectroscopy (EDX) analysis was performed with a scanning electron microscope (SEM, Zeiss, Leo 1430 VP, Carl Zeiss, Oberkochen, Germany). The crystal structure of the samples was determined from X-ray diffraction patterns recorded in the range of 2θ = 20°–90°, using an X-ray diffractometer (X'Pert Pro, Panalytical, Almelo, The Netherlands) with Cu Kα radiation. The crystallite size was calculated based on the Scherrer formula. Raman spectra were measured with a micro-Raman spectrometer (Senterra, Bruker Optik, Billerica, MA, USA) with a 532 nm excitation laser.

The UV-Vis absorbance spectra were registered on a SHIMADZU (UV-2600) UV-VIS Spectrophotometer (SHIMADZU, Kioto, Japan) equipped with an integrating sphere. The measurements were carried out in the wavelength range of 300–800 nm, the baseline was determined with barium sulfate as the reference, and the scanning speed was 250 nm/min at room temperature. The photoluminescence (PL) spectra were recorded at room temperature with a LS-50B Luminescence Spectrometer equipped with a Xenon discharge lamp as an excitation source and a R928 photomultiplier (HAMAMATSU, Hamamatsu, Japan) as detector. The excitation radiation (300 nm) was directed onto the surface of the samples at an angle of 90°.

3.4. Measurement of Photocatalytic Activity

The photocatalytic activity of the as-prepared NTs was analyzed, for the first time, in the purification of air from toluene, which was used as a model pollutant. The photodegradation experiments were carried out in a stainless-steel reactor with a volume of ca. 35 cm^3. The reactor included a quartz window, two valves, and a septum. The light source consisting of an array of 25 LEDs (λ_{max} = 375, 415 and 465 nm, Optel, Opole, Poland) was located above the sample. The anodized foil was placed at the bottom side of the reactor and it was closed with the quartz window. A gas mixture (200 ppmv) was passed through the reactor during 1 min, the valves were then closed, and the reactor was kept in the dark for 30 min in order to achieve the equilibrium. Before starting the irradiation, a reference toluene sample was taken. The concentration was determined by using a gas chromatograph (TRACE 1300, Thermo Scientific, Waltham, MA, USA), equipped with an ionization flame detector (FID) and an Elite-5 capillary column. The samples (200 µL) were dosed with a gas-tight syringe for 10 min.

4. Conclusions

In summary, V_2O_5-TiO_2 mixed oxide layers were successfully synthesized through the one-step anodization of Ti-V alloys in a fluoride-containing EG-based electrolyte. The obtained layers exhibited a sponge-like and nanotubular structure with highly enhanced optical and visible-light-photocatalytic properties, in contrast with pristine TiO_2 NTs. The photoactivity of these anodically-obtained composites was evaluated for the first time in the degradation of toluene (200 ppmv) in the gas phase under visible light, with a twenty-five-LED array as the irradiation source (λ_{max} = 465 nm). All of the V_2O_5-TiO_2 samples were reported as photoactive and the initial degradation reaction rate was in the range of 4.50–7.08 × 10^{-2} µmol·dm^{-3}·min^{-1}. The visible light harvesting was attributed to the presence of the narrow-bandgap V_2O_5 species in the matrix of the V_2O_5-TiO_2 composites. A morphological study was also reported and the vanadium content in the alloy was found as the key factor limiting the self-ordering of the electrochemically prepared thin layers. The highest photoactivity under visible light (465 nm) was observed in the presence of the $Ti_{85}V_{15}$_40V_5% sample. This sample not only exhibited the highest absorbance intensity at about 500 nm, but also reported the highest diameter (103 nm), the optimum length (0.9 µm), and the largest crystallite size (36 nm) among all of the modified samples. EDX analysis revealed that the vanadium content in this sample was equal to 9.08 wt %. In sum, the photocatalytic properties of these highly efficient nanocomposites, obtained through the most suitable method (electrochemical technique), permit new insights into the exploitation of industrially oriented applications, for instance, photocatalytic devices for air purification. The presented materials are photoactive under a low powered light source, and thus, the use of low cost light-emitting diodes (LEDs) as an irradiation source can significantly reduce the cost of photocatalytic air treatment processes, which is consistent with the principles of green chemistry.

Acknowledgments: This research was financially supported by the Polish National Science Center (research grant, Ordered TiO_2/M_xO_y nanostructures obtained by electrochemical method; contract no. NCN 2014/15/B/ST5/00098).

Author Contributions: A.Z.-M. supervised and directed the project; A.Z.-M. and P.M. conceived the concept; M.C.N.-M., P.M., M.P.K., G.K., A.M., M.K. and G.S. performed the experiments; M.C.N.-M. and P.M. analyzed the data; M.C.N.-M. and P.M. contributed reagents/materials/analysis tools; M.C.N-M., P.M, P.J.E.-M. and A.Z.-M. wrote the paper.

Conflicts of Interest: The authors declare no conflict of interest.

References

1. Ghicov, A.; Schmidt, B.; Kunze, J.; Schmuki, P. Photoresponse in the visible range from Cr doped TiO_2 nanotubes. *Chem. Phys. Lett.* **2007**, *433*, 323–326. [CrossRef]

2. Kubacka, A.; Fernandez-Garcia, M.; Colon, G. Advanced nanoarchitectures for solar photocatalytic applications. *Chem. Rev.* **2012**, *112*, 1555–1614. [CrossRef] [PubMed]

3. Schneider, J.; Bahnemann, D.; Ye, J.; Puma, G.L.; Dionysiou, D.D. *Photocatalysis: Fundamentals and Perspectives*; Royal Society of Chemistry: Cambridge, UK, 2016.

4. Dionysiou, D.D.; Puma, G.L.; Ye, J.; Schneider, J.; Bahnemann, D. *Photocatalysis: Applications*; Royal Society of Chemistry: Cambridge, UK, 2016.

5. Pichat, P. *Photocatalysis: Fundamentals, Materials and Potential*; MDPI: Basel, Switzerland, 2016.

6. Colmenares Quintero, J.C.; Xu, Y.-J. *Heterogeneous Photocatalysis: From Fundamentals to Green Applications*; Springer: Berlin/Heidelberg, Germany, 2016; Volume 8, p. 416.

7. Fattakhova-Rohlfing, D.; Zaleska, A.; Bein, T. Three-dimensional titanium dioxide nanomaterials. *Chem. Rev.* **2014**, *114*, 9487–9558. [CrossRef] [PubMed]

8. Daghrir, R.; Drogui, P.; Robert, D. Modified TiO_2 for environmental photocatalytic applications: A review. *Ind. Eng. Chem. Res.* **2013**, *52*, 3581–3599. [CrossRef]

9. Ahmed, S.; Rasul, M.G.; Brown, R.; Hashib, M.A. Influence of parameters on the heterogeneous photocatalytic degradation of pesticides and phenolic contaminants in wastewater: A short review. *J. Environ. Manag.* **2011**, *92*, 311–330. [CrossRef] [PubMed]

10. Pichat, P. *Photocatalysis and Water Purification: From Fundamentals to Recent Applications*; John Wiley & Sons: Weinheim, Germany, 2013.

11. Kang, D.; Kim, T.W.; Kubota, S.R.; Cardiel, A.C.; Cha, H.G.; Choi, K.S. Electrochemical synthesis of photoelectrodes and catalysts for use in solar water splitting. *Chem. Rev.* **2015**, *115*, 12839–12887. [CrossRef] [PubMed]

12. Highfield, J. Advances and recent trends in heterogeneous photo(electro)-catalysis for solar fuels and chemicals. *Molecules* **2015**, *20*, 6739–6793. [CrossRef] [PubMed]

13. Ye, M.; Zheng, D.; Wang, M.; Chen, C.; Liao, W.; Lin, C.; Lin, Z. Hierarchically structured microspheres for high-efficiency rutile TiO_2-based dye-sensitized solar cells. *ACS Appl. Mater. Interfaces* **2014**, *6*, 2893–2901. [CrossRef] [PubMed]

14. Low, J.; Cheng, B.; Yu, J. Surface modification and enhanced photocatalytic CO_2 reduction performance of Tio_2: A review. *Appl. Surf. Sci.* **2017**, *392*, 658–686. [CrossRef]

15. Wang, R.; Hashimoto, K.; Fujishima, A.; Chikuni, M.; Kojima, E.; Kitamura, A.; Shimohigoshi, M.; Watanabe, T. Light-induced amphiphilic surfaces. *Nature* **1997**, *388*, 431–432. [CrossRef]

16. Pichat, P. Self-cleaning materials based on solar photocatalysis. In *New and Future Developments in Catalysis: Solar Photocatalysis*; Suib, S.L., Ed.; Elsevier: Amsterdam, The Netherlands, 2013; Volume 7, pp. 167–190.

17. Li, Z.; Luo, W.; Zhang, M.; Feng, J.; Zou, Z. Photoelectrochemical cells for solar hydrogen production: Current state of promising photoelectrodes, methods to improve their properties, and outlook. *Energy Environ. Sci.* **2013**, *6*, 347–370. [CrossRef]

18. Pfitzner, A.; Dankesreiter, S.; Eisenhofer, A.; Cherevatskaya, M. Heterogeneous semiconductor photocatalysis. In *Chemical Photocatalysis*; König, B., Ed.; De Gruyter: Berlin, Germany, 2013.

19. Skinner, D.E.; Colombo, D.P.; Cavaleri, J.J.; Bowman, R.M. Femtosecond investigation of electron trapping in semiconductor nanoclusters. *J. Phys. Chem.* **1995**, *99*, 7853–7856. [CrossRef]

20. Banerjee, S.; Pillai, S.C.; Falaras, P.; O'Shea, K.E.; Byrne, J.A.; Dionysiou, D.D. New insights into the mechanism of visible light photocatalysis. *J. Phys. Chem. Lett.* **2014**, *5*, 2543–2554. [CrossRef] [PubMed]

21. Wang, Z.; Ma, W.; Chen, C.; Zhao, J. Sensitization of titania semiconductor: A promising strategy to utilize visible light. In *Photocatalysis and Water Purification*; Pichat, P., Ed.; Wiley-VCH: Weinheim, Germany, 2013; pp. 199–240.

22. Choi, W.; Termin, A.; Hoffmann, M.R. The role of metal ion dopants in quantum-sized TiO_2: Correlation between photoreactivity and charge carrier recombination dynamics. *J. Phys. Chem.* **1994**, *98*, 13669–13679. [CrossRef]

23. Asahi, R.; Morikawa, T.; Ohwaki, T.; Aoki, K.; Taga, Y. Visible-light photocatalysis in nitrogen-doped titanium oxides. *Science* **2001**, *293*, 269–271. [CrossRef] [PubMed]

24. Li, D.; Haneda, H.; Labhsetwar, N.K.; Hishita, S.; Ohashi, N. Visible-light-driven photocatalysis on fluorine-doped TiO_2 powders by the creation of surface oxygen vacancies. *Chem. Phys. Lett.* **2005**, *401*, 579–584. [CrossRef]

25. Mazierski, P.; Nischk, M.; Gołkowska, M.; Lisowski, W.; Gazda, M.; Winiarski, M.J.; Klimczuk, T.; Zaleska-Medynska, A. Photocatalytic activity of nitrogen doped TiO_2 nanotubes prepared by anodic oxidation: The effect of applied voltage, anodization time and amount of nitrogen dopant. *Appl. Catal. B Environ.* **2016**, *196*, 77–88. [CrossRef]

26. Mazierski, P.; Lisowski, W.; Grzyb, T.; Winiarski, M.J.; Klimczuk, T.; Mikołajczyk, A.; Flisikowski, J.; Hirsch, A.; Kołakowska, A.; Puzyn, T.; et al. Enhanced photocatalytic properties of lanthanide-TiO_2 nanotubes: An experimental and theoretical study. *Appl. Catal. B Environ.* **2017**, *205*, 376–385. [CrossRef]

27. Chatterjee, D.; Mahata, A. Demineralization of organic pollutants on the dye modified TiO_2 semiconductor particulate system using visible light. *Appl. Catal. B Environ.* **2001**, *33*, 119–125. [CrossRef]

28. Hirai, T.; Suzuki, K.; Komasawa, I. Preparation and photocatalytic properties of composite CdS nanoparticles–titanium dioxide particles. *J. Colloid Interface Sci.* **2001**, *244*, 262–265. [CrossRef]

29. Zielińska-Jurek, A.; Zaleska, A. Ag/Pt-modified TiO_2 nanoparticles for toluene photooxidation in the gas phase. *Catal. Today* **2014**, *230*, 104–111. [CrossRef]

30. Nischk, M.; Mazierski, P.; Wei, Z.; Siuzdak, K.; Kouame, N.A.; Kowalska, E.; Remita, H.; Zaleska-Medynska, A. Enhanced photocatalytic, electrochemical and photoelectrochemical properties of TiO_2 nanotubes arrays modified with Cu, AgCu and Bi nanoparticles obtained via radiolytic reduction. *Appl. Surf. Sci.* **2016**, *387*, 89–102. [CrossRef] [PubMed]

31. Pichat, P. Surface-properties, activity and selectivity of bifunctional powder photocatalysts. *New J. Chem.* **1987**, *11*, 135–140.

32. Wu, Z.; Dong, F.; Liu, Y.; Wang, H. Enhancement of the visible light photocatalytic performance of C-doped TiO_2 by loading with V_2O_5. *Catal. Commun.* **2009**, *11*, 82–86. [CrossRef]

33. Wang, Y.; Zhang, J.; Liu, L.; Zhu, C.; Liu, X.; Su, Q. Visible light photocatalysis of V_2O_5/TiO_2 nanoheterostructures prepared via electrospinning. *Mater. Lett.* **2012**, *75*, 95–98. [CrossRef]

34. Yang, X.; Ma, F.; Li, K.; Guo, Y.; Hu, J.; Li, W.; Huo, M.; Guo, Y. Mixed phase titania nanocomposite codoped with metallic silver and vanadium oxide: New efficient photocatalyst for dye degradation. *J. Hazard. Mater.* **2010**, *175*, 429–438. [CrossRef] [PubMed]

35. Wang, Y.; Su, Y.R.; Qiao, L.; Liu, L.X.; Su, Q.; Zhu, C.Q.; Liu, X.Q. Synthesis of one-dimensional TiO_2/V_2O_5 branched heterostructures and their visible light photocatalytic activity towards rhodamine b. *Nanotechnology* **2011**, *22*, 225702. [CrossRef] [PubMed]

36. Li, B.; Xu, Y.; Rong, G.; Jing, M.; Xie, Y. Vanadium pentoxide nanobelts and nanorolls: From controllable synthesis to investigation of their electrochemical properties and photocatalytic activities. *Nanotechnology* **2006**, *17*, 2560–2566. [CrossRef] [PubMed]

37. Fei, H.-L.; Zhou, H.-J.; Wang, J.-G.; Sun, P.-C.; Ding, D.-T.; Chen, T.-H. Synthesis of hollow V_2O_5 microspheres and application to photocatalysis. *Solid State Sci.* **2008**, *10*, 1276–1284. [CrossRef]

38. Shahid, M.; Rhen, D.S.; Shakir, I.; Patole, S.P.; Yoo, J.B.; Yang, S.-J.; Kang, D.J. Facile synthesis of single crystalline vanadium pentoxide nanowires and their photocatalytic behavior. *Mater. Lett.* **2010**, *64*, 2458–2461. [CrossRef]

39. Qiu, G.; Dharmarathna, S.; Genuino, H.; Zhang, Y.; Huang, H.; Suib, S.L. Facile microwave-refluxing synthesis and catalytic properties of vanadium pentoxide nanomaterials. *ACS Catal.* **2011**, *1*, 1702–1709. [CrossRef]

40. Shen, T.F.R.; Lai, M.-H.; Yang, T.C.K.; Fu, I.P.; Liang, N.-Y.; Chen, W.-T. Photocatalytic production of hydrogen by vanadium oxides under visible light irradiation. *J. Taiwan Inst. Chem. Eng.* **2012**, *43*, 95–101. [CrossRef]

41. Aslam, M.; Ismail, I.M.; Salah, N.; Chandrasekaran, S.; Qamar, M.T.; Hameed, A. Evaluation of sunlight induced structural changes and their effect on the photocatalytic activity of V_2O_5 for the degradation of phenols. *J. Hazard. Mater.* **2015**, *286*, 127–135. [CrossRef] [PubMed]

42. Xie, L.; Liu, P.; Zheng, Z.; Weng, S.; Huang, J. Morphology engineering of V_2O_5/TiO_2 nanocomposites with enhanced visible light-driven photofunctions for arsenic removal. *Appl. Catal. B Environ.* **2016**, *184*, 347–354. [CrossRef]

43. Choi, S.; Lee, M.-S.; Park, D.-W. Photocatalytic performance of TiO_2/V_2O_5 nanocomposite powder prepared by DC arc plasma. *Curr. Appl. Phys.* **2014**, *14*, 433–438. [CrossRef]

44. Zhu, K.; Neale, N.R.; Miedaner, A.; Frank, A.J. Enhanced charge-collection efficiencies and light scattering in dye-sensitized solar cells using oriented TiO_2 nanotubes arrays. *Nano Lett.* **2007**, *7*, 69–74. [CrossRef] [PubMed]

45. Lee, K.; Mazare, A.; Schmuki, P. One-dimensional titanium dioxide nanomaterials: Nanotubes. *Chem. Rev.* **2014**, *114*, 9385–9454. [CrossRef] [PubMed]

46. Macak, J.M.; Schmuki, P. Anodic growth of self-organized anodic TiO_2 nanotubes in viscous electrolytes. *Electrochim. Acta* **2006**, *52*, 1258–1264. [CrossRef]

47. Pichat, P. Are TiO_2 nanotubes worth using in photocatalytic purification of air and water? *Molecules* **2014**, *19*, 15075–15087. [CrossRef] [PubMed]

48. Albu, S.P.; Ghicov, A.; Macak, J.M.; Schmuki, P. 250 μm long anodic TiO_2 nanotubes with hexagonal self-ordering. *Phys. Status Solidi–R.* **2007**, *1*, R65–R67. [CrossRef]

49. Khudhair, D.; Bhatti, A.; Li, Y.; Hamedani, H.A.; Garmestani, H.; Hodgson, P.; Nahavandi, S. Anodization parameters influencing the morphology and electrical properties of TiO_2 nanotubes for living cell interfacing and investigations. *Mater. Sci. Eng. C* **2016**, *59*, 1125–1142. [CrossRef] [PubMed]

50. Nischk, M.; Mazierski, P.; Gazda, M.; Zaleska, A. Ordered TiO_2 nanotubes: The effect of preparation parameters on the photocatalytic activity in air purification process. *Appl. Catal. B Environ.* **2014**, *144*, 674–685. [CrossRef]

51. Mazierski, P.; Nadolna, J.; Lisowski, W.; Winiarski, M.J.; Gazda, M.; Nischk, M.; Klimczuk, T.; Zaleska-Medynska, A. Effect of irradiation intensity and initial pollutant concentration on gas phase photocatalytic activity of TiO_2 nanotube arrays. *Catal. Today* **2016**, *284*, 19–26. [CrossRef]

52. Bashirom, N.; Razak, K.A.; Yew, C.K.; Lockman, Z. Effect of fluoride or chloride ions on the morphology of ZrO_2 thin film grown in ethylene glycol electrolyte by anodization. *Procedia Chem.* **2016**, *19*, 611–618. [CrossRef]

53. Tsuchiya, H.; Schmuki, P. Self-organized high aspect ratio porous hafnium oxide prepared by electrochemical anodization. *Electrochem. Commun.* **2005**, *7*, 49–52. [CrossRef]

54. Jha, H.; Hahn, R.; Schmuki, P. Ultrafast oxide nanotube formation on TiNb, TiZr and TiTa alloys by rapid breakdown anodization. *Electrochim. Acta* **2010**, *55*, 8883–8887. [CrossRef]

55. Yang, Y.; Kim, D.; Schmuki, P. Electrochromic properties of anodically grown mixed V_2O_5-TiO_2 nanotubes. *Electrochem. Commun.* **2011**, *13*, 1021–1025. [CrossRef]

56. Yang, Y.; Kim, D.; Yang, M.; Schmuki, P. Vertically aligned mixed V_2O_5-TiO_2 nanotube arrays for supercapacitor applications. *Chem. Commun.* **2011**, *47*, 7746–7748. [CrossRef] [PubMed]

57. Paramasivam, I.; Nah, Y.C.; Das, C.; Shrestha, N.K.; Schmuki, P. WO_3/TiO_2 nanotubes with strongly enhanced photocatalytic activity. *Chemistry* **2010**, *16*, 8993–8997. [CrossRef] [PubMed]

58. Ning, X.; Wang, X.; Yu, X.; Li, J.; Zhao, J. Preparation and capacitance properties of Mn-doped TiO_2 nanotube arrays by anodisation of Ti-Mn alloy. *J. Alloys Compd.* **2016**, *658*, 177–182. [CrossRef]

59. Allam, N.K.; Deyab, N.M.; Abdel Ghany, N. Ternary Ti-Mo-Ni mixed oxide nanotube arrays as photoanode materials for efficient solar hydrogen production. *Phys. Chem. Chem. Phys.* **2013**, *15*, 12274–12282. [CrossRef] [PubMed]

60. Luo, B.; Yang, H.; Liu, S.; Fu, W.; Sun, P.; Yuan, M.; Zhang, Y.; Liu, Z. Fabrication and characterization of self-organized mixed oxide nanotube arrays by electrochemical anodization of Ti-6Al-4V alloy. *Mater. Lett.* **2008**, *62*, 4512–4515. [CrossRef]

61. Mazierski, P.; Malankowska, A.; Kobyłański, M.; Diak, M.; Kozak, M.; Winiarski, M.J.; Klimczuk, T.; Lisowski, W.; Nowaczyk, G.; Zaleska-Medynska, A. Photocatalytically active TiO_2/Ag_2O nanotube arrays interlaced with silver nanoparticles obtained from the one-step anodic oxidation of Ti-Ag alloys. *ACS Catal.* **2017**, *7*, 2753–2764. [CrossRef]

62. Ding, J.; Huang, Z.; Zhu, J.; Kou, S.; Zhang, X.; Yang, H. Low-temperature synthesis of high-ordered anatase TiO_2 nanotube array films coated with exposed {001} nanofacets. *Sci. Rep.* **2015**, *5*, 17773. [CrossRef] [PubMed]

63. Nevárez-Martínez, M.C.; Mazierski, P.; Kobylański, M.; Szczepańska, G.; Trykowski, G.; Malankowska, A.; Kozak, M.; Espinoza-Montero, P.J.; Zaleska-Medynska, A. Self-organized TiO_2-MnO_2 nanotube arrays for efficient photocatalytic degradation of toluene. *Molecules* **2017**, *22*, 564. [CrossRef] [PubMed]

64. Macak, J.M.; Tsuchiya, H.; Ghicov, A.; Yasuda, K.; Hahn, R.; Bauer, S.; Schmuki, P. TiO_2 nanotubes: Self-organized electrochemical formation, properties and applications. *Curr. Opin. Solid State Mater. Sci.* **2007**, *11*, 3–18. [CrossRef]

65. Wood, G.C.; Khoo, S.W. The mechanism of anodic oxidation of alloys. *J. Appl. Electrochem.* **1971**, *1*, 189–206. [CrossRef]

66. Cox, B. 186. Complex fluorides. Part IV. The structural chemistry of complex fluorides of the general formula ABF_6. *J. Chem. Soc.* **1956**, 876–878. [CrossRef]

67. Wu, J.C.S.; Chen, C.-H. A visible-light response vanadium-doped titania nanocatalyst by sol-gel method. *J. Photochem. Photobiol. A Chem.* **2004**, *163*, 509–515. [CrossRef]

68. Nishanthi, S.T.; Subramanian, E.; Sundarakannan, B.; Padiyan, D.P. An insight into the influence of morphology on the photoelectrochemical activity of TiO_2 nanotube arrays. *Sol. Energy Mater. Sol. Cell* **2015**, *132*, 204–209. [CrossRef]

69. Tang, H.; Berger, H.; Schmid, P.E.; Lévy, F. Optical properties of anatase (TiO_2). *Solid State Commun.* **1994**, *92*, 267–271. [CrossRef]

70. Knorr, F.J.; Mercado, C.C.; McHale, J.L. Trap-state distributions and carrier transport in pure and mixed-phase TiO_2: Influence of contacting solvent and interphasial electron transfer. *J. Phys. Chem. C* **2008**, *112*, 12786–12794. [CrossRef]

71. Chin, S.; Park, E.; Kim, M.; Jurng, J. Photocatalytic degradation of methylene blue with TiO_2 nanoparticles prepared by a thermal decomposition process. *Powder Technol.* **2010**, *201*, 171–176. [CrossRef]

72. Zhang, W.; He, Y.; Zhang, M.; Yin, Z.; Chen, Q. Raman scattering study on anatase TiO_2 nanocrystals. *J. Phys. D Appl. Phys.* **2000**, *33*, 912. [CrossRef]

73. Wu, Z.; Dong, F.; Zhao, W.; Wang, H.; Liu, Y.; Guan, B. The fabrication and characterization of novel carbon doped TiO_2 nanotubes, nanowires and nanorods with high visible light photocatalytic activity. *Nanotechnology* **2009**, *20*, 235701. [CrossRef] [PubMed]

74. Borbón-Nuñez, H.A.; Dominguez, D.; Muñoz-Muñoz, F.; Lopez, J.; Romo-Herrera, J.; Soto, G.; Tiznado, H. Fabrication of hollow TiO_2 nanotubes through atomic layer deposition and MWCNT templates. *Powder Technol.* **2017**, *308*, 249–257. [CrossRef]

75. Sethi, D.; Jada, N.; Tiwari, A.; Ramasamy, S.; Dash, T.; Pandey, S. Photocatalytic destruction of escherichia coli in water by V_2O_5/TiO_2. *J. Photochem. Photobiol. B* **2015**, *144*, 68–74. [CrossRef] [PubMed]

76. Ohsaka, T.; Izumi, F.; Fujiki, Y. Raman spectrum of anatase, TiO_2. *J. Raman Spectrosc.* **1978**, *7*, 321–324. [CrossRef]

77. Kim, Y.S.; Song, M.Y.; Park, E.S.; Chin, S.; Bae, G.N.; Jurng, J. Visible-light-induced bactericidal activity of vanadium-pentoxide (V_2O_5)-loaded TiO_2 nanoparticles. *Appl. Biochem. Biotechnol.* **2012**, *168*, 1143–1152. [CrossRef] [PubMed]

Sample Availability: Samples of the compounds are available from the authors.

Article

Self-Organized TiO_2–MnO_2 Nanotube Arrays for Efficient Photocatalytic Degradation of Toluene

María C. Nevárez-Martínez [1,2], Marek P. Kobylański [3], Paweł Mazierski [3,*], Jolanta Wółkiewicz [4], Grzegorz Trykowski [4], Anna Malankowska [3], Magda Kozak [3], Patricio J. Espinoza-Montero [2] and Adriana Zaleska-Medynska [3,*]

[1] Facultad de Ingeniería Química y Agroindustria, Escuela Politécnica Nacional, Ladrón de Guevara E11-253, P.O. Box 17-01-2759, Quito 170525, Ecuador; ma.cristina.nevarez@gmail.com

[2] Centro de Investigación y Control Ambiental "CICAM", Departamento de Ingeniería Civil y Ambiental, Facultad de Ingeniería Civil y Ambiental, Escuela Politécnica Nacional, Ladrón de Guevara E11-253, P.O. Box 17-01-2759, Quito 170525, Ecuador; patricio.espinoza@epn.edu.ec

[3] Department of Environmental Technology, Faculty of Chemistry, University of Gdansk, 80-308 Gdansk, Poland; marek.kobylanski@phdstud.ug.edu.pl (M.P.K.); anna.malankowska@ug.edu.pl (A.M.); magda.kozak@ug.edu.pl (M.K.)

[4] Faculty of Chemistry, Nicolaus Copernicus University, 87-100 Torun, Poland; jolanta.wolkiewicz@umk.pl (J.W.); tryki@umk.pl (G.T.)

* Correspondences: pawel.mazierski@phdstud.ug.edu.pl (P.M.); adriana.zaleska@ug.edu.pl (A.Z.-M.); Tel.: +48-58-523-52-29 (P.M.); +48-58-523-52-20 (A.Z.-M.)

Academic Editor: Pierre Pichat
Received: 3 February 2017; Accepted: 28 March 2017; Published: 31 March 2017

Abstract: Vertically oriented, self-organized TiO_2–MnO_2 nanotube arrays were successfully obtained by one-step anodic oxidation of Ti–Mn alloys in an ethylene glycol-based electrolyte. The as-prepared samples were characterized by scanning electron microscopy (SEM), energy-dispersive X-ray spectroscopy (EDX), UV-Vis absorption, photoluminescence spectroscopy, X-ray diffraction (XRD), and micro-Raman spectroscopy. The effect of the applied potential (30–50 V), manganese content in the alloy (5–15 wt. %) and water content in the electrolyte (2–10 vol. %) on the morphology and photocatalytic properties was investigated for the first time. The photoactivity was assessed in the toluene removal reaction under visible light, using low-powered LEDs as an irradiation source (λ_{max} = 465 nm). Morphology analysis showed that samples consisted of auto-aligned nanotubes over the surface of the alloy, their dimensions were: diameter = 76–118 nm, length = 1.0–3.4 μm and wall thickness = 8–11 nm. It was found that the increase in the applied potential led to increase the dimensions while the increase in the content of manganese in the alloy brought to shorter nanotubes. Notably, all samples were photoactive under the influence of visible light and the highest degradation achieved after 60 min of irradiation was 43%. The excitation mechanism of TiO_2–MnO_2 NTs under visible light was presented, pointing out the importance of MnO_2 species for the generation of e^- and h^+.

Keywords: TiO_2–MnO_2 nanotubes; visible light induced photocatalysis; alloys; toluene degradation; anodization

1. Introduction

TiO_2-based photocatalysis is an effective technique for pollutant removal from both gas and liquid phase [1–7]. In fact, applications of TiO_2 are not limited only to photodegradation reactions, but it offers the facility to drive many others such as organic synthesis [8], water splitting [9,10], disinfection [11], CO_2 reduction [10,12], self-cleaning or antimicrobial surfaces [13,14], and dye-sensitized solar cells [10,15,16]. Due to the environmentally-friendly nature of TiO_2, its chemical and biological inertness, low cost, availability, and excellent photoactivity, this semiconductor material has become of great interest [17]. Nevertheless, some drawbacks as the rapid charge recombination of the photogenerated electrons and holes, and the wide bandgap (3.0 eV for rutile and 3.2 eV for anatase), which restricts photoabsorption to only ultraviolet region (ca. 5% of solar spectrum), need to be overcome in order to extend the practical application of TiO_2 photocatalysts for solar or interior light driven photoreactions at large scale [18].

Bandgap engineering in addition to tuning strategies have been studied over last decades with a common aim: shifting the absorption wavelength range of TiO_2 to the visible region. Since 1980s, TiO_2 has been modified by platinization [19–21]. So far, numerous approaches as ion (either cation or anion) doping [22–25], coupling with a narrower-bandgap semiconductor [26], with noble metals [27,28], with either organic or inorganic dyes [29] have been presented by a large number of research groups. Synthesizing composites with oxide semiconductors has become a promising way to enhance the photoactivity of TiO_2 by promoting the absorption of visible light and inhibiting the fast recombination of charge carriers [18,30]. Recent studies have focused on TiO_2–MnO_2 system due to the MnO_2 features as non-toxicity and earth abundance. These composites have been used mainly for capacitance applications [31,32], and despite the narrow bandgap of MnO_2 (0.26–2.7 eV), which could allow the absorption of visible and theoretically even infrared light [33–38], there exist just few reports in literature about the application of this system in photocatalysis. Xue, et al. [39] synthesized mesoporous MnO_2/TiO_2 nanocomposite, photoactive for the visible light-driven degradation of MB. They attributed the improved photocatalytic efficiency to the effective separation of photogenerated electrons and holes.

However, the industrial usage of photocatalysts is still in need of improvements to maximize the overall efficiency which also depends on mass and charge transfer processes. Therefore, TiO_2 nanostructures like zero-dimensional (nanoparticles), one-dimensional (nanowires, rods, and tubes), two dimensional (layers and sheets), and three dimensional (hierarchical spheres) have been widely synthesized, and used [40]. Since the discovery of carbon nanotubes in 1991 by Iijima [41], 1D morphologies as nanotubes (NTs) have become attractive materials due to the efficient separation of charge carriers, shape selectivity in chemical processes, high surface area to volume ratio, high electron mobility, mechanical strength [40,42], and high photoactivity in air purification [43]. Many approaches as sol-gel, template assisted, hydro/solvothermal and electrochemical have achieved to prepare TiO_2 NTs. Among these techniques, the electrochemical anodization of a suitable metal or alloy is the simplest, cheapest, and the most direct to grow self-highly-organized-nanotube arrays under specific electrochemical conditions which permit to control the properties of the fabricated NTs [44]. It was reported previously that MnO_2-TiO_2 NTs composite could be successfully formed by one-step anodic oxidation of Ti–Mn alloy [22,37]. Mohapatra, et al. [45] synthesized ordered arrays of mixed oxide NTs by anodization of Ti/Mn alloys, under ultrasonication in the presence of a fluoride-containing ethylene glycol solution. They pointed out that before calcination, the as-formed NTs showed a stoichiometry of $(Ti,Mn)O_2$, while annealing at 500 °C resulted in formation of nanotubes composed of anatase and rutile phases of TiO_2 and Mn_2O_3. Ning, Wang, Yu, Li, and Zhao [32] electrochemically prepared mixed oxide NTs from Ti–Mn alloys which showed enhanced capacitive properties compared with those of pristine TiO_2 NTs.

Herein, this work aims to anodically grow TiO_2–MnO_2 NTs in a fluoride-containing ethylene glycol-based electrolyte, and their application in the photodegradation of a model gaseous pollutant. According to our best knowledge, photocatalytic properties of nanotubes made of titania and manganese oxide mixtures have been investigated in this work for the first time. Moreover, parameters as the applied voltage (30–50 V), manganese content in the alloy (5–15 wt. %), and water content (2–10 vol. %) in the electrolyte have been also studied for the first time to analyze their effect on the morphology and photoactivity of the obtained NT arrays. Photodegradation tests in the gas phase were conducted with toluene as the model pollutant, and a possible mechanism of visible-light driven decomposition over the TiO_2–MnO_2 NTs was proposed as well.

2. Results and Discussion

2.1. Morphology and Growth Mechanism

One-step anodization processes were conducted for 60 min to synthesize pristine TiO_2 and TiO_2–MnO_2 nanotube layers from technical grade Ti sheets and Ti–Mn alloys under specific conditions, which are summarized in Table 1. SEM technique was used to analyze the effect of the applied voltage (30, 40 and 50 V), manganese content in the alloy (5, 10 and 15 wt. %) and water content in the electrolyte (2, 5 and 10 vol. %) on the morphology of the as-prepared samples. Figure 1 shows the top and cross-sectional SEM images which indicate that all synthesized nanotubes were uniform and vertically oriented. Pristine TiO_2 NTs presented smooth and uniform walls while TiO_2–MnO_2 NTs had ripples on their walls, which was also observed by Mohapatra, et al. [45] in samples anodized from Ti-8Mn alloys. It is well known that the dimensions of the nanotubes can be easily tuned by changing the preparation parameters [25]. Length and diameter increased with increasing the applied voltage, starting from $d = 81 \pm 9$ nm and $l = 1.5 \pm 0.1$ μm (Ti_30V); and reaching values of $d = 120 \pm 12$ nm and $l = 16.2 \pm 0.2$ μm (Ti_50V) for pristine TiO_2 NTs. The influence of anodization voltage was studied keeping constant the manganese content in the alloy (10 wt. %) and the water content in the electrolyte solution (2 vol. %). This way, dimensions of samples synthesized from $Ti_{90}Mn_{10}$ alloy also were bigger as the applied potential was higher, starting from $d = 76 \pm 9$ nm and $l = 1.0 \pm 0.1$ μm ($Ti_{90}Mn_{10}$_30V) and rising to $d = 118 \pm 4$ nm and $l = 2.8 \pm 0.1$ μm ($Ti_{90}Mn_{10}$_50V). Similar behavior of morphological results were reported by Macak, et al. [46]. They performed a systematic study of the factors influencing the two-step anodization of Ti foils in ammonium fluoride-containing glycerol/water mixtures. They prepared NT layers with diameters in the range of 20–300 nm for the potentials 2–40 V, while the thickness of the NT layers, tube length, was in the range of 150 nm up to 3 μm. This dependence of the dimensions, diameter, and length, with the applied potential is in well agreement with the present work. However, herein, the electrolyte media (EG-based) favored longer tubes in the case of pristine TiO_2 NTs. Furthermore, a complementary discussion about the anodization voltage effect on the diameter of NT arrays has been reported by Macak, et al. [47]. They stated that, particularly for TiO_2 NTs, the diameter strongly depends on the applied potential and electrolyte media, and consequently a wide variety of nanotube diameters can be obtained.

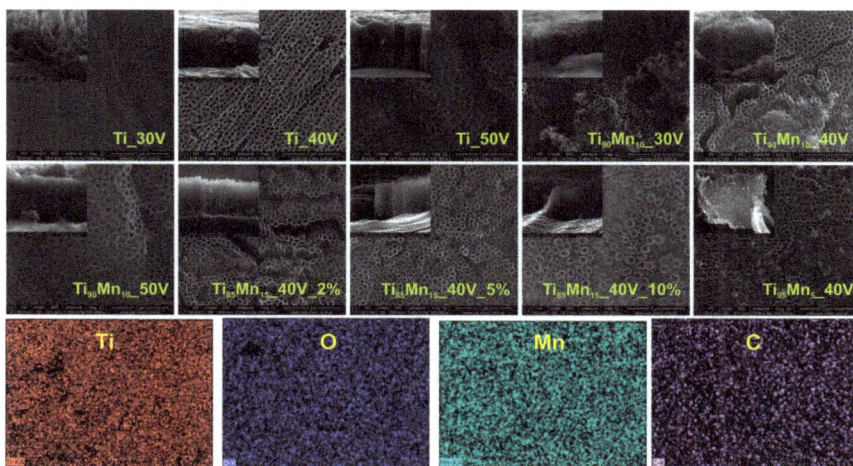

Figure 1. Top-view and cross-sectional SEM images of pristine TiO_2 and TiO_2–MnO_2 NTs (the effect of applied voltage, manganese content in the Mn/Ti alloy, and water content in the electrolyte on the morphology of formed nanotubes) and EDX mapping of the $Ti_{90}Mn_{10}$_30V sample.

Table 1. Sample labels, preparation conditions, and selected properties of pristine TiO$_2$ and TiO$_2$–MnO$_2$ nanotubes.

Sample Label	Preparation Parameters Electrolyte, Applied Voltage	External Diameter (nm)	Tube Length (µm)	Wall Thickness (nm)	Average Crystallite Size (nm)	EDX Analysis			
						Ti (wt. %)	Mn (wt. %)	C (wt. %)	O (wt. %)
Ti_30V	EG 98% (v/v), H$_2$O 2% (v/v), NH$_4$F 0.09 M, 30 V	81 ± 9	1.5 ± 0.1	10 ± 2	33	71.47	0	0.19	28.34
Ti_40V	EG 98% (v/v), H$_2$O 2% (v/v), NH$_4$F 0.09 M, 40 V	100 ± 7	5 ± 0.4	13 ± 2	34	66.73	0	0.03	33.24
Ti_50V	EG 98% (v/v), H$_2$O 2% (v/v), NH$_4$F 0.09 M, 50 V	120 ± 12	16.2 ± 0.2	18 ± 3	38	67.69	0	0.03	32.28
Ti$_{90}$Mn$_{10}$_30V	EG 98% (v/v), H$_2$O 2% (v/v), NH$_4$F 0.09 M, 30 V	76 ± 9	1 ± 0.1	8 ± 3	31	76.15	8.91	0.01	14.83
Ti$_{90}$Mn$_{10}$_40V	EG 98% (v/v), H$_2$O 2% (v/v), NH$_4$F 0.09 M, 40 V	92 ± 8	1.5 ± 0.1	9 ± 3	32	82.73	7.77	0.01	9.51
Ti$_{90}$Mn$_{10}$_50V	EG 98% (v/v), H$_2$O 2% (v/v), NH$_4$F 0.09 M, 50 V	118 ± 4	2.8 ± 0.1	9 ± 2	34	68.79	6.46	0.03	24.72
Ti$_{85}$Mn$_{15}$_40V_2%	EG 98% (v/v), H$_2$O 2% (v/v), NH$_4$F 0.09 M, 40 V	94 ± 11	1.3 ± 0.1	9 ± 2	31	77.20	11.14	0.01	11.67
Ti$_{85}$Mn$_{15}$_40V_5%	EG 95% (v/v), H$_2$O 5% (v/v), NH$_4$F 0.09 M, 40 V	90 ± 7	1.3 ± 0.1	9 ± 2	35	79.94	12.40	0.01	7.66
Ti$_{85}$Mn$_{15}$_40V_10%	EG 90% (v/v), H$_2$O 10% (v/v), NH$_4$F 0.09 M, 40 V	115 ± 8	1.1 ± 0.1	11 ± 2	34	61.76	9.11	1.18	27.95
Ti$_{95}$Mn$_5$_40V	EG 98% (v/v), H$_2$O 2% (v/v), NH$_4$F 0.09 M, 40 V	94 ± 8	3.4 ± 0.3	9 ± 1	32	70.89	2.10	0.03	27.00

The samples fabricated at 40 V from $Ti_{85}Mn_{15}$ alloy in electrolytes with different water content (2–10 vol. %) reported smaller length (1.1–1.3 µm) than that of the analogous non-modified (Ti_40V, 5.0 ± 0.4 µm). As it was mentioned in previous works [44,46], the increase of water in the electrolyte, provoked the increase in the formation of ripples in the tube walls. The sample prepared from the alloy with 5 wt. % of Mn showed the longest modified nanotubes (3.4 ± 0.3 µm), presumably due to the low content of Mn in the alloy which allowed a better stabilization of the nanotube matrix by TiO_2 species. Detailed information is displayed in Table 1. As it can be seen, all TiO_2–MnO_2 NTs were shorter, with smaller wall thickness than their pristine analog. Their length decreased with increasing the manganese content in the alloy. This could be attributed to the increase in the dissolution rate in phases with higher manganese content [32,45].

Table 1 also presents the results from EDX analysis which is in accordance with the composition of the alloys and no elements different from Ti, Mn, C, and O were found. Figure 1 presents also the EDX mapping of a selected sample where all elements are well dispersed and thus, there was not aggregation of Ti and Mn which guaranteed chemically homogeneous nanotube arrays.

A possible growth mechanism was proposed in Figure 2, based on the obtained results from SEM images of the sample $Ti_{95}Mn_5$_40V anodized during 4, 15 and 60 min and information provided in literature. It is possible to observe that the current density-time curves recorded for TiO_2–MnO_2 NTs resemble those corresponding to pristine TiO_2 NTs. The characteristic exponential decay of current density during the first stage indicates the formation of the oxide layer composed of TiO_2 and MnO_2 [45], Progressively, the chemical etching induces the apparition of initial random pits in the mixed oxide layer due to its dissolution through the formation of the fluoride complexes $[TiF_6]^{2-}$ and $[MnF_6]^{2-}$ [48,49]. Consequently, the resistive field decreases, allowing the current density to increase along the second stage. Finally, throughout the third stage, an equilibrium is established between oxidation and chemical dissolution, leading to the self-organized nanotube growth under steady state conditions [47] allowing the auto-alignment of the nanotubes.

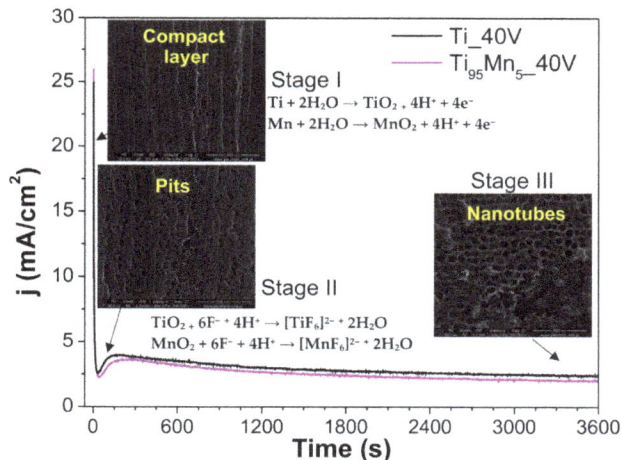

Figure 2. Proposed growth mechanism of TiO_2–MnO_2 NTs.

2.2. Structural Properties

The XRD patterns of the as-obtained NTs are presented in Figure 3. As it can be seen, obtained pristine and TiO_2–MnO_2 NTs consisted mainly of pure anatase TiO_2, while the peaks of Ti came from Ti substrate. Five common planes of anatase were found, namely (101), (004), (200), (105) and (211). The intensity of anatase diffraction peaks increased with increasing the anodization potential as a result of thicker NT layer. It was possible to observe just one characteristic peak ascribed to MnO_2 at about 58° [50,51]. The absence of any other band corresponding to the signature of MnO_2 can be related to the small content and good dispersion of manganese oxide in the TiO_2 NT layer, as it was mentioned in previous reports [32,39,52]. However, the constant diffraction peak positions indicate that the structure of TiO_2 was not changed through the anodization of Ti–Mn alloy.

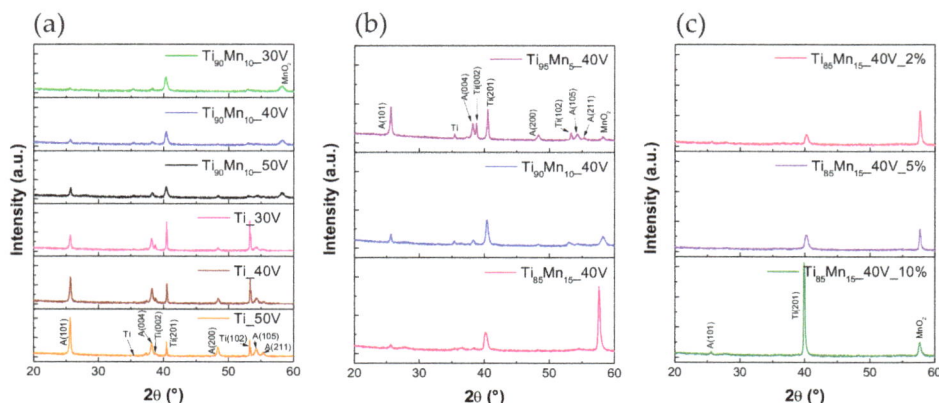

Figure 3. XRD spectra of pristine TiO_2 and TiO_2–MnO_2 NTs. Effect of (**a**) anodization potential; (**b**) manganese content in the alloy; and (**c**) water content in the electrolyte.

The calculated average crystallite size for pristine and modified TiO_2 NTs is summarized in Table 1. The average crystallite size was calculated using the Scherrer equation, based on (101) diffraction peak. The largest crystallite size was observed for pristine TiO_2 NTs and varied from 33 (30 V) to 38 nm (50 V). Among Ti–Mn series, crystallite sizes tended to be smaller than those of pristine TiO_2 NTs. This can be correlated to the wall thickness, as mentioned above, wall thickness of TiO_2–MnO_2 NTs was smaller than that of pristine TiO_2 NTs, thus there is less space to allow the growth of grain.

To further analyze the structure of the synthesized photocatalysts, micro-Raman spectroscopy was performed using a 532 nm laser as excitation light. Figure 4 displays the recorded spectra of pristine TiO_2 and TiO_2–MnO_2 NTs. As it can be seen, the spectra of the samples obtained from alloys with 5 and 10% of Mn mainly presented the signature peaks of anatase phase which are sharper in the spectra of pristine TiO_2 NTs. These peaks at approximately 150, 396, 515, and 636 cm^{-1} can be attributed to the E_g (TiO_2 symmetry), B_{1g} (O–Ti–O bending), A_{1g} + B_{1g} (T–O stretching), and E_g modes of anatase as it was exposed in previous reports [53]. The presence of MnO_2 in these samples decreased the intensity and broadened the anatase bands. On the other hand, the characteristic peaks of MnO_2 at around 521 and 644 cm^{-1}, assigned to the stretching mode of octahedral MnO_6 [54], overlapped the anatase peak at 636 cm^{-1} in the spectra of the samples prepared from $Ti_{85}Mn_{15}$ alloys, making it broaden to a range of 575–650 cm^{-1} [52]. These spectra also showed week bands at about 260 and 420 cm^{-1} originated from the bending modes of the metal–oxygen chain of Mn–O–Mn in the MnO_2 octahedral lattice [55–57].

2.3. Optical Properties

Figure 5 shows the absorption spectra of pristine TiO_2 and TiO_2–MnO_2 NTs. All the samples synthesized from the Ti–Mn alloy exhibited absorption in the full visible range due to the presence of MnO_2 as it was previously reported for TiO_2 NTs coated by MnO_2 [58]. The absorption band edge of pure TiO_2 NTs at about 400 nm registered a red-shift at about 500 nm which is easier to appreciate in samples prepared from alloys with 15% of Mn. This was also observed in the case of mesoporous structured MnO_2/TiO_2 nanocomposites [39]. As it was stated by Ding, et al. [59],

TiO$_2$–MnO$_2$ NTs could be used for solar-light driven photocatalysis owing to their absorption in the UV and visible region.

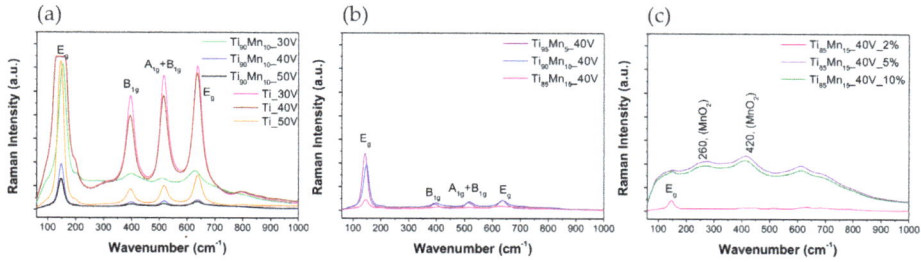

Figure 4. Raman spectra of pristine TiO$_2$ and TiO$_2$–MnO$_2$ NTs. Effect of (**a**) anodization potential; (**b**) manganese content in the alloy; and (**c**) water content in the electrolyte.

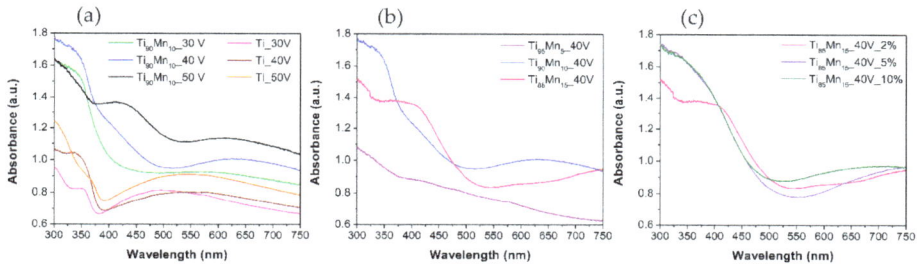

Figure 5. UV-Vis spectra of pristine TiO$_2$ and TiO$_2$–MnO$_2$ NTs. Effect of (**a**) anodization potential; (**b**) manganese content in the alloy; and (**c**) water content in the electrolyte.

Figure 6 shows photoluminescence (PL) spectra of both: pristine TiO$_2$ and TiO$_2$–MnO$_2$ NTs. Four emission peaks were detected among all series of photocatalysts. First one, at approximately 420 nm can be ascribed to the existence of self-trapped excitons from TiO$_6$$^{8-}$ octahedron, while the second and third peaks at 450 and 485 nm are associated with the presence of surface defects and oxygen vacancies. The last peak at approximately 525 nm is associated with radiative recombination of charge carriers [60,61].

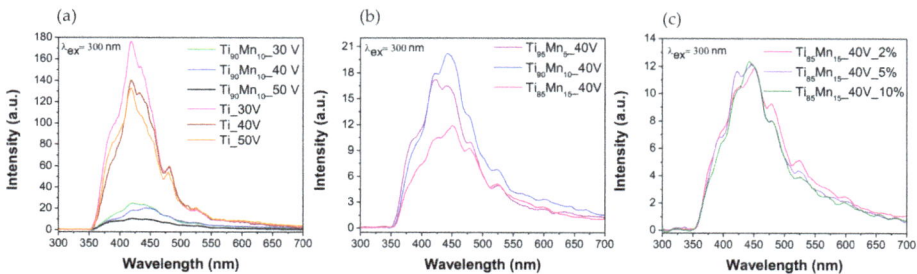

Figure 6. Photoluminescence spectra of pristine TiO$_2$ and TiO$_2$–MnO$_2$ NTs. Effect of (**a**) anodization potential; (**b**) manganese content in the alloy; and (**c**) water content in the electrolyte.

2.4. Photocatalytic Performance

The photoactivity of the prepared samples was tested in the visible-light-driven photodegradation of toluene (200 ppmv) from an air mixture. The irradiation source consisted of a LED array with λ_{max} = 465 nm. The effect of anodization voltage, manganese content in the alloy and water content in the electrolyte was systematically studied. Figure 7 presents the degradation curves in the presence of obtained NT photocatalysts and a reference curve in the absence of any photocatalyst, to test photolysis. It is clearly showed that in the reference curve, degradation was not achieved. Pristine TiO_2 NTs exhibited insignificant toluene removal (about 5%) while all of the samples were photoactive towards the degradation of the model pollutant. Figure 7a shows that the highest degradation after 60 min of irradiation was achieved in the presence of the $Ti_{90}Mn_{10}\_30V$ sample (43%). The samples anodized from $Ti_{90}Mn_{10}$ alloys at 40 V and 50 V reported similar toluene removal, 28% and 33% respectively. The results displayed in Figure 7b indicate that the manganese content in the alloy inversely affected the photoactivity, the higher the manganese content in the alloy was, the less degradation was achieved. This way, samples prepared from alloys with 5, 10 and 15 wt. % of manganese reached a degradation of 29%, 28%, and 24%, respectively. This was also observed by Xue, Huang, Wang, Wang, Gao, Zhu, and Zou [39] in the dye-mediated photodegradation of MB under visible light in the presence of mesoporous MnO_2/TiO_2 nanocomposites. They attributed the lower degradation to the accumulation of MnO_2 on the surface of TiO_2 which increased the transfer rate of photogenerated electrons within MnO_2, overall weakening the effect of improving the photoactivity. The photoactivity of the samples from the series with different water content in the electrolyte was similar between each other (Figure 7c), the highest toluene removal (28%) was accomplished by the sample with 10% of water in the electrolyte. The other two samples exhibited 24% of toluene removal. The kinetic parameters of each photocatalyst are included in Table 2.

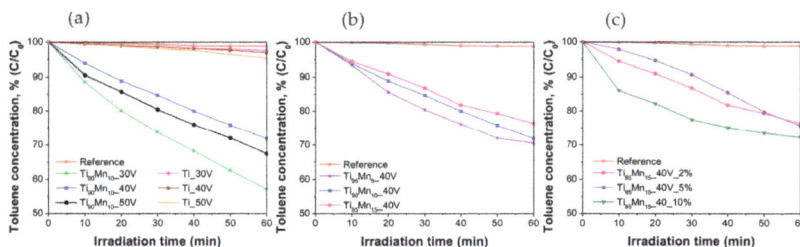

Figure 7. Photoactivity of pristine TiO_2 and TiO_2–MnO_2 NTs in gas phase degradation of toluene under Vis light irradiation (λ_{max} = 465 nm). Effect of (a) applied voltage; (b) manganese content in the alloy, and (c) water content in the electrolyte.

Table 2. Initial reaction rate and reaction rate constant for the gas phase degradation of toluene (200 ppmv) under Vis light irradiation (25-LED array, λ_{max} = 465 nm, irradiation intensity = 14.5 mW·cm^{-2}) in the presence of pristine TiO_2 and TiO_2–MnO_2 NTs.

Sample Label	Photocatalytic Toluene Degradation	
	Initial Reaction Rate $\times 10^2$ (μmol·dm^{-3}·min^{-1})	Reaction Rate Constant $\times 10^3$ (min^{-1})
Ti_30V	0.37 ± 0.09	0.42 ± 0.10
Ti_40V	0.43 ± 0.09	0.49 ± 0.10
Ti_50V	0.64 ± 0.04	0.72 ± 0.04
$Ti_{90}Mn_{10}\_30V$	8.54 ± 0.53	9.57 ± 0.59
$Ti_{90}Mn_{10}\_40V$	4.97 ± 0.30	5.57 ± 0.33
$Ti_{90}Mn_{10}\_50V$	6.04 ± 0.08	6.77 ± 0.09
$Ti_{85}Mn_{15}\_40V\_2\%$	4.18 ± 0.77	4.69 ± 0.87
$Ti_{85}Mn_{15}\_40V\_5\%$	3.79 ± 0.43	4.24 ± 0.48
$Ti_{85}Mn_{15}\_40V\_10\%$	5.84 ± 1.61	6.54 ± 1.81
$Ti_{95}Mn_5\_40V$	5.76 ± 0.12	6.45 ± 0.14

The highest initial reaction rate ($8.54 \pm 0.53 \times 10^{-2}$ $\mu mol \cdot dm^{-3} \cdot min^{-1}$) and reaction rate constant ($9.57 \pm 0.59 \times 10^{-3}$ min^{-1}) were observed for the toluene degradation over the $Ti_{90}Mn_{10}\_30V$ sample and they were more than 23 times higher compared with those of pristine TiO_2 NTs obtained by anodization at 30 V ($0.37 \pm 0.09 \times 10^{-2}$ $\mu mol \cdot dm^{-3} \cdot min^{-1}$ and $0.42 \pm 0.10 \times 10^{-3}$ min^{-1}).

As shown in Table 2, the most photoactive sample, $Ti_{90}Mn_{10}\_30V$, was used to analyze the effect of the irradiation wavelength (λ_{max} = 375, 415 and 465 nm) in the same degradation reaction. As evident from Figure 8a, the maximum toluene removal (43%) was reached under 465 nm while under 375 nm (UV light) and 415 nm (25% and 20% of degradation, respectively) the sample was less active. This can be explained by a synergistic effect of MnO_2 and TiO_2 in the NT matrix. As it was formerly reported [58], MnO_2 has lower photoactivity than TiO_2 under UV light irradiation, and thereby, this narrow-bandgap semiconductor reduced the overall photoactivity of the composite in this wavelength range because of a synergistic effect in the composite. This behavior under UV light was also exposed by Xue, et al. [39] who indicated that the transferring of photoexcited electrons (generated in TiO_2) within MnO_2 can correspond to an internal dissipation able to suppress the photocatalytic activity. On the other hand, Xu, et al. [58] reported improved visible light-photoactivity for NTs electrodeposited with MnO_2. Therefore, we can conclude that the presence of MnO_2 in TiO_2 NTs favored the conditions for the degradation of toluene in gas phase under visible light (longer wavelength of irradiation) owing to the ability of MnO_2 species to absorb visible light irradiation and promote the enhancement of the charge transfer rate.

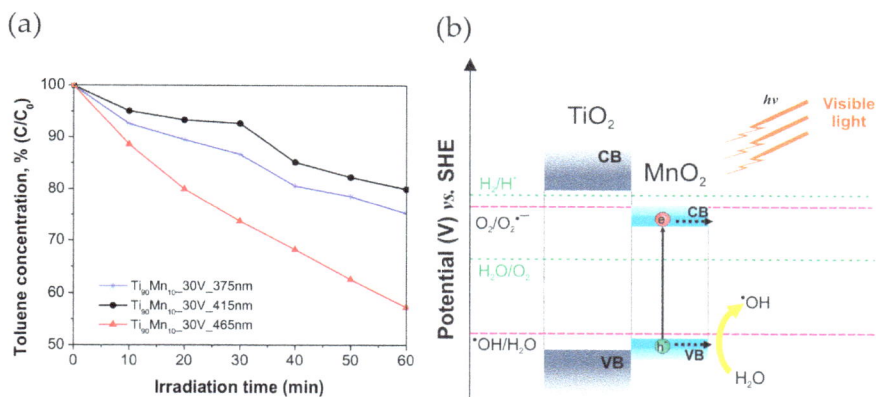

Figure 8. (**a**) Photoactivity of $Ti_{90}Mn_{10}\_30V$ sample in gas phase degradation of toluene under different wavelengths of irradiation (λ_{max} = 375, 415, 465 nm) and (**b**) possible excitation mechanism of TiO_2–MnO_2 NTs under Vis light irradiation.

Additionally, a possible excitation mechanism of TiO_2–MnO_2 NTs under Vis light was proposed and diagrammed in Figure 8b. The conduction band and valence band edge values of MnO_2 were calculated to be 0.57 and 2.34 eV, respectively [34]. Thus, it is likely that photogenerated holes from the valence band (VB) of MnO_2 could be involved in the formation of hydroxyl radicals ($^{\bullet}OH$), while electrons from the CB of MnO_2 can participate indirectly in the degradation of toluene, considering that the potential of photogenerated electrons is not high enough to generate other reactive oxygen species, such as $O_2^{\bullet-}$, H_2O_2, and $HO_2^{\bullet}$ radicals.

3. Materials and Methods

3.1. Materials

Acetone, isopropanol, and methanol were purchased from P.P.H. "STANLAB" Sp. J. (Lublin, Poland), while ethylene glycol (EG) from CHEMPUR and ammonium fluoride from ACROS ORGANICS. Technical grade Ti foils and Ti–Mn alloys with 5, 10 and 15 wt. % of manganese content were provided by HMW-Hauner Metallische Werkstoffe (Röttenbach, Germany). Deionized (DI) water with conductivity of 0.05 µS was used to prepare all aqueous solutions.

3.2. Synthesis of Pristine TiO$_2$ and TiO$_2$–MnO$_2$ Nanotubes

Ti foils as well as Ti–Mn alloys were ultrasonically cleaned in acetone, isopropanol, methanol, and deionized water for 10 min, respectively. Then, foils were dried in an air stream. The anodization processes were carried out at room temperature, in an electrochemical cell consisting of a platinum mesh as counter electrode, and the Ti foils or the Ti–Mn alloy (2.5 cm × 2.5 cm) as working electrode. A reference electrode of Ag/AgCl connected to a digital multimeter (BRYMEN BM857a) was used to control and record information about the actual potential and current on the alloy. The anodization was conducted in an electrolyte composed of EG, water and NH$_4$F 0.09 M, during 60 min with a voltage in the range of 30–50 V applied with a programmable DC power supply (MANSON SDP 2603). Three electrolyte solutions with different water content were used (volume ratios of EG:water of 98:2, 95:5 and 90:10) The obtained samples were rinsed with deionized water, sonicated in deionized water (1 min), dried in air (80 °C for 24 h), and calcined (450 °C, heating rate 2 °C/min) for 1 h.

3.3. Characterization of Pristine TiO$_2$ and TiO$_2$–MnO$_2$ Nanotubes

The morphology of synthesized pristine TiO$_2$ and TiO$_2$–MnO$_2$ nanotubes was determined by using scanning electron microscopy (SEM, FEI QUANTA 3D FEG, FEI Company, Brno, Czech Republic). Energy-dispersive X-ray spectroscopy (EDX) analysis were performed with a scanning electron microscope (SEM, Zeiss, Leo 1430 VP, Carl Zeiss, Oberkochen, Germany) coupled to an energy-dispersive X-ray fluorescence spectrometer (EDX) Quantax 200 with the XFlash 4010 (Bruker AXS, Karlsruhe, Germany) detector. The crystal structure of the samples was determined from X-ray diffraction patterns recorded in the range of 2θ = 20°–90°, using an X-ray diffractometer (X'Pert Pro, Panalytical, Almelo, The Netherlands) with Cu Kα radiation. The crystallite size was calculated based on the Scherrer formula. Raman spectra were measured with a micro-Raman spectrometer (Senterra, Bruker Optik, Billerica, MA, USA) with a 532 nm excitation laser. The UV-Vis absorbance spectra were registered with the UV-VIS Spectrophotometer, SHIMADZU UV-2600, in the wavelength range of 300–800 nm equipped with an integrating sphere. The baseline was determined with barium sulfate as reference, the scanning speed was 250 nm/min at room temperature. The photoluminescence (PL) spectra were recorded at room temperature with a LS-50B Luminescence Spectrometer equipped with a Xenon discharge lamp, as an excitation source, and a R928 photomultiplier as detector. The excitation radiation (300 nm) was directed on the surface of the samples at an angle of 90°.

3.4. Measurement of Photocatalytic Activity

Photocatalytic activity of the as-prepared NTs was analyzed, for the first time, in the purification of air from toluene which was used as a model pollutant. The photodegradation experiments were carried out in a stainless steel reactor of a volume of ca. 35 cm^3. The reactor included a quartz window, two valves and a septum. The light source consisting of an array of 25 LEDs (λ$_{max}$ = 375, 415 and 465 nm, Optel, Opole, Poland) was located above the sample. The anodized foil was placed at the bottom side of the reactor and it was closed with the quartz window. A gas mixture (toluene, 200 ppmv) was passed through the reactor for 1 min, then the valves were closed and the reactor was kept in dark for 30 min in order to achieve the equilibrium. Before starting the irradiation, a reference toluene sample was taken. The concentration was determined by using a gas chromatograph (TRACE 1300, Thermo Scientific, Waltham, MA, USA), equipped with an ionization flame detector (FID) and an Elite-5 capillary column. The samples (200 µL) were dosed with a gas-tight syringe each 10 min. Irradiation intensity was measured by an optical power meter (HAMAMATSU, C9536-01, Hamamatsu, Japan) and reached 14.7, 14.1 and 14.5 mW/cm^2 for LEDs with λ_{max} = 375, 415 and 465 nm, respectively.

4. Conclusions

The analysis of the effect of applied potential, manganese content in the alloy and water content in the electrolyte on the morphology and visible-light photocatalytic activity of TiO$_2$–MnO$_2$ NTs obtained from one-step anodic oxidation of Ti–Mn alloys in a fluoride-containing EG-based electrolyte was reported here for the first time. All fabricated samples were described as vertically-oriented, self-organized nanotube arrays with a diameter of 76–115 nm and length of 1–3.4 µm. Diameter and length were directly influenced by the applied voltage while the manganese content led to obtain shorter tubes than those prepared from Ti sheets. The as-prepared TiO$_2$–MnO$_2$ arrays exhibited improved optical and photocatalytic properties in comparison with those of pristine TiO$_2$ NTs. The photoactivity assessment was carried out towards the degradation of toluene (200 ppmv) in gas phase under Vis light irradiation (LEDs, λ_{max} = 465 nm). The highest degradation after 60 min of irradiation corresponded to 43% and the initial reaction rate reached values of 3.79–8.54 × 10^{-2} µmol·dm^{-3}·min^{-1}. A wavelength dependence exploration was performed as well, MnO$_2$ modified NTs showed the highest activity under visible light irradiation and therefore, a possible mechanism of excitation was presented. These findings suggest that TiO$_2$–MnO$_2$ mixed oxide nanotube arrays, activated by low-powered LEDs, could be a promising material for air purification systems. Moreover, the electrochemical approach is a successful way to obtain these highly-organized nanostructures from Ti–Mn alloys. Consequently, the industrially-oriented application of photocatalysis for air treatment using LEDs, as a low-cost and suitable irradiation source, follows the trends of green chemistry and environmentally friendly performance.

Acknowledgments: This research was financially supported by the Polish National Science Center (research grant, Ordered TiO$_2$/MxOy nanostructures obtained by electrochemical method; contract No. NCN 2014/15/B/ST5/00098).

Author Contributions: A.Z.-M. supervised and directed the project; A.Z.-M. and P.M. conceived the concept; M.C.N.-M., P.M., M.P.K., A.M., M.K., J.W., G.T. performed the experiments; M.C.N.-M., P.M., and M.K. analyzed the data; M.C.N.-M., P.M., and M.P.K. contributed reagents/materials/analysis tools; M.N.-M., P.M., P.J.E.-M. and A.Z.-M. wrote the paper.

Conflicts of Interest: The authors declare no conflict of interest.

References

1. Gaya, U.I.; Abdullah, A.H. Heterogeneous photocatalytic degradation of organic contaminants over titanium dioxide: A review of fundamentals, progress and problems. *J. Photochem. Photobiol. C Photochem. Rev.* **2008**, *9*, 1–12.

2. Pichat, P. *Photocatalysis and Water Purification: From Fundamentals to Recent Applications*; John Wiley & Sons: Hoboken, NJ, USA, 2013.

3. Schneider, J.; Bahnemann, D.; Ye, J.; Puma, G.L.; Dionysiou, D.D. *Photocatalysis: Fundamentals and Perspectives*; Royal Society of Chemistry: Cambridge, UK, 2016.

4. Dionysiou, D.D.; Puma, G.L.; Ye, J.; Schneider, J.; Bahnemann, D. *Photocatalysis: Applications*; Royal Society of Chemistry: Cambridge, UK, 2016.

5. Pichat, P. *Photocatalysis: Fundamentals, Materials and Potential*; MDPI: Basel, Switzerland, 2016.

6. Colmenares Quintero, J.C.; Xu, Y.-J. *Heterogeneous Photocatalysis: From Fundamentals to Green Applications*; Springer: Berlin/Heidelberg, Germany, 2016; Volume VIII, p. 416.

7. Mazierski, P.; Nadolna, J.; Lisowski, W.; Winiarski, M.J.; Gazda, M.; Nischk, M.; Klimczuk, T.; Zaleska-Medynska, A. Effect of irradiation intensity and initial pollutant concentration on gas phase photocatalytic activity of TiO_2 nanotube arrays. *Catal. Today* **2017**, *284*, 19–26. [CrossRef]

8. Zavahir, S.; Xiao, Q.; Sarina, S.; Zhao, J.; Bottle, S.; Wellard, M.; Jia, J.; Jing, L.; Huang, Y.; Blinco, J.P.; et al. Selective oxidation of aliphatic alcohols using molecular oxygen at ambient temperature: Mixed-valence vanadium oxide photocatalysts. *ACS Catal.* **2016**, *6*, 3580–3588. [CrossRef]

9. Kang, D.; Kim, T.W.; Kubota, S.R.; Cardiel, A.C.; Cha, H.G.; Choi, K.S. Electrochemical synthesis of photoelectrodes and catalysts for use in solar water splitting. *Chem. Rev.* **2015**, *115*, 12839–12887. [CrossRef] [PubMed]

10. Highfield, J. Advances and recent trends in heterogeneous photo(electro)-catalysis for solar fuels and chemicals. *Molecules* **2015**, *20*, 6739–6793. [CrossRef] [PubMed]

11. Sethi, D.; Jada, N.; Tiwari, A.; Ramasamy, S.; Dash, T.; Pandey, S. Photocatalytic destruction of escherichia coli in water by V_2O_5/TiO_2. *J. Photochem. Photobiol. B* **2015**, *144*, 68–74. [CrossRef] [PubMed]

12. Low, J.; Cheng, B.; Yu, J. Surface modification and enhanced photocatalytic CO_2 reduction performance of TiO_2: A review. *Appl. Surf. Sci.* **2017**, *392*, 658–686. [CrossRef]

13. Wang, R.; Hashimoto, K.; Fujishima, A.; Chikuni, M.; Kojima, E.; Kitamura, A.; Shimohigoshi, M.; Watanabe, T. Light-induced amphiphilic surfaces. *Nature* **1997**, *388*, 431–432. [CrossRef]

14. Pichat, P. Self-cleaning materials based on solar photocatalysis. In *New and Future Developments in Catalysis: Solar Photocatalysis*; Suib, S.L., Ed.; Elsevier: Amsterdam, The Netherlands, 2013; Volume 7 "Solar catalysis", pp. 167–190.

15. Ye, M.; Zheng, D.; Wang, M.; Chen, C.; Liao, W.; Lin, C.; Lin, Z. Hierarchically structured microspheres for high-efficiency rutile TiO_2-based dye-sensitized solar cells. *ACS Appl. Mater. Interfaces* **2014**, *6*, 2893–2901. [CrossRef] [PubMed]

16. Wang, Z.; Ma, W.; Chen, C.; Zhao, J. Sensitization of titania semiconductor: A promising strategy to utilize visible light. In *Photocatalysis and Water Purification*; Pichat, P., Ed.; Wiley-VCH: Weinheim, Germany, 2013; pp. 199–240.

17. Zhang, H.; Banfield, J.F. Structural characteristics and mechanical and thermodynamic properties of nanocrystalline TiO_2. *Chem. Rev.* **2014**, *114*, 9613–9644. [CrossRef] [PubMed]

18. Banerjee, S.; Pillai, S.C.; Falaras, P.; O'Shea, K.E.; Byrne, J.A.; Dionysiou, D.D. New insights into the mechanism of visible light photocatalysis. *J. Phys. Chem. Lett.* **2014**, *5*, 2543–2554. [CrossRef] [PubMed]

19. Sato, S.; White, J.M. Photodecomposition of water over Pt/TiO_2 catalysts. *Chem. Phys. Lett.* **1980**, *72*, 83–86. [CrossRef]

20. Ohtani, B.; Osaki, H.; Nishimoto, S.; Kagiya, T. A novel photocatalytic process of amine N-alkylation by platinized semiconductor particles suspended in alcohols. *J. Am. Chem. Soc.* **1986**, *108*, 308–310. [CrossRef]

21. Pichat, P. Surface-properties, activity and selectivity of bifunctional powder photocatalysts. *New J. Chem.* **1987**, *11*, 135–140.

22. Choi, W.; Termin, A.; Hoffmann, M.R. The role of metal ion dopants in quantum-sized TiO_2: Correlation between photoreactivity and charge carrier recombination dynamics. *J. Phys. Chem.* **1994**, *98*, 13669–13679. [CrossRef]

23. Asahi, R.; Morikawa, T.; Ohwaki, T.; Aoki, K.; Taga, Y. Visible-light photocatalysis in nitrogen-doped titanium oxides. *Science* **2001**, *293*, 269–271. [CrossRef] [PubMed]

24. Mazierski, P.; Lisowski, W.; Grzyb, T.; Winiarski, M.J.; Klimczuk, T.; Mikołajczyk, A.; Flisikowski, J.; Hirsch, A.; Kołakowska, A.; Puzyn, T.; et al. Enhanced photocatalytic properties of lanthanide-TiO_2 nanotubes: An experimental and theoretical study. *Appl. Catal. B Environ.* **2017**, *205*, 376–385. [CrossRef]

25. Mazierski, P.; Nischk, M.; Gołkowska, M.; Lisowski, W.; Gazda, M.; Winiarski, M.J.; Klimczuk, T.; Zaleska-Medynska, A. Photocatalytic activity of nitrogen doped TiO_2 nanotubes prepared by anodic oxidation: The effect of applied voltage, anodization time and amount of nitrogen dopant. *Appl. Catal. B Environ.* **2016**, *196*, 77–88. [CrossRef]

26. Ouyang, J.; Chang, M.; Li, X. CdS-sensitized ZnO nanorod arrays coated with TiO_2 layer for visible light photoelectrocatalysis. *J. Mater. Sci.* **2012**, *47*, 4187–4193. [CrossRef]

27. Diak, M.; Grabowska, E.; Zaleska, A. Synthesis, characterization and photocatalytic activity of noble metal-modified TiO_2 nanosheets with exposed {001} facets. *Appl. Surf. Sci.* **2015**, *347*, 275–285. [CrossRef]

28. Nischk, M.; Mazierski, P.; Wei, Z.; Siuzdak, K.; Kouame, N.A.; Kowalska, E.; Remita, H.; Zaleska-Medynska, A. Enhanced photocatalytic, electrochemical and photoelectrochemical properties of TiO_2 nanotubes arrays modified with Cu, AgCu and Bi nanoparticles obtained via radiolytic reduction. *Appl. Surf. Sci.* **2016**, *387*, 89–102. [CrossRef] [PubMed]

29. Chatterjee, D.; Mahata, A. Demineralization of organic pollutants on the dye modified TiO_2 semiconductor particulate system using visible light. *Appl. Catal. B Environ.* **2001**, *33*, 119–125. [CrossRef]

30. Paramasivam, I.; Nah, Y.C.; Das, C.; Shrestha, N.K.; Schmuki, P. WO_3/TiO_2 nanotubes with strongly enhanced photocatalytic activity. *Chemistry* **2010**, *16*, 8993–8997. [CrossRef] [PubMed]

31. Zhou, H.; Zhang, Y. Electrochemically self-doped TiO_2 nanotube arrays for supercapacitors. *J. Phys. Chem. C* **2014**, *118*, 5626–5636. [CrossRef]

32. Ning, X.; Wang, X.; Yu, X.; Li, J.; Zhao, J. Preparation and capacitance properties of Mn-doped TiO_2 nanotube arrays by anodisation of Ti–Mn alloy. *J. Alloys Compd.* **2016**, *658*, 177–182. [CrossRef]

33. Islam, A.K.M.F.U.; Islam, R.; Khan, K.A. Studies on the thermoelectric effect in semiconducting MnO_2 thin films. *J. Mater. Sci. Mater. Electron.* **2005**, *16*, 203–207. [CrossRef]

34. Zhao, J.; Nan, J.; Zhao, Z.; Li, N.; Liu, J.; Cui, F. Energy-efficient fabrication of a novel multivalence Mn_3O_4–MnO_2 heterojunction for dye degradation under visible light irradiation. *Appl. Catal. B Environ.* **2017**, *202*, 509–517. [CrossRef]

35. Pinaud, B.A.; Chen, Z.; Abram, D.N.; Jaramillo, T.F. Thin films of sodium birnessite-type MnO_2: Optical properties, electronic band structure, and solar photoelectrochemistry. *J. Phys. Chem. C* **2011**, *115*, 11830–11838. [CrossRef]

36. Chen, Z.; Jaramillo, T.F.; Deutsch, T.G.; Kleiman-Shwarsctein, A.; Forman, A.J.; Gaillard, N.; Garland, R.; Takanabe, K.; Heske, C.; Sunkara, M.; et al. Accelerating materials development for photoelectrochemical hydrogen production: Standards for methods, definitions, and reporting protocols. *J. Mater. Res.* **2011**, *25*, 3–16. [CrossRef]

37. Sherman, D.M. Electronic structures of iron(III) and manganese(IV) (hydr)oxide minerals: Thermodynamics of photochemical reductive dissolution in aquatic environments. *Geochim. Cosmochim. Acta* **2005**, *69*, 3249–3255. [CrossRef]

38. Sakai, N.; Ebina, Y.; Takada, K.; Sasaki, T. Photocurrent generation from semiconducting manganese oxide nanosheets in response to visible light. *J. Phys. Chem. B* **2005**, *109*, 9651–9655. [CrossRef] [PubMed]

39. Xue, M.; Huang, L.; Wang, J.Q.; Wang, Y.; Gao, L.; Zhu, J.H.; Zou, Z.G. The direct synthesis of mesoporous structured MnO_2/TiO_2 nanocomposite: A novel visible-light active photocatalyst with large pore size. *Nanotechnology* **2008**, *19*, 185604. [CrossRef] [PubMed]

40. Roy, P.; Berger, S.; Schmuki, P. TiO_2 nanotubes: Synthesis and applications. *Angew. Chem. Int. Ed. Engl.* **2011**, *50*, 2904–2939. [CrossRef] [PubMed]

41. Iijima, S. Helical microtubules of graphitic carbon. *Nature* **1991**, *354*, 56–58. [CrossRef]

42. Kubacka, A.; Fernandez-Garcia, M.; Colon, G. Advanced nanoarchitectures for solar photocatalytic applications. *Chem. Rev.* **2012**, *112*, 1555–1614. [CrossRef] [PubMed]

43. Pichat, P. Are TiO_2 nanotubes worth using in photocatalytic purification of air and water? *Molecules* **2014**, *19*, 15075–15087. [CrossRef] [PubMed]

44. Lee, K.; Mazare, A.; Schmuki, P. One-dimensional titanium dioxide nanomaterials: Nanotubes. *Chem. Rev.* **2014**, *114*, 9385–9454. [CrossRef] [PubMed]

45. Mohapatra, S.K.; Raja, K.S.; Misra, M.; Mahajan, V.K.; Ahmadian, M. Synthesis of self-organized mixed oxide nanotubes by sonoelectrochemical anodization of Ti–8Mn alloy. *Electrochim. Acta* **2007**, *53*, 590–597. [CrossRef]

46. Macak, J.M.; Hildebrand, H.; Marten-Jahns, U.; Schmuki, P. Mechanistic aspects and growth of large diameter self-organized TiO_2 nanotubes. *J. Electroanal. Chem.* **2008**, *621*, 254–266. [CrossRef]

47. Macak, J.M.; Tsuchiya, H.; Ghicov, A.; Yasuda, K.; Hahn, R.; Bauer, S.; Schmuki, P. TiO_2 nanotubes: Self-organized electrochemical formation, properties and applications. *Curr. Opin. Solid State Mater. Sci.* **2007**, *11*, 3–18. [CrossRef]

48. Helmholz, L.; Russo, M.E. Spectra of manganese(IV) hexafluoride ion ($MnF_6^=$) in environments of O_h and D_{3d} symmetry. *J. Chem. Phys.* **1973**, *59*, 5455–5470. [CrossRef]

49. Pourbaix, M. *Atlas of Electrochemical Equilibria in Aqueous Solutions*; Pergamon Press: New York, NY, USA, 1966.

50. Devaraj, S.; Munichandraiah, N. Effect of crystallographic structure of MnO_2 on its electrochemical capacitance properties. *J. Phys. Chem. C* **2008**, *112*, 4406–4417. [CrossRef]

51. Li, J.; Chen, J.; Ke, R.; Luo, C.; Hao, J. Effects of precursors on the surface Mn species and the activities for NO reduction over MnO_x/TiO_2 catalysts. *Catal. Commun.* **2007**, *8*, 1896–1900. [CrossRef]

52. Liao, J.-Y.; Higgins, D.; Lui, G.; Chabot, V.; Xiao, X.; Chen, Z. Multifunctional TiO_2–C/MnO_2 core–double-shell nanowire arrays as high-performance 3D electrodes for lithium ion batteries. *Nano Lett.* **2013**, *13*, 5467–5473. [CrossRef] [PubMed]

53. Borbón-Nuñez, H.A.; Dominguez, D.; Muñoz-Muñoz, F.; Lopez, J.; Romo-Herrera, J.; Soto, G.; Tiznado, H. Fabrication of hollow TiO_2 nanotubes through atomic layer deposition and MWCNT templates. *Powder Technol.* **2017**, *308*, 249–257. [CrossRef]

54. Jana, S.; Basu, S.; Pande, S.; Ghosh, S.K.; Pal, T. Shape-selective synthesis, magnetic properties, and catalytic activity of single crystalline β-MnO_2 nanoparticles. *J. Phys. Chem. C* **2007**, *111*, 16272–16277. [CrossRef]

55. Wei, M.; Konishi, Y.; Zhou, H.; Sugihara, H.; Arakawa, H. Synthesis of single-crystal manganese dioxide nanowires by a soft chemical process. *Nanotechnology* **2005**, *16*, 245–249. [CrossRef] [PubMed]

56. Jana, S.; Pande, S.; Sinha, A.K.; Sarkar, S.; Pradhan, M.; Basu, M.; Saha, S.; Pal, T. A green chemistry approach for the synthesis of flower-like Ag-doped MnO_2 nanostructures probed by surface-enhanced raman spectroscopy. *J. Phys. Chem. C* **2009**, *113*, 1386–1392. [CrossRef]

57. Luo, J.; Zhu, H.T.; Fan, H.M.; Liang, J.K.; Shi, H.L.; Rao, G.H.; Li, J.B.; Du, Z.M.; Shen, Z.X. Synthesis of single-crystal tetragonal α-MnO_2 nanotubes. *J. Phys. Chem. C* **2008**, *112*, 12594–12598. [CrossRef]

58. Xu, X.; Zhou, X.; Li, X.; Yang, F.; Jin, B.; Xu, T.; Li, G.; Li, M. Electrodeposition synthesis of MnO_2/TiO_2 nanotube arrays nanocomposites and their visible light photocatalytic activity. *Mater. Res. Bull.* **2014**, *59*, 32–36. [CrossRef]

59. Ding, S.; Liyong, W.; Shaoyan, Z.; Qiuxiang, Z.; Yu, D.; Shujuan, L.; Yanchao, L.; Quanying, K. Hydrothermal synthesis, structure and photocatalytic property of nano-TiO_2–MnO_2. *Sci. China Ser. B* **2003**, *46*, 542. [CrossRef]

60. Tang, H.; Berger, H.; Schmid, P.E.; Lévy, F. Optical properties of anatase (TiO_2). *Solid State Commun.* **1994**, *92*, 267–271. [CrossRef]

61. Knorr, F.J.; Mercado, C.C.; McHale, J.L. Trap-state distributions and carrier transport in pure and mixed-phase TiO_2: Influence of contacting solvent and interphasial electron transfer. *J. Phys. Chem. C* **2008**, *112*, 12786–12794. [CrossRef]

Sample Availability: Samples of the compounds are available from the authors.

molecules

MDPI

Review

Photoactive Hybrid Catalysts Based on Natural and Synthetic Polymers: A Comparative Overview

Juan Carlos Colmenares * and Ewelina Kuna

Institute of Physical Chemistry, Polish Academy of Sciences, Kasprzaka 44/52, 01-224 Warsaw, Poland; ekuna@ichf.edu.pl
* Correspondence: jcarloscolmenares@ichf.edu.pl; Tel.: +48-22-343-3215

Academic Editor: Pierre Pichat
Received: 18 March 2017; Accepted: 7 May 2017; Published: 12 May 2017

Abstract: In the present review, we would like to draw the reader's attention to the polymer-based hybrid materials used in photocatalytic processes for efficient degradation of organic pollutants in water. These inorganic–organic materials exhibit unique physicochemical properties due to the synergistic effect originating from the combination of individual elements, i.e., photosensitive metal oxides and polymeric supports. The possibility of merging the structural elements of hybrid materials allows for improving photocatalytic performance through (1) an increase in the light-harvesting ability; (2) a reduction in charge carrier recombination; and (3) prolongation of the photoelectron lifetime. Additionally, the great majority of polymer materials exhibit a high level of resistance against ultraviolet irradiation and improved corrosion resistance. Taking into account that the chemical and environmental stability of the hybrid catalyst depends, to a great extent, on the functional support, we highlight benefits and drawbacks of natural and synthetic polymer-based photocatalytic materials and pay special attention to the fact that the accessibility of synthetic polymeric materials derived from petroleum may be impeded due to decreasing amounts of crude oil. Thus, it is necessary to look for cheap and easily available raw materials like natural polymers that come from, for instance, lignocellulosic wastes or crustacean residues to meet the demand of the "plastic" market.

Keywords: photoactive hybrid materials; photocatalyst; biopolymers; synthetic polymers; water/air detoxification; metal oxides

1. Introduction

In recent years, the interest in photocatalysis as a green and eco-friendly method for pollution remediation [1,2], energy conversion [3,4] or chemical synthesis [5] has been increasing. However, the proficient application of photocatalysis is restricted to the fabrication of advanced nanostructured materials consisting of various types of semiconductors [6–11]. Current research in this field mainly concentrates on the development of hybrid materials containing photoactive metal oxides [9–13], quantum dots, or perovskites [14,15] to enhance the efficiency of photocatalytic systems [16]. In order to understand the main restrictions of photocatalytic performance, we have to know the main photochemical and photophysical mechanisms leading that process (Figure 1) [16–20]. It is worth mentioning that correlation between semiconductors and light plays a crucial role in the case of photocatalysis [12,16]. The absorption of photons with a specific energy allows for excitation of electrons from the valence band to the conduction band, producing hole–electron pairs responsible for redox processes. However, oxidation and reduction processes can be inhibited due to charges and radical recombination or back electron transfer processes [21,22]. Consequently, the employed photonic power is much higher than desirable rate of the desired chemical transformation, which is the main limitation of photocatalytic performance [23].

Significant features which can extensively improve photocatalytic efficiency and light absorption capability, are strongly correlated to size and structure of photocatalytic materials, their specific surface area, or crystalline phase [7,24]. Taking into consideration the fact that the required physicochemical characteristics are a consequence of various features of the main components and interaction between them [25,26], the enhancement of photocatalytic performances could be obtained through an organic–inorganic hybrid complex made of a semiconductor and suitable support [25]. Multiple material combinations provide synergistic effects that are able to create and improve properties of nanomaterials [27–29] which are beneficial for enhancing the efficiency of photocatalytic reactions. The synergistic effect originating from the combination of individual elements is clearly evident in the case of polymer-based hybrid photocatalysts. A great majority of polymer materials exhibits a high level of resistance against ultraviolet irradiation and improved corrosion resistance as well as environmental stability [30,31]. Compared with semiconductor oxides, the great majority of polymeric supports are chemically inert, mechanically stable, inexpensive and easily available. Additionally, the hydrophobic nature of polymer gives an advantage to congregate the organic pollutants on the surface and raise the efficiency of adsorption and subsequent degradation reaction rates [32]. Therefore, in recent years polymer-based hybrid materials have been emerging as promising device in the photocatalytic field.

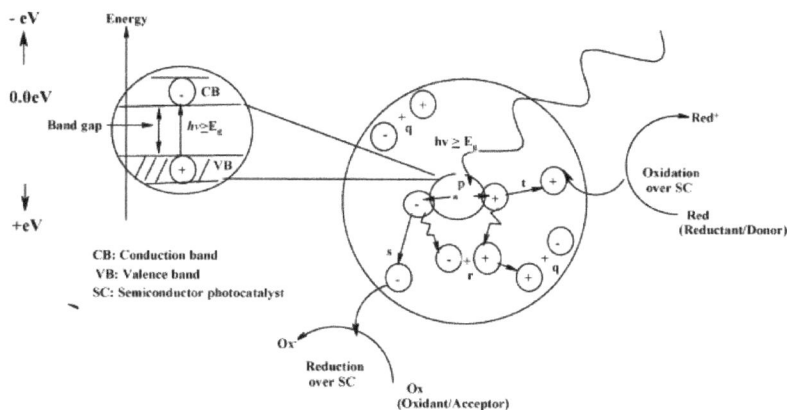

Figure 1. Photochemical and photophysical processes over photon-activated semiconductors, where: (p) is photogeneration of electron/hole pair, (q) is surface recombination, (r) is recombination in the bulk, (s) is diffusion of acceptor and reduction on the surface of semiconductor and (t) is oxidation is oxidation of donor on the surface of semiconductor particle. Reprinted from [31] with permission from Elsevier.

Nowadays, plastic production is based mainly on feedstock derived from petroleum refineries. A wide range of synthetic or semisynthetic polymerization products are obtained from oil and gas, which undergo chemical processing [33]. As inputs for plastic manufacturing, refinery olefins (mainly propylene and less quantities of ethylene or butylenes) are produced from alkane transformations [34]. Unfortunately, there are difficulties with the discovery of new oil deposits, and some sedimentary basins that contain crude oil have already been explored. The oilfields which still have not been explored are located in inaccessible regions of the world [35]. Taking into consideration the fact that the plastic production is based mainly on feedstock derived from oil refinery, it is necessary to look for cheap and easily available raw materials like highly abundant biopolymers in nature. In this state-of-the-art review, we focus on synthetic and natural polymers, and highlight the main benefits and limitations coming from polymer materials which could be used as support for the fabrication of photocatalytic hybrid materials.

2. Synthetic and Natural Polymers

According to IUPAC (International Union of Pure and Applied Chemistry) nomenclature, the word "polymer" refers to substances composed of macromolecules with high relative molecular mass. However, this term could be also be applied to polymer substances, polymer blends or polymer molecules [36]. Additionally, polymers include a wide class of materials which can be grouped according to source, functionalities, structure, thermal behavior, polymerization mechanism or preparation techniques (Figure 2) [37,38]. In this mini-review, we mainly focus on synthetic and natural polymers, which can find applications in photocatalysis, taking into consideration factors related to the photocatalytic properties, including stability, biodegradability, and biocompatibility with inorganic materials as well as following recent progress on the synthesis of hybrid materials.

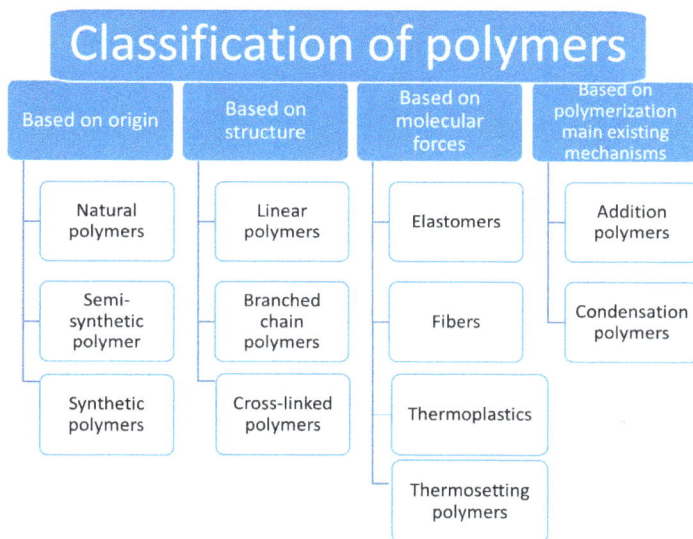

Figure 2. Classification of polymers according to [37].

Synthetic and semisynthetic polymers originating from crude oil have had a huge impact on modern science and technology due to their physicochemical properties. In many cases, chemical, physical and biological resistance play crucial roles in the selection of a desirable polymer for determined function. However, with respect to most synthetic materials, the affected time-resistant properties of polymeric wastes can lead to the release of toxic degradation products during decomposition which is not acceptable from an eco-friendly point of view [38]. Owing to concerns about the natural environment and shortages of non-renewable sources, the interest in polymers derived from natural sources like starch, lignocellulose or proteins is still increasing [34]. Biodegradable polymers that can be obtained from renewable resources (Figure 3) have emerged as environmentally friendly substitutes for non-biodegradable materials. It is worth mentioning that some of them exhibit similar or superior mechanical properties to petroleum-based polymers [39]. However, in many cases they possess inferior physical feature in terms of stability and strength and lots of them require high-cost production [33]. Additionally, in comparison with synthetic polymers, several natural polymers cannot be processed into a wide range of shapes, due to the fact that the high processing temperature destroys their structure [40]. Consequently, the design of new eco-friendly and highly efficient and stable photocatalytic biopolymer hybrid materials is challenging.

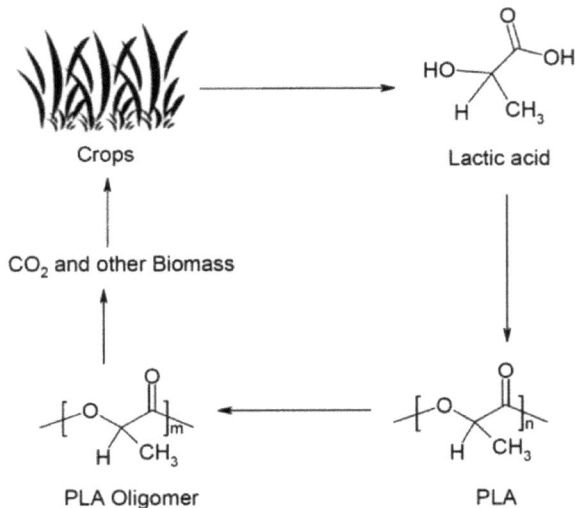

Figure 3. Life cycle of polylactide (PLA), an example of biodegradable synthetic polymers. Reprinted from [41] with permission from Elsevier.

2.1. Photocatalytic Hybrid Materials Based on Synthetic Polymers

Various types of synthetic polymer shave been reported as photocatalytic supports in the literature, namely: polyethylene (PE) [42], polypropylene (PP) [43], polystyrene (PS) [44], polyethylene terephthalate (PET) [45], polyvinyl chloride (PVC) [46], polyvinyl alcohol (PVA) [47], polycarbonate (PC) [48] and so on. To our best knowledge, the first attempt to produce polymer hybrid materials was made in 1995 by Tennakone [49]. Titanium oxide with polyethylene films as support was used for the photocatalytic decomposition of phenol with a high degradation ratio (50% after 2.5 h of illumination). Further, experimental studies on polypropylene non-woven with zinc oxide nanorods indicate that this kind of photocatalytic materials exhibited not only excellent catalytic activity but also high stability [50,51]. Hence, these hybrid materials can be successfully used for water treatment processes by acting as photocatalysts and filters at the same time [52]. Additionally, the synergetic effect of the combination of metal oxide and polymers allows for protection of the polypropylene fiber against surface cracks and limits the well-known photocorrosion process of zinc oxide [52,53]. Similar photoactive hybrid materials based on polybutylene terephthalate (PBT) polymer fiber mats were used for photocatalytic dye degradation. These studies confirmed that the catalyst supported on the polymer mat could be reused without a particular recovery step [54]. They also pointed out the fact that the combination of proper fabrication methods allows for better photocatalytic performance (Table 1, Entry 1) [55]. Another example of synthetic polymer hybrid materials, which have applications in water treatment processes, are polyethersulfone or polyvinylidene fluoride membranes with various types of metal oxides (e.g., titanium, zinc or chromium) displaying good antifouling performance, including photo-catalysis, self-cleaning, and filterability properties [55,56]. Our special attention gives merit to hybrid materials based on conjugated organic polymers (COPs) like polyaniline (PANI) [57] (Table 1, Entry 2), poly(pyrrole) (PPy) [58], polythiophene (PT) [59], polyacetylene (PA) [60], poly(methyl methacrylate) (PMMA) [61], polythiopene (PT) [62], polyparaphenylene (PPP) [63], polyparaphenylenevenylene (PPV), poly(3,4-ethylenedioxythiophene) (PEDOT) [63] or poly(O-phenylenediamine) (POPD)) [64]. The conjugated organic polymers are mostly p-type semiconductors, due to their electrical and optical properties. Specifically, their high electron mobility or high photon absorption coefficient under visible spectra has attracted increasing interest for photocatalytic applications, e.g., degradation of pollutants or hydrogen generation by water splitting [65]. In terms of water treatment processes, another interesting perspective solution offered by polymeric support is the possibility of fabricating a floatable photocatalyst, the concept of which is shown in Figure 4. These kinds of materials are able to maximize illumination utilization and oxygenation processes of the photocatalyst by approaching the air/water interface, which in the end can result in higher rates of radical formation and oxidation efficiencies [66].

Polymeric supports possess different morphologies and can exist in the form of sheets [67], nanospheres [68], or nanoparticles [69]. Of course in all these cases, polymer materials contribute to an increase the photocatalytic activity of inorganic–organic materials. However, contact surface area of the hybrid photocatalyst, which has a significant influence on their activity, is lower for fiber polymeric supports. Selected examples of catalysts based on polymeric fibers with high photocatalytic activity are shown in Table 1.

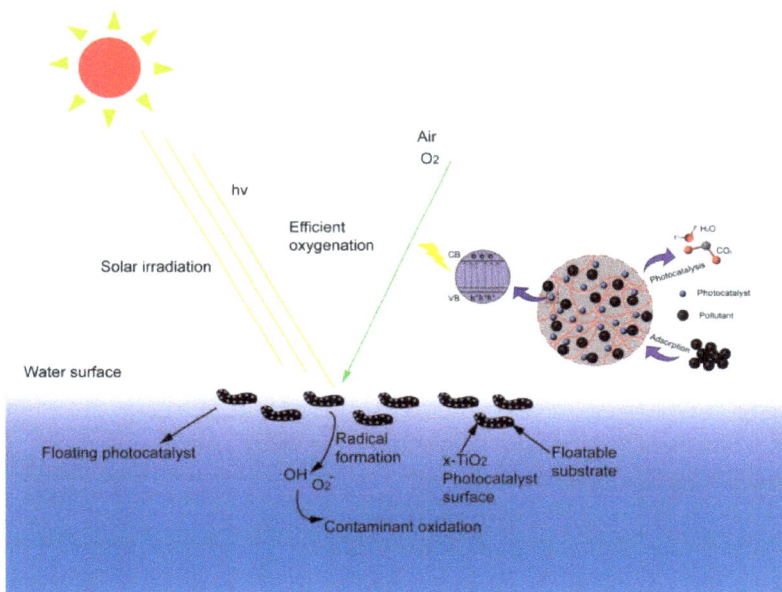

Figure 4. Schematic representation of a floating photocatalyst (CB: conduction band; VB: valence band). Reproduced from reference [70] by permission of the Royal Society of Chemistry.

It is worth mentioning that in the open literature a new class of hybrid materials, represented by coordination polymers (CPs) and composed of metal clusters with organic ligands (Figure 5), can be used in photocatalysis. These crystalline materials possess high dispersion of active sites, tuneable adsorption properties, appropriate pore size and topology [71]. Additionally, the effective solar light absorption properties can be obtained through modifying the composition of the metal cations and organic linkers. The possibility to connect the light-harvesting and catalytic components allows for conversion of solar energy to chemical energy by artificial photosynthesis [72]. Materials based on coordination polymers provide crucial information about synergistic effects derived from multiple elements of hybrid materials and allow for understanding the fundamental principles about light harvesting and energy transfer phenomena, schematically explained in Figure 6. In spite of some examples of CP-based heterogeneous photocatalytic systems [73], until now the CP-based soluble complexes have mainly been used in homogenous catalytic processes which is a serious limitation for broad industrial application due mostly to the problem of photocatalyst filtration after the process.

Figure 5. The coordination polymers which consist of the branch of metal organic framework and metal organic complex. Adapted and modified with permission from [72] Copyright (2012) American Chemical Society.

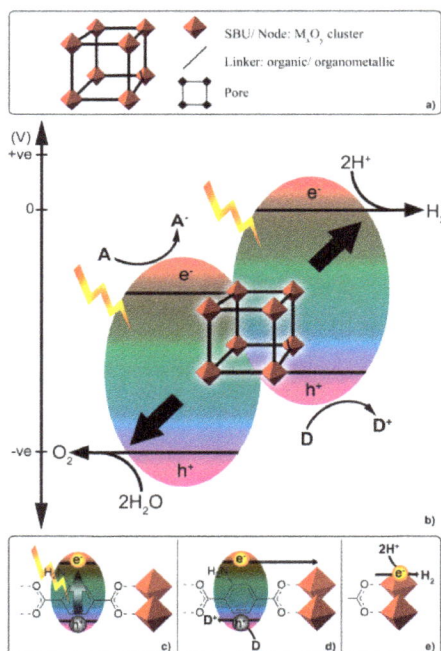

Figure 6. (**a**) the components of the MOF structure; (**b**) conceptual schematic for photo-catalyzed water oxidation or reduction using a MOF in the presence of an acceptor or donor; (**c**) light harvesting accomplished by an organic linker; (**d**) generation of a charge separated state and quenching of h$^+$ by a donor; (**e**) electron transfer to the metal oxide node and subsequent proton reduction (SBU: secondary building units). Reproduced from reference [73] with permission of The Royal Society of Chemistry.

Table 1. Selected photocatalytic hybrid materials based on synthetic polymers used for degradation of organic contaminants.

Entry	Polymer Hybrid Materials	Target Contaminant	Light Source	Fabrication Method	Photocatalytic Behavior	Ref.
1	ZnO nanorods on polybutylene terephthalate (PBT) polymer fiber mats	Azo organic dye (acid red 40)	Ultraviolet (UV) radiation in the range of 320–390 nm providing 79 mW/cm^2 of energy flux.	Thin films formed by low temperature vapor phase atomic layer deposition (ALD) and hydrothermal growth of ZnO nanorod crystals on a seed layer.	Degradation ratio ~90% of the dye within 2 h. The combination of ALD and hydrothermal method allow to obtain the best performance of the photocatalyst and may be also used for other crystal growth systems, such as TiO$_2$, Fe$_2$O$_3$, SnO$_2$ and V$_2$O$_5$, where high area and ready solution access are desired.	[55]
2	ZnO nanoparticles on wool and polyacrylonitrile (PANI) fibers	Methylene blue (MB) and eosin yellowish (EY) dye	High-pressure mercury lamp covers illumination spectrum ranging from ultraviolet to visible (200–800 nm).	Impregnation of polymeric fibers using sol-gel process at ambient temperature. ZnO-sol is based on the method described in the literature with minor changes in details.	There is 77% MB dye degradation after 6 h upon ZnO/PANI and 80% upon ZnO/wool fibers, which is 4-fold more in comparison to bare fibers. Similar results of degradation were obtained for EY dye, where the degradation ratios equal 64% and 50%, respectively.	[57]
3	CeO$_2$–ZnO-polyvinylpyrrolidone (PVP)	Rhodamine B (RhB)	UV lamp (8 W) with emission wavelengths at 254 nm.	The electrospinning technique was followed by thermal treatment to obtain CeO$_2$–ZnO nanofibers. The nonwoven mat was prepared from the precursor solution of PVP/Ce(NO$_3$)$_3$/Zn(CH$_3$COO)$_2$.	After 3 h of irradiation, only 17.4% and 82.3% of Rhodamine B was decomposed catalyzed by pure CeO$_2$ and ZnO fibers, respectively, whereas almost 98% was decomposed applying the CeO$_2$–ZnO-composite fibers.	[74]
4	ZnO nanowires on polyethylene (PP)	Methylene blue (MB)	UV light source (6 W)	ZnO nanowires were grown from seed ZnO nanoparticles affixed onto the commercially available fibers by hydrothermal method.	After 2.5 h of irradiation, ZnO/polyethylene fibers degraded 83% of the MB, whereas the fibers without ZnO degrade only 32%. 24% of MB was found undergo self-degradation under the same UV light without using polyethylene fibers.	[75]
5	ZnO/SnO$_2$-polyvinylpyrrolidone (PVP)	Rhodamine B	High-pressure mercury lamp (50 W) with main emission wavelength at 313 nm.	A simple combination method of sol-gel process and electrospinning technique. The electrospun composite nanofibers was obtained by the precursor solution of PVP/ZnCl$_2$/SnCl$_2$.	After 50 min, the degradation efficiency of RhB was equal to 75, 35, and 85% for ZnO, SnO$_2$, and TiO$_2$ fibers, respectively. However, the time for complete decolorization of dye solution over the ZnO/SnO$_2$-nanofibers was 30 min.	[76]
6	Reduced graphene oxide/titanium dioxide filter (RGO/TiO$_2$) and reduced graphene oxide/zinc oxide filter (RGO/ZnO) on polypropylene(PP) porous filter	Methylene blue (MB)	Halogen lamp (150 W)	The polypropylene (PP) porous filter was incorporated with reduced graphene oxide (RGO) and metal oxides via a simple hydrothermal approach.	The combination of RGO and the metal oxide compounds on the filters shows more than 70% of MB adsorption in 20 min compared with those consisting of individual materials, degradation after 120 min 99%.	[50]

2.2. Photocatalytic Hybrid Materials Based on Natural Polymers

Natural polymers derived from renewable resources or waste products can also serve as a desirable organic support for inorganic semiconductor metal oxides [77–79]. Polysaccharides, lignin, cellulose, hemicellulose, chitin, chitosan, starch or xylan, possess excellent sustainability, biodegradability and can be used as abundant industrial raw feed stock to synthesize photocatalytic hybrid materials [80–82]. Natural polymers (see structures in Figure 7) are mainly obtained by extractions from wood, plants or even residues of living organisms [83], that make them more attractive in comparison with synthetic polymers. Natural fibers are frequently used as a reinforcing composite for producing hybrid materials because they exhibit advantages, like recyclability and eco-friendliness, over their synthetic counterparts [84]. Additionally, natural fibers possess a higher volume fraction, and thus a larger loading capacity [85]. For these reasons they are widely used to produce composite materials, especially in the field of photocatalysis. For instance, depositing titanium dioxide on cellulose fiber surface allows for obtaining hybrid materials with a high degradation ratio of organic compounds like organic dye or phenolic contaminants [86] (Table 2, Entry 1). Yu et al. obtained cellulose-templated TiO_2/Ag nanosponge composites with enhanced photocatalytic activities for the degradation of RhB; the synthetic procedures for this material are shown in Figure 8 [86]. The polymeric nanocomposite membranes with cellulose fibers can be also used for gas separation processes (e.g., hydrogen recovering, nitrogen generation or carbon dioxide separation) [87,88]. However, due to the fact that cellulose consists of monosaccharide units (Figure 7a), it is hydrophilic and exhibits a rather poor interaction with most of the non-polar compound. Many efforts have been done to obtain uniform dispersion of the fibers within the matrices. Furthermore, it is worth noting that plant fibers like cellulose possess relatively low processing temperatures (<200 °C), and for this reason, researchers have used low-temperature techniques (e.g., sol-gel method, hydrothermal method, dip coating method, etc.) to coat natural fibers [89]. Despite this, in the open literature we can find successfully completed examples of many attempts to modify surface properties of natural fibers (see Table 2).

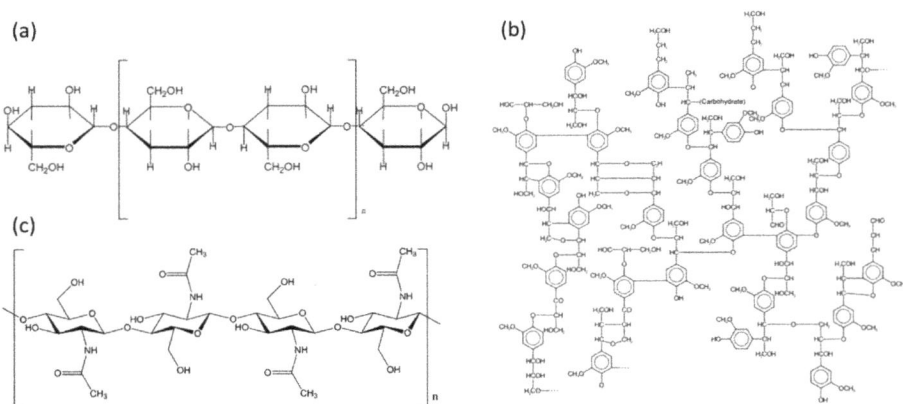

Figure 7. Structure of most common natural polymers: cellulose (**a**), lignin (**b**) and chitin (**c**). Reprinted from references [83] with permission from Elsevier.

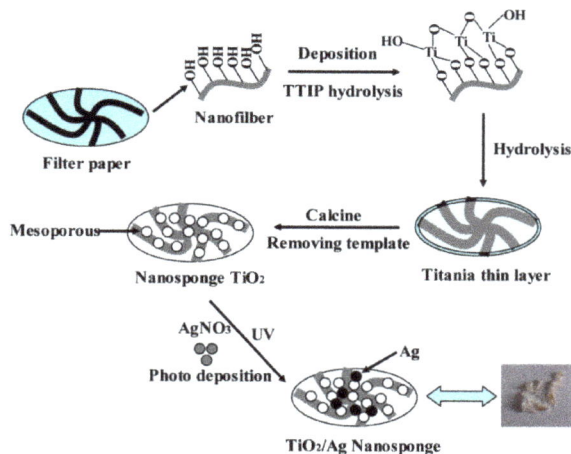

Figure 8. Schematic illustration of the synthetic approach for TiO$_2$/Ag nanosponge materials (TTIP: titanium *iso*-propoxide). Reprinted with permission from [86]. Copyright (2012) American Chemical.

In contrast to hemicellulose and cellulose, lignin possesses rather a complex structure due to the different types of linkages connecting the phenylpropanoid-based units [90] (Figure 7b) and is seen as a product with little intrinsic value but potentially, for instance, can serve as a stable biopolymeric support for hybrid materials [91]. It should be noted that lignin can be used for the synthesis of functional hybrid materials with antimicrobial properties [92] as well as adsorptive properties for the removal of inorganic compounds from aqueous solutions [93]. However, the stability of biopolymers based on lignin under photocatalytic conditions is not high. Some critical review reports pointed out that photocatalytic methods can be used also to degradate lignin and lignin-based phenolic compounds [94]. From this point of view, it is interesting to study if lignin, which consists of phenylpropanoid-based structures, can be used as a support for the fabrication of stable hybrid photocatalysts with the aim of being applied for mineralization of organic contaminants in water.

Chitin (Figure 7c) represents the next most plentiful natural polysaccharide and is source of chitosan which is obtained by partial deacetylation of chitin. Due to the presence of various functional groups, chitosan can be applied as an adsorbent for the removal of different kinds of pollutants [95,96]. Additionally, this polymer can be used in the wide range of form, e.g., hydrogels [97], nanofiber mats [98], and nano-beads as well as powders [99]. Chitosan may have a high specific surface area as a bead or fiber, and thus exhibits strong adsorption capacities. However, all of these forms depend on the fabrication method. The formation of a highly specific surface area requires mainly specialized methods like electrospinning [100]. In the open literature, one can find the description of many other facile methods to obtain hybrid catalysts like those based on titanium dioxide [100], zinc oxide [101] (Table 2), and niobium oxide [102] among others, combined with chitosan, which can be used for water treatment processes. However, chitosan, as well as other polysaccharides have some limitations associated with wastewater treatment application, namely: they possess high swelling capacities and low resistance, especially in extreme wastewater conditions (e.g., acid medium), which may result in significant leaching. Consequently, the catalyst based on these natural polymers cannot be stable and may promote the expansion of organic matter in wastewater [103].

Table 2. Selected photocatalytic hybrid materials based on natural polymers used for degradation of organic contaminants.

Entry	Polymer Hybrid Materials	Target Contaminant	Light Source	Fabrication Method	Photocatalytic Behavior	Ref.
1	Titanium dioxide (TiO_2) immobilized in cellulose matrix	Phenol	UV (6 W) light at wavelength of 254 nm was used. The mean light intensity equal to 0.56 mW/cm^2.	Composite films have been prepared via a sol-gel method.	The composite films exhibited high degradation ratio (90% after 2 h of irradiation) without remarkable loss of photocatalytic activity after three times.	[85]
2	ZnAc/cellulose acetate (CA) composite nanofibers	Rhodamine B and phenol	Ultraviolet lamps (PHILIPS 365 nm) as the irradiation source.	Electrospinning technique in combination with calcination.	Almost 100% of Rhodamine B and 85% phenol (after 24 h) was decomposed in the presence of TiO_2/ZnO composite nanofibers under mild conditions.	[104]
3	ZnO/cellulose hybrid nanofibers	Methylene blue (MB) and eosin yellowish (EY) dye	Tungsten lamp (500 W) was used as the visible light source.	A novel method that combines electrospinning and solvothermal techniques	Nearly 50% of Rhodamine B was decomposed after 24 h of irradiation under visible light.	[105]
4	Photoactive TiO_2 films on cellulose fibers	Methylene blue (MB) and heptane-extracted bitumen fraction (BF) containing a mixture of heavy aromatic hydrocarbons	Reproducible solar light (50 mW/cm^2).	Sol-gel method.	The degradation ratio of MB reached 90% after 20 h and 90% for BF fraction after 9 h without loss of activity after three illumination cycles.	[106]
5	Rice-straw-derived hybrid TiO_2–SiO_2 structures	Methylene blue (MB)	UV-A (8 W) lamps (300–450 nm) providing an irradiation power flux of 2.0 mW/cm^2.	Impregnation method.	The photocatalytic decomposition of methylene blue after 90 min obtained was 100%.	[107]
6	Chitosan (CS)-encapsulated TiO_2 nanohybrid	Methylene blue (MB)	UV light at a wavelength of 365 nm.	Nanohybrid materials was prepared by chemical precipitation method.	The catalyst showed high photocatalytic activity of 90% degradation after 3 h of irradiation and without losing photocatalytic activity after five recycle tests.	[100]
7	Fe_3O_4/chitosan/TiO_2 nanocomposites	Methylene blue (MB)	Illumination with UV light.	Facile and low-cost method by solvents thermal reduction.	The degradation rate of methyl blue was 93% after 30 min for Fe_3O_4/CTS/TiO_2 nanocomposites.	[108]

3. Summary and Future Perspectives

The selected studies in this short overview serve as clear examples that both natural and synthetic polymers can be successfully used in the field of hybrid innovative materials for heterogeneous photocatalysis in the context of pollutant degradation. The organic–inorganic hybrid materials exhibit significantly better photocatalytic properties than the separate components, due to the synergistic effect coming from the intrinsic properties of a photoactive semiconductor and polymers. Several key advantages can be expected from polymeric support, such as: (a) an increase of the specific surface area which consequently allows for adsorption of higher amounts of target pollutants [109–111], and (b) an improvement of the photocatalytic performance by promoting reduction of the charge carriers recombination and prolongation of the photoelectron lifetime [112]. In this mini-review, the highlighted benefits and drawbacks of natural and synthetic polymer-based photocatalytic materials will exponentially increase in importance due to the fact that accessibility of synthetic polymeric materials derived from oil, gas and carbon (non-renewable sources) will be impeded because of the decreasing amount of fossil resources. Thus, this fact turns on the alarm to look for cheap and easily available raw materials like bio-polymers that come from sources such as lignocellulosic wastes or crustacean residues to cover the increasing demand of the market for plastics. Currently, the scientific world indicates that polymer materials are the key promising components of the next generation of photocatalytic hybrid materials for water and air treatment processes [113,114]. However, there are still some limitations on this topic (Table 3) that should be studied extensively.

Table 3. The main pros/cons of using synthetic polymers and biopolymers for photocatalytic hybrid materials.

	Synthetic Polymers	Biopolymers
Availability	Decreasing	High
Physicochemical resistance	High	Low
Thermal stability	High	Low
Large-scale applications	Possible	Difficult
Environmental-friendly	No	Yes
Cost of production	Low	High
Sustainability	Low	High

Acknowledgments: Juan C. Colmenares is very grateful for the partial support from the National Science Centre in Poland within Sonata Bis Project No. 2015/18/E/ST5/00306.

Author Contributions: The authors contributed equally to this work.

Conflicts of Interest: The authors declare no conflict of interest. The founding sponsors had no role in the design of the study; in the collection, analyses, or interpretation of data; in the writing of the manuscript, and in the decision to publish the results.

References

1. Pichat, P.; Oills, D. Photocatalytic Treatment of Water: Irradiance Influences. In *Photocatalysis and Water Purification: From Fundamentals to Recent Applications*, 1st ed.; Pichat, P., Ed.; Wiley-VCH Verlag GmbH & Co., KGaA: Weinheim, Germany, 2013; pp. 311–333.
2. George, C.; Beeldens, A.; Barmpas, F.; Doussin, J.F.; Manganelli, G.; Herrmann, H.; Kleffmann, J.; Mellouki, A. Impact of photocatalytic remediation of pollutants on urban air quality. *Front. Environ. Sci. Eng.* **2016**, *10*. [CrossRef]
3. Yu, C.; Zhou, W.; Yu, J.C.; Liu, H.; Wei, L. Design and fabrication of heterojunction photocatalysts for energy conversion and pollutant degradation. *Chin. J. Catal.* **2014**, *35*, 1609–1618. [CrossRef]
4. Peng, Y.; Shang, L.; Bian, T.; Zhao, Y.; Zhou, C.; Yu, H.; Tunga, C.; Zhang, T. Flower-like CdSe ultrathin nanosheet assemblies for enhanced visible-light-driven photocatalytic H_2 production. *Chem. Commun.* **2015**, *51*, 4677–4680. [CrossRef] [PubMed]

5. Corrigan, N.; Shanmugam, S.; Xu, J.; Boyer, C. Photocatalysis in organic and polymer synthesis. *Chem. Soc. Rev.* **2016**, *45*, 6165–6212. [CrossRef] [PubMed]

6. Penga, Y.; Shanga, L.; Cao, Y.; Wang, Q.; Zhao, Y.; Zhou, C.; Bian, T.; Wu, L.Z.; Tung, C.H.; Zhang, T. Effects of surfactants on visible-light-driven photocatalytic hydrogen evolution activities of $AgInZn_7S_9$ nanorods. *Appl. Surf. Sci.* **2015**, *358*, 485–490. [CrossRef]

7. Zhao, Y.; Chen, G.; Bian, T.; Zhou, C.; Waterhouse, G.I.N.; Wu, L.Z.; Tung, C.H.; Smith, L.J.; O'Hare, D.; Zhang, T. Defect-Rich Ultrathin ZnAl-Layered Double Hydroxide Nanosheets for Efficient Photoreduction of CO_2 to CO with Water. *Adv. Mater.* **2015**, *27*, 7823–7831. [CrossRef]

8. Yu, H.; Shi, R.; Zhao, Y.; Bian, T.; Zhao, Y.; Zhou, C.; Waterhouse, G.I.N.; Wu, L.Z.; Tung, C.H.; Zhang, T. Alkali-Assisted Synthesis of Nitrogen Deficient Graphitic Carbon Nitride with Tunable Band Structures for Efficient Visible-Light-Driven Hydrogen Evolution. *Adv. Mater.* **2017**, *29*, 1605148. [CrossRef] [PubMed]

9. Colmenares, J.C.; Varma, R.S.; Lisowski, P. Sustainable hybrid photocatalysts: Titania immobilized on carbon materials derived from renewable and biodegradable resources. *Green Chem.* **2016**, *18*, 5736–5750. [CrossRef]

10. Colmenares, J.C.; Kuna, E.; Lisowski, P. Synthesis of Photoactive Materials by Sonication: Application in Photocatalysis and Solar Cells. *Top. Curr. Chem.* **2016**, *374*. [CrossRef] [PubMed]

11. Colmenares, J.C.; Xu, Y.J. Heterogeneous Photocatalysis: From Fundamentals to Green Applications. In *Heterogeneous Photocatalysis*, 1st ed.; He, L.N., Rogers, R.D., Su, D., Tundo, P., Zhang, C., Eds.; Springer: Berlin/Heidelberg, Germany, 2016. [CrossRef]

12. Arora, A.K.; Jaswal, V.S.; Singh, K.; Singh, R. Applications of Metal/Mixed Metal Oxides as Photocatalyst: (A Review). *Orient. J. Chem.* **2016**, *32*, 2035–2042. [CrossRef]

13. Zhao, Y.; Jia, X.; Waterhouse, G.I.N.; Wu, L.Z.; Tung, C.H.; O'Hare, D.; Zhang, T. Layered Double Hydroxide Nanostructured Photocatalysts for Renewable Energy Production. *Adv. Energy Mater.* **2016**, *6*. [CrossRef]

14. Zhang, Z.; Zheng, T.; Li, X.; Xu, J.; Zeng, H. Progress of Carbon Quantum Dots in Photocatalysis Applications. *Part. Part. Syst. Charact.* **2016**, *33*, 447–588. [CrossRef]

15. Wang, W.; Tadéa, M.O.; Shao, Z. Research progress of perovskite materials in photocatalysis- and photovoltaics-related energy conversion and environmental treatment. *Chem. Soc. Rev.* **2015**, *44*, 5371–5408. [CrossRef] [PubMed]

16. Khan, M.M.; Adil, S.F.; Al-Mayouf, A. Metal oxides as photocatalysts. *J. Saudi Chem. Soc.* **2015**, *19*, 462–464. [CrossRef]

17. Ahmad, R.; Ahmad, Z.; Khan, A.U.; Mastoi, N.R.; Aslam, M.; Kim, J. Photocatalytic systems as an advanced environmental remediation: Recent developments, limitations and new avenues for applications. *J. Environ. Chem. Eng.* **2016**, *4*, 4143–4164. [CrossRef]

18. Schneider, J.; Bahnemann, D.; Ye, J.; Puma, G.L.; Dionysiou, D.D. *Photocatalysis, Vol. 1: Fundamentals and Perspectives; Vol. 2: Applications*, 1st ed.; RSC Publishing: Cambridge, UK, 2016; pp. 1–936.

19. Pichat, P. *Photocatalysis: Fundamentals, Materials and Potential*, 1st ed.; MDPI: Basel, Switzerland, 2016; pp. 3–650.

20. Nosaka, Y.; Nosaka, A. *Introduction to Photocatalysis: From Basic Science to Applications*, 1st ed.; RSC Publishing: Cambridge, UK, 2016; pp. 1–272.

21. Portela, R.; Hernández-Alonso, M.D. *Design of Advanced Photocatalytic Materials for Energy and Environmental Applications*, 1st ed.; Coronado, J., Fresno, F., Hernández-Alonso, M.D., Portela, R., Eds.; Springer: London, UK, 2013; pp. 35–36.

22. Chaturvedi, S.; Dave, P.N. Environmental Application of Photocatalysis. *Mater. Sci. Forum* **2012**, *734*, 273–294. [CrossRef]

23. Jiang, L.; Wang, Y.; Feng, C. Application of Photocatalytic Technology in Environmental Safety. *Procedia Eng.* **2012**, *45*, 993–997. [CrossRef]

24. Nakataa, K.; Fujishimaa, A. TiO_2 photocatalysis: Design and applications. *J. Photochem. Photobiol. C Photochem. Rev.* **2012**, *13*, 169–189. [CrossRef]

25. Sanchez, C.; Ribot, F.; Lebeau, B. Molecular design of hybrid organic-inorganic nanocomposites synthesized via sol-gel chemistry. *J. Mater. Chem.* **1999**, *9*, 35–44. [CrossRef]

26. Singh, S.; Mahalingam, H.; Singh, P.K. Polymer-supported titanium dioxide photocatalysts for environmental remediation: A review. *Appl. Catal. A* **2013**, *462–463*, 178–195. [CrossRef]

27. Vilatela, J.J.; Eder, D. Nanocarbon composites and hybrids in sustainability: A review. *ChemSusChem* **2012**, *12*, 456–478. [CrossRef] [PubMed]

28. Thakur, R.S.; Chaudhary, R.; Singh, C. Fundamentals and applications of the photocatalytic treatment for the removal of industrial organic pollutants and effects of operational parameters: A review. *J. Renew. Sustain. Energy Rev.* **2010**, *2*, 042701. [CrossRef]

29. Peng, Y.; Shang, L.; Cao, Y.; Waterhouse, G.I.N.; Zhou, C.; Bian, T.; Wu, L.Z.; Tunga, C.H.; Zhang, T. Copper(I) cysteine complexes: Efficient earth-abundant oxidation co-catalysts for visible light-driven photocatalytic H_2 production. *Chem. Commun.* **2015**, *51*, 12556–12599. [CrossRef] [PubMed]

30. Reddy, K.R.; Hassan, M.; Gomes, V.G. Hybrid nanostructures based on titanium dioxide for enhanced photocatalysis. *Appl. Catal. A Gen.* **2015**, *489*, 1–16. [CrossRef]

31. Gaya, U.I.; Abdullah, A.H. Heterogeneous photocatalytic degradation of organic contaminants over titanium dioxide: A review of fundamentals, progress and problems. *J. Photochem. Photobiol. C Photochem. Rev.* **2008**, *9*, 1–12. [CrossRef]

32. Ghosh, T.; Oh, W.C. Carbon Based Titania Photocatalysts. *Asian J. Chem.* **2012**, *24*, 5419–5423.

33. Muggeridge, A.; Cockin, A.; Webb, K.; Frampton, H.; Collins, I.; Moulds, T.; Salino, P. Recovery rates, enhanced oil recovery and technological limits. *Philos. Trans. R. Soc. A* **2013**, *372*, 20120320. [CrossRef] [PubMed]

34. How Much Oil Is Used to Make Plastic? Available online: http://www.eia.gov/tools/faqs/faq.cfm?id=34& t=6 (accessed on 25 April 2016).

35. Gervet, B. *The Use of Crude Oil in Plastic Making Contributes to Global Warming*; Department of Civil and Environmental Engineering Luleå University of Technology: Luleå, Sweden, May 2007.

36. Compendium of Polymer Terminology and Nomenclature. Available online: https://www.iupac.org/ cms/wp-content/uploads/2016/01/Compendium-of-Polymer-Terminology-and-Nomenclature-IUPAC-Recommendations-2008.pdf (accessed on 19 January 2009).

37. Olatunji, O. *Classification of Natural Polymers*; Springer: Basel, Switzerland, 2015; pp. 1–17.

38. Luckachan, G.E.; Pillai, C.K.S. Biodegradable Polymers—A Review on Recent Trends and Emerging Perspectives. *J. Polym. Environ.* **2011**, *19*, 637–676. [CrossRef]

39. Jiang, L.; Zhang, J. Biodegradable and Biobased Polymers. In *Handbook of Biopolymers and Biodegradable Plastics: Properties, Processing and Applications*, 1st ed.; Ebnesajjad, S., Andrew, W., Eds.; Elsevier: Oxford, UK, 2013; pp. 109–124.

40. Sionkowska, A. Current research on the blends of natural and synthetic polymers as new biomaterials: Review. *Prog. Polym. Sci.* **2011**, *36*, 1254–1276. [CrossRef]

41. Gupta, A.P.; Kumar, V. New emerging trends in synthetic biodegradable polymers—Polylactide: A critique. *Eur. Polym. J.* **2007**, *43*, 4053–4074. [CrossRef]

42. Yu, Z.; Mielczarski, E.; Mielczarski, J.; Laub, C.; Buffat, P.; Klehm, U.; Albers, P.; Lee, K.; Kulike, A.; Kiwi-Minerska, L.; et al. Preparation, stabilization and characterization of TiO_2 on thin polyethylene films (LDPE). Photocatalytic applications. *Water Res.* **2007**, *41*, 862–874.

43. Ma, S.; Meng, J.; Li, J.; Zhang, Y.; Ni, L. Synthesis of catalytic polypropylene membranes enabling visible-light-driven photocatalytic degradation of dyes in water. *J. Membr. Sci.* **2014**, *453*, 221–229. [CrossRef]

44. Zan, L.; Tian, L.; Liu, Z.; Peng, Z. A new polystyrene–TiO_2 nanocomposite film and its photocatalytic degradation. *Appl. Catal. A Gen.* **2004**, *264*, 237–242.

45. Taylor, D.M.; Lewis, T.J. Electrical conduction in polyethylene terephthalate and polyethylene films. *J. Phys. D Appl. Phys.* **1971**, *4*, 1346–1354. [CrossRef]

46. Wang, D.; Shi, L.; Luo, Q.; Li, X.; An, J. An efficient visible light photocatalyst prepared from TiO_2 and polyvinyl chloride. *J. Mater. Sci.* **2012**, *47*, 2136–2145. [CrossRef]

47. Araújo, V.D.; Tranquilin, R.L.; Motta, F.V.; Paskocimas, C.A.; Bernardi, M.I.B.; Cavalcante, L.S.; Andres, J.; Longo, E.; Bomio, M.R.D. Effect of polyvinyl alcohol on the shape, photoluminescence and photocatalytic properties of $PbMoO_4$ microcrystals. *Mater. Sci. Semicond. Process.* **2014**, *26*, 425–430. [CrossRef]

48. Fateh, R.; Dillert, R.; Bahnemann, D. Preparation and characterization of transparent hydrophilic photocatalytic TiO_2/SiO_2 thin films on polycarbonate. *Langmuir* **2015**, *29*, 3730–3739. [CrossRef] [PubMed]

49. Tennakone, K.; Tilakaratne, C.T.K.; Kottegoda, I.R.M. Photocatalytic degradation of organic contaminants in water with TiO_2 supported on polyethene films. *J. Photochem. Photobiol. A Chem.* **1995**, *87*, 177–179. [CrossRef]

50. Ariffin, S.N.; Lima, H.N.; Jumeri, F.A.; Zobir, M.; Abdullah, A.H.; Ahmad, M.; Ibrahim, N.A.; Huang, N.M.; Teo, P.S.; Muthoosamy, K.; et al. Modification of polypropylene filter with metal oxide and reduced graphene oxide for water treatment. *Ceram. Int.* **2014**, *40*, 6927–6936. [CrossRef]

51. Li, M.; Li, G.; Fan, Y.; Jiang, J.; Ding, Q.; Dai, X.; Mai, K. Effect of nano-ZnO-supported 13X zeolite on photo-oxidation degradation and antimicrobial properties of polypropylene random copolymer. *Polym. Bull.* **2014**, *71*, 2981–2997. [CrossRef]

52. Colmenares, J.C.; Kuna, E.; Jakubiak, S.; Michalski, J.; Kurzydłowski, K. Polypropylene nonwoven filter with nanosized ZnO rods: Promising hybrid photocatalyst for water purification. *Appl. Catal. B Environ.* **2015**, *170–171*, 273–282. [CrossRef]

53. Sakthivel, S.; Neppolian, B.; Shankar, M.V.; Arabindoo, B.; Palanichamy, M.; Murugesan, V. Solar photocatalytic degradation of azo dye: Comparison of photocatalytic efficiency of ZnO and TiO$_2$. *Sol. Energy Mater. Sol. Cells* **2003**, *77*, 65–82. [CrossRef]

54. Preparation and Characterization of Composite PES/Nanoparticle Membranes. Available online: https://web.wpi.edu/Pubs/E-project/Available/E-project-042313--223351/unrestricted/Sheppard_MQP_Report.pdf (accessed on 25 April 2013).

55. Gong, B.; Peng, Q.; Na, J.S.; Parsons, G.N. Highly active photocatalytic ZnO nanocrystalline rods supported on polymer fiber mats: Synthesis using atomic layer deposition and hydrothermal crystal growth. *Appl. Catal. A Gen.* **2011**, *407*, 211–216. [CrossRef]

56. Hong, J.; He, Y. Polyvinylidene fluoride ultrafiltration membrane blended with nano-ZnO particle for photo-catalysis self-cleaning. *Desalination* **2014**, *332*, 67–75. [CrossRef]

57. Moafi, H.F.; Shojaie, A.F.; Zanjanchi, M.A. Semiconductor-assisted self-cleaning polymeric fibers based on zinc oxide nanoparticles. *J. Appl. Polym. Sci.* **2011**, *121*, 3641–3650. [CrossRef]

58. Sun, L.; Shi, Y.; Li, B.; Li, X.; Wang, Y. Preparation and characterization of polypyrrole/TiO$_2$ nanocomposites by reverse microemulsion polymerization and its photocatalytic activity for the degradation of methyl orange under natural light. *Polym. Compos.* **2013**, *34*, 1076–1080. [CrossRef]

59. Ansari, M.O.; Khan, M.M.; Ansari, S.A.; Cho, M.H. Polythiophene nanocomposites for photodegradation applications: Past, present and future. *J. Saudi Chem. Soc.* **2015**, *19*, 494–504. [CrossRef]

60. Aizawa, M.; Watanabe, S.; Shinohara, H.; Shirakawa, H. Photodoping of polyacetylene films. *J. Chem. Soc. Chem. Commun.* **1985**, *2*, 62–63. [CrossRef]

61. Yang, M.; Dan, Y. Preparation of poly(methyl methacrylate)/titanium oxide composite particles via in-situ emulsion polymerization. *J. Appl. Polym. Sci.* **2006**, *101*, 4056–4063. [CrossRef]

62. Zhang, Z.; Zheng, T.; Xu, J.; Zeng, H. Polythiophene/Bi$_2$MoO$_6$: A novel conjugated polymer/nanocrystal hybrid composite for photocatalysis. *J. Mater. Sci.* **2016**, *51*, 3846–3853. [CrossRef]

63. Wang, X.; Maeda, K.; Thomas, A.; Takanabe, K.; Xin, G.; Carlsson, J.M.; Domen, K.; Antonietti, M. A metal-free polymeric photocatalyst for hydrogen production from water under visible light. *Nat. Mater.* **2009**, *8*, 76–80. [CrossRef] [PubMed]

64. Ghosh, S.; Kouame, N.A.; Remita, S.; Ramos, L.; Goubard, F.; Aubert, P.H.; Dazzi, A.; Deniset-Besseau, A.; Remita, H. Visible-light active conducting polymer nanostructures with superior photocatalytic activity. *Sci. Rep.* **2015**, *5*. [CrossRef] [PubMed]

65. Ullah, H.; Tahir, A.A.; Mallick, T.K. Polypyrrole/TiO$_2$ composites for the application of photocatalysis. *Sens. Actuators B Chem.* **2017**, *241*, 1161–1169. [CrossRef]

66. Magalhães, F.; Mourab, F.C.C.; Lago, R.M. TiO$_2$/LDPE composites: A new floating photocatalyst for solar degradation of organic contaminants. *Desalination* **2011**, *276*, 266–271. [CrossRef]

67. Naskar, S.; Pillay, A.S.; Chanda, M. Photocatalytic degradation of organic dyes in aqueous solution with TiO$_2$ nanoparticles immobilized on foamed polyethylene sheet. *J. Photochem. Photobiol. A Chem.* **1998**, *3*, 257–264. [CrossRef]

68. Jin, L.; Wu, H.; Morbidelli, M. Synthesis of Water-Based Dispersions of Polymer/TiO$_2$ Hybrid Nanospheres. *Nanomaterials* **2015**, *5*, 1454–1468. [CrossRef] [PubMed]

69. Nabid, M.R.; Golbabaee, M.; Moghaddam, A.B.; Dinarvand, R.; Sedghi, R. Polyaniline/TiO$_2$ Nanocomposite: Enzymatic Synthesis and Electrochemical Properties. *Int. J. Electrochem. Sci.* **2008**, *3*, 1117–1126.

70. Wang, X.; Wang, X.; Wang, W.; Zhao, J. Enhanced visible light photocatalytic activity of a floating photocatalyst based on B-N-codoped TiO$_2$ grafted on expanded perlite. *RSC Adv.* **2015**, *5*, 41385–41392. [CrossRef]

71. Kan, W.Q.; Liu, B.; Yang, J.; Liu, Y.Y.; Ma, J.F. A Series of Highly Connected Metal–Organic Frameworks Based on Triangular Ligands and d10 Metals: Syntheses, Structures, Photoluminescence, and Photocatalysis. *Cryst. Growth Des.* **2012**, *12*, 2288–2298. [CrossRef]

72. Wang, C.; Xie, Z.; deKrafft, K.E.; Lin, W. Light-Harvesting Cross-Linked Polymers for Efficient Heterogeneous Photocatalysis. *ACS Appl. Mater. Interfaces* **2012**, *4*, 2288–2294. [CrossRef] [PubMed]

73. Meyer, K.; Ranocchiari, M.; Bokhoven, J.A. Metal organic frameworks for photo-catalytic water splitting. *Energy Environ. Sci.* **2015**, *8*, 1923–1937. [CrossRef]

74. Li, C.; Chen, R.; Zhang, X.; Shu, S.; Xiong, J.; Zheng, Y.; Dong, W. Electrospinning of CeO₂–ZnO composite nanofibers and their photocatalytic property. *Mater. Lett.* **2011**, *65*, 1327–1330. [CrossRef]

75. Baruah, S.; Thanachayanont, C.; Dutta, J. Growth of ZnO nanowires on nonwoven polyethylene fibers. *Sci. Technol. Adv. Mater.* **2008**, *9*, 025009. [CrossRef] [PubMed]

76. Zhang, Z.; Shao, C.; Li, X.; Zhang, L.; Xue, H.; Wang, C.; Liu, Y. Electrospun Nanofibers of ZnO−SnO₂ Heterojunction with High Photocatalytic Activity. *J. Phys. Chem. C* **2010**, *114*, 7920–7925. [CrossRef]

77. Gao, M.; Peh, C.K.G.; Onga, W.L.; Ho, G.W. Green chemistry synthesis of a nanocomposite graphene hydrogel with three-dimensional nano-mesopores for photocatalytic H₂ production. *RSC Adv.* **2013**, *3*, 13169–13177. [CrossRef]

78. Petzold, J.C.; Herrington, T.M. The measurement of concentration and stability of weak acids and a study of the ionisation of polysaccharides using buffer capacity curves. *Macromol. Chem. Phys.* **2003**, *192*, 1741–1748. [CrossRef]

79. Zhang, G.; Shen, X.; Yang, Y. Facile Synthesis of Monodisperse Porous ZnO Spheres by a Soluble Starch-Assisted Method and Their Photocatalytic Activity. *J. Phys. Chem. C* **2011**, *115*, 7145–7152. [CrossRef]

80. Hamdi, A.; Boufi, S.; Bouattour, S. Phthalocyanine/chitosan-TiO₂ photocatalysts: Characterization and photocatalytic activity. *Appl. Surf. Sci.* **2015**, *339*, 128–136. [CrossRef]

81. Benabid, F.Z.; Zouai, F. Natural polymers: Cellulose, chitin, chitosan, gelatin, starch, carrageenan, xylan and dextran. *Alger. J. Nat. Prod.* **2016**, *4*, 348–357.

82. Fu, S.; Song, P.; Liu, X. Thermal and flame retardancy properties of thermoplastics/natural fiber biocomposites. In *Advanced High Strength Natural Fibre Composites in Construction*, 1st ed.; Gwen, J., Ed.; Wood Publishing: Duxford, UK, 2017; pp. 479–508. [CrossRef]

83. Thakura, V.K.; Thakurb, M.K. Processing and characterization of natural cellulose fibers/thermoset polymer composites. *Carbohydr. Polym.* **2014**, *109*, 102–117. [CrossRef] [PubMed]

84. Li, H.; Fu, S.; Peng, L. Surface modification of cellulose fibers by layer-by-layer self-assembly of lignosulfonates and TiO₂ nanoparticles: Effect on photocatalytic abilities and paper properties. *Fibers Polym.* **2013**, *14*, 1794–1802. [CrossRef]

85. Zeng, J.; Liu, S.; Cai, J.; Zhang, L. TiO₂ Immobilized in Cellulose Matrix for Photocatalytic Degradation of Phenol under Weak UV Light Irradiation. *J. Phys. Chem. C* **2010**, *114*, 7806–7811. [CrossRef]

86. Yu, D.H.; Yu, X.; Wang, C.; Liu, X.C.; Xing, Y. Synthesis of Natural Cellulose-Templated TiO₂/Ag Nanosponge Composites and Photocatalytic Properties. *ACS Appl. Mater. Interfaces* **2012**, *4*, 2781–2788. [CrossRef] [PubMed]

87. Ahmadizadegan, H. Surface modification of TiO₂ nanoparticles with biodegradable nanocellolose and synthesis of novel polyimide/cellulose/TiO₂ membrane. *J. Colloid Interface Sci.* **2017**, *491*, 390–400. [CrossRef] [PubMed]

88. Foruzanmehr, M.R.; Vuillaume, P.Y.; Robert, M.; Elkoun, S. The effect of grafting a nano-TiO₂ thin film on physical and mechanical properties of cellulosic natural fibers. *Mater. Des.* **2015**, *85*, 671–678. [CrossRef]

89. Horvath, A.L. Solubility of structurally complicated materials: I. wood. *J. Phys. Chem.* **2006**, *35*, 77–92. [CrossRef]

90. Ten, E.; Vermerris, W. Functionalized Polymers from Lignocellulosic Biomass: State of the Art. *Polymers* **2013**, *5*, 600–642. [CrossRef]

91. Dong, Y.Y.; Li, S.M.; Ma, M.G.; Zhao, J.J.; Sun, R.C.; Wang, S.P. Environmentally-friendly sonochemistry synthesis of hybrids from lignocelluloses and silver. *Carbohydr. Polym.* **2014**, *102*, 445–452. [CrossRef] [PubMed]

92. Klapiszewski, Ł.; Siwińska-Stefańska, K.; Kołodyńska, D. Preparation and characterization of novel TiO₂/lignin and TiO₂-SiO₂/lignin hybrids and their use as functional biosorbents for Pb(II). *Chem. Eng. J.* **2017**, *314*, 169–181. [CrossRef]

93. Kansala, S.K.; Singh, M.; Sud, D. Studies on TiO₂/ZnO photocatalysed degradation of lignin. *J. Hazard. Mater.* **2008**, *153*, 412–417. [CrossRef] [PubMed]

94. Pandey, K.K. A note on the influence of extractives on the photo-discoloration and photo-degradation of wood. *Polym. Degrad. Stab.* **2005**, *87*, 375–379. [CrossRef]

95. Chen, X.; Chew, S.L.; Kerton, F.M.; Yan, N. Direct conversion of chitin into a *N*-containing furan derivative. *Green Chem.* **2014**, *16*, 2204–2212. [CrossRef]

96. Tang, H.; Zhou, W.; Zhang, L. Adsorption isotherms and kinetics studies of malachite green on chitin hydrogels. *J. Hazard. Mater.* **2012**, *209–210*, 218–225. [CrossRef] [PubMed]
97. Naseri, N.; Kristiina, O.; Aji, P.M. Electrospun Chitosan Nanofiber Random Mats Reinforced with Chitin and Cellulose Nanocrystals for Wound Dressing Application. 2016. Available online: http://www.diva-portal. org/smash/record.jsf?pid=diva2%3A1011676&dswid=-5388 (accessed on 11 May 2017).
98. Muzzarelli, R.A. Biomedical exploitation of chitin and chitosan via mechano-chemical disassembly, electrospinning, dissolution in imidazolium ionic liquids, and supercritical drying. *Mar. Drugs* **2010**, *9*, 1510–1533. [CrossRef] [PubMed]
99. Kamal, T.; Anwar, Y.; Khan, S.B.; Tariq Saeed Chania, M.; Asiria, A.B. Dye adsorption and bactericidal properties of TiO_2/chitosan coating layer. *Carbohydr. Polym.* **2016**, *148*, 153–160. [CrossRef] [PubMed]
100. Haldorai, Y.; Shim, J.J. Novel chitosan-TiO_2 nanohybrid: Preparation, characterization, antibacterial, and photocatalytic properties. *Polym. Compos.* **2014**, *35*, 327–333. [CrossRef]
101. Salehi, R.; Arami, M.; Mahmoodi, N.M.; Bahrami, H.; Khorramfar, S. Novel biocompatible composite (Chitosan-zinc oxide nanoparticle): Preparation, characterization and dye adsorption properties. *Colloids Surf. B Biointerfaces* **2010**, *80*, 86–93. [CrossRef] [PubMed]
102. Torres, J.D.; Faria, E.A.; De Souza, J.R.; Prado, A.G.S. Preparation of photoactive chitosan–niobium (V) oxide composites for dye degradation. *J. Photochem. Photobiol. A Chem.* **2006**, *182*, 202–206. [CrossRef]
103. Sabar, S.; Nawi, M.A.; Ngah, W.S.W. Photocatalytic removal of Reactive Red 4 dye by immobilised layer-by-layer TiO_2/cross-linked chitosan derivatives system. *Desalin. Water Treat.* **2014**, *57*, 5851–5857. [CrossRef]
104. Liu, H.; Yang, J.; Liang, J.; Huang, J.; Tang, C. ZnO Nanofiber and Nanoparticle Synthesized through Electrospinning and Their Photocatalytic Activity under Visible Light. *J. Am. Chem. Soc.* **2008**, *91*, 1287–1291. [CrossRef]
105. Ye, S.; Zhang, D.; Liu, J.; Zhou, J. ZnO Nanocrystallites/Cellulose Hybrid Nanofibers Fabricated by Electrospinning and Solvothermal Techniques and Their Photocatalytic Activity. *J. Appl. Polym. Sci.* **2011**, *121*, 1757–1764. [CrossRef]
106. Uddin, M.J.; Cesano, F.; Bonino, F.; Bordiga, S.; Spoto, G.; Scarano, D.; Zecchina, A. Photoactive TiO_2 films on cellulose fibres: Synthesis and characterization. *J. Photochem. Photobiol. A Chem.* **2007**, *189*, 286–294. [CrossRef]
107. Ge, Y.; Xiang, Y.; He, Y.; Ji, M.; Song, G. Preparation of Zn-TiO_2/RH/Fe_3O_4 composite material and its photocatalytic degradation for the dyes in wastewater. *Desalin. Water Treat.* **2016**, *57*, 9837–9844. [CrossRef]
108. Choia, C.; Hwanga, K.J.; Kimb, Y.J.; Kimb, G.; Parkc, J.Y.; Sungho, J. Rice-straw-derived hybrid TiO2–SiO2 structures with enhanced photocatalytic properties for removal of hazardous dye in aqueous solutions. *Nano Energy* **2016**, *20*, 76–83. [CrossRef]
109. Lee, M.; Chen, B.Y.; Walter Den, W. Chitosan as a Natural Polymer for heterogeneous Catalysts Support: A Short Review on Its Applications. *Appl. Sci.* **2015**, *5*, 1272–1283. [CrossRef]
110. Luo, L.; Yang, L.C.; Xiao, M.; Bian, L.; Yuan, B.; Liu, Y.; Jiang, F.; Pan, X. A novel biotemplated synthesis of TiO_2/wood charcoal composites for synergistic removal of bisphenol A by adsorption and photocatalytic degradation. *Chem. Eng. J.* **2015**, *262*, 1275–1283. [CrossRef]
111. Ohtani, N.; Tonoi, M. Improved Photoluminescence Lifetime of Organic Emissive Materials Embedded in Organic-Inorganic Hybrid Thin Films Fabricated by Sol-Gel Method Using Tetraethoxysilane. *Mol. Cryst. Liq. Cryst.* **2014**, *599*, 132–138. [CrossRef]
112. Yan, S.C.; Lv, S.B.; Li, Z.S.; Zouabd, Z.G. Organic-inorganic composite photocatalyst of g-C(3)N(4) and TaON with improved visible light photocatalytic activities. *Dalton Trans.* **2010**, *39*, 1488–1491. [CrossRef] [PubMed]
113. Corma, A.; Navarro, M.T.; Rey, F.; Ruiz, V.R.; Sabater, M.J. Direct synthesis of a photoactive inorganic-organic mesostructured hybrid material and its application as a photocatalyst. *ChemPhysChem* **2010**, *10*, 1084–1089. [CrossRef] [PubMed]
114. Foruzanmehr, M.R.; Vuillaum, P.Y.; Elkoun, S.; Robert, M. Physical and mechanical properties of PLA composites reinforced by TiO_2 grafted flax fibers. *Mater. Des.* **2016**, *106*, 295–304. [CrossRef]

molecules

MDPI

Review

Self-Sterilizing Sputtered Films for Applications in Hospital Facilities

Sami Rtimi *, Stefanos Giannakis and Cesar Pulgarin *

Group of Advanced Oxidation Processes, Swiss Federal Institute of Technology, EPFL-SB-ISIC-GPAO, Station 6, CH-1015 Lausanne, Switzerland; stefanos.giannakis@epfl.ch
* Correspondence: sami.rtimi@epfl.com (S.R.); cesar.pulgarin@epfl.ch (C.P.); Tel.: +41-216-936-150 (S.R.)

Received: 4 May 2017; Accepted: 23 May 2017; Published: 28 June 2017

Abstract: This review addresses the preparation of antibacterial 2D textile and thin polymer films and 3D surfaces like catheters for applications in hospital and health care facilities. The sputtering of films applying different levels of energy led to the deposition of metal/oxide/composite/films showing differentiated antibacterial kinetics and surface microstructure. The optimization of the film composition in regards to the antibacterial active component was carried out in each case to attain the fastest antibacterial kinetics, since this is essential when designing films avoiding biofilm formation (under light and in the dark). The antimicrobial performance of these sputtered films on *Staphylococcus aureus* (MRSA) and *Escherichia coli* (*E. coli*) were tested. A protecting effect of TiO_2 was found for the release of Cu by the TiO_2-Cu films compared to films sputtered by Cu only. The Cu-released during bacterial inactivation by TiO_2-Cu was observed to be much lower compared to the films sputtered only by Cu. The FeOx-TiO_2-PE films induced *E. coli* inactivation under solar or under visible light with a similar inactivation kinetics, confirming the predominant role of FeOx in these composite films. By up-to-date surface science techniques were used to characterize the surface properties of the sputtered films. A mechanism of bacteria inactivation is suggested for each particular film consistent with the experimental results found and compared with the literature.

Keywords: antibacterial surfaces; light; metal oxides; coatings; magnetron sputtering

1. Introduction

Hospital acquired infections (HAIs) are on the rise in Europe, infecting 5–7% of hospital patients staying beyond 10 days with the consequent high cost related to the long and costly hospital residence time [1]. Thirty years ago, Domek et al. [2] reported the inactivation of coliform bacteria by Cu, later Keevil's group reported the Cu inactivation of *MRSA* [3] and *E. coli* [4]. The Cu-ions released from coated surfaces were reported to induce a strong biocidal effect below the cytotoxic levels accepted for mammalian cells. This consideration validates the use of Cu-immobilized devices in bloodstream infections [5]. The disinfection in some cases seems to proceed via an oligodynamic effect due to the low amounts (in the ppb/ppm range) of Cu or Ag released by these surfaces [6]. Cu-ions have been shown to complex proteins and break hydrogen bonds within the DNA opening the double helix [7].

Antimicrobial coatings are being investigated to prepare implants and medical devices [8–11]. Recently, research groups reported Chemical Vapor Deposition (CVD) of Cu-titania films being applied in single or multilayered coatings [8–13]. Innovative films against MRSA infections have been recently reported for packaging materials [14], plastics [15] and stethoscopes [16]. Boyce et al. [17] found MRSA contamination up to 65% in the hospital staff gloves and uniforms due to the contact in hospital-infected rooms/surfaces. Later, Bhalla et al. [18] showed that hospital workers frequently had infected gloves/uniforms in variable concentrations with the bacteria/fungi available in the hospital facilities. Bacteria invade, adhere and form biofilms that tightly glue to the surface (catheters or other medical devices) [19]. For this reason, biofilm formation has to be precluded. Catheters impregnated with antibiotics/antiseptics or both showed a short-term effect lasting only a few days. This is due to the rapid release of the antibacterial agent. This complicated in hospital settings by the increased bacterial resistance to many antibiotics affecting patients during long-hospital residence times [20,21].

Studies have shown rapid killing of bacterial cells when exposed to Cu-surfaces but the mechanism of the Cu-MRSA killing is still controversial [6–13]. The Cu-antimicrobial action seems to comprise the cellular metabolism damaging the cell DNA [22]. A recent study has shown that the uptake of Cu-ions by MRSA was fast and damaged the cell DNA, but the mechanism of this uptake remains controversial and more work is needed in this direction [1,22].

Recently, Heidenau et al. [10,11] demonstrated that in Ag-Cu amalgams, Cu in extremely low amounts inactivate more effectively bacteria compared to other metals. The Ag-Cu films were found to present surfaces with a high in vitro compatibility. Recent work in our laboratory with Cu-sputtered surfaces induced a faster kinetic bacterial inactivation [23,24] compared to Ag-sputtered surfaces [25,26]. TiO_2 surfaces in the dark are ineffective against bacterial infection, but Cu-TiO_2 surfaces introduce antibacterial action in the dark on medical implants [27,28]. TiO_2 has been reported to increase the adhesion of Cu on glass and other surfaces avoiding leaching during disinfection and therefore assumes the role of a protective additive hindering the Cu-release.

In the past few years, the continuous exposure to antibiotics over long times has led to increased antibiotic resistance of bacteria. However, only a few pathogens display resistance to Ag and Cu and combinations thereof [1,6,29,30]. Cu-Ag films show long-operational lifetime, which is not the case for antibiotics/antiseptics rapidly detaching from the film surface. This is important when antibacterial films are used in connection to bloodstream. In this review, we focus on the preparation of innovative antibacterial coatings on 2D surfaces (polyethylene or polyurethane films) and on 3D complex shape medical devices (intravascular catheters). The sputtering of uniform, adhesive coatings with Cu and Ag on low thermal resistant fabrics like hospital textiles and thin polymer films are reviewed as well as material related to TiO_2, TiO_2-Cu (both on polyester), FeOx and FeOx-TiO_2 (both on polyethylene films) [30–35].

2. Antibacterial Coatings: From Conventional to up to Date Methods

Conventional deposition techniques e.g., sol-gel or films prepared via the colloidal route enable the synthesis of materials with a high thermal resistance since a temperature of a few hundred degrees has to be applied to anneal the active antibacterial component to the particular substrate. Commercial sol-gel methods have reported TiO_2 and other thin films on heat resistant substrates [35].

Photo-induced reactions by finely dispersed Ag-nanoparticles (Ag-NPs) of ~2 nm in size have been reported to inactivate *E. coli* and *Staphylococcus aureus*. This dispersion also showed photo-switchable behavior involving a transition from an initial hydrophobic surface towards a super-hydrophilic surface within the bacterial inactivation time under light irradiation. The contact angle (and surface energies) was followed under light and in the dark. The reversible process to reestablish the initial hydrophobicity (dark storage of the sample after bacterial inactivation under light) was seen to happen within long times (>24 h) [35].

Estimates for the effective bacteria reduction by Ag-coated surfaces suggest that these may be able to reduce the contamination in hospitals and public places [36–39]. The level found for these bacteria in many hospitals is higher than the allowed level for the healthcare rooms. For example, the contamination of 10^5 CFU/cm^2 was observed in a diabetic wound dressing. Nevertheless, near the patient, a microbial density of about 10^2 CFU/cm^2 was found. The use of catalytic/photocatalytic textiles (for beddings, curtains, lab-coats ...) can drastically decrease the bacterial propagation [38–42].

The major limitations towards the use of sol-gel photoactive surfaces are related to (i) the lack of uniformity of these sol-gel films; (ii) the non-reproducible preparation observed from different batches with the same composition and (iii) the lack of mechanical stability and adhesion to the substrate [24–26,40]. The sputtering methods described in this review have been applied to obtain robust antibacterial films on cotton, polyester, polyethylene and polyurethane thin films [34,35,40].

During the last four decades, the sputtering of surfaces for industries in the aviation, car and machine tool sectors was used to protect the surfaces against corrosion by sputtering micrometer thick coatings of Cr-Fe to avoid the corrosion of the Fe/Fe_2O_3 layers. Nowadays, thin coatings are used for many purposes such as to avoid surface corrosion, to serve as anti-reflective and to acquire self-cleaning and self-sterilizing properties. Recently sputtering in the presence of O_2 have been carried out to deposit thin metal oxides on non-heat resistant substrates like textiles at temperatures <130 °C by DCMS, DCPMS [41] and HIPIMS [42]. During the last decade, Kelly et al. has addressed the preparation of antibacterial thin films by magnetron sputtering (DCMS, DCMP and HIPIMS). They characterized the film surface properties and corrosion and correlated with the antibacterial activities [43–46]. Several recent studies report that the antibacterial activity of metal/oxide films is a function of the dispersion, size and thickness of the metal/oxide coatings [5,8,12,13,30,44,46–50]. Diverse thicknesses in the TiO_2 coatings led to a complete different Cu-dispersion, Cu nanoparticulate size and bacterial inactivation kinetics as recently reported [12,29,40,49,51,58]. During the last years, we investigated the bactericidal activity on Ag, Cu, and other metal oxides sputtered on polymers and textile fabrics. The antibacterial kinetics, microstructure and surface properties have been reported in details [40–42,50–53].

3. Photocatalytic Coated Polyester Showing Duality in the *E. coli* and MRSA Inactivation under Actinic Indoor Light

Recently, some laboratories have reported the preparation of antibacterial Ag, Cu and TiO_2 coatings on glass and polymer films by depositing the metal/oxides by CVD and sputtering techniques [8,29,30,32,44–46]. The direct current magnetron sputtering (DCMS) co-deposition of TiO_2 and Cu films leading to uniform, adhesive and robust layers on polyester (PES) at temperatures not exceeding 120–130 °C has been reported recently [29,30]. There is a need for innovative active coatings showing a fast bacterial inactivation kinetics and a high adhesion to the substrate. In this way, the formation of toxic biofilms spreading bacteria/virus/fungi in hospital facilities leading to increased HAIs may be precluded.

Recent studies report the preparation of TiO_2, Cu and TiO_2/Cu films by sol-gel methods inducing significant photo-induced bacterial inactivation of films deposited on glass surfaces from Ti-chloride/ethyl acetate annealed at 500 °C [40,47–50]. Our laboratory has reported *Escherichia coli* (*E. coli*) inactivation on TiO_2/Cu sequentially sputtered (deposited one after the other starting with an under layer of TiO_2 then an upper layer of Cu/CuOx). Moreover, the co-sputtering of TiO_2-Cu (simultaneous deposition) on textiles leading to *E. coli* and methicillin-resistant *Staphylococcus aureus* (MRSA) reduction was reported [43]. The later reports the differential effect of actinic and visible light (400–700 nm) on the bacterial reduction kinetics on *E. coli* as a MRSA.

The TiO_2-Cu-PES microstructure is shown by TEM in Figure 1. The denser Cu-clusters presented diameters between 16 and 20 nm while the TiO_2 clusters presented smaller sizes from 5 and up to 10 nm. The TiO_2-Cu coatings were 120 to 160 nm thick. This is equivalent to 500 to 800 atomic layers (being each layer 0.2 nm thick) each containing 10^{15} atoms/cm^2 [51]. It showed close contact between the TiO_2 and Cu-nanoparticles and was uniform in its microstructure [32,35,50].

The bacterial inactivation by TiO_2-Cu-PES under actinic light irradiation and in the dark was evaluated as a function of the amount of TiO_2 and Cu sputtered on the substrate [41]. The deposition time of TiO_2-Cu (co-sputtering) was optimized to determine the most suitable amount of TiO_2 and Cu on the inactivating *E. coli*. In the dark, *E. coli* bacterial inactivation proceeds within 120 min on TiO_2-Cu-PES. The mechanism of TiO_2-Cu(CuO) mediated *E. coli* inactivation under light irradiation has been reported to involve interfacial charge transfer (IFCT) [29,40]. Bacterial reduction in the dark as shown in Figure 2, trace 6) and seems to proceed through a mechanism where O_2 (air) reacts with the Cu^0/Cu-ions. Figure 2, trace 1) shows the complete bacterial reduction under visible light within 30 min for TiO_2-Cu samples co-sputtered for 3 min. Within 3 min, a sufficient amount of TiO_2

and Cu was sputtered on the PES leading to the most suitable number of exposed catalytic sites leading to the fastest *E. coli* inactivation.

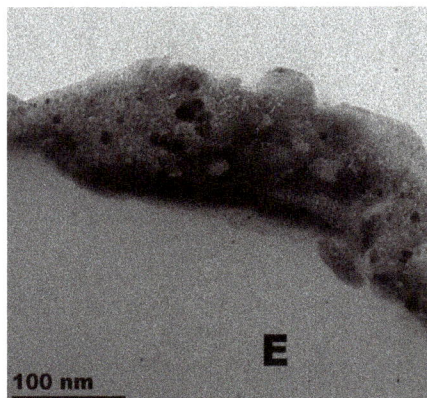

Figure 1. Transmission electron microscopy (TEM) of TiO_2-Cu co-sputtered for 3 min on PES. "E" stands for the epoxide required to embed the sample during the sample preparation/cutting for the TEM image [32].

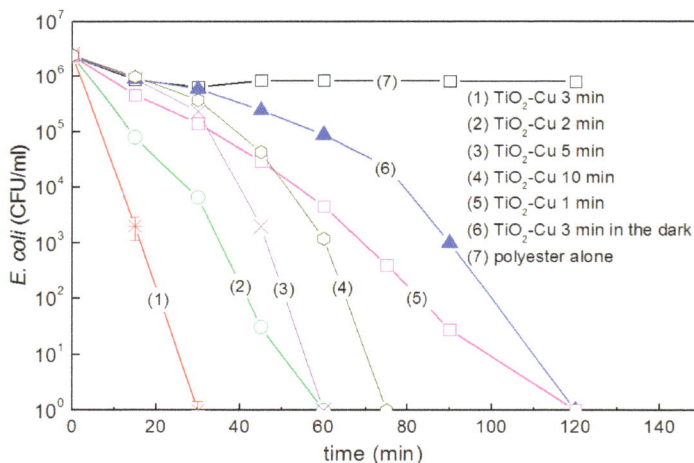

Figure 2. *E. coli* inactivation on TiO_2-Cu co-sputtered for different times on PES as indicated in the traces: (1) 3 min, (2) 2 min, (3) 5 min, (4) 10 min, (5) 1 min (6) 3 min in the dark and (7) PES-alone. The bacterial reduction under light irradiation used a lamp Philips Master-18W/865 (4.65 mW/cm^2), Error bars: standard deviation (*n* = 5%).

Co-sputtering for short times (1–2 min) did not deposit the necessary amount of TiO_2 and Cu to induce a fast bacterial inactivation. Co-sputtering TiO_2-Cu for 5 and 10 min (Figure 2) led to longer bacterial inactivation kinetics compared to TiO_2-Cu (3 min). This is due to the inward charge diffusion of the generated charges in the TiO_2 under band–gap irradiation [29,51]. In addition, longer sputtering times facilitate the TiO_2 inter-particle growth decreasing the TiO_2 contact surface with bacteria [52]. Figure 2, trace 6) shows complete bacterial reduction in the dark. Bacterial inactivation takes place in the dark possibly through an oligodynamic effect as recently reported by S. Rtimi [41,52,53]. The effect of Cu on bacteria has been associated to reactions blocking the of proteins/enzymes regulating the respiratory chain [32,41,52].

The electronic transfer between the TiO_2/Cu and *E. coli* depends on the length of the charge diffusion in the TiO_2/Cu layers. The diffusion of the charges induced by band-gap irradiation is a function of the TiO_2 and Cu particle size and shape [54,55]. The interfacial distances between TiO_2 and Cu/CuO on the polyester surface ranges from 5 nm and up. The IFCT, as shown in Figure 2, proceeds with a quantum efficiency depending on the light intensity and the nanoparticulate size [56–58]. Quantum size effects occur in particles with sizes ~10 nm and about 10^4 atoms or smaller [59]. The surface composition and properties of the TiO_2-CuO play a role in the charge transfer. The bacterial inactivation kinetics depends on the film (i) surface defects; (ii) surface imperfections; and (iii) dangling bonds on the edge of this composite. In TiO_2-Cu composite, the charge recombination in nanoparticles is short due to their small particle size. The small particle size decreases the space for charge separation. In addition, the semiconductor space charge layer in both the TiO_2 and CuO further decreases the potential depth available for the charge injection at the TiO_2-Cu hetero-junction. This in turn, decreases the energy difference between TiO_2 and Cu, which is not favorable for the charge injection [29,52]. The conduction band of CuO at -0.30 V vs. SCE (pH 7) is at a more negative potential than the potential required for one electron oxygen reduction [52]. Furthermore, the Cu^{2+} can also react with O_2^-:

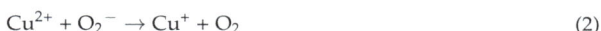

$$O_2 + H^+ + e^- \rightarrow HO_2^{\bullet} \ -0.22 \text{ V} \tag{1}$$

$$Cu^{2+} + O_2^- \rightarrow Cu^+ + O_2 \tag{2}$$

The irradiation with solar simulated light induces the transfer of the e^- and h^+ from TiO_2 to CuO as shown in Figure 3 below. Charge transfer from ad-atoms to TiO_2 under light has been investigated during the last few decades [40]. The potential energy levels of the TiO_2 conduction band (cb) and TiO_2 valence band (vb) lie above the CuO(cb) and CuO(vb) as shown in Figure 3. The partial recombination of e^-/h^+ in TiO_2 is hindered by the charge injection into CuO. The interfacial charge transfer (IFCT) between the TiO_2 and the CuO(vb) of $+1.4$ eV to the TiO_2(vb) at 2.5 eV vs. SCE, pH 0, and proceeds with a considerable driving force due to the large potential energy difference between these two valence band levels.

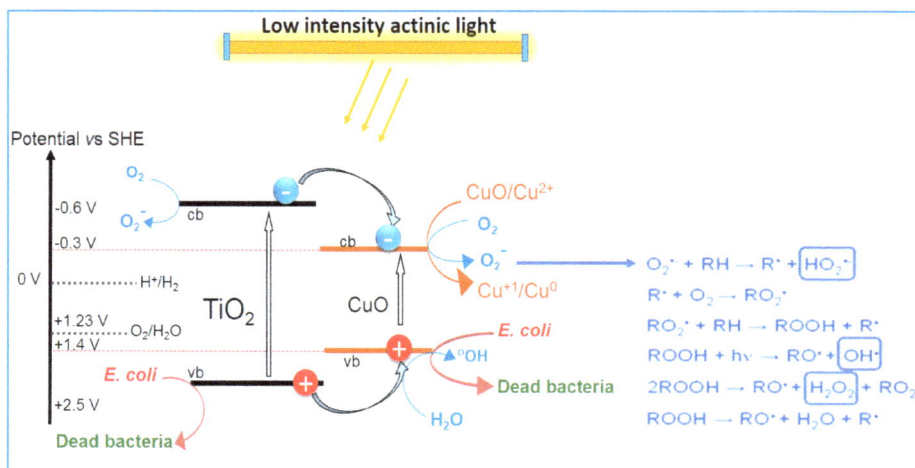

Figure 3. Diagram suggested for of bacterial inactivation under solar simulated light photocatalyzed by TiO_2/Cu films sputtered on polyester (PES) [52].

A model for the charge transfer between TiO_2 and CuO under solar light (UV-Vis) has been suggested [40,52] with respect to previous reports [56–58] and presented in Figure 3. The *E. coli* inactivation proceeded within a few minutes [29,52]. These TiO_2(vb) holes react with the surface-OH groups of the TiO_2 releasing OH-radicals. The CuO nanoparticles on the TiO_2 can be reduced to Cu_2O by the charges generated in the TiO_2 under light and can later re-oxidize to CuO as investigated by X-ray photoelectron spectroscopy (XPS) in recent report [52]. The TiO_2/Cu co-sputtered samples showed that TiO_2 plays a stabilizing effect on the Cu-release from the co-sputtered surfaces during bacterial inactivation compared to Cu deposed individually [29,52]. A low amount of Cu-released in the ppb range inactivated both Gram-positive and Gram-negative bacteria, possibly through an oligodynamic effect as observed in dark runs but needing longer times [53,59].

4. Nanostructured Fe-Oxides for Self-Sterilizing through an Oligodynamic Effect: Surface Properties

Iron oxide nanoparticles (NP's) are of considerable interest due to their wide applications in fields such as magnetic storage, medicine, chemical industries, catalytic materials and water purification [60]. The synthesis of Fe_2O_3, FeO and Fe_3O_4 involve routes including precipitation, sol-gel, hydrothermal, dry vapor deposition, has been carried out by way of micro-emulsion, electro-deposition and sonochemistry [61,62].

To avoid the time, work and reagents needed to separate the products from reactions catalyzed by suspensions at the end of water detoxification processes involving nanoparticles, metal/metal oxide coatings were prepared. Polymer-based films have been applied in protective coatings of medical devices [63,64], thin-films and bactericide/self-cleaning surfaces [64–66]. Films grafted by colloids weakly adhered to the substrate, have shown to be not entirely reproducible and can be wiped out of the polymer surface [49,66,67].

Polyethylene (PE) thin film is a flexible low cost polymer resistant to corrosion and withstands up to 120 °C for short times were coated/sputtered with FeOx [49,68]. Due to its low surface energy, the PE limits the adhesion of particles on its surface. In order to bind a higher amount of catalytic species on the PE surface suitable anchor groups have been used to attain acceptable catalysts loadings leading to the degradation of pollutants/bacteria occurring with a satisfactory kinetics. Surface pretreatment was necessary to increase the number of oxidative sites, hydrophilicity, and surface-roughness necessary for better FeOx bonding [69]. The polyethylene fabrics were pretreated in the cavity of an RF-plasma unit (13.56 MHz, 100 W, Harrick Corp., Ithaca, NY, USA) at a pressure of 1 torr. The topmost PE-layers of 2 nm (~10 atomic layers) were RF-plasma pretreated for 15 min. Oxygen RF-produced plasma reacts with the PE surface to induce groups like C-O, C=O, O-C=O, C-O-O- on the PE surface. This pretreatment introduces hydrophilic groups on the PE-surface and breaks the intermolecular PE- and the H-H bonds segmenting the PE-fibers [35,70]. The slightly positive FeOx binds the negatively charged pretreated PE (containing the overall negative carboxylic groups) through electrostatic interaction and chelation/complexation [68]. FeOx was sputtered from a target 5 cm diameter (Kurt Lesker, East Sussex, UK) positioned at 10 cm from the target by direct current magnetron sputtering (DCMS) on PE. The PE consists of highly branched low crystalline semi-transparent film with the formula $H(CH_2$-$CH_2)_n H$ (ET3112019, Goodfellow, Huntingdon, UK). The PE-FeOx mediated bacterial reduction was determined on *Escherichia coli* (*E. coli K12* ATCC23716) on 2 cm by 2 cm PE-FeOx samples under solar simulated light (52 mW/cm^2, ~0.8×10^{16} photons/s) for utilization in environmental cleaning [68,71].

Bacterial inactivation under low intensity solar simulated light on PE-FeOx sputtered films in $Ar + O_2$ atmosphere proceeded with an acceptable kinetics [68]. The PE-FeOx films avoid the use of heavy metals whose discharge into the environment is not desired or admissible by pertinent sanitary regulations. The bacterial inactivation kinetics was attributed to the redox processes occurring on the surface. The regeneration of the surface initial catalytic states was reported to happen by simply washing the surface with NaOH. The PE-FeOx properties like surface polarity, roughness and stability were described in details [68,71]. Figure 4 presents the *E. coli* inactivation under solar simulated light irradiation for PE-samples pretreated with Rf-plasma and Fe-sputtered between 30 s and 150 s. It is readily seen that Fe-sputtering for 60 s (trace 1) led to the faster bacterial reduction time. The FeOx film thickness (42 nm equivalent to 210 atomic layers) led to the shortest bacterial reduction time [68]. If one atomic layer is ~0.2 nm thick and including 10^{15} atoms/cm^2, the Fe deposition rate can be estimated as 3.5×10^{15} atoms/cm^2 × s. Sputtering for 30 s (Figure 4, trace 4) did not deposit enough FeOx on the PE (0.040 wt % Fe$_2$O$_3$/weight PE). The longer bacterial reduction time shown in Figure 4, traces 4 and 5) for samples sputtered for 120 and 150s showing higher loadings > 0.084 wt % Fe$_2$O$_3$/weight PE. This was probably due to: (i) the increase in layer thickness leading to the bulk inward diffusion of the charge carriers [59,68], (ii) the increased size of the FeOx at longer sputtering times leading to cluster agglomeration [68,71] and (iii) the increase of the Fe-metal content with respect to FeOx. The bacterial reduction on PE-FeOx films does not proceed like on PE-alone under simulated solar irradiation.

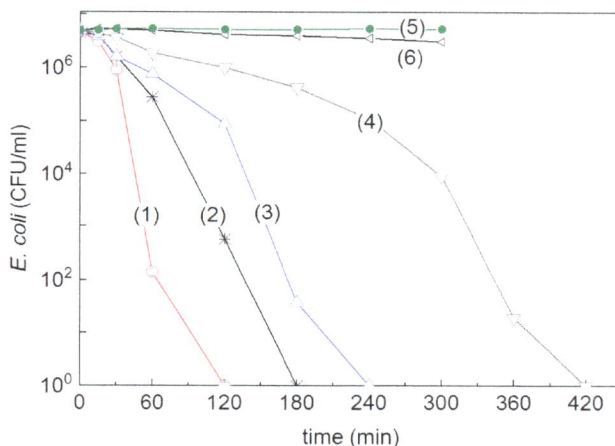

Figure 4. *E. coli* reduction on PE-FeOx pre-treated surfaces and sputtered for (1) 60 s; (2) 120 s; (3) 150 s; (4) 30 s and illuminated with solar simulated light of 52 mW/cm^2; (5) PE-FeOx sputtered for 60 s in the dark and (6) Un-sputtered PE under solar simulated light; Error bars: standard deviation (*n* = 5%) [68].

PE-FeOx mediated bacterial reduction was investigated and remained stable up to five cycles (one cycle = one bacterial inactivation run + one water washing run) [68]. The Fe-leached out in ppb amounts during *E. coli* bacterial reduction was determined by inductively coupled mass spectrometry (ICP-MS) and showed a small release of Fe-ions to the environment at below toxicity levels. PE-FeOx induced stable re-cycling during bacterial inactivation trials [68]. Evidence of ppb amounts of Fe was observed in the PE-FeOx mediated bacterial inactivation suggesting an oligodynamic effect. Until now, only heavy meals as Ag, Pt and Pd have been reported to induce bacterial inactivation through the oligodynamic effect, but, this introduces undesirable/detrimental metals into the environment [53]. Fe$_2$O$_3$ colloids have been reported to leach consistent amounts of Fe during the degradation of pollutants in aqueous suspensions. Fe$_2$O$_3$ is a stable n-type semiconductor responding to visible light up to 500 nm with a band-gap of 2.2 eV able to separate the electrons at a potential ~0.1 eV and holes at ~2.3 eV as a function of the applied light (pH 6–7) [68].

By XPS analysis, the PE-FeOx surface atomic concentration and the changes in the Fe-oxidation states during the bacterial inactivation period were followed. The initial Fe(III) in Fe$_2$O$_3$ at 712.2 eV decreases from 80.0% (time zero) to 53.0% after the disinfection period. Fe in the form of Fe$_3$O$_4$ at 713.6 eV and Fe(II) with peaks at 709.7 eV was detected before the bacterial disinfection [68]. After disinfection, Fe(III), FeO(II/III) and Fe(II) peaks were: 711.4, 708.6 and 713.8 eV, respectively.

These changes point out to the shift in the oxidation states in the Fe-oxides during bacterial reduction. The catalysis deriving the bacterial reduction contains three different FeO_x oxide species each one offering a different potential couple and its own surface potential (eigenvalue). The high turnover of the biological material on the photocatalyst surface avoids the accumulation of residual intermediates during bacterial inactivation as observed for the lack of C1s and N2p peaks after the bacterial inactivation on the PE-FeOx surface [68,70].

The electrostatic attraction between the bacteria and PE-FeOx is a dominant effect at distances below four Angstroms, polarizing strongly the interaction between the PE-FeOx within this short distance. A mechanism of reaction was recently suggested taking into consideration the bacterial inactivation dynamics and Equations (3)–(6) below. The generation of highly oxidative radicals (ROS) by the O_2 (air) reduction under the solar simulated light was observed concomitantly to the Fe(III)/Fe(II) reduction during bacterial oxidation leading to CO_2 and a small amount of mineral trace residues:

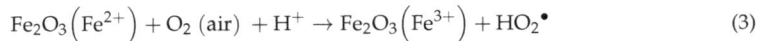

$$Fe_2O_3\left(Fe^{2+}\right) + O_2 \text{ (air) } + H^+ \rightarrow Fe_2O_3\left(Fe^{3+}\right) + HO_2^\bullet \tag{3}$$

The $HO_2^\bullet$ radical would convert to O_2^- at the biocompatible pH 6–7 through Equation (4):

$$HO_2^\bullet \Leftrightarrow H^+ + O_2^- \quad pKa = 4.8 \tag{4}$$

In the presence of Fe-ions, the $HO_2^\bullet$ decomposes at much slower kinetics compared to the fast reaction between $O_2^- + Fe^{3+}$, it is suggested to follow reactions (5) and (6) below:

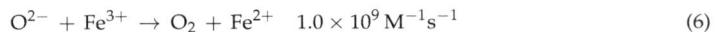

$$HO_2^\bullet + Fe^{3+} \rightarrow Fe^{2+} + H^+ + O_2 \quad 3.3 \times 10^5 \, M^{-1}s^{-1} \tag{5}$$

$$O^{2-} + Fe^{3+} \rightarrow O_2 + Fe^{2+} \quad 1.0 \times 10^9 \, M^{-1}s^{-1} \tag{6}$$

The reaction between PE-FeOx and the bacteria cell wall involves diffusion of the metal-ions and absorption/translocation of the metal-ions on the bacterial cell bilayer. Electrostatic and Van der Waals interactions and controlled diffusion of FeOx at the interface with *E. coli* cells drive the interaction between the bacteria and the catalyst/photocatalyst surfaces [68].

5. Coupling of TiO_2 and Fe-Oxide: Innovative Preparation for Self-Sterilizing Surfaces

Binary-oxides semiconductors due to their optical absorption and semiconductor behavior have been widely used for environmental decontamination purposes like FeO_x-TiO_2. These binary oxides play an important role in pollutants and bacterial abatement involving redox processes [72,73]. Supported photocatalyst films, for self-cleaning and self-sterilizing purposes have recently been developed. This works involves the grafting of narrow band-gap semiconductors increase the light absorption in the visible region [68,72–76]. Work of this kind involves the selection of the meta/oxide components taking into consideration factors like (i) the surface spectral properties; (ii) the transients generated under femto-second laser pulses induced in the visible light region (545 nm /25 fs); (iii) the bacterial inactivation kinetics of *E. coli* under low intensity solar/visible light and effect of different irradiation intensities and (iv) the mechanism of electron transfer from the FeOx used as photosensitizer to the low lying TiO_2 trapping states as recently reported [73].

Femto-second ultra-fast kinetics is a powerful tool to detect and register the initial charge separation within very short times when PE-FeOx-TiO_2 is irradiated under visible light [8]. Femto-second kinetics pulses in the visible range (545 nm/25 fs) to photo-induce and identify the short-lived transients. These transients are shown in Figure 5 at different pulse delay times as a function of wavelength [73,75]. FeOx absorbs laser pulses in the visible leading to the separation of the cb (e−) and vb(h+) or excited d-d states. Figure 5 also shows that the spectral features are about the same for different delays. An increase in the pulse delay up to 500 ps, leads to a decrease in absorption bands. Additional experiments showed that the main part of the absorption amplitude at 600 nm decays within 25 ps ($t_{\frac{1}{2}}$ ~25 ps). Mid-gap Fe d-d states have been suggested as the main trapping sites with lifetimes of a few hundred picoseconds. The d-d states are ascribed to local excitons in the FeOx matrix [73]. In Figure 5, the electron trapping process was initiated at times of 150 fs and compete with electron-hole recombination. The absorption band with a maximum at 600 nm was assigned to the cb(e−) spectrum of the FeOx-TiO$_2$ i.e., an IFCT process occurring at the heterojunction [75,76]. These ultra short-lived transients in the visible region lead later to longer-lived intermediates in the minute range able to inactivate bacteria under band-gap continuous irradiation [73–75].

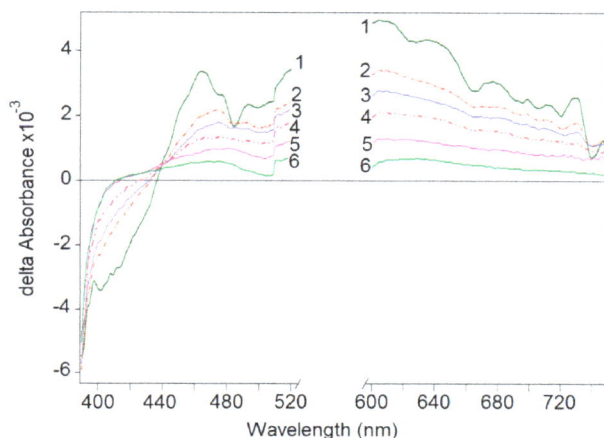

Figure 5. Transient spectra of the FeOx-TiO$_2$-PE as a function of wavelength after femtosecond laser pulse 25 fs (544 nm). Time delays: (1) 150 fs; (2) 500 fs; (3) 1 ps; (4) 3 ps; (5) 10 ps; (6) 500 ps [73].

6. Conclusions

The preparation of TiO$_2$, TiO$_2$-TiO$_2$, FeOx and FeOx-TiO$_2$ thin films and their dynamics have been reviewed. The bacterial interactions with the film surface in the dark and under light conditions were discussed. The reactivity of TiO$_2$-Cu towards bacteria seems to proceed through mechanisms that are still controversial. TiO$_2$-Cu films obtained by sputtering were described as well as their surface properties. The mechanism of bacterial inactivation by Cu, TiO$_2$ and FeOx possibly involve an oligodynamic effect. Innovative PE-FeOx composites/coatings may have a potential application in disinfection for biomedical devices favored by the low Fe-cytotoxicity and high Fe-biocompatibility compared to heavy metals like Ag, Pt, and Au. The co-sputtered FeOx-TiO$_2$-PE films show also a potential to improve the removal of pathogens and prevent biofilm formation under sun/visible light.

Acknowledgments: We thank the EPFL and Swiss National Science Foundation (SNF) Project (200021-143283/1) for financial support.

Conflicts of Interest: Authors declare no conflict of interest.

References

1. Soe, M.M.; Gould, C.V.; Pollock, D.; Edwards, J. Targeted Assessment for Prevention of Healthcare-Associated Infections: A New Prioritization Metric. *Infect. Control Hosp. Epidemiol.* **2015**, *36*, 12. [CrossRef] [PubMed]

2. Domek, M.J.; Chevalier, M.W.; Cameron, S.C.; McFeters, G.A. Evidence for the role of copper in the injury process of coliform bacteria in drinking water. *Appl. Environ. Microbiol.* **1984**, *48*, 289–293. [PubMed]

3. Noyce, J.O.; Michels, H.; Keevil, C.W. Potential use of copper surfaces to reduce survival of epidemic meticillin-resistant *Staphylococcus aureus* in the healthcare environment. *J. Hosp. Infect.* **2006**, *63*, 289–297. [CrossRef] [PubMed]

4. Wilks, S.A.; Michels, H.; Keevil, C.W. The survival of *Escherichia coli* O157 on a range of metal surfaces. *Int. J. Food Microbiol.* **2005**, *105*, 445–454. [CrossRef] [PubMed]

5. Noimak, S.; Dunnill, Ch.; Wilson, M.; Parkin, I.P. The role of surfaces in catheter-associated infections. *Chem. Soc. Rev.* **2009**, *38*, 3435–3448. [CrossRef] [PubMed]

6. Thurman, R.B.; German, C.P. The molecular mechanisms of copper and silver ion disinfection of bacteria and viruses. *Crit. Rev. Environ. Control* **1989**, *18*, 259–315. [CrossRef]

7. Borkow, G.; Gabbay, J. Copper as abiocidal tool. *Curr. Med. Chem.* **2005**, *12–13*, 2163–2170. [CrossRef]

8. Page, K.; Wilson, M.; Parkin, I. Antimicrobial surfaces and their potential in reducing the role of the inanimate environment in the incidence of hospital-acquired infections. *J. Mater. Chem.* **2009**, *19*, 3818–3831. [CrossRef]

9. Monroe, D. Looking for chinks in the armor of bacterial biofilms. *PLoS Biol.* **2007**, *5*, e307. [CrossRef] [PubMed]

10. Haenle, M.; Fritshe, M.; Zietz, C.; Bader, R.; Heidenau, F. An extended spectrum bactericidal titanium dioxide (TiO$_2$) coating for metallic implants: In vitro effectiveness against MRSA and mechanical properties. *J. Mater. Sci.* **2011**, *22*, 381–387. [CrossRef] [PubMed]

11. Heidenau, F.; Mittelmeier, W.; Detsch, R.; Haenle, M.; Stenzel, F.; Ziegler, G.; Gollwitzer, H. A novel antibacterial Titania coating: Metal ion toxicity and in vitro surface colonization. *J. Mater. Sci.* **2005**, *16*, 883–889. [CrossRef] [PubMed]

12. Yates, H.M.; Brook, L.A.; Ditta, I.B.; Evans, P.; Foster, H.A.; Sheel, D.W.; Steele, A. Photo-induced self-cleaning and biocidal behviour of titania and copper oxide multilayers. *J. Photochem. Photobiol. A* **2008**, *197*, 187–205. [CrossRef]

13. Foster, H.A.; Sheel, P.; Sheel, W.D.; Evans, P.; Varghese, S.; Rutschke, N.; Yates, M.H. Antimicrobial activity of titatnia/silver and titania/copper films prepared by CVD. *J. Photochem. Photobiol. A* **2010**, *216*, 283–289. [CrossRef]

14. Frise, A.P. Choosing disinfectants. *J. Hosp. Infect.* **1999**, *43*, 255–264. [CrossRef]

15. Neely, A.N.; Maley, M.P. Survival of Enterococci and Staphylococci on Hospital Fabrics and Plastic. *J. Clin. Microbiol.* **2000**, *38*, 724–726. [PubMed]

16. Cohen, H.A.; Amir, J.; Matalon, A.; Mayan, R.; Beni, S.; Barzilai, A. Stethoscopes and otoscopes-a potential vector of infection? *Fam. Pract.* **1997**, *14*, 446–449. [CrossRef] [PubMed]

17. Boyce, J.; Potter-Bynoe, G.; Chevenert, C.; King, T. Environmental Contamination Due to Methicillin-Resistant *Staphylococcus aureus* Possible Infection Control Implications. *Infect. Control Hosp. Epidemiol.* **1997**, *18*, 622–627. [CrossRef] [PubMed]

18. Bhalla, A.; Pultz, N.; Gries, D.; Ray, A.; Eckstein, E.; Aron, D.; Donskey, C. Acquisition of Nosocomial Pathogens on Hands after Contact With Environmental Surfaces Near Hospitalized Patients. *Infect. Control Hosp. Epidemiol.* **2004**, *25*, 164–167. [CrossRef] [PubMed]

19. Ding, X.; Liu, Z.; Su, J.; Yan, D. Human serum inhibits adhesion and biofilm formation in Candida albicans. *BMC Microbiol.* **2014**, *14*, 80. [CrossRef] [PubMed]

20. Vogt, S.; Kühn, K.-D.; Ege, W.; Pawlik, K.; Schnabelrauch, M. Novel Polylactide-Based Release Systems for Local Antibiotic Therapies. *Materialwiss. Werkstofftech.* **2003**, *34*, 1041–1047. [CrossRef]

21. Ginalska, G.; Osinska, M.; Uryniak, A.; Urbanik-Sypniewska, T.; Belcarz, A.; Rzeski, W.; Wolski, A. Antibacterial activity of gentamicin-bonded gelatin-sealed polyethylene terephtalate vascular prostheses. *Eur. J. Vasc. Endovasc. Surg.* **2005**, *29*, 419–424. [CrossRef] [PubMed]

22. Weaver, L.; Noyce, J.O.; Michels, H.T.; Keevil, C.W. Potential action of copper surfaces on meticillin-resistant *Staphylococcus aureus*. *J. Appl. Microbiol.* **2010**, *109*, 2200–2205. [CrossRef] [PubMed]

23. Torres, A.; Ruales, C.; Pulgarin, C.; Aimable, A.; Bowen, P.; Kiwi, J. Enhanced Inactivation of E. coli by RF-plasma Pretreated Cotton/CuO (65 m^2/g) under Visible Light. *ACS Appl. Mater. Interfaces* **2010**, *2*, 2547–2552. [CrossRef] [PubMed]

24. Rio, L.; Kusiak, E.; Kiwi, J.; Pulgarin, C.; Trampuz, A.; Bizini, A. Comparative methods to evaluate the bactericidal activity of copper-sputtered surfaces against methicillin-resistant *Staphylococcus aureus*. *J. Appl. Microbiol.* **2012**, *78*, 8176–8182. [CrossRef] [PubMed]

25. Mejia, M.I.; Restrepo, G.; Marin, J.M.; Sanjines, R.; Pulgarin, C.; Mielczarski, E.; Mielczarski, J.; Kiwi, J. Magnetron-Sputtered Ag-Modified Cotton Textiles Active in the Inactivation of Airborne Bacteria. *ACS Appl. Mater. Interfaces* **2010**, *2*, 230–235. [PubMed]

26. Baghriche, O.; Kiwi, J.; Pulgarin, C.; Sanjines, R. Antibacterial Ag–ZrN surfaces promoted by subnanometric ZrN-clusters deposited by reactive pulsed magnetron sputtering. *J. Photochem. Photobiol. A* **2012**, *229*, 39. [CrossRef]

27. Polak, M.; Ohl, A.; Quaas, M.; Lukowski, G.; Lüthen, F.; Weltmann, K.-D.; Schröder, K. Oxygen and Water Plasma-Immersion Ion Implantation of Copper into Titanium for Antibacterial Surfaces of Medical Implants. *Adv. Eng. Mater.* **2010**, *12*, B511–B518. [CrossRef]

28. Finke, B.; Polak, M.; Hempel, F.; Schroeder, K.; Lukowski, G.; Müller, W.D.; Weltmann, K.D. Electrochemical Assessment of Cu-PIII Treated Titanium Samples for Antimicrobial Surfaces. *Mater. Sci. Forum* **2011**, *706–709*, 478–483. [CrossRef]

29. Ritmi, S.; Baghriche, O.; Sanjines, R.; Pulgarin, C.; Bensimon, M.; Lavanchy, J.-C.; Kiwi, J. Growth of TiO$_2$/Cu Films by HIPIMS for Accelerated Bacterial Loss of Viability. *Surf. Coat. Technol.* **2013**, *232*, 804. [CrossRef]

30. Baghriche, O.; Rtimi, S.; Pulgarin, C.; Sanjines, R.; Kiwi, J. Innovative TiO$_2$/Cu surfaces inactivating bacteria < 5 min under low intensity visible/actinic light. *ACS Appl. Mater. Interfaces* **2012**, *4*, 5234–5240. [PubMed]

31. Rtimi, S.; Sanjines, R.; Pulgarin, C.; Kiwi, J. Accelerated *Esherichia coli* inactivation in the Dark on Uniform Copper Flexible Surfaces. *Biointerphases* **2014**, *9*, 029012. [CrossRef] [PubMed]

32. Rtimi, S.; Ballo, M.; Pulgarin, C.; Entenza, J.; Bizzini, A.; Kiwi, J. Duality in the *Escherichia coli* and Methicillin Resistant *Staphylococcus aureus* Reduction Mechanism under Actinic Light on Innovative Co-sputtered Surfaces. *Appl. Catal. A* **2015**, *498*, 4185–4191. [CrossRef]

33. Rtimi, S.; Sanjines, R.; Pulgarin, C.; Kiwi, J. Quasi-Instantaneous Bacterial Inactivation on Cu-Ag Nano-particulate 3D-Catheters in the Dark and Under Light: Mechanism and Dynamics. *ACS Appl. Mater. Interfaces* **2016**, *8*, 47–55. [CrossRef] [PubMed]

34. Rtimi, S.; Sanjines, R.; Pulgarin, C.; Kiwi, J. Microstructure of Cu-Ag Uniform Nanoparticulate Films on Polyurethane 3D-catheters: Surface Properties. *ACS Appl. Mater. Interfaces* **2016**, *8*, 56–64. [CrossRef] [PubMed]

35. Rtimi, S.; Sanjines, R.; Andrzejczuk, M.; Pulgarin, C.; Kulik, A.; Kiwi, J. Innovative transparent non-scattering TiO$_2$ bactericide thin films inducing increased *E. coli* cell wall fluidity. *Surf. Coat. Technol.* **2014**, *254*, 333–343. [CrossRef]

36. Zhang, H.; Liu, P.; Liu, X.; Zhang, S.; Yao, X.; An, T.; Amal, R.; Zhao, H. Fabrication of Highly Ordered TiO$_2$ Nanorod/Nanotube Adjacent Arrays for Photo-electrochemical Applications. *Langmuir* **2010**, *26*, 11226–11232. [CrossRef] [PubMed]

37. Gunawan, C.; Teoh, W.Y.; Marquis, C.P.; Amal, R. Induced Adaptation of *Bacillus* sp. to Antimicrobial Nanosilver. *Small* **2013**, *9*, 3554–3560. [CrossRef] [PubMed]

38. Kramer, A.; Schwebke, I.; Kampf, G. How long do nosocomial pathogens pathogens persist on inanimate surfaces? A systematic reviews. *BMC Infect. Dis.* **2006**, *6*, 130–139. [CrossRef] [PubMed]

39. Dancer, S.J. The role of environmental cleaning in the control of hospital-acquired infections. *J. Hosp. Infect.* **2009**, *73*, 378–385. [CrossRef] [PubMed]

40. Rtimi, S.; Kiwi, J.; Pulgarin, C. Accelerated Antibacterial Inactivation on 2D Cu-Titania Surfaces: Latest Developments and Critical Issues. *Coatings* **2017**, *7*, 20. [CrossRef]

41. Rtimi, S.; Sanjines, R.; Pulgarin, C.; Kiwi, J. Effect of Light and Oxygen on repetitive bacterial inactivation on uniform, adhesive, robust and stable Cu-polyester surfaces. *J. Adv. Oxid. Technol.* **2017**, *20*, 20160178 . [CrossRef]

42. Rtimi, S.; Giannakis, S.; Bensimon, M.; Pulgarin, C.; Sanjines, R.; Kiwi, J. Supported TiO$_2$ films deposited at different energies: Implications of the surface compactness on the catalytic kinetics. *Appl. Catal. B* **2016**, *191*, 42–52. [CrossRef]

43. Kelly, P.J.; Arnell, R.D. Magnetron sputtering: A review of recent developments and applications. *Vacuum* **2000**, *65*, 159–172. [CrossRef]

44. Farahani, N.; Kelly, P.J.; West, G.; Ratova, M.; Hill, C.; Vishnyakov, V. Photocatalytic activity of reactively sputtered and directly sputtered titania coatings. *Thin Solid Films* **2011**, *520*, 91464–91469. [CrossRef]

45. Ratova, M.; West, G.T.; Kelly, P.J. Optimisation of HIPIMS photocatalytic titania coatings for low temperature deposition. *Surf. Coat. Technol.* **2014**, *250*, 7–13. [CrossRef]

46. Fisher, L.; Ostovapour, S.; Kelly, P.; Whitehead, K.A.; Cooke, K.; Storgards, E.; Verran, J. Molybdenum doped titanium dioxide photocatalytic coatings for use as hygienic surfaces: The effect of soiling and antimicrobial activity. *Biofouling* **2014**, *30*, 911–919. [CrossRef] [PubMed]

47. Qiu, X.; Miyauchi, M.; Sunada, K.; Minoshima, M.; Liu, M.; Lu, Y.; Li, D.; Shimodaira, Y.; Hosogi, Y.; Kuroda, Y.; et al. Hybrid CuxO/TiO$_2$ Nanocomposites As Risk-Reduction Materials in Indoor Environments. *ACS Nano* **2012**, *6*, 1609–1618. [CrossRef] [PubMed]

48. Sunada, K.; Watanabe, T.; Hashimoto, K. Bactericidal Activity of Copper-Deposited TiO$_2$ Thin Film under Weak UV Light Illumination. *Environ. Sci. Technol.* **2003**, *37*, 4785–4789. [CrossRef] [PubMed]

49. Zhang, L.; Dillert, R.; Bahnemann, D.; Vormoor, M. Photo-induced hydrophilicity and self-cleaning: Models and reality. *Energy Environ. Sci.* **2012**, *5*, 7491–7507. [CrossRef]

50. Rtimi, S.; Pulgarin, C.; Sanjines, R.; Kiwi, J. Innovative semi-transparent nanocomposite films presenting photo-switchable behavior and leading to a reduction of the risk of infection under sunlight. *RSC Adv.* **2013**, *3*, 16345–16348. [CrossRef]

51. Rtimi, S.; Baghriche, O.; Sanjines, R.; Pulgarin, C.; Bensimon, M.; Kiwi, J. TiON and TiON-Ag sputtered surfaces leading to bacterial inactivation under indoor actinic light. *J. Photochem. Photobiol. A* **2013**, *256*, 52–63. [CrossRef]

52. Rtimi, S.; Pulgarin, C.; Robyr, M.; Aybush, A.; Shelaev, I.; Gostev, F.; Nadtochenko, V.; Kiwi, J. Insight into the catalyst/photocatalyst microstructure presenting the same composition but leading to a variance in bacterial reduction under indoor visible light. *Appl. Catal. B* **2017**, *208*, 135–147. [CrossRef]

53. Rtimi, S.; Pascu, M.; Sanjines, R.; Pulgarin, C.; Ben-Simon, M.; Bouas, A.; Lavanchy, J.-C.; Kiwi, J. ZrNO–Ag co-sputtered surfaces leading to *E. coli* inactivation under actinic light: Evidence for the oligodynamic effect. *Appl. Catal. B* **2013**, *138–139*, 113–121. [CrossRef]

54. Rossnagel, S.; Hopwood, J. Magnetron sputter deposition with high levels of metal ionization. *Appl. Phys. Lett.* **1993**, *63*, 3285–3287. [CrossRef]

55. Ehasarian, P.A. High-power impulse magnetron sputtering and its applications. *Pure Appl. Chem.* **2010**, *82*, 1247–1258. [CrossRef]

56. Irie, H.; Kamiya, K.; Shibanuma, T.; Miura, S.; Tryck, D.; Yo koyama, T.; Hashimoto, K. Visible light sensitive Cu(II)-grafted TiO$_2$ photocatalysts: Activities and X-ray absorption fine structure analyses. *J. Phys. Chem. C* **2009**, *113*, 10671–10766. [CrossRef]

57. Irie, H.; Miura, S.; Kamiya, K.; Hashimoto, K. Efficient visible light-sensitive photocatalysts: Grafting Cu(II) onto TiO$_2$ and WO$_3$ photocatalysts. *Chem. Phys. Lett.* **2008**, *457*, 202–205. [CrossRef]

58. Ishiguro, H.; Yao, Y.; Nakano, R.; Hara, M.; Sunada, K.; Hashimoto, K.; Kajioka, J.; Fujishima, A.; Kubota, Y. Photocatalytic activity of Cu^{2+}/TiO$_2$-coated cordierite foam inactivates bacteriophages and Legionella pneumophila. *Appl. Catal. B* **2013**, *129*, 56–61. [CrossRef]

59. Rtimi, S.; Pulgarin, C.; Baghriche, O.; Kiwi, J. Accelerated inactivation obtained by HIPIMS sputtering on low cost surfaces with concomitant reduction in the metal-semiconductor content. *RSC Adv.* **2013**, *3*, 13127–13130. [CrossRef]

60. Mohapatr, M.; Anand, S. Synthesis and applications of nano-structured iron oxides/hydroxides—A review. *Int. J. Eng. Sci. Technol.* **2010**, *2*, 127–146. [CrossRef]

61. Fujishima, A.; Zhang, X.; Tryk, D.A. TiO$_2$ photocatalysis and related surface phenomena. *Surf. Sci. Rep.* **2008**, *63*, 515–582. [CrossRef]

62. Daoud, W. *Self-Cleaning Materials and Surfaces*; John Wiley and Sons Ltd.: Chichester, UK, 2013.

63. Wu, W.; He, Q.; Jiang, C. Magnetic Iron Oxide Nanoparticles: Synthesis and Surface Functionalization Strategies. *Nanoscale Res. Lett.* **2008**, *3*, 397–415. [CrossRef] [PubMed]

64. Naka, K.; Narita, A.; Tanaka, H.; Chujo, Y.; Morita, M.; Inubushi, T.; Nishimura, I.; Hiruta, J.; Shibayama, H.; Koga, M.; et al. Biomedical applications of imidazolium cation-modified iron oxide nanoparticles. *Polym. Adv. Technol.* **2008**, *19*, 1421–1429. [CrossRef]

65. Levy, M.; Wilhelm, C.; Siaugue, J.-M.; Horner, O.; Bacri, J.-C.; Gazeau, F. Magnetically induced hyperthermia: Size-dependent heating power of γ-Fe$_2$O$_3$ nanoparticles. *J. Phys.* **2008**, *20*, 204133. [CrossRef] [PubMed]

66. Mukh-Qasem, R.; Gedanken, R. Sonochemical synthesis of stable hydrosol of Fe$_3$O$_4$ nanoparticles. *J. Colloid Interface Sci.* **2005**, *284*, 489–494. [CrossRef] [PubMed]

67. Zhou, L.; He, B.; Huang, J. One-Step Synthesis of Robust Amine- and Vinyl-Capped Magnetic Iron Oxide Nanoparticles for Polymer Grafting, Dye Adsorption, and Catalysis. *ACS Appl. Mater. Interfaces* **2013**, *5*, 8678–8685. [CrossRef] [PubMed]

68. Rtimi, S.; Pulgarin, C.; Sanjines, R.; Kiwi, J. Novel FeOx–polyethylene transparent films: Synthesis and mechanism of surface regeneration. *RSC Adv.* **2015**, *5*, 80203–80211. [CrossRef]

69. Kever, K.; Imlay, J.A. Superoxide accelerates DNA damage by elevating free-iron levels. *Proc. Natl. Acad. Sci. USA* **1996**, *93*, 13635–13640.

70. Baghriche, O.; Rtimi, S.; Pulgarin, C.; Roussel, C.; Kiwi, J. RF-plasma pretreatment of surfaces leading to TiO_2 coatings with improved optical absorption and OH-radical production. *Appl. Catal. B* **2013**, *130–131*, 65–72. [CrossRef]

71. Rtimi, S.; Gulin, A.; Sanjines, R.; Pulgarin, C.; Nadtochenko, V.; Kiwi, J. Innovative self-sterilizing transparent Fe-phosphate polyethylene films under visible light. *RSC Adv.* **2016**, *6*, 77066–77074. [CrossRef]

72. Rtimi, S.; Robyr, M.; Pulgarin, C.; Lavanchy, J.-C.; Kiwi, J. A New Perspective in the Use of FeOx-TiO_2 Photocatalytic Films: Indole Degradation in the Absence of Fe-Leaching. *J. Catal.* **2016**, *342*, 184–192. [CrossRef]

73. Rtimi, S.; Sanjines, R.; Kiwi, J.; Pulgarin, C.; Bensimon, M.; Khmel, I.; Nadtochenko, V. Innovative photocatalyst (FeOx-TiO_2): Transients induced by Femtosecond laser leading to bacterial inactivation under visible light. *RSC Adv.* **2015**, *5*, 101751–101759. [CrossRef]

74. Rtimi, S. Indoor Light Enhanced Photocatalytic Ultra-Thin Films on Flexible Non-Heat Resistant Substrates Reducing Bacterial Infection Risks. *Catalysts* **2017**, *7*, 57. [CrossRef]

75. Rtimi, S.; Pulgarin, C.; Nadtochenko, V.A.; Gostev, F.E.; Shelaev, I.V.; Kiwi, J. FeOx-TiO_2 Film with Different Microstructures Leading to Femtosecond Transients with Different Properties: Biological Implications under Visible Light. *Sci. Rep.* **2016**, *6*, 30113. [CrossRef] [PubMed]

76. Rtimi, S.; Sanjines, R.; Pulgarin, C.; Houas, A.; Lavanchy, J.-C.; Kiwi, J. Coupling of narrow and wide band-gap semiconductors on uniform films active in bacterial disinfection under low intensity visible light: Implications of the interfacial charge transfer (IFCT). *J. Hazard. Mater.* **2013**, *260*, 860–868. [CrossRef] [PubMed]

molecules

MDPI

Article

Photocatalytic and Adsorption Performances of Faceted Cuprous Oxide (Cu$_2$O) Particles for the Removal of Methyl Orange (MO) from Aqueous Media

Weng Chye Jeffrey Ho [1,*], Qiuling Tay [1], Huan Qi [1], Zhaohong Huang [2], Jiao Li [3] and Zhong Chen [1,*]

[1] School of Materials Science and Engineering, Nanyang Technological University, Singapore 639798, Singapore; qltay@ntu.edu.sg (Q.T.); QIHU0002@ntu.edu.sg (H.Q.)
[2] Singapore Institute of Manufacturing Technology, 71 Nanyang Drive, Singapore 638075, Singapore; zhhuang@SIMTech.a-star.edu.sg
[3] School of Materials Science and Engineering, Shandong University of Technology, Zibo 255049, Shandong Province, China; haiyan9943@163.com
* Correspondence: jeffreychemical@gmail.com (W.C.J.H.); aszchen@ntu.edu.sg (Z.C.);
  Tel.: +65-96464025 (W.C.J.H.); +65-67904256 (Z.C.)

Academic Editor: Pierre Pichat
Received: 25 February 2017; Accepted: 19 April 2017; Published: 23 April 2017

Abstract: Particles of sub-micron size possess significant capacity to adsorb organic molecules from aqueous media. Semiconductor photocatalysts in particle form could potentially be utilized for dye removal through either physical adsorption or photo-induced chemical process. The photocatalytic and adsorption capabilities of Cu$_2$O particles with various exposed crystal facets have been studied through separate adsorption capacity test and photocatalytic degradation test. These crystals display unique cubic, octahedral, rhombic dodecahedral, and truncated polyhedral shapes due to specifically exposed crystal facet(s). For comparison, Cu$_2$O particles with no clear exposed facets were also prepared. The current work confirms that the surface charge critically affects the adsorption performance of the synthesized Cu$_2$O particles. The octahedral shaped Cu$_2$O particles, with exposed {111} facets, possess the best adsorption capability of methyl orange (MO) dye due to the strongest positive surface charge among the different types of particles. In addition, we also found that the adsorption of MO follows the Langmuir monolayer mechanism. The octahedral particles also performed the best in photocatalytic dye degradation of MO under visible light irradiation because of the assistance from dye absorption. On top of the photocatalytic study, the stability of these Cu$_2$O particles during the photocatalytic processes was also investigated. Cu(OH)$_2$ and CuO are the likely corrosion products found on the particle surface after the photocorrosion in MO solution. By adding hole scavengers in the solution, the photocorrosion of Cu$_2$O was greatly reduced. This observation confirms that the photocatalytically generated holes were responsible for the photocorrosion of Cu$_2$O.

Keywords: cuprous oxide (Cu$_2$O); adsorption; photocatalytic performance; methyl orange (MO); dye degradation; photocorrosion

---

1. Introduction

One of the main issues that human society faces in the 21st century is pollution of water and air caused by the increasing industrial activities and human consumption of natural resources including minerals, coal, petroleum and their derivatives. Pollution poses serious threat to not only human health, but also marine life, which in turns affects livelihood of human being through the food chains.

Membrane technology provides an effective way to extract clean water from polluted water, however the equipment and processing cost is high, as the technology requires a large amount of electricity and space. Alternative cost-effective approaches are needed to not only remove, but also, ideally, chemically decompose the organic pollutants in wastewater. Semiconductor photocatalysts offer such a possibility.

Since the discovery of hydrogen production through photocatalytic reactions by Fujishima and Honda [1] in 1972, enormous efforts have been devoted to its development [2,3]. Besides hydrogen evolution, photocatalysts are also capable of purifying wastewater, simply by breaking down the organic molecules using chemical reactions energized by the renewable solar light [4–6]. Devising suitable semiconductor photocatalysts for the decomposition of organic compounds using solar energy has been a major topic in the research community in recent decades. Substantial progress has been made, and some of the recent development has been summarized in [7–12].

Nanoparticulated photocatalysts offer clear advantages in this regard, as they possess a larger specific surface area for photo absorption and more active sites for the catalytic reactions [13]. If the surface of these particles is properly engineered, these particles can also physically adsorb the pollutant molecules from wastewater [14]. Therefore, semiconductors in particle form can be potentially applied to remove organic pollutants from aqueous solution through either chemical (photocatalytic) reactions, or physical adsorption, or both. It has been demonstrated that photocatalysis and adsorption can be applied synergistically to speed up the removal of pollutants from aqueous media [4,6,14]. This also implies that, when assessing the photocatalytic activity based on pollutant removal from the aqueous solution (typically by measuring the waste molecule concentration in water), care has to be taken to evaluate the effectiveness from photocatalytic action and adsorption separately.

Cuprous oxide (Cu_2O), due to its relatively narrow band gap among semiconductor metal oxides, has received lots of attention in recent years as visible light photocatalyst [15–18]. Particularly, faceted single crystal Cu_2O particles have been synthesized from Cu salts using different type of reducing agents. Their surface, electrical and catalytic properties have been studied. It has been acknowledged that particles with different exposed facets possess different physical and chemical properties, thus the simple solution process for crystal facet-control provides an attractive means to improve the material's properties. However, thus far the reported photocatalytic activity among different Cu_2O facets differs depending on the type of organic pollutants used as well as the experimental conditions. Huang et al. [15] prepared cubic, cuboctahedra, truncated octahedral, octahedral, and multipod structured Cu_2O nanocrystals and compared their photocatalytic activities. They found that octahedra and hexapods with the {111} exposed facets are catalytically most active in photocatalytically decomposing negatively charged molecules such as methyl orange (MO), while cubes with only the {100} faces are not photocatalytically active. In one of the following papers from the same group [16], rhombic dodecahedra exposing only the {110} facets were reported to exhibit the best photocatalytic activity among cubic, face-raised cubic, edge- and corner-truncated octahedral, all-corner-truncated rhombic dodecahedral, {100}-truncated rhombic dodecahedral, and rhombic dodecahedral Cu_2O particles. In both cases, the authors attributed the activity to the high number of surface copper atoms on the exposed facets [15,16].

When nano or sub-micron sized particles are used, it is important to study both adsorption and photocatalytic performances in order to reach a comprehensive conclusion. In the current work, we also chose the commonly used organic dye, MO, as the model pollutant. Cu_2O particles with different types of exposed facets were synthesized and compared for their adsorption and visible light photocatalytic activities. Since Cu_2O is well known for its instability under photocatalytic testing condition [19], comparison was also made on the photo stability of these particles. The mechanism for the photocorrosion was investigated based on corrosion product analysis.

2. Results and Discussion

2.1. Morphology and Crystal Structure of the Synthesized Cu_2O and Surface Area

The average size of the Cu_2O particles with standard deviation is summarized in Table 1. When 0.01 M of NaOH was used, porous Cu_2O particles of about 0.8 μm in size were obtained (Figure 1a). The particles consisted of agglomerated (non-separable) smaller nanoparticles with mean size around 40 nm. The small nanoparticles have no specific exposed facets, and the agglomerated cluster has an equal axial shape. For convenience, this morphology is denoted as spherical Cu_2O particles. When NaOH concentration increased from 0.01 to 0.03 M, the particle was transformed into cubic shape with six {100} facets exposed (Figure 1b). To form the cubic shape, the growth rate in the <100> direction must be the slowest among all other growth directions. This was achieved by an increased NaOH concentration together with the addition of ascorbic acid. The obtained cubic Cu_2O particles show uniform size and sharp edges.

Table 1. Size and specific surface area of the synthesized Cu_2O particles.

	Spherical	Cubic	Octahedral	Rhombic Dodecahedral	Truncated
Size (μm)	0.80 ± 0.15	0.75 ± 0.37	0.65 ± 0.08	0.80 ± 0.157	5.0 ± 0.37
Surface area (m^2/g)	3.95	1.13	1.13	0.34	Not available

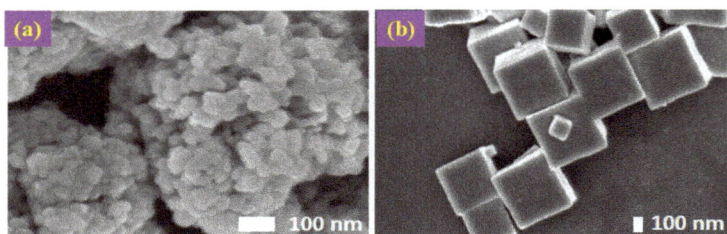

Figure 1. Field Emission Scanning Electron Microscopy (FESEM) images of Cu_2O particles: (**a**) Cu_2O of agglomerated spherical shape with no clear exposed facets; and (**b**) Cu_2O of cubic shape with exposed {100} facets.

Figure 2 shows the Field Emission Scanning Electron Microscopy (FESEM) image of well-defined, smooth and uniform octahedral shaped Cu_2O particles. Each particle has eight exposed {111} facets. The synthesis method was similar to the formation of the cubic shape, except that ascorbic acid was replaced by a stronger reducing agent, hydrazine hydrate.

Figure 2. FESEM images of octahedral shaped Cu_2O particles exposing the {111} facets.

Figure 3 shows flat and smooth surfaces of rhombic dodecahedral Cu_2O with 12 {110} facets. Both oleic acid and glucose were used in the synthesis. Apparently, they are able to reduce the growth along the <110> direction, resulting the exposed facets.

Figure 3. FESEM images of particles in rhombic dodecahedral shape with exposed {110} facets.

Figure 4 shows the truncated polyhedral Cu_2O particles with 26 facets. Among them, eight are the {111} facets (indicated in red) and six are the {100} facets (indicated in black). The remaining 12 are the {110} facets (indicated in pink). Due to the high concentration of the precursors used and longer cooling time (1.5 h), the size of the smooth and uniform truncated polyhedral Cu_2O particle was seen to be relatively large (5 μm) compared to the other particles. It was observed that by lowering the concentration of the precursors and shortening the cooling time, some irregular cubic shaped Cu_2O particles were obtained with an average size of 2 μm (not shown here). Therefore, both factors, the concentration and the cooling rate, are important to synthesize the truncated polyhedral Cu_2O particles.

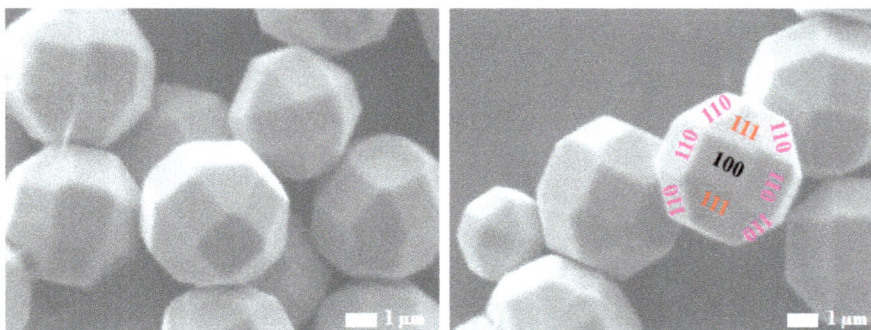

Figure 4. FESEM images of the truncated polyhedral Cu_2O particles, exposing eight {111} facets, six {100} facets and 12 {110} facets.

Figure 5 shows the nitrogen adsorption curves for the Cu_2O samples. The Brunauer–Emmett–Teller (BET) surface area is summarized in Table 1. Not surprisingly, the spherical shaped Cu_2O yields the highest surface area. The one for the truncated polyhedral Cu_2O could not be measured, due to the large size of the particles exceeding the detection limit of the equipment (~2 μm).

The crystal structure of various shaped Cu_2O synthesized by the wet chemical method is confirmed by the X-ray diffraction (XRD) pattern shown in Figure 6. The diffraction peaks labeled with

"★" matches well with the standard data of Cu_2O (JCPDS No. 05-0667), confirming the formation of a single cubic phase structure. All the peaks were perfectly indexed to crystalline Cu_2O not only in their peak positions (2θ values of $29.60°$, $36.52°$, $42.44°$, $61.40°$ and $73.55°$), but also in their relative intensity. It was observed that the XRD peaks for all the faceted samples were relatively sharp, due to the good crystallinity of the crystals. However, the crystallinity for the spherical shaped Cu_2O particles was poor, as indicated by the weak and broad peaks.

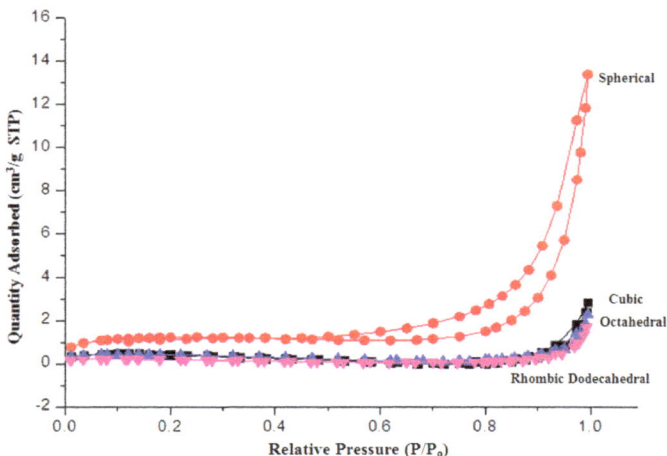

Figure 5. Isotherm N_2 adsorption of Cu_2O particles. STP: Standard Temperature and Pressure.

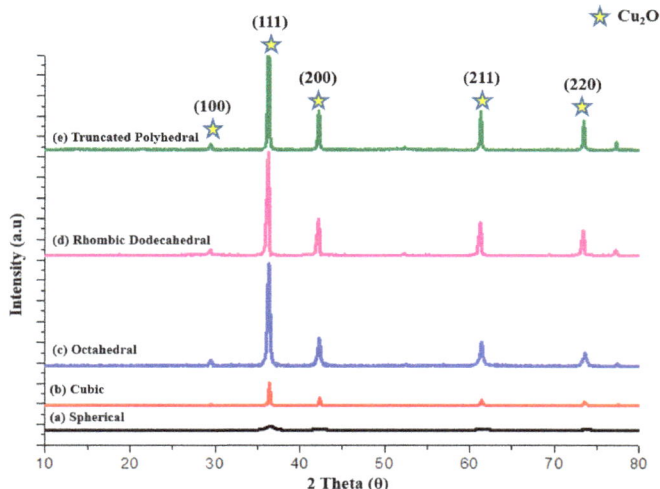

Figure 6. The X-ray diffraction (XRD) patterns of Cu_2O particles: (**a**) spherical; (**b**) cubic; (**c**) octahedral; (**d**) rhombic dodecahedral; and (**e**) truncated polyhedral.

2.2. FTIR and Zeta Potential Analyses

The Fourier Transform Infrared (FTIR) spectra for all the five samples synthesized are shown in Figure 7. Though the spectrum was run from 400 to 4000 cm^{-1}, the bands from metal oxide are generally below 1000 cm^{-1} [20]. For Cu$_2$O, the bands that lie at 1130, 798 and 621 cm^{-1} were attributed to the stretching vibration of Cu–O in Cu$_2$O. Thus, the results obtained have proven the existence of Cu$_2$O. No bands related to CuO, which would appear at 588, 534 and 480 cm^{-1} [21], were detected. The use of oleic acid may lead to carboxylate (C=O stretch) on the surface of the particles which would show in the FTIR spectra at 1710 cm^{-1}. Based on Figure 7, the peak for the carboxylate could not be detected across all the synthesized Cu$_2$O particles. This clearly proves that the surfaces of the Cu$_2$O particles are ligand free.

The zeta potentials for all the Cu$_2$O samples were measured and the results are tabulated in Table 2. The surface charge of the spherical, cubic and octahedral shaped Cu$_2$O was positively charged (indicated by the "+" sign) while the surfaces of the rhombic dodecahedral and truncated polyhedral Cu$_2$O were negatively charged (indicated by the "−" sign). The surface charge is related to the surface bonding state on the exposed facets. Our observation that the octahedral particles with exposed {111} planes possess the most positive surface charge indicates that there are probably more un-coordinated Cu bonds on the particle surfaces. On the other hand, the surface charge of the rhombic dodecahedral Cu$_2$O particles consisting of 12 {110} planes was found to be negative, suggesting the O ions are likely to be dominant on the exposed surface. This result is contrary to some previous reports [16,18] that claim the rhombic dodecahedral particles are positively charged based on visualizing the atomic models of Cu$_2$O. Strictly speaking, this approach (cutting and visualizing a particular crystal model) cannot deduce the information on particle surface charge.

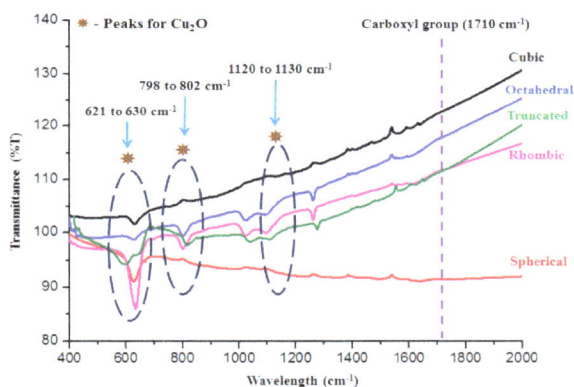

Figure 7. The Fourier Transform Infrared (FTIR) spectra of the faceted Cu$_2$O particles.

Table 2. Zeta potentials of the faceted Cu$_2$O.

Zeta Potential (mV)	Spherical (+)	Cubic (+)	Octahedral (+)	Rhombic Dodecahedral (−)	Truncated (−)
Test 1	5.93	6.70	19.7	−15.3	−4.59
Test 2	5.76	7.63	20.9	−15.1	−2.99
Test 3	5.83	6.25	23.1	−15.3	−3.41
Test 4	4.92	4.51	31.8	−17.6	−2.68
Test 5	3.33	5.96	30.6	−18.9	−1.45
Test 6	5.79	5.63	31.5	−18.1	−1.75
Test 7	5.79	5.88	33.0	−18.3	−5.41
Test 8	5.53	5.27	31.9	−7.84	−3.93
Test 9	4.51	6.31	31.3	−7.84	−1.66
Test 10	3.33	5.16	32.9	−7.78	−3.24
Range	3.33 to 5.93	4.51 to 7.63	19.7 to 33.0	−7.68 to −18.9	−1.45 to −4.59

2.3. The Adsorption Performance of Cu$_2$O Particles

The adsorption performance of the Cu$_2$O particles was evaluated using solution of MO, a negatively charged dye (Figure 8). It was observed that Cu$_2$O particles with positive surface charges (spherical, cubic and octahedral) were able to adsorb MO due to the electrostatic force. The adsorption was almost complete after 300 min for the octahedral Cu$_2$O. Both the rhombic dodecahedral and truncated Cu$_2$O particles possess negative charges, and there was nearly no adsorption due to electrostatic repulsion. Based on the zeta potential results shown earlier (Table 2), it was observed that the octahedral sample was more positively charged, followed by the cubic and then the spherical structures. The sequence agrees well with the adsorption performance despite the difference in the specific area: although the spherical sample possessed a higher surface area, the adsorption was observed to be slower and no further adsorption could be seen after 360 min. Hence, it is proven that the surface charge critically affects the adsorption performance of the synthesized Cu$_2$O particles.

As the adsorption performance of octahedral sample was evidently the best, it is of interest to further understand the adsorption mechanism and capacity of this material. Figure 9 presents the adsorption isotherm of the Cu$_2$O octahedral shaped sample.

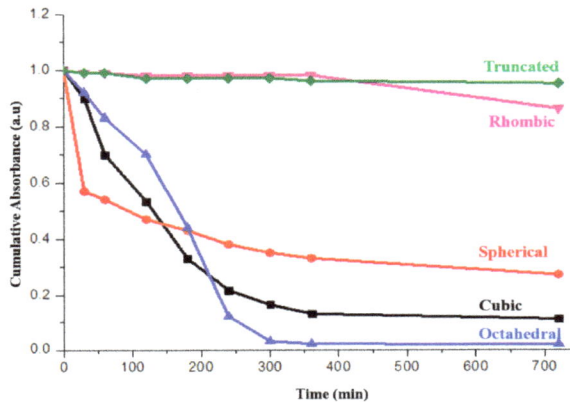

Figure 8. Adsorption performance of the Cu$_2$O (spherical, cubic, octahedral, rhombic dodecahedral and truncated polyhedral) particles in 20 ppm methyl orange (MO) solution.

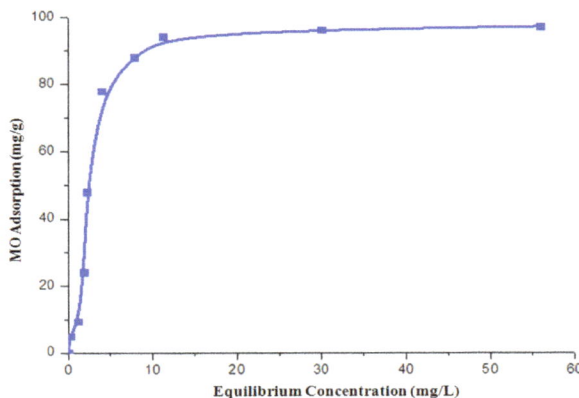

Figure 9. The adsorption isotherms of the octahedral shaped Cu$_2$O.

The curve is fitted into the well-known Freundlich (Equation (1)) and Langmuir (Equation (2)) models:

$$\text{Freundlich}: \quad q = K_f C^{\frac{1}{n}} \tag{1}$$

$$\text{Langmuir}: q = \frac{q_m KC}{1 + KC} \tag{2}$$

where q (mg/g) is the amount of adsorbed MO, C the concentration of MO at equilibrium, q_m (mg/g) the maximum adsorption capacity, K_f and n are Freundlich constants and K is the Langmuir constant. The amount of MO adsorbed onto the photocatalyst, q, is given by:

$$q = \frac{(C_o - C_f)V}{M} \tag{3}$$

where C_o and C_f represent the initial and final concentration of MO in the solution, V is the volume of MO solution (L) and M is the mass of photocatalyst added (g). The Freundlich and Langmuir models can be linearized as:

$$\text{Freundlich}: \ln q_e = \ln K_F + \left(\frac{1}{n}\right) \ln C_e \tag{4}$$

$$\text{Langmuir}: \frac{1}{q_e} = \frac{1}{q_{max} K_L C_e} + \frac{1}{q_{max}} \tag{5}$$

If the adsorption isotherm exhibits Langmuir behavior, it indicates monolayer adsorption. In contrast, a good fit into the Freundlich model indicates a heterogeneous surface binding. The results of fitting the isotherm curves to Freundlich and Langmuir models are summarized in Table 3. The fitting is clearly better for the Langmuir model. On the other hand, the R^2 value is less than 0.8 when the data is fitted with the Freundlich model. This illustrates that the adsorption of MO by the octahedral Cu_2O is governed by the monolayer adsorption. The MO adsorbed (q_m) by the octahedral shaped Cu_2O sample was found to be 96.42 mg/g (experimentally) and 66.0 mg/g (theoretically). The values are comparable with some reported adsorbents such as rice husk (40.58 mg/g) [22] and raw date pits (80.29 mg/g) [23].

Table 3. Adsorption isotherm parameters fitted to Langmuir and Freundlich models.

Adsorbent	Langmuir				Freundlich		
	q_m (exp) (mg/g)	K_L (L/mg)	q_{m1} (mg/g)	R^2	K_F (mg^{1-n}L^{-n}g^{-1})	n	R^2
Octahedral	96.42	0.40	66	0.86	17.3	1.79	0.76

2.4. The Photocatalytic Performance of Cu_2O Particles

The photocatalytic activity of the as-prepared Cu_2O samples was evaluated by degradation of MO under visible light irradiation. Photolysis test was also carried out as a reference. Water loss due to evaporation was calibrated when reporting the dye concentration.

Figure 10 shows the photocatalytic performance. When assessing the photocatalytic activity, reference has to be made with the adsorption experiment as shown in Figure 8. Through comparison, it is evident that the octahedral and cubic shaped Cu_2O particles demonstrated substantial photocatalytic degradation of MO under visible light irradiation. The demonstrated good activity is attributed to the adsorption of the dye to the surface before the photo degradation occurs.

The rhombic dodecahedral and truncated Cu_2O showed limited activity despite the fact that their negative surface charge repels the dye molecules. Huang et al. [24] have attributed the poor photocatalytic activity of the rhombic dodecahedral shaped Cu_2O to the surface residues from the oleic acid used during the synthesis. However, as discussed earlier, in the current study, no FTIR peaks related to oleic acid could be found. Therefore, our current work confirms that the rhombic dodecahedral Cu_2O indeed possesses some degree of photocatalytic activity.

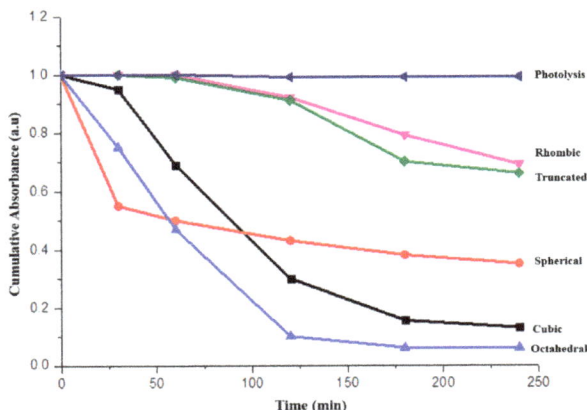

Figure 10. The photocatalytic performance of the faceted Cu_2O (spherical, cubic, octahedral, rhombic dodecahedral and truncated polyhedral) particles under visible light irradiation.

No obvious dye degradation was observed for spherical particles once the adsorption curve is plotted together with the photocatalytic one, as shown in Figure 11. The poor crystallinity could be the main reason, which has resulted in the fast recombination of the photo-generated electron–hole pairs.

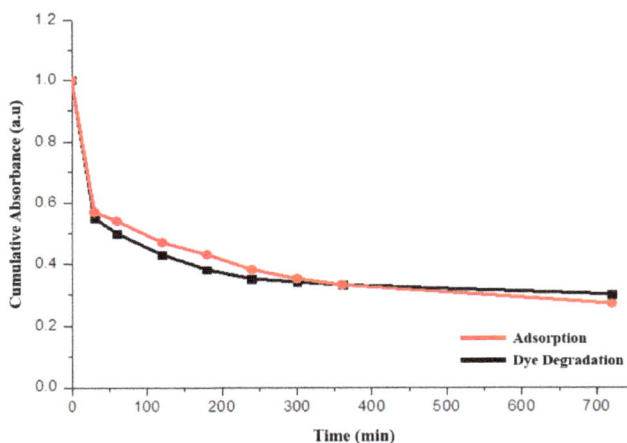

Figure 11. Comparison of adsorption (red line) and dye degradation (black line) performances of the spherical Cu_2O particles.

2.5. Band Gap, Band Edge Positions and Proposed Photodegradation Mechanism

To determine the flatband potential, Mott–Schottky graph was obtained by measuring the apparent capacitance as a function of potential under depletion condition at the semiconductor–electrolyte junction based on:

$$\frac{1}{C_{sc}^2} = \frac{2}{e\varepsilon\varepsilon_0 N}\left(E - E_{fb} - \frac{kT}{e}\right) \qquad (6)$$

where C_{sc} is the capacitance of the space charged region, e the electron charge (1.602×10^{-19} C), ε the diaelectric constant of the semiconductor, ε_0 the permittivity of free space (8.85×10^{-14} F cm^{-1}), N the donor density, E the applied potential, E_{fb} the flatband potential, k the Boltzmann constant (1.38×10^{-23} J K^{-1}), and T the absolute temperature. Extrapolation of the linear plot to the applied potential axis leads to the value for E_{fb}. Table 4 summaries the valence band potential position, the type of semiconductor, and the optical bandgap values that were determined by the optical absorption measurement.

Table 4. The tabulated valence band position and optical bandgap for the Cu_2O particles.

Particle	Valence Band Position vs. NHE (eV)	Type	Optical Bandgap (eV)
Spherical	+0.547	p-type	1.88
Cubic	+0.597	p-type	2.00
Octahedral	+0.727	p-type	1.93
Rhombic Dodecahedral	+0.547	p-type	1.96
Truncated Polyhedral	+0.537	p-type	1.90

Figure 12 shows the band edge positions (with reference to NHE at pH ~7) and band gap energy of different Cu_2O samples. The measured bandgap value only slightly varies from 1.9 to 2.0 eV, and the largest difference in the valence band position is about 0.19 eV. The difference in the band gap energy and band potential positions could be due to the size and the specific exposed facet(s) [25]. It was reported that the main oxidation species for MO degradation are $^\bullet$OH radicals under our experimental pH condition [26]. In general, there are two routes that the $^\bullet$OH radicals are generated, viz., through the photo-generated holes in the valence band or the photo-generated electrons in the conduction band. However, none of the particles are able to produce $^\bullet$OH radicals directly from the photo-generated holes as the required semiconductor valence band potential is +1.58 eV for pH = 7 (+1.99 eV vs. NHE [27]), far more anodic than valence band potentials of all these Cu_2O particles. Therefore, the mechanism of $^\bullet$OH radical generation should be through the photogenerated electrons from the conduction band. The electrons first react with adsorbed oxygen to generate $O_2^{-\bullet}$ and H_2O_2, which then react to form $^\bullet$OH radicals [25].

In addition to the photocatalytic action discussed above, there is another possible dye degradation mechanism through the so-called photo-assisted degradation. Under such mechanism, the electrons generated by the dyes after light adsorption are injected to the conduction band of the photocatalyst. The electrons generated by the dye itself, rather than by the photocatalyst in a typical photocatalytic reaction, will cause chemical destruction of the dye molecules [28]. The LUMO (lowest unoccupied molecular orbital) energy of MO is around -1.636 eV with reference to NHE at ~pH 7 [29], so it is possible for the electrons to be injected from MO to the conduction band of Cu_2O, leading to the photo-assisted degradation of MO itself. Because the energy gap between the LUMO and HOMO (highest occupied molecular orbital) of MO is about 2.35 eV [29], which is larger than the band gap of the Cu_2O samples (1.9–2.0 eV), it becomes impossible to choose a light illumination that only excites the dye but not the photocatalyst—such experiment would be able to differentiate the photocatalytic action vs. the photo-assisted degradation. However, based on the fact that the spherical samples did not degrade MO, it is reasonable to deduce that the photo-assisted degradation is not likely to be a main contributing mechanism since it would not be affected by the charge recombination in the photocatalyst. In other words, if the photo-assisted degradation does exist, it should be manifested through the spherical particles.

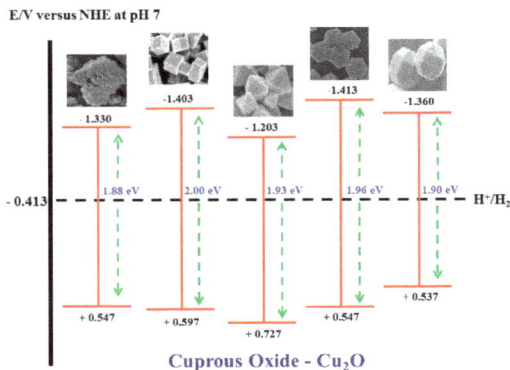

E/V versus NHE at pH 7

Cuprous Oxide - Cu$_2$O

Figure 12. The band edge potential positions of the faceted Cu$_2$O particles.

2.6. Stability of the Synthesized Cu$_2$O Particles

The stability of Cu$_2$O is determined via the solar light illumination of intensity 100 mW/cm^2 for 12 h when the particles were dispersed in MO solution. The samples were collected and analyzed using scanning electron microcopy (SEM), X-ray photoelectron microcopy (XPS) and FTIR. Comparison was made with 25% methanol (v/v) added to the solution to understand the source of the photocorrosion.

Figure 13 compares the morphology of the faceted Cu$_2$O particles at the as-synthesized state (column 1), after light illumination in the (MO + methanol) solution (column 2), and after light illumination in the MO solution (column 3). Comparing the first and the third columns, it is observed that the spherical shaped (Figure 14a) Cu$_2$O particles do not seem to have suffered much photocorrosion. In relation to nearly zero photocatalytic activity as reported earlier, it strongly suggests that corrosion of the Cu$_2$O is related to photo-generated species.

In the case of the cubic and octahedral shaped Cu$_2$O particles (Figure 13b,c), small amount of precipitates could be observed on the surface. For the rhombic dodecahedral Cu$_2$O particles (Figure 13d), the precipitates could be seen on the surface of the particles. On the other hand, the truncated polyhedral Cu$_2$O (Figure 13e) particles suffered from minimum photocorrosion as only a very small amount of precipitates could be found on the surface.

The addition of methanol, a hole scavenger, in the MO solution has clearly alleviated the severity of photocorrosion (comparing columns 2 and 3). This suggests that the photocatalytically generated holes are responsible for the photocorrosion of Cu$_2$O.

To determine the corrosion mechanism, we took the octahedral sample as a representative and exposed it under the solar light irradiation for an extended period of 120 h. After that, XPS and FTIR were employed in the analysis of the corrosion products on the particle surface. XRD was also explored but the intensity of the corrosion products was not strong enough for a firm identification.

Possible photocorrosion products of Cu$_2$O include Cu, Cu(OH)$_2$ and CuO. As the binding energies for Cu^0 and Cu^{1+} are 932.67 eV and 932.6 eV, respectively [30], it is difficult to determine the phase of the photocorrosion products through Cu 2p$_{3/2}$ scan. As such, XPS scan for the Auger peak of Cu LMM is needed to differentiate between Cu^0 and Cu^{1+}. The Auger LMM energy for Cu^0 and Cu^{1+} lie in 334.95 eV and 336.80 eV respectively [31]. Figure 14 shows the Auger LMM spectra of octahedral shaped Cu$_2$O particles which have Auger LMM energy of 341 eV. This Auger energy generated corresponds to the Auger energy of Cu^{1+}. Figure 15 shows the Cu 2p$_{3/2}$ XPS spectra of the Cu$_2$O particles. The main peaks at 932.7 and 935.4 eV correspond to the binding energies of Cu$_2$O and Cu(OH)$_2$ respectively [31,32]. Lastly, the satellite peaks on the higher binding energy, 943.6 eV, indicates the presence of an unfilled Cu 3d shell and thus confirms the existence of Cu^{2+} on the sample surface.

Figure 13. Column 1—FESEM images of as-synthesized Cu_2O: (**a1**) spherical; (**b1**) cubic; (**c1**) octahedral; (**d1**) rhombic dodecahedral; and (**e1**) truncated polyhedral. Column 2—FESEM images of Cu_2O in mixture of MO and methanol solution under solar light illumination for 12 h: (**a2**) spherical; (**b2**) cubic; (**c2**) octahedral; (**d2**) rhombic dodecahedral; and (**e2**) truncated polyhedral. Column 3—FESEM images of Cu_2O in pure MO solution under solar light illumination for 12 h: (**a3**) spherical; (**b3**) cubic; (**c3**) octahedral; (**d3**) rhombic dodecahedral; and (**e3**) truncated polyhedral.

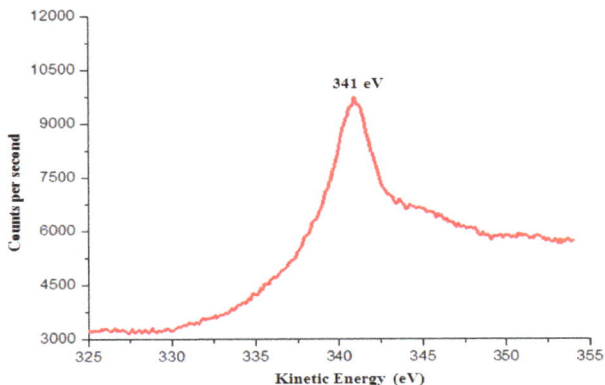

Figure 14. The kinetic energy (Auger LMM) of the photocorroded Cu_2O particles.

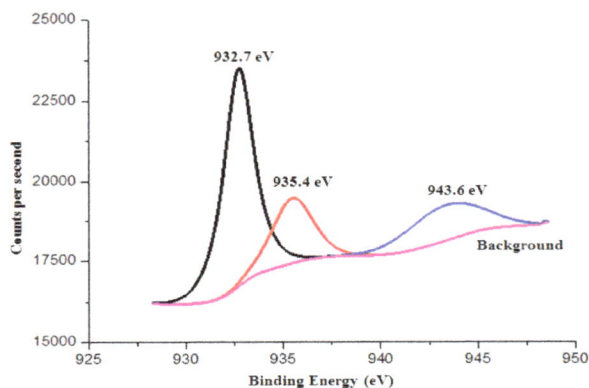

Figure 15. The binding energy of the photocorroded Cu_2O particles.

The XPS confirms that the top surface (4–8 nm) of the Cu_2O particles contain $Cu(OH)_2$ precipitates. To explore possible photocorrosion products beneath the top layer, FTIR measurement was used. As shown in Figure 16a, the broad bands centered at 3436 and 1638 cm^{-1} are attributed to the O–H stretching and bending modes of water [16,33]. The peaks located at 631, 809 and 1156 cm^{-1} are attributed to the stretching vibration of Cu–O in Cu_2O.

When zooming into the details in the range of 400 to 800 cm^{-1}, the adsorption band at 530 cm^{-1} associated with CuO is revealed (Figure 16b). The low intensity suggests that the amount of CuO is relatively small.

Based on the above analyses, the photocorrosion precipitates formed on the surface of Cu_2O particles are mainly $Cu(OH)_2$ and a small amount of CuO. This finding agrees with our early analysis that the photo-generated electrons are responsible for the MO degradation, while the holes are left to oxidize Cu^+ to Cu^{2+} state. The presence of the photocorrosion products will hinder the photocatalytic activity of Cu_2O as they block the light from reaching the Cu_2O surface. Adding hole scavenger has alleviated the photocorrosion, and this observation also confirms that the holes are responsible for the photocorrosion of Cu_2O.

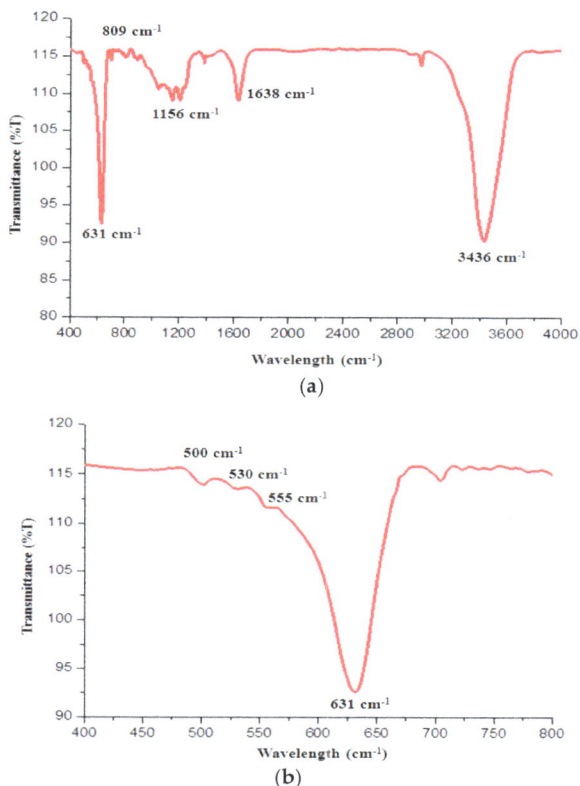

Figure 16. (**a**) FTIR spectrum of the photocorroded Cu_2O particles; and (**b**) zoom-in FTIR showing the existence of CuO.

3. Materials and Methods

3.1. Synthesis of Cu_2O Particles

Copper(II) acetate monohydrate ($C_4H_6CuO_4 \cdot H_2O$, 99.0%) was purchased from Fluka (Singapore). Copper(II) sulfate pentahydrate ($CuSO_4 \cdot 5H_2O$, 98.0%), L-ascorbic acid ($C_6H_8O_6$, reagent grade), hydrazine hydrate ($H_4N_2 \cdot xH_2O$, reagent grade 50-60%), oleic acid ($C_{18}H_{34}O_2$, technical grade 90.0%), and D-(+)-Glucose ($C_6H_{12}O_6$, 99.5%) were obtained from Sigma Aldrich (Singapore). Sodium hydroxide pellets (NaOH, 99.0%) were acquired from Schedelco (Singapore). Absolute ethanol (C_2H_5OH, 99.0%) was purchased from Merck (Singapore). Absolute methanol (CH_3OH, analytical reagent grade) was purchased from Fisher Scientific (Singapore). Hexane (C_6H_{14}, reagent grade) was obtained from Riverbank Chemical Pte Ltd (Singapore). All chemicals were used as received without further purification.

3.1.1. Spherical Cu_2O Particles

Copper(II) acetate (Cu(OAc)$_2$) (0.01 mol) was dissolved into 20 mL of deionized water under constant stirring at 500 rpm. Forty milliliters of 0.01 M NaOH was then added to the solution followed by 0.01 M of ascorbic acid (20 mL). After stirring for 30 min, the solution was centrifuged at 6000 rpm for 3 min and then washed with deionized water and ethanol solution. The powder collected was then dried in a vacuum oven at 60 °C for 6 h.

3.1.2. Cubic Cu$_2$O Particles

Copper(II) acetate (Cu(OAc)$_2$) (0.01 mol) was dissolved in 20 mL of deionized water under constant stirring at 500 rpm. Forty milliliters NaOH solution (0.03 M) was then added, followed by 20 mL of 0.01 M ascorbic acid. After stirring for 30 min, the solution was centrifuged at 6000 rpm for 3 min followed by washing with deionized water and ethanol solution. The powder collected was then dried in a vacuum oven at 60 °C for 6 h.

3.1.3. Octahedral Cu$_2$O Particles

Copper(II) acetate (Cu(OAc)$_2$) (0.01 mol) was dissolved in 20 mL of deionized water under constant stirring at 500 rpm. Forty milliliters NaOH solution (0.03 M) was then added to the solution followed by addition of 40 mL hydrazine hydrate solution (0.05 M). After 45 min of stirring, the solution was centrifuged at 6000 rpm for 3 min and washed with deionized water and ethanol solution. The powder collected was dried in a vacuum oven at 60 °C for 6 h.

3.1.4. Rhombic Dodecahedral Cu$_2$O Particles

Copper(II) sulfate (CuSO$_4$·5H$_2$O) (0.25 g) was dissolved in 40 mL deionized water. When the powder was fully dissolved, 5 mL oleic acid and 20 mL absolute ethanol were added to the solution. The solution was vigorously stirred at 700 rpm for 30 min before being heated at 90 °C. Twenty milliliters NaOH solution (0.08 M) was then added to the CuSO$_4$ solution under constant stirring for 5 min at 90 °C. Afterwards, 3.42 g of glucose, which was dissolved in 30 mL of heated deionized water, was added to the solution. The final solution was left to stir at 90 °C for 45 min. After that, the solution was centrifuged and washed with hexane (10 times) and ethanol (3 times). The powder collected was then dried in a vacuum oven at 60 °C for 6 h.

3.1.5. Truncated Polyhedral Cu$_2$O Particles

Copper(II) acetate (Cu(OAc)$_2$) (0.015 mol) was dissolved in 40 mL deionized water at 70 °C. After continuous stirring at 500 rpm for 5 min, 10 mL of 9.0 M NaOH and 0.6 g glucose were added into the solution. The solution continued to be stirred at 70 °C for about 60 min before being cooled down to room temperature. The powder was washed and centrifuged with ethanol (6 times) and deionized water (3 times). Finally, the powder was dried in a vacuum oven at 60 °C for 6 h.

3.2. Materials Characterization

Crystal structure was identified by X-ray diffraction (XRD) using a Shimadzu LabX-6000 diffractometer (Shimadzu Corporation, Tokyo, Japan) with Cu Kα radiation (λ = 1.54178 Å). A step size of 0.02° over 2θ ranging from 10° to 80° was used with a scanning rate at 2.33° per minute. The accelerating voltage and emission current were 40 kV and 30 mA, respectively.

The morphology of the samples was examined by field emission scanning electron microscopy (FESEM, JEOL JSM-7600F, JEOL Ltd., Tokyo, Japan) and transmission electron microscopy (TEM, JEOL JEM-2010, JEOL Ltd., Tokyo, Japan). The specific surface areas were evaluated using a Micromeritics ASAP 2020 (Micromeritics Instrument Corporation, Norcross, GA, USA) surface analyzer based on the BET theory. The samples were outgassed under vacuum and heated to 100 °C before the test.

The surface chemical analysis was carried out by X-ray photoelectron spectroscopy (XPS) using a VG ESCALab 220i-XL system (Thermo Scientific, Waltham, MA, USA). Mg Kα X-ray (hν = 1253.6 eV) from twin anode X-ray gun was employed using a large area lens mode for analysis with photoelectron takeoff angle of 90° with respect to surface plane. The maximum analysis depth is in the range of 4–8 nm. Survey spectra were acquired for elemental identification while high-resolution spectra were acquired for chemical state identification and surface composition calculation. For chemical state analysis, a spectral deconvolution was performed by a curve-fitting procedure based on Lorentzians broadened by Gaussian using the manufacturer's standard software. The error of binding energy is estimated to be within 0.2 eV.

FTIR was carried out in a Perkin Elmer Instruments Spectrum GX FTIR spectrometer (PerkinElmer, Waltham, MA, USA). Synthesized particles were mixed with standard KBr particles and pressed into thin pellets. The spectral range was 400–4000 cm^{-1} and a total of 40 scans were recorded at a resolution of 4 cm^{-1} averaging each spectrum.

Zeta potential was measured by a Mavern Nanosizer system (Malvern Instruments Ltd, Malvern, UK). Particles (5–10 mg) were dispersed in deionized water (~pH 7) and sonicated for 5 min. The equilibrium time was set at 120 s and each sample was run for 10 times to obtain the average value of the surface charge.

Optical absorption of bulk powders was measured on a Perkin Elmer Lambda 900 UV-Visible spectrometer (PerkinElmer, Waltham, MA, USA) in the diffuse reflectance spectroscopy mode over the spectra ranging from 250 to 800 nm. The optical diffuse reflectance spectrum of Cu_2O was measured on a Shimadzu 2550 UV-vis-NIR spectrometer (Shimadzu Corporation, Tokyo, Japan) using $BaSO_4$ as a reference standard. The bandgap of Cu_2O was calculated using the Kubelka–Munk function.

The flat band potentials were measured by impedance spectroscopy based on the Mott–Schottky plots. To prepare for the test samples, 15 mg of the as-synthesized Cu_2O powder was sonicated in 1 mL of ethanol to obtain a homogeneous mixture. The Cu_2O suspension was then drop-casted on a conductive fluorine-tin oxide (FTO) glass substrate with adhesive tapes acting as spacers attached on the four edges. The substrate was then dried at 80 °C, and the adhesive tape attached on the top side of the substrate was removed. Electrical contact was formed by first applying silver paint on the top uncoated area of FTO, and then sticking a conductive copper tape onto the dried silver paint. Three electrodes were used for the impedance measurements which include the working electrode (the Cu_2O film), a Pt counter electrode, and a reference electrode (Ag/AgCl, saturated in KCl). A 0.1 M Na_2SO_4 solution was used as the electrolyte. The measurements were carried out by a Gamry electrochemical impedance spectrometer, and the potential was systemically varied between +0 V and +2.0 V with frequency of 50 Hz.

3.3. Photocatalytic and Adsorption Experiments

Methyl orange (MO) was chosen as a representative dye to test the photocatalytic degradation activity and the adsorption capacity of the prepared Cu_2O samples. Solution of the dye was prepared by dissolving the dye in deionized water at 20 ppm concentration, and the solution pH was around 6.7. One hundred mg of the powder samples were dispersed in 50 mL MO solution for the dye degradation test under visible light irradiation. A 100 mL glass beaker, wrapped with aluminum foil on its side wall, was used as the reactor with light shone from the top. The irradiation source comes from a solar simulator equipped with a 300 W Xe-lamp (HAL-320, Asahi Spectra Co., Ltd., Kita-ku, Japan). Super cold filter (YSC0750) was used to provide visible light ranging from 420 to 700 nm. The light intensity was around 50 mW/cm^2. The amount of photocatalyst used was chosen through a few trial runs based on their degradation speed.

An adsorption isotherm test was carried out in the dark to prevent the potential photocatalytic degradation of MO under light. The equilibrium adsorption isotherm was determined using various concentrations of MO (10, 20, 30, 50 and 80 ppm). For each test, 100 mg of adsorbent was added to 50 mL MO solution. After 48 h, the equilibrium concentration was measured.

To determine the adsorption capacity, 100 mg of synthesized powder was added to 50 mL MO solution of various concentrations (10, 20, 30, 50 and 80 ppm). Stirring was applied throughout the duration of the test (1440 min) at a speed of 400 rpm in dark. At different intervals, the sample was collected, centrifuged and measured using the UV-visible spectrometer to determine the dye concentration.

3.4. Photocatalytic Stability Study

Two separate tests were carried out to measure the photocatalytic stability of Cu_2O particles and to determine the possible root cause of the instability. In the first experiment, 50 mg powder was dispersed in 50 mL MO solution with 10 ppm concentration. For the second experiment, 50 mg of the powder was dispersed in 40 mL of MO solution (10 ppm) mixed with 10 mL of absolute methanol (purity > 99%). Both were constantly stirred under a solar simulator equipped with a 300 W Xe-lamp (HAL-320, Asahi Spectra Co., Ltd., Kita-ku, Japan) for 12 h under the intensity of 100 mW/cm^2. An AM 1.5 G filter (400 to 1100 nm) was used.

4. Conclusions

In this paper, we have studied the adsorption and photocatalytic performances of the various faceted Cu_2O samples for MO removal from its solution. The adsorption capability of Cu_2O particles was found to be mainly determined by the surface charge. The octahedral shaped Cu_2O particles with exposed {111} facets performed the best due to its most positive surface charges. The adsorption of the octahedral shaped Cu_2O was found to follow the Langmuir monolayer mechanism.

The spherical shaped Cu_2O without clearly defined facets did not display photocatalytic activity. All the faceted samples showed different degree of photocatalytic activities. The octahedral shaped Cu_2O particles with exposed {111} facets performed the best in photocatalytic degradation of MO under visible light. The photo-generated electrons are responsible for the degradation of MO solution, while the photo-generated holes attack Cu_2O, causing photocorrosion. It was observed that the corrosion precipitates are mainly $Cu(OH)_2$, together with a small amount of CuO. These photocorrosion products hinder the photocatalytic activity of Cu_2O and thus shorten the service life of Cu_2O particles. The addition of hole scavengers such as methanol has shown to have alleviated the corrosion attack on Cu_2O.

Acknowledgments: Financial support from Ministry of Education (grant RG15/16) and Singapore National Research Foundation through the Singapore-Berkeley Initiative for Sustainable Energy (SINBERISE) CREATE Programme is gratefully acknowledged. Authors are grateful for helpful discussion with Wei Chen, Xin Zhao, and Pushkar Kanhere.

Author Contributions: Weng Chye Jeffrey Ho and Zhong Chen conceived the research plan. Weng Chye Jeffrey Ho carried out the various experiments with the assistance from Qiuling Tay and Huan Qi. Weng Chye Jeffrey Ho and Zhong Chen analyzed the experiment results and wrote the manuscript after discussion with Zhaohong Huang and Jiao Li.

Conflicts of Interest: The authors declare no conflict of interest.

References

1. Fujishima, A.; Honda, K. Electrochemical Photolysis of Water at a Semiconductor Electrode. *Nature* **1972**, *238*, 37–38. [CrossRef] [PubMed]
2. Tay, Q.; Liu, X.; Tang, Y.; Jiang, Z.; Sum, T.C.; Chen, Z. Enhanced Photocatalytic Hydrogen Production with Synergistic Two-Phase Anatase/Brookite TiO_2 Nanostructures. *J. Phys. Chem. C* **2013**, *117*, 14973–14982. [CrossRef]
3. Jiang, Z.; Tang, Y.; Tay, Q.; Zhang, Y.; Malyi, O.I.; Wang, D.; Deng, J.; Lai, Y.; Zhou, H.; Chen, X.; et al. Understanding the Role of Nanostructures for Efficient Hydrogen Generation on Immobilized Photocatalysts. *Adv. Energy Mater.* **2013**, *3*, 1368–1380. [CrossRef]
4. Tang, Y.; Jiang, Z.; Tay, Q.; Deng, J.; Lai, Y.; Gong, D.; Dong, Z.; Chen, Z. Visible-light plasmonic photocatalyst anchored on titanate nanotubes: A novel nanohybrid with synergistic effects of adsorption and degradation. *RSC Adv.* **2012**, *2*, 9406–9414. [CrossRef]
5. Tang, Y.; Wee, P.; Lai, Y.; Wang, X.; Gong, D.; Kanhere, P.D.; Lim, T.; Dong, Z.; Chen, Z. Hierarchical TiO_2 Nanoflakes and Nanoparticles Hybrid Structure for Improved Photocatalytic Activity. *J. Phys. Chem. C* **2012**, *116*, 2772–2780. [CrossRef]

6. Cheng, Y.H.; Huang, Y.; Kanhere, P.D.; Subramaniam, V.P.; Gong, D.; Zhang, S.; Highfield, J.; Schreyer, M.; Chen, Z. Dual-Phase Titanate/Anatase with Nitrogen Doping for Enhanced Degradation of Organic Dye under Visible Light. *Chem. A Eur. J.* **2011**, *17*, 2575–2578. [CrossRef] [PubMed]

7. Pichat, P. (Ed.) *Photocatalysis and Water Purification*; Wiley-VCH: Weinheim, Germany, 2013.

8. Schneider, J.; Bahnemann, D.; Ye, J.; Puma, G.L.; Dionysiou, D.D. (Eds.) *Photocatalysis: Fundamentals and Perspectives*; RSC: London, UK, 2016.

9. Dionysiou, D.D.; Puma, G.L.; Ye, J.; Schneider, J.; Bahnemann, D. (Eds.) *Photocatalysis: Applications*; RSC: London, UK, 2016.

10. Colmenares, J.C.; Xu, Y.-J. (Eds.) *Heterogeneous Photocatalysis*; Springer: Berlin, Germany, 2016.

11. Pichat, P. (Ed.) *Photocatalysis: Fundamentals, Materials and Potential*; MDPI: Basel, Switzerland, 2016.

12. Nosaka, Y.; Nosaka, A. *Introduction to Photocatalysis from Basic Science to Applications*; RSC: London, UK, 2016.

13. Zhang, Y.Y.; Jiang, Z.L.; Huang, J.Y.; Lim, L.Y.; Li, W.L.; Deng, J.Y.; Gong, D.G.; Tang, Y.X.; Lai, Y.K.; Chen, Z. Titanate and Titania Nanostructured Materials for Environmental and Energy Applications: A Review. *RSC Adv.* **2015**, *5*, 79479–79510. [CrossRef]

14. Tang, Y.X.; Gong, D.G.; Lai, Y.K.; Shen, Y.Q.; Zhang, Y.Y.; Huang, Y.Z.; Tao, J.; Lin, C.J.; Dong, Z.L.; Chen, Z. Hierarchical Layered Titanate Microspherulite: Formation by Electrochemical Spark Discharge Spallation and Application in Aqueous Pollutant Treatment. *J. Mater. Chem.* **2010**, *20*, 10169–10178. [CrossRef]

15. Kuo, C.-H.; Huang, M.H. Morphologically controlled synthesis of Cu_2O nanocrystals and their properties. *Nano Today* **2010**, *5*, 106–116. [CrossRef]

16. Huang, W.-C.; Lyu, L.-M.; Yang, Y.-C.; Huang, M.H. Synthesis of Cu_2O Nanocrystals from Cubic to Rhombic Dodecahedral Structures and Their Comparative Photocatalytic Activity. *J. Am. Chem. Soc.* **2012**, *134*, 1261–1267. [CrossRef] [PubMed]

17. Chai, C.; Peng, P.; Wang, X.; Li, K. Cuprous oxide microcrystals via hydrothermal approach: Morphology evolution and photocatalytic properties. *Cryst. Res. Technol.* **2015**, *5*, 299–303. [CrossRef]

18. Huang, M.H.; Rej, S.; Hsu, S.-C. Facet-dependent properties of polyhedral nanocrystals. *Chem. Commun.* **2014**, *50*, 1634–1644. [CrossRef] [PubMed]

19. Qi, H.; Wolfe, J.; Fichou, D.; Chen, Z. Cu_2O Photocathode for Low Bias Photoelectrochemical Water Splitting Enabled by NiFe-Layered Double Hydroxide Co-Catalyst. *Sci. Rep.* **2016**, *6*, 30882. [CrossRef] [PubMed]

20. Yu, Y.; Zhang, L.; Wang, J.; Yang, Z.; Long, M.; Hu, N.; Zhang, Y. Preparation of hollow porous Cu_2O microspheres and photocatalytic activity under visible light irradiation. *Nanoscale Res. Lett.* **2012**, *7*, 347. [CrossRef] [PubMed]

21. Singhal, A.; Pai, M.R.; Rao, R.; Pillai, K.T.; Lieberwirth, I.; Tyagi, A.K. Copper(I) Oxide Nanocrystals–One Step Synthesis, Characterization, Formation Mechanism, and Photocatalytic Properties. *Eur. J. Inorg. Chem.* **2013**, 2640–2651. [CrossRef]

22. Vadivelan, V.; Kumar, K.V. Equilibrium, kinetics, mechanism, and process design for the sorption of methylene blue onto rice husk. *J. Colloid Interface Sci.* **2005**, *286*, 90–100. [CrossRef] [PubMed]

23. Banat, F.; Al-Asheh, S.; Al-Makhadmeh, L. Evaluation of the use of raw and activated date pits as potential adsorbents for dye containing waters. *Process Biochem.* **2003**, *39*, 193–202. [CrossRef]

24. Hua, Q.; Cao, T.; Bao, H.; Jiang, Z.; Huang, W. Crystal-Plane-Controlled Surface Chemistry and Catalytic Performance of Surfactant-Free Cu_2O Nanocrystals. *ChemSusChem* **2013**, *6*, 1966–1972. [CrossRef] [PubMed]

25. Huang, L.; Peng, F.; Yu, H.; Wang, H. Preparation of cuprous oxides with different sizes and their behaviors of adsorption, visible-light driven photocatalysis and photocorrosion. *Solid State Sci.* **2009**, *11*, 129–138. [CrossRef]

26. Niu, P. Photocatalytic Degradation of Methyl Orange in Aqueous TiO_2 Suspensions. *Asian J. Chem.* **2013**, *25*, 1103–1106.

27. Liang, Y.; Lin, S.; Liu, L.; Hu, J.; Cui, W. Oil-in-water self-assembled Ag@AgCl QDs sensitized Bi2WO6: Enhanced photocatalytic degradation under visible light irradiation. *Appl. Catal. B Environ.* **2015**, *164*, 192–203. [CrossRef]

28. Zhang, F.; Zhao, J.; Zang, L.; Shen, T.; Hidaka, H.; Pelizzetti, E.; Serpone, N. Photoassisted degradation of dye pollutants in aqueous TiO_2 dispersions under irradiation by visible light. *J. Mol. Catal. A Chem.* **1997**, *120*, 173–178. [CrossRef]

29. Chang, X.; Gondal, M.A.; Al-Saadi, A.A.; Ali, M.A.; Shen, H.; Zhou, Q.; Zhang, J.; Du, M.; Liu, Y.; Ji, G. Photodegradation of Rhodamine B over unexcited semiconductor compounds of BiOCl and BiOBr. *J. Colloid Interface Sci.* **2012**, *377*, 291–298. [CrossRef] [PubMed]

30. Bi, F.; Ehsan, M.F.; Liu, W.; He, T. Visible-Light Photocatalytic Conversion of Carbon Dioxide into Methane Using Cu_2O/TiO_2 Hollow Nanospheres. *Chin. J. Chem.* **2015**, *33*, 112–118. [CrossRef]

31. Zhang, Y.X.; Huang, M.; Li, F.; Wen, Z.Q. Controlled Synthesis of Hierarchical CuO Nanostructures for Electrochemical Capacitor Electrodes. *Int. J. Electrochem. Sci.* **2013**, *8*, 8645–8661.

32. Hernandez, J.; Wrschka, P.; Oehrlein, G.S. Surface Chemistry Studies of Copper Chemical Mechanical Planarization. *J. Electrochem. Soc.* **2011**, *148*, G389–G397. [CrossRef]

33. Parveen, M.F.; Umapathy, S.; Dhanalakshmi, V.; Anbarasan, R. Synthesis and characterization of nano sized $Cu(OH)_2$ and its surface catalytic effect on poly(vinyl alcohol). *Indian J. Sci.* **2013**, *3*, 111–117.

Sample Availability: Samples of the Cu_2O particles are not available from the authors, but can be synthesized following the described procedures.

molecules

MDPI

Article

Sulfur-Doped Carbon Nitride Polymers for Photocatalytic Degradation of Organic Pollutant and Reduction of Cr(VI)

Yun Zheng [1], Zihao Yu [1], Feng Lin [1], Fangsong Guo [1], Khalid A. Alamry [2], Layla A. Taib [2], Abdullah M. Asiri [2,3] and Xinchen Wang [1,*]

[1] State Key Laboratory of Photocatalysis on Energy and Environment, College of Chemistry, Fuzhou University, Fuzhou 350002, China; zhengyun1101@163.com (Y.Z.); yuzihao15@163.com (Z.Y.); linfeng13@mails.ucas.ac.cn (F.L.); guofangsong@outlook.com (F.G.)

[2] Chemistry Department, Faculty of Science, King Abdulaziz University, Jeddah 21589, Saudi Arabia; k_alamry@yahoo.com (K.A.A.); laylataib@gmail.com (L.A.T.); aasiri2@kau.edu.sa (A.M.A.)

[3] Center of Excellence for Advanced Materials Research (CEAMR), Faculty of Science, King Abdulaziz University, Jeddah 21589, Saudi Arabia

*   Correspondence: xcwang@fzu.edu.cn; Tel.: +86-591-8392-0097

Academic Editor: Pierre Pichat
Received: 18 February 2017; Accepted: 29 March 2017; Published: 1 April 2017

Abstract: As a promising conjugated polymer, binary carbon nitride has attracted extensive attention as a metal-free and visible-light-responsive photocatalyst in the area of photon-involving purification of water and air. Herein, we report sulfur-doped polymeric carbon nitride microrods that are synthesized through thermal polymerization based on trithiocyanuric acid and melamine (TM) supramolecular aggregates. By tuning the polymerization temperature, a series of sulfur-doped carbon nitride microrods are prepared. The degradation of Rhodamine B (RhB) and the reduction of hexavalent chromium Cr(VI) are selected as probe reactions to evaluate the photocatalytic activities. Results show that increasing pyrolysis temperature leads to a large specific surface area, strong visible-light absorption, and accelerated electron-hole separation. Compared to bulk carbon nitride, the highly porous sulfur-doped carbon nitride microrods fabricated at 650 °C exhibit remarkably higher photocatalytic activity for degradation of RhB and reduction of Cr(VI). This work highlights the importance of self-assembly approach and temperature-control strategy in the synthesis of photoactive materials for environmental remediation.

Keywords: carbon nitride; self-assembly; photocatalysis; pollutant degradation; Cr(VI) reduction

1. Introduction

With the rapid advancement of urbanization and industrialization, a series of environmental issues has come about owing to the excessive industrial contaminations containing toxic organic pollutants and poisonous metal ions [1]. Semiconductor-mediated photocatalysis technology has been regarded as the most promising strategy for environmental treatment of pollutants and heavy metal ions [2,3]. To date, various semiconductors, such as metal oxides, nitrides and sulphides, have been developed and further utilized as photocatalysts [4–6]. However, the low quantum efficiency, expensive raw materials and activity instability raise the threshold of practical applications of photocatalytic technology. Therefore, the development of photocatalytically polymeric materials with abundant materials and ease modifications take more considerations.

Heptazine-based polymer melon, also denoted as graphitic carbon nitride (g-C_3N_4) for simplicity, has drawn significant attention in the past few years attributed to its unique chemical, electronic and (photo)catalytic properties [7,8]. g-C_3N_4 has been reported to be an attractive photocatalyst for water splitting, pollutant purification, CO_2 reduction, organic synthesis, bacterial inactivation, etc. [9,10]. However, carbon nitride polymers synthesized via traditional routes usually possess limited surface area, moderate visible light absorption, high recombination rate of charge and carrier, and inefficient photocatalytic activity [11–13]. Recently, a variety of nanostructures have been created by the nanocasting methods [14,15], using traditional hard-templating approach to tune the texture and photocatalytic performance [16–18]. Meanwhile, soft-templating approaches are newly emerging, which endow the synthetic process with simple and the morphological tuning diversiform [19,20]. However, the molecules of soft templates are easily decomposed, which restrains the polycondensation of precursors during the thermal treatment and the structure optimization of carbon nitride. Therefore, the development of an effective pathway for the structure modification of polymeric carbon nitride is an urgent task.

The supramolecular system based on small molecule self-assembly is a relatively new and fascinating area in material science [21–23]. For carbon nitride, thermal polymerization of hydrogen-bonded supramolecular aggregates stemming from trithiocyanuric acid-melamine (TM) and cyanuric acid-melamine complex have been synthesized [24–28]. By controlling the reaction conditions, the chemical and electron structure can be effectively tailored, and further affecting the photocatalytic activity of carbon nitride [29,30]. In addition, the supramolecular aggregates can be also modified via copolymerization, element doping, composite, heterojunction, and salt-melt method [31–33]. Such modification methods are usually relied on the additional chemicals (e.g., organic co-monomer, inorganic salt and acid) in the starting process, which may lead to an inhomogeneous mixture [34–38]. As far as we know, the utilization of a facile temperature-control protocol for the self-assembly synthesis of carbon nitride photocatalysts without adding extra additives is less explored.

In this work, by employing TM supramolecular aggregates as the starting material, nanostructured carbon nitride is prepared via direct thermal condensation in Ar flow. The pre-organization of trithiocyanuric acid and melamine at molecular level intrinsically tunes the microstructure and morphology of target materials. By controlling the condensation temperature of TM, a series of sulfur-doped carbon nitride microrods with diverse properties and photocatalytic activities are obtained. The effects of pyrolysis temperature on the condensation degree, morphology and photocatalytic performances of the resultant carbon nitride are analyzed. In addition, the photocatalytic activity tests and the corresponding mechanism about dyes decomposing and Cr(VI) reduction are also carried out and discussed. Our findings will open up new opportunities for future design and development of polymeric materials for environmental remediation and energy conversion.

2. Results and Discussion

2.1. Characterization of TM Supramolecular Aggregates

TM supramolecular aggregate is synthesized from a simplified way by mixing an equimolar mixture of trithiocyanuric acid and melamine in CH_3OH/H_2O solution at room temperature (Scheme 1), which is different from hydrothermal method in the previous literature [39–42]. As shown in Figure 1a, TM cocrystal exhibits a totally different X-ray diffraction (XRD) pattern compared with pristine raw materials. The peaks at 12.3°, 13.1°, and 18.4° are ascribed to the in-planar packing, while the peak at 24.6° with a *d*-spacing of 0.362 nm is attributed to graphite-like stacking of individual two-dimensional sheets [40–42]. Fourier transform infrared spectroscopy (FT-IR) spectra (Figure 1b) indicate that the functional groups of both trithiocyanuric acid and melamine are retained in the TM adduct. Additionally, benefiting from the hydrogen-bonding interaction between trithiocyanuric acid and melamine, the triazine ring vibration of melamine is shifted to a lower wave-number from 814 to 781 cm^{-1}, and the N–H stretching vibration of melamine is shifted to a lower wave-number from 3468 to 3420 cm^{-1} with enhanced intensity [40–42]. These results confirmed that hydrogen-bonded TM supramolecular aggregates are formed.

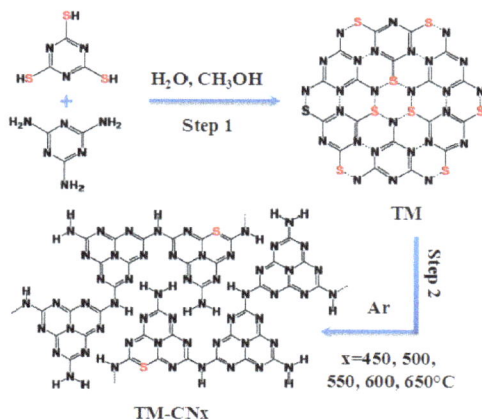

Scheme 1. Schematic illustration of the formation process of TM-CNx.

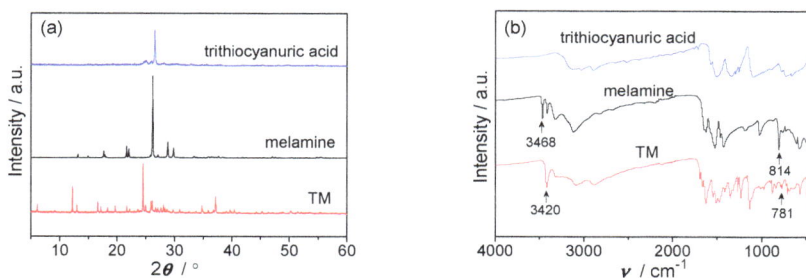

Figure 1. (**a**) XRD patterns and (**b**) FT-IR spectra of trithiocyanuric acid, melamine and TM.

2.2. Morphology of TM-CNx

The TM supramolecular aggregates are then heated at various temperature (450, 500, 550, 600, and 650 °C) in Ar gas to induce polymerization. The resultant carbon nitride products are abbreviated as TM-CNx, where x is the thermal polymerization temperature. As the condensation temperature rose (450 to 650 °C), the yields of products were decreased. It is worth noting that almost no product can be obtained at the condensation temperature of 700 °C due to the complete decomposition of TM.

To analyze the morphology of the as-prepared samples, the characterizations including scanning electron microscopy (SEM) and transmission electron microscopy (TEM) were performed. Figure 2 shows a smooth rod-like morphology for TM aggregates with the average width and length of 1 μm and 5–10 μm, respectively. After pyrolysis, the TM-CN550 sample preserves integrated rod morphology, whereas the TM-CN650 sample possesses a loose rod-like appearance with abundant pores and channels. The variation of nanostructure from TM to TM-CNx can be attributed to the thermal-driven rearrangement of some atoms as well as the breaking and reformation of some chemical bonds in TM precursor. Increasing the temperature from 550 °C to 650 °C leads to more pores due to the sulfur species elimination. TEM images of TM-CN650 (Figure 3) reveal a microrod morphology composed of spatially interconnected nanosheets and enormous macrospores. The elemental-mapping image of TM-CN650 shows that two major elements of C and N (as well as trace amount of S) are homogeneously distributed over the entire nanoarchitecture.

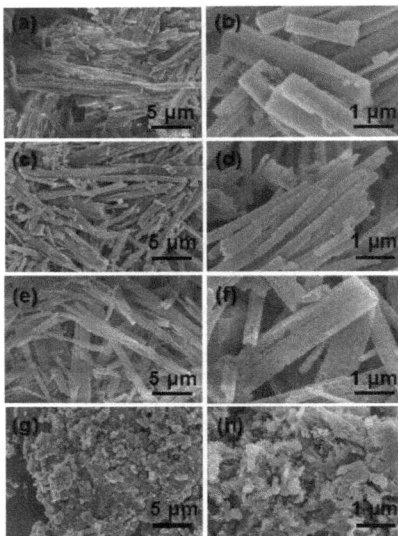

Figure 2. SEM images of (**a**,**b**) TM, (**c**,**d**) TM-CN550, (**e**,**f**) TM-CN650 and (**g**,**h**) bulk g-C_3N_4.

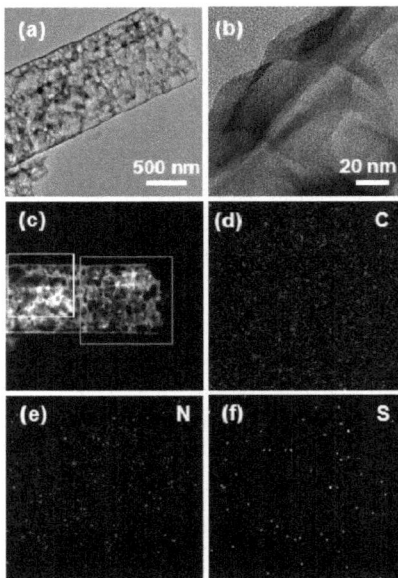

Figure 3. (**a**,**b**) TEM images and (**c**–**f**) elemental mapping images of TM-CN650.

2.3. Texture and Chemical Structure of TM-CNx

The porous structure and surface area of TM-CN650 were studied by N_2 sorption measurements (Figure 4a and Table 1). A characteristic type-IV isotherm with an H3 hysteresis loop was observed in the N_2 sorption isotherms of TM-CN650, suggesting that the macroporous structure is formed by the accumulation of carbon nitride sheet. It indicates that TM-CN650 has a Brunauer–Emmett–Teller (BET) surface area of 72 m^2 g^{-1}, which is much higher than that of bulk g-C$_3$N$_4$ (3 m^2 g^{-1}) and other TM-CNx synthesized at lower temperatures. When the temperature is decreased from 650 to 450 °C, the BET surface area of TM-CNx is remarkably reduced, and the pore volume is lowered. These results clearly proved that the variation of temperature greatly changes the texture of carbon nitride.

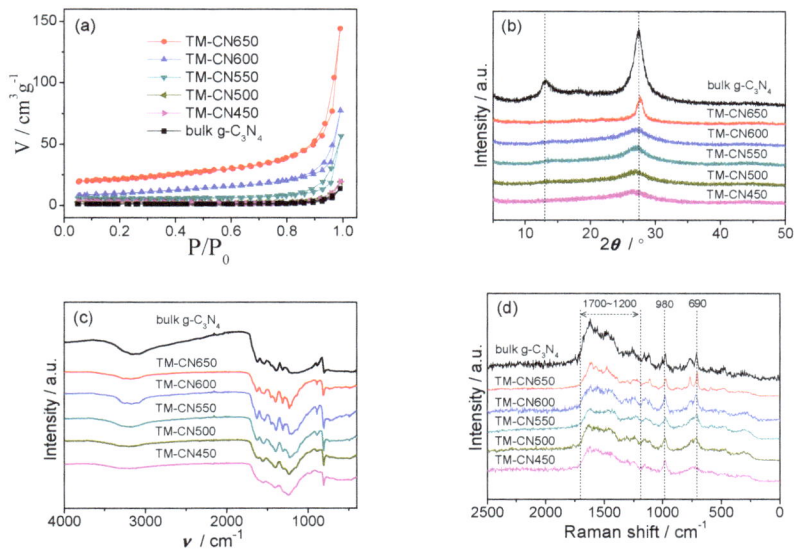

Figure 4. (**a**) N$_2$ adsorption-desorption isotherms, (**b**) XRD patterns, (**c**) FT-IR spectra and (**d**) Raman spectra of TM-CNx and bulk g-C$_3$N$_4$.

Table 1. The surface area, pore volume, and elemental composition of TM-CNx samples.

Entry	Samples	Surface Area (m^2 g^{-1})	Pore Volume (cm^3 g^{-1})	C (wt. %)	N (wt. %)	H (wt. %)	S (wt. %)	Molar ratio C/N (%)
1	bulk g-C$_3$N$_4$	3	0.01	34.1	58.6	2.0	-	0.68
2	TM-CN450	8	0.01	32.1	56.4	2.3	<0.5	0.66
3	TM-CN500	9	0.01	33.1	57.3	2.1	<0.5	0.67
4	TM-CN550	17	0.01	33.4	57.1	2.3	<0.5	0.68
5	TM-CN600	35	0.04	32.8	55.6	2.2	<0.5	0.69
6	TM-CN650	72	0.06	32.8	55.8	2.2	<0.5	0.69

The crystal and chemical structure of the as-prepared TM-CNx polymers were carefully investigated by XRD, FT-IR, UV-Raman, and X-ray photoelectron spectroscopy (XPS) measurements. As shown in Figure 4b, all TM-CNx samples presented similar XRD patterns, which is an indication of heptazine-based polymeric melon. The characteristic diffraction peaks of TM-CNx at ca. 27° are assigned to the periodic in-plane tri-*s*-triazine stacking and the interlayer structural aromatic packing, and also ascribed to the (002) plane of the graphitic layer structures. As reaction temperature increases from 450 °C to 650 °C, the interlayer stacking peaks of TM-CNx samples shift from 26.6° to 27.4°, corresponding to the decrease of interlayer distance from 0.335 to 0.326 nm. This phenomenon suggests an improvement of crystallinity and a decreased interlayer distance, thus proving an enhanced condensation degree and more regular stacking structure of carbon nitride with increasing heating temperature [39]. Compared to typical XRD pattern of bulk g-C$_3$N$_4$ material, these TM-CNx samples exhibit relatively weaker intensities at 27°, and the peak at ca. 13° as the (100) peak is not evident in TM-CNx samples. This result suggests the reduced order degree of chemical structure owing to the incomplete molecular polymerization as well as the existence of one-dimensional nanostructure with increasing pyrolysis temperature.

FT-IR measurement (Figure 4c) was also performed to investigate the formation of carbon nitride. Both bulk g-C_3N_4 and TM-CNx polymers displayed similar FT-IR vibration modes. The peaks at 810 cm^{-1} and 1200–1600 cm^{-1} were assigned to the characteristic breathing and stretching vibration modes of aromatic C–N heterocycles, identifying the existence of triazine units [43]. The broad and weak bands at 2900–3300 cm^{-1} were typical signals of N–H or O–H vibrations, which were ascribed to the uncondensed amino groups as the surface terminal groups and the absorbed H_2O molecules.

To probe the chemical structure of TM-CNx, Raman spectra were carried out over the catalysts pressed on glass slides (Figure 4d). With a 325 nm UV laser excitation, an intense, broad, asymmetric peak in the range of 1200–1700 cm^{-1} can be ascribed to C–N stretching vibrations, which resembles the "G" and "D" band profiles for structurally disordered graphitic carbon-based materials. The existence of a heptazine ring structure is further confirmed by the two sharp peaks appeared at 690 and 980 cm^{-1} [44]. The former peak at 690 cm^{-1} is a doubly degenerate mode for in-plane bending vibrations of the heptazine linkages, while the latter peak at 980 cm^{-1} is assigned to the symmetric N-breathing mode of heptazine units. No band is observed between 2000 and 2500 cm^{-1}, since there is no triply bonded C≡N units or N=C=N groups within the carbon nitride framework. These features are observed for all of the TM-CNx catalysts.

Additional proof for the formation of polymeric carbon nitride was obtained by XPS measurements of TM-CN650 (Figure 5). The weak peak at 284.6 eV was attributed to the sp^2-hybridized carbon (C–C) of standard carbon [43]. The strong peak at binding energy of 288.0 eV was determined as the sp^2-bonded carbon atoms in the heterocycle (N=C–N) of aromatic carbon nitride, which was related to the major skeleton carbon in the triazine-based heterocycle [43]. The high resolution of N 1s spectra could be deconvoluted into four peaks at 398.4, 399.5, 400.7 and 404.2 eV, respectively [43]. These peaks were corresponded to the sp^2 bonded nitrogen (C–N=C), the tertiary nitrogen (N–C$_3$) groups, the surface uncondensed amino groups (C–N–H), and the charging effects or positive charge localization in the heterocycles, respectively [43]. The first two nitrogen together with the sp^2-bonded carbon (N–C=N) constitute the triazine-based heterocyclic ring (C$_6$N$_7$) units [43]. Weak signal of S 2p can be detected in TM-CN650 sample, proving the existence of sulfur species in carbon nitride structure. The weak peak of S 2p at 165.0 eV is ascribed to the generation of S-N bonds by replacing lattice carbon with sulfur, while the peak at 168.2 eV is determined as the formation of sulfur oxide during the thermal condensation. According to the XPS spectra of TM-CN650, the atomic concentration ratio of S/C is calculated to be 1:1627 based on the method reported by Huang et al. [45]. Thus, it can be concluded that sulfur has been doped into the structure of carbon nitride.

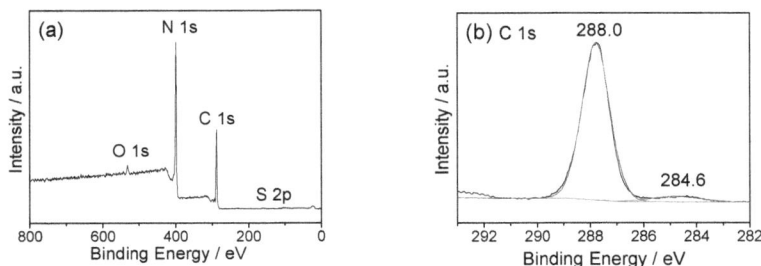

Figure 5. *Cont.*

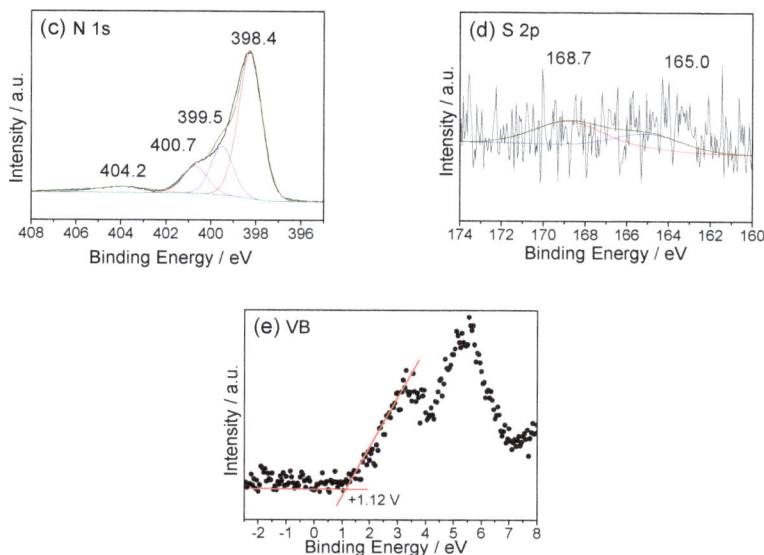

Figure 5. (**a**) XPS survey spectra, (**b**) C 1s spectra, (**c**) N 1s spectra, (**d**) S 2p spectra, (**e**) VB XPS spectra of TM-CN650.

To further confirm the element composition of TM-CNx, the elemental analysis was performed. As shown by the elemental analysis result (Table 1), the C/N molar ratios of TM-CNx under different temperatures remain in the range of 0.66–0.69 without any major fluctuations, similar to that of bulk g-C_3N_4 (0.68). All samples feature 2 wt. % hydrogen (Table 1), suggesting that these materials are indeed polymeric melon rather than fully-condensed phase of covalent carbon nitride materials. Tiny sulfur can be found in TM-CNx. It can be concluded from these results that the structure of TM-CNx are sulfur-doped and heptazine-based melon polymers.

2.4. Optical Property and Band Structure of TM-CNx

The optical properties of the TM-CNx samples were investigated by UV-Vis diffuse reflectance (DRS) spectra (Figure 6a and Table 2). All TM-CNx samples feature typical semiconductor-like absorption. Considerable improved light-harvesting capability and gradual bathochromic shift of optical absorption edges are found for TM-CNx samples prepared under increasing condensation temperature. By incresing the condensation temperature, the absorption band edges of TM-CNx samples red-shifted from 442 to 680 nm, and the electronic band gap narrowed from 2.81 to 1.82 eV for TM-CN450 to TM-CN650. The modified light-absorption property from temperature processing is primarily attributed to the emergence of structural distortion and the activation of more n→π* transitions, as well as the improved π-electron delocalization and inter-planar packing towards J-type aggregates in the conjugated system [39]. The increase of visible light absorption ability and the decrease of band gap contributed to the capturing of more visible photons, which is beneficial for enhancing the photocatalytic activity. Combined with the band gap of 1.82 eV and the valence-band (VB) potential of 1.12 eV (determined by the VB XPS spectra in Figure 5e), the conduction band (CB) level of TM-CN650 is calculated to be −0.70 eV.

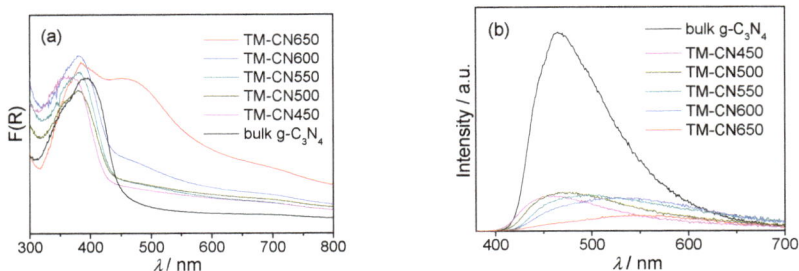

Figure 6. (**a**) UV-Vis DRS spectra and (**b**) PL spectra of TM-CNx samples.

Table 2. The absorption band edge, band gap energy, and photocatalytic activity of TM-CNx samples.

Entry	Samples	Absorption Band Edge (nm)	Band Gap Energy (eV)	k_{RhB} (min^{-1})	$k_{Cr(VI)}$ (min^{-1})
1	bulk g-C₃N₄	460	2.70	0.0152	0.0036
2	TM-CN450	442	2.81	0.0047	0.0020
3	TM-CN500	452	2.74	0.0352	0.0066
4	TM-CN550	456	2.72	0.0496	0.0181
5	TM-CN600	463	2.68	0.1031	0.0204
6	TM-CN650	680	1.82	0.2283	0.1287

The charge-carriers separation/recombination rates were next investigated by room temperature photoluminescence (PL) spectra under excitation wavelength of 370 nm (Figure 6b). All the fluorescence of TM-CNx are quenched in comparison with bulk g-C₃N₄, indicating the lower exciton energy and the improved electron-hole separation in TM-CNx. Additionally, the PL intensities of the emission peaks were greatly decreased when the temperature were increased, which illustrates that raising temperature is effective for suppressing the rapid charge carrier recombination. The quenching of emission intensity also suggests that the relaxation of a portion of photocarriers occurs via a non-radiative pathway, presumably due to charge transfer of electrons and holes to new localized/surface states [46]. Moreover, when the temperature is varied from 450 to 650 °C, an obvious red-shift of the emission peak from 460 to 560 nm was observed for TM-CNx, which further certified the decreased band-gap energy of TM-CNx samples.

2.5. Photocatalytic Activity of TM-CNx in Degradation of RhB

The photocatalytic activities of the TM-CNx samples were evaluated by degradation of Rhodamine B (RhB) under visible light (λ > 420 nm) irradiation. The photodegradation process was recorded by the temporal evolution of the spectra of supernants at different reaction times. Figure 7 displays the changes of the RhB concentration versus the reaction time over the TM-CNx catalysts and the corresponding first-order kinetics plot by the equation of $\ln(C_0/C) = kt$, where C_0 and C are the RhB concentrations in solution at times 0 and t, respectively, and k is the apparent first-order rate constant. Based on the result of control experiments, RhB was not degraded under dark conditions, and RhB is also stable under visible light if there is no photocatalyst involved. As shown in Figure 7a, the TM-CNx samples exhibit accelerated degradation ability and higher photocatalytic activity with increasing condensation temperature. An optimal activity was found for TM-CN650 with the superior degradation rate of 97% after 15 min. Furthermore, TM-CN650 shows much better photocatalytic activity than bulk g-C₃N₄, urea derived carbon nitride (CNU), and commercial Degussa P25 (Figure 7b). Figure 7c and Table 2 show the first-order rate constant k (min^{-1}) measurements for RhB degradation of the TM-CNx samples. The measured k value of TM-CN650 is 0.2283 min^{-1}, which is almost 14 times higher than that of bulk g-C₃N₄ (0.0152 min^{-1}).

The stability of TM-CN650 photocatalyst was investigated by recycling the photocatalyst for RhB degradation under visible-light irradiation (Figure 8). The high degradation capability of TM-CN650 was maintained without a significant decrease in six consecutive experiments. The XRD, FT-IR, and Raman analysis also proved that the crystal and chemical structure of TM-CN650 remain unchanged after the photocatalytic reaction. These results proved the good stability of TM-CN650 in the photocatalytic reaction.

To further investigate the possible photodegradation mechanism, the effects of various radical scavengers and N_2 purging on the degradation of RhB over TM-CN650 were examined (Figure 7d). The photodegradation of RhB was repeated with modification by adding benzoquinone (BQ), tert-butyl alcohol (TBA), and ammonium oxalate (AO) as the scavengers of superoxide radical ($\cdot O_2^-$), hydroxyl radical ($\cdot OH$) and hole (h^+), respectively. To suppress the capture of photo-induced electrons by oxygen to generate $\cdot O_2^-$, N_2 bubbling was also applied to remove the oxygen molecules dissolved in the RhB aqueous solution. After 15 min of reaction, the removal efficiency of RhB are 20%, 34%, 53% and 83% in the presence of BQ, N_2 bubbling, AO and TBA, respectively. It was observed that the degradation rate of RhB was depressed under an N_2 atmosphere. The addition of BQ almost completely hindered the decomposition of RhB. These results confirmed that $\cdot O_2^-$ was the major oxidation species during degradation of RhB in the TM-CN650/RhB system. The degrading rate of RhB also decreased obviously with the addition of AO, indicating that photogenerated hole oxidation plays a role in the degradation of RhB. Additionally, the degradation rate was slightly reduced by the addition of TBA (an efficient trap of $\cdot OH$), which was important but did not result in complete quenching of the photodegradation reaction. The formation of reactive oxygen species over the TM-CN650 catalyst under visible light irradiation was further probed by a 5,5-dimethyl-1-pyrroline-*N*-oxide (DMPO) spin-trapping electron paramagnetic resonance (EPR) technique. DMPO/O_2^- adducts were clearly observed when the TM-CN650 catalyst was exposed to visible light irradiation (Figure 7e), whereas no detectable $\cdot OH$ signals can be observed in experiments carried out under darkness or visible light irradiation (Figure 7f). These results demonstrated the fact that $\cdot OH$ is involved but not exclusively and that the major reactive species are $\cdot O_2^-$ and holes in RhB degradation with a TM-CN650 catalyst.

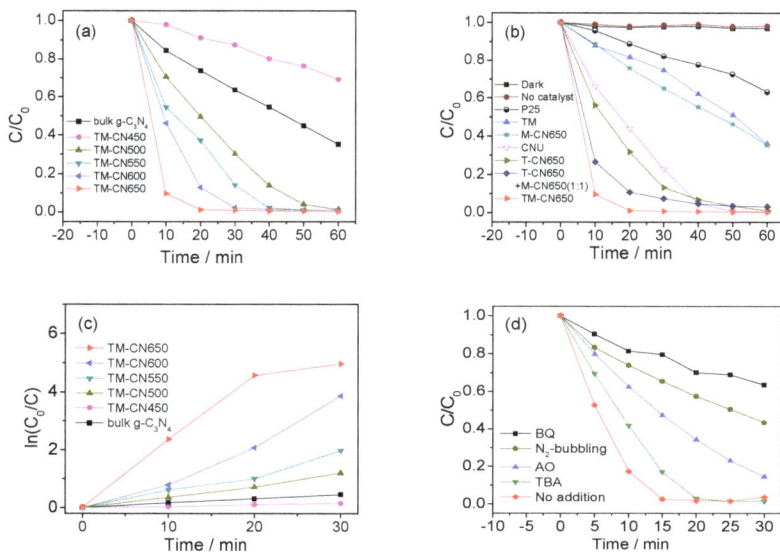

Figure 7. *Cont.*

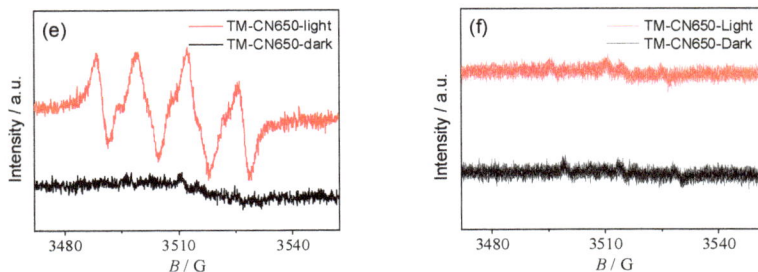

Figure 7. (**a**,**b**) Concentration changes of RhB as a function of irradiation time with different catalyst under visible light irradiation; (**c**) first-order rate constant k (min^{-1}) of TM-CNx; (**d**) effect of quencher additive and N_2 purging on the photocatalytic activity of TM-CN650 in RhB degradation. DMPO spin-trapping EPR spectra of TM-CN650 sample with visible light irradiation ($\lambda > 400$ nm) for the detection of (**e**) DMPO-·O_2^- and (**f**) DMPO-·OH.

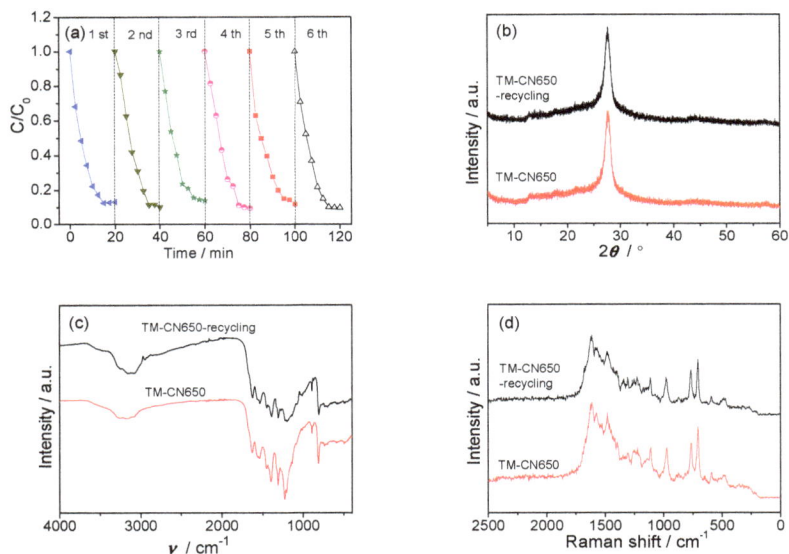

Figure 8. (**a**) cycling runs of RhB degradation in TM-CNx under visible light irradiation; (**b**) XRD patterns, (**c**) FT-IR spectra, and (**d**) Raman spectra of TM-CN650 and recycled TM-CN650 after photocatalytic degradation.

2.6. Photocatalytic Activity of TM-CNx in Reduction of Cr(VI)

In order to understand the visible light photocatalytic activity of the as-synthesized samples for Cr(VI) reduction, a test reaction was carried out for the reduction of 5 mg/L of Cr (VI) solution with the catalyst and ammonium oxalate (as hole scavenger) in nitrogen atmosphere under visible-light irradiation ($\lambda > 400$ nm). Figure 9 displays the changes of Cr (VI) concentration versus the reaction time over the TM-CNx catalysts. The apparent rate constant for the reduction of Cr(VI) is calculated by the following equation: $k = \ln(C_0/C)/t$, where C_0 and C are the Cr (VI) concentrations in solution at times 0 and t, respectively, and k is the apparent first-order rate constant. As shown in Figure 9b and Table 2, the apparent reaction rate constant and photocatalytic activity of TM-CNx for reduction of Cr(VI) followed the order of TM-CN650 > TM-CN600 > TM-CN550 > TM-CN500 > bulk g-C_3N_4 > TM-CN450.

Apparently, TM-CN650 exhibits the best photocatalytic activity in reduction of Cr(VI). Furthermore, the results of recycled experiment reveal that the TM-CN650 sample has good photostability, since there is no noticeable decrease in the removal ratio over four recycling tests (Figure 9d).

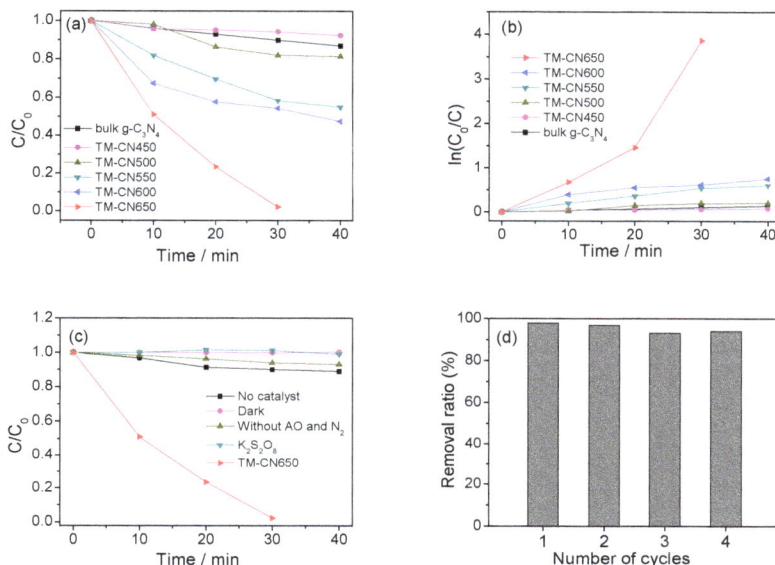

Figure 9. (**a**) concentration changes of Cr(VI) as a function of irradiation time with different catalyst under visible light irradiation; (**b**) first-order rate constant k (min^{-1}) of TM-CNx in photocatalytic reduction of Cr(VI); (**c**) control experiments of photocatalytic reduction of Cr(VI); and (**d**) the reusability of TM-CN650 for the reduction of Cr(VI) after 40 min of irradiation.

Some control experiments were performed to certify the possible mechanism of Cr(VI) reduction. As can be seen from Figure 9c, the reduction of Cr(VI) hardly occurs in the absence of light or photocatalyst. No significant reaction of Cr(VI) is observed without AO and N_2, even when the reaction is operated in the presence of catalyst with irradiation. The reduction of Cr(VI) to Cr(III) by photogenerated electrons can be described by the following equation:

$$Cr_2O_7^{2-} + 14H^+ + 6e^- \rightarrow 2Cr(III) + 7H_2O \tag{1}$$

According to Equation (1), Cr (VI) was reduced to Cr(III) in the presence of TM-CN650 and AO upon purging with N_2 under visible light irradiation. For the photocatalytic reduction of Cr(VI), AO as a hole scavenger can capture photo-induced holes (h^+) of TM-CN650, and thus mitigate the recombination of photogenerated carriers of TM-CN650. Furthermore, the superoxide radicals ($\cdot O_2^-$), which is formed by the transformation of photogenerated electrons to the absorbed O_2, is significantly suppressed under N_2 bubbling. Accordingly, the introduction of AO and purging with N_2 atmosphere in the reaction solution play crucial roles in the photocatalytic reduction of Cr(VI). Additionally, the control experiments for the photoreduction of Cr(VI) under visible light illumination have been performed by using $K_2S_2O_8$ as a scavenger for photogenerated electrons. A remarkably reduced activity was noticed by adding $K_2S_2O_8$, further confirming that the reduction reaction of Cr(VI) is driven by the photoexcited electrons of TM-CN650 [47].

2.7. Mechanism of Photocatalytic Degradation of RhB and Reduction of Cr(VI) over TM-CN650

According to the above discussions, a possible mechanism for photocatalytic degradation of RhB and reduction of Cr(VI) by TM-CN650 is proposed and depicted in Figure 10. With visible light irradiation, a photon is absorbed by the TM-CN650 semiconductor, and an electron is excited from the VB to the CB, generating a positive hole in the VB and an electron in the CB. The CB potential (-0.7 eV) of TM-CN650 is negative enough to $E(O_2/\cdot O_2^-)$ (-0.046 eV vs. normal hydrogen electrode (NHE)) [48,49], and O_2 adsorbed on the surface of TM-CN650 can be reduced to $\cdot O_2^-$ by the electrons left in the CB via one electron reducing reaction. It is noted that $\cdot O_2^-$ is a relatively mild oxidant that results in partial oxidation (the cleavage of the RhB chromophore structure) instead of mineralization [50,51]. According to the VB potential ($+1.1$ eV) of TM-CN650, hydroxyl groups or water molecules adsorbed on the surface of TM-CN650 can not be directly oxidized to $\cdot OH$ radicals ($+1.99$ eV vs. NHE). The generation of $\cdot OH$ via photogenerated electron-induced multistep reduction of O_2 ($O_2 + e \rightarrow \cdot O_2^-$, $\cdot O_2^- + e^- + 2H^+ \rightarrow H_2O_2$, $H_2O_2 + e^- \rightarrow \cdot OH + OH^-$) is involved but not exclusively in the current photocatalytic system [51,52]. The photodegradation of RhB by TM-CN650 catalyst is mainly attributed to the partial oxidation of RhB by $\cdot O_2^-$ and photogenerated hole oxidation of RhB. Different from the degradation of RhB, the photocatalytic reduction of Cr(VI) is directly reduced to Cr(III) by photo-generated electrons (e^-) of TM-CN650. Benefiting from the one-dimensional nanostructure, high surface area, and appropriate band structure of TM-CN650, the recombination of photoinduced electron-hole pairs are inhibited and the lifetime of photo-induced charge carriers are prolonged, thereby contributing to the enhancement of overall photocatalytic activity in an aqueous phase [53–55].

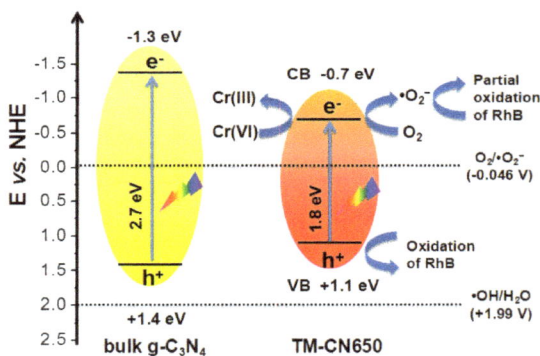

Figure 10. Schematic illustration of the mechanism of photocatalytic degradation of RhB and reduction of Cr(VI) over TM-CN650.

3. Materials and Methods

3.1. Chemicals

Melamine, Rhodamine B, ammonium oxalate, potassium dichromate ($K_2Cr_2O_7$), potassium persulfate ($K_2S_2O_8$), methanol, and tert-butyl alcohol were purchased from China Sinopharm Chemical Reagent Co. Ltd (Shanghai, China). Trithiocyanuric acid and benzoquinone were purchased from Aladdin Industrial Corporation (Shanghai, China). All of the chemicals were of analytical grade and were used without further purification.

3.2. Synthesis of TM

The synthetic process was performed in a 250 mL beaker containing a 10×30 mm magnetic stirring bar. Trithiocyanuric acid (0.71 g, 4 mmol) and melamine (0.50 g, 4 mmol) were added with

120 mL deionized water and 25 L methanol. Then, the mixture was sonicated for 10 min, followed by stirring at 500 r for 10 min and at 250 r for 12 h. The as-obtained mixture was filtered to recover the solid precipitate, washed with deionized water (200 mL), and dried in an oven at 70 °C.

3.3. Preparation of TM-CNx from TM

TM-CNx samples were synthesized by heating the TM precursor at a certain temperature (x = 450, 500, 550, 600, and 650) for 2 h with the heating rate of 4.6 °C/min under flowing Ar gas (200 mL/min).

3.4. Preparation of Bulk g-C₃N₄

Bulk g-C_3N_4 sample was prepared by heating melamine (10 g) at 550 °C for 4 h with the heating rate of 2.3 °C/min in the air, and followed by grounding into power.

3.5. Preparation of CNU

Carbon nitride derived from urea (CNU) was prepared by heating urea (10 g) at 600 °C for 2 h with the heating rate of 4.6 °C/min in the air.

3.6. Preparation of T-CN650

T-CN650 sample was synthesized by heating trithiocyanuric acid (5 g) at 650 °C for 2 h with the heating rate of 4.6 °C/min under flowing Ar gas (200 mL/min).

3.7. Preparation of M-CN650

M-CN650 sample was synthesized by heating melamine (5 g) at 650 °C for 2 h with the heating rate of 4.6 °C/min under flowing Ar gas (200 mL/min).

3.8. Characterization

XRD measurements were collected on a Bruker D8 Advance diffractometer (Billerica, MA, USA) with Cu Ka1 radiation (λ = 1.5406 Å). FT-IR spectra were collected with a thermo Nicolet Nexus 670 FT-IR spectrometer (Waltham, MA, USA), and the samples were mixed with KBr at a concentration of ca. 0.2 wt. %. UV-Raman scattering measurements were performed with a multichannel modular triple Raman system (Renishaw Co., Wotton-under-Edge, Gloucestershire, UK) with confocal microscope at room temperature using a 325 nm laser. XPS data were obtained on a Thermo Scientific ESCALAB250 instrument (Waltham, MA, USA) with a monochromatized Al Kα line source (200 W). All binding energies were referenced to the C 1s peak at 284.6 eV of surface adventitious carbon. Elemental analysis results were collected from Elementar Vario EL (Langenselbold, Germany). SEM measurement was conducted using Hitachi SU-8010 and S4800 Field Emission Scanning Electron Microscopes (Chiyoda, Tokyo, Japan). TEM measurement was obtained using a FEI TECNAIG2F20 instrument (Hillsboro, OR, USA). N_2 adsorption-desorption isotherms were performed using a 3020 micromeritics tristarII surface area and porosity analyzer (Atlanta, GA, USA). The UV-Vis absorption spectra were measured on Shimadzu UV-1780 (Kyoto, Japan) and Varian Cary 50 UV/Vis spectrophotometer (Palo Alto, CA, USA). The UV-Vis DRS spectra were measured on Cary 5000 Scan UV-Vis-NIR system from Agilent Technologies (Santa Clara, CA, USA). Photoluminescence spectra were recorded on an Edinburgh FI/FSTCSPC 920 spectrophotometer (Livingston, West Lothian, UK). The EPR technique was used to detect ·$O_2{}^-$ or ·OH radicals spin-trapped by DMPO in methanol or water, respectively. The signals were collected by a Bruker model A300 spectrometer (Bruker Instruments, Inc., Billerica, MA, USA) with the settings of center field (3512 G), microwave frequency (9.86 GHz), and power (20 mW).

3.9. Photocatalytic Degradation of RhB

An aqueous dispersion of catalyst and RhB dye was prepared for the photocatalytic test by dispersing 20 mg of catalyst powder to 80 mL RhB solution (10^{-5} mol/L). This reaction dispersion was magnetically stirred in the dark for ca. 30 min prior to irradiation to establish the adsorption/desorption equilibrium of the dye on the catalyst surface. The mixture was then irradiated under a 300 W Xe-lamp with cut-off filter to produce visible light irradiation ($\lambda > 420$ nm) at room temperature. At given irradiation time intervals, specimens (2 mL) were taken from the dispersion, and centrifuged at 10,000 rpm for 10 min to separate the catalyst particles. The concentration of aqueous RhB was determined using a Shimadzu UV-1780 UV-vis spectrophotometer at 554 nm by measuring its absorbance. The RhB degradation was calculated by the Lambert–Beer equation. Photoactivities for RhB in the dark in the presence of the photocatalyst and under visible-light irradiation in the absence of the photocatalyst were also evaluated.

3.10. Photocatalytic Activity for Reduction of Cr(VI)

In a typical measurement, a 300 W Xe arc lamp (PLS-SXE 300, Beijing Perfect Light Co., Ltd., Beijing, China) with a UV cut-off filter to eliminate light of wavelength $\lambda < 400$ nm was used as the light source. The power density of the light source used in the measurement is ca. 700 mW/cm^2. In addition, 10 mg of the sample, 50 mg hole scavenger (ammonium oxalate), and 40 mL of the Cr(VI) aqueous solution (5 mg·L^{-1}, which was based on Cr in a dilute $K_2Cr_2O_7$ solution) were added in a quartz vial. After the photocatalyst was dispersed in the solution with an ultrasonic bath for 10 min, the above suspension was bubbled with nitrogen gas (100 mL/min) to remove air and was stirred in the dark for 1 h, in order to ensure the establishment of adsorption–desorption equilibrium between the sample and the reactant before being exposed to visible light illumination. During the process of the reaction, 4 mL of sample solution was collected at a certain time interval and centrifuged to remove the catalyst completely at 10,000 rpm for 10 min. The Cr(VI) content in the supernatant solution was analyzed colorimetrically at 540 nm using the diphenylcarbazide method [56,57] on a Varian UV-vis spectrophotometer (Cary-50, Varian Co., Palo Alto, CA, USA).

4. Conclusions

In summary, one-dimensional sulfur-doped carbon nitride nanostructures have been prepared by thermal polymerization of trithiocyanuric acid-melamine supramolecular aggregates under Ar atmosphere at different temperatures. The acidic sulfur-containing gases generated from the thermal decomposition of precursors greatly affect the condensation/polymerization process, thus modifying both the texture and electronic structure of carbon nitride polymers [39,44]. Increasing the heating temperature not only causes a structure peeling effect to increase the porosity/surface area, but also induces structural distortion and the activation of more n→π* transitions to modify the band structure of carbon nitride polymers [39,44]. These integrative factors synergistically contribute to the overall photoactivity improvement of porous sulfur-doped carbon nitride microrods for degradation of RhB and reduction of Cr(VI). The porous rod-like sulfur-doped carbon nitrides synthesized at 650 °C demonstrates the best photocatalytic activity for degradation of RhB and reduction of Cr(VI), presumably owing to the unique electronic structure, strong light harvesting ability, high surface area, and accelerated separation rate of photo-induced electron-hole. Our findings not only significantly improve the understanding on the role of condensation temperature in the property of carbon nitride polymers, but also open the possibilities for the facile synthesis of nanostructured materials to manipulate their activity for photon-involving purification of water and air.

Acknowledgments: This work is financially supported by the National Basic Research Program of China (2013CB632405), the National Natural Science Foundation of China (21425309 and 21761132002), and the 111 Project.

Author Contributions: Xinchen Wang conceived of the project and supervised the research work. Yun Zheng performed most of the experiments and analyzed the data; Zihao Yu performed the FT-IR experiments; Feng Lin, Fangsong Guo, Khalid A. Alamry, Layla A. Taib, and Abdullah M. Asiri contributed to the analysis and discussion on the characterization results. All authors contributed to the writing of the paper and approved the final version of the manuscript.

Conflicts of Interest: The authors declare no conflict of interest.

References

1. Pichat, P. (Ed.) *Photocatalysis and Water Purification: From Fundamentals to Recent Applications*; Wiley-VCH: Weinheim, Germany, 2013; pp. 1–406.

2. Schneider, J.; Bahnemann, D.; Ye, J.; Puma, G.L.; Dionysiou, D.D. (Eds.) *Photocatalysis: Fundamentals and Perspectives*; Royal Society of Chemistry: London, UK, 2016; Volume 1.

3. Dionysiou, D.D.; Puma, G.L.; Ye, J.; Schneider, J.; Bahnemann, D. (Eds.) *Photocatalysis: Applications*; Royal Society of Chemistry: London, UK, 2016; Volume 2.

4. Pichat, P. (Ed.) *Photocatalysis: Fundamentals, Materials and Potential*; MDPI: Basel, Switzerland, 2016.

5. Colmenares Quintero, J.C.; Xu, Y. (Eds.) *Heterogeneous Photocatalysis: From Fundamentals to Green Applications*; Springer: Berlin, Germany, 2016.

6. Nosaka, Y.; Nosaka, A. (Eds.) *Introduction to Photocatalysis: From Basic Science to Applications*; Royal Society of Chemistry: London, UK, 2016.

7. Yang, C.; Ma, B.C.; Zhang, L.; Lin, S.; Ghasimi, S.; Landfester, K.; Zhang, K.A.I.; Wang, X.C. Molecular engineering of conjugated polybenzothiadiazoles for enhanced hydrogen production by photosynthesis. *Angew. Chem. Int. Ed.* **2016**, *55*, 9202–9206. [CrossRef] [PubMed]

8. Zhang, G.G.; Lan, Z.A.; Lin, L.H.; Lin, S.; Wang, X.C. Overall water splitting by Pt/g-C_3N_4 photocatalysts without using sacrificial agents. *Chem. Sci.* **2016**, *7*, 3062–3066. [CrossRef]

9. Wang, X.C.; Maeda, K.; Thomas, A.; Takanabe, K.; Xin, G.; Carlsson, J.M.; Domen, K.; Antonietti, M. A metal-free polymeric photocatalyst for hydrogen production from water under visible light. *Nat. Mater.* **2009**, *8*, 76–80. [CrossRef] [PubMed]

10. Pan, Z.M.; Zheng, Y.; Guo, F.S.; Niu, P.P.; Wang, X.C. Decorating CoP and Pt nanoparticles on graphitic carbon nitride nanosheets to promote overall water splitting by conjugated polymers. *ChemSusChem* **2017**, *10*, 87–90. [CrossRef] [PubMed]

11. Zhang, G.G.; Lan, Z.A.; Wang, X.C. Conjugated polymers: Catalysts for photocatalytic hydrogen evolution. *Angew. Chem. Int. Ed.* **2016**, *55*, 15712–15727. [CrossRef] [PubMed]

12. Wu, X.; Fang, S.; Zheng, Y.; Sun, J.; Lv, K. Thiourea-modified TiO_2 nanorods with enhanced photocatalytic activity. *Molecules* **2016**, *21*, 181. [CrossRef] [PubMed]

13. Liang, S.; Zhou, Z.; Wu, X.; Zhu, S.; Bi, J.; Zhou, L.; Liu, M.; Wu, L. Constructing a MoS_2 QDs/CdS core/shell flowerlike nanosphere hierarchical heterostructure for the enhanced stability and photocatalytic activity. *Molecules* **2016**, *21*, 213. [CrossRef] [PubMed]

14. Zheng, D.D.; Cao, X.N.; Wang, X.C. Precise formation of a hollow carbon nitride structure with a Janus surface to promote water splitting by photoredox catalysis. *Angew. Chem. Int. Ed.* **2016**, *55*, 11512–11516. [CrossRef] [PubMed]

15. Zheng, Y.; Lin, L.H.; Ye, X.J.; Guo, F.S.; Wang, X.C. Helical graphitic carbon nitrides with photocatalytic and optical activities. *Angew. Chem. Int. Ed.* **2014**, *53*, 11926–11930. [CrossRef] [PubMed]

16. Long, B.H.; Zheng, Y.; Lin, L.H.; Alamry, K.A.; Asiri, A.M.; Wang, X.C. Cubic Mesoporous Carbon Nitride Polymers with Large Cage Type Pores for Visible Light Photocatalysis. *J. Mater. Chem. A* **2017**. [CrossRef]

17. Li, X.; Zhang, J.; Chen, X.; Fischer, A.; Thomas, A.; Antonietti, M.; Wang, X.C. Condensed graphitic carbon nitride nanorods by nanoconfinement: Promotion of crystallinity on photocatalytic conversion. *Chem. Mater.* **2011**, *23*, 4344–4348. [CrossRef]

18. Li, X.H.; Wang, X.C.; Antonietti, M. Mesoporous g-C_3N_4 nanorods as multifunctional supports of ultrafine metal nanoparticles: Hydrogen generation from water and reduction of nitrophenol with tandem catalysis in one step. *Chem. Sci.* **2012**, *3*, 2170–2174. [CrossRef]

19. Wang, Y.; Wang, X.C.; Antonietti, M.; Zhang, Y. Facile one-pot synthesis of nanoporous carbon nitride solids by using soft templates. *ChemSusChem* **2010**, *3*, 435–439. [CrossRef] [PubMed]

20. Yan, H. Soft-templating synthesis of mesoporous graphitic carbon nitride with enhanced photocatalytic H_2 evolution under visible light. *Chem. Commun.* **2012**, *48*, 3430–3432. [CrossRef] [PubMed]

21. Aida, T.; Meijer, E.W.; Stupp, S.I. Functional supramolecular polymers. *Science* **2012**, *335*, 813–817. [CrossRef] [PubMed]

22. Li, L.; Zhao, Y.; Antonietti, M.; Shalom, M. New organic semiconducting scaffolds by supramolecular preorganization: Dye intercalation and dye oxidation and reduction. *Small* **2016**, *12*, 6090–6097. [CrossRef] [PubMed]

23. Xu, J.; Zhu, J.; Yang, X.; Cao, S.; Yu, J.; Shalom, M.; Antonietti, M. Synthesis of organized layered carbon by self-templating of dithiooxamide. *Adv. Mater.* **2016**, *28*, 6727–6733. [CrossRef] [PubMed]

24. Seto, C.T.; Whitesides, G.M. Self-assembly based on the cyanuric acid-melamine lattice. *J. Am. Chem. Soc.* **1990**, *112*, 6409–6411. [CrossRef]

25. Seto, C.T.; Whitesides, G.M. Molecular self-assembly through hydrogen bonding: Supramolecular aggregates based on the cyanuric acid-melamine lattice. *J. Am. Chem. Soc.* **1993**, *115*, 905–916. [CrossRef]

26. Mathias, J.P.; Simanek, E.E.; Zerkowski, J.A.; Seto, C.T.; Whitesides, G.M. Structural preferences of hydrogen-bonded networks in organic solution-the cyclic $CA_3 \cdot M_3$ "rosette". *J. Am. Chem. Soc.* **1994**, *116*, 4316–4325. [CrossRef]

27. Ishida, Y.; Chabanne, L.; Antonietti, M.; Shalom, M. Morphology control and photocatalysis enhancement by the one-pot synthesis of carbon nitride from preorganized hydrogen-bonded supramolecular precursors. *Langmuir* **2014**, *30*, 447–451. [CrossRef] [PubMed]

28. Shalom, M.; Guttentag, M.; Fettkenhauer, C.; Inal, S.; Neher, D.; Llobet, A.; Antonietti, M. In situ formation of heterojunctions in modified graphitic carbon nitride: Synthesis and noble metal free photocatalysis. *Chem. Mater.* **2014**, *26*, 5812–5818. [CrossRef]

29. Xu, J.; Brenner, T.J.K.; Chabanne, L.; Neher, D.; Antonietti, M.; Shalom, M. Liquid-based growth of polymeric carbon nitride layers and their use in a mesostructured polymer solar cell with Voc exceeding 1 V. *J. Am. Chem. Soc.* **2014**, *136*, 13486–13489. [CrossRef] [PubMed]

30. Guo, S.; Deng, Z.; Li, M.; Jiang, B.; Tian, C.; Pan, Q.; Fu, H. Phosphorus-doped carbon nitride tubes with a layered micro-nanostructure for enhanced visible-light photocatalytic hydrogen evolution. *Angew. Chem. Int. Ed.* **2016**, *55*, 1830–1834. [CrossRef] [PubMed]

31. Yang, X.; Tang, H.; Xu, J.; Antonietti, M.; Shalom, M. Silver phosphate/graphitic carbon nitride as an efficient photocatalytic tandem system for oxygen evolution. *ChemSusChem* **2015**, *8*, 1350–1358. [CrossRef] [PubMed]

32. Zou, X.; Silva, R.; Goswami, A.; Asefa, T. Cu-doped carbon nitride: Bio-inspired synthesis of H_2-evolving electrocatalysts using graphitic carbon nitride (g-C_3N_4) as a host material. *Appl. Surf. Sci.* **2015**, *357*, 221–228. [CrossRef]

33. Cui, Q.; Xu, J.; Wang, X.; Li, L.; Antonietti, M.; Shalom, M. Phenyl-modified carbon nitride quantum dots with distinct photoluminescence behavior. *Angew. Chem. Int. Ed.* **2016**, *55*, 3672–3676. [CrossRef] [PubMed]

34. Guo, Y.; Li, J.; Yuan, Y.; Li, L.; Zhang, M.; Zhou, C.; Lin, Z. A Rapid microwave-assisted thermolysis route to highly crystalline carbon nitrides for efficient hydrogen generation. *Angew. Chem. Int. Ed.* **2016**, *55*, 14693–14697. [CrossRef] [PubMed]

35. Xu, J.; Brenner, T.J. K.; Chen, Z.; Neher, D.; Antonietti, M.; Shalom, M. Upconversion-agent induced improvement of g-C_3N_4 photocatalyst under visible light. *ACS Appl. Mater. Interfaces* **2014**, *6*, 16481–16486. [CrossRef] [PubMed]

36. Shalom, M.; Inal, S.; Fettkenhauer, C.; Neher, D.; Antonietti, M. Improving carbon nitride photocatalysis by supramolecular preorganization of monomers. *J. Am. Chem. Soc.* **2013**, *135*, 7118–7121. [CrossRef] [PubMed]

37. Jun, Y.S.; Lee, E.Z.; Wang, X.; Hong, W.H.; Stucky, G.D.; Thomas, A. From melamine-cyanuric acid supramolecular aggregates to carbon nitride hollow spheres. *Adv. Funct. Mater.* **2013**, *23*, 3661–3667. [CrossRef]

38. Li, Y.; Yang, L.; Dong, G.; Ho, W. Mechanism of NO photocatalytic oxidation on g-C_3N_4 was changed by Pd-QDs modification. *Molecules* **2016**, *21*, 36. [CrossRef] [PubMed]

39. Chen, Y.; Wang, B.; Lin, S.; Zhang, Y.; Wang, X. Activation of n→π* transitions in two-dimensional conjugated polymers for visible light photocatalysis. *J. Phys. Chem. C* **2014**, *118*, 29981–29989. [CrossRef]

40. Ranganathan, A.; Pedireddi, V.R.; Rao, C.N.R. Hydrothermal synthesis of organic channel structures: 1:1 hydrogen-bonded adducts of melamine with cyanuric and trithiocyanuric acids. *J. Am. Chem. Soc.* **1999**, *121*, 1752–1753. [CrossRef]

41. Fan, Q.; Liu, J.; Yu, Y.; Zuo, S.; Li, B. A simple fabrication for sulfur doped graphitic carbon nitride porous rods with excellent photocatalytic activity degrading RhB dye. *Appl. Surf. Sci.* **2017**, *391*, 360–368. [CrossRef]

42. Feng, L.; Zou, Y.; Li, C.; Gao, S.; Zhou, L.; Sun, Q.; Fan, M.; Wang, H.; Wang, D.; Li, G.; Zou, X. Nanoporous sulfur-doped graphitic carbon nitride microrods: A durable catalyst for visible-light-driven H_2 evolution. *Int. J. Hydrog. Energy* **2014**, *39*, 15373–15379. [CrossRef]

43. Lan, Z.A.; Zhang, G.G.; Wang, X.C. A facile synthesis of Br-modified g-C_3N_4 semiconductors for photoredox water splitting. *Appl. Catal. B* **2016**, *192*, 116–125. [CrossRef]

44. Long, B.H.; Lin, J.L.; Wang, X.C. Thermally-induced desulfurization and conversion of guanidine thiocyanate into graphitic carbon nitride catalysts for hydrogen photosynthesis. *J. Mater. Chem. A* **2014**, *2*, 2942–2951. [CrossRef]

45. Wang, Y.; Lin, F.; Peng, J.; Dong, Y.; Li, W.; Huang, Y. A robust bilayer nanofilm fabricated on copper foam for oil-water separation with improved performances. *J. Mater. Chem. A* **2016**, *4*, 10294–10303. [CrossRef]

46. Sun, J.; Xu, J.; Grafmueller, A.; Huang, X.; Liedel, C.; Algara-Siller, G.; Willinger, M.; Yang, C.; Fu, Y.; Wang, X.C.; et al. Self-assembled carbon nitride for photocatalytic hydrogen evolution and degradation of p-nitrophenol. *Appl. Catal. B* **2017**, *205*, 1–10. [CrossRef]

47. Liang, R.; Jing, F.; Shen, L.; Qin, N.; Wu, L. MIL-53(Fe) as a highly efficient bifunctional photocatalyst for the simultaneous reduction of Cr(VI) and oxidation of dyes. *J. Hazard. Mater.* **2015**, *287*, 364–372. [CrossRef] [PubMed]

48. Fujishima, A.; Rao, T.N.; Tryk, D.A. Titanium dioxide photocatalysis. *J. Photochem. Photobiol. C* **2000**, *1*, 1–21. [CrossRef]

49. He, Y.; Cai, J.; Li, T.; Wu, Y.; Lin, H.; Zhao, L.; Luo, M. Efficient degradation of RhB over $GdVO_4$/g-C_3N_4 composites under visible-light irradiation. *Chem. Eng. J.* **2013**, *215–216*, 721–730. [CrossRef]

50. Sawyer, D.T.; Valentine, J.S. How super is superoxide? *Acc. Chem. Res.* **1981**, *14*, 393–400. [CrossRef]

51. Cui, Y.; Ding, Z.; Liu, P.; Antonietti, M.; Fu, X.; Wang, X. Metal-free activation of H_2O_2 by g-C_3N_4 under visible light irradiation for the degradation of organic pollutants. *Phys. Chem. Chem. Phys.* **2012**, *14*, 1455–1462. [CrossRef] [PubMed]

52. Yan, S.C.; Li, Z.S.; Zou, Z.G. Photodegradation of rhodamine B and methyl orange over boron-doped g-C_3N_4 under visible light irradiation. *Langmuir* **2010**, *26*, 3894–3901. [CrossRef] [PubMed]

53. Zheng, Y.; Pan, Z.M.; Wang, X.C. Advances in photocatalysis in China. *Chin. J. Catal.* **2013**, *34*, 524–535. [CrossRef]

54. Zheng, Y.; Lin, L.H.; Wang, B.; Wang, X.C. Graphitic carbon nitride polymers toward sustainable photoredox catalysis. *Angew. Chem. Int. Ed.* **2015**, *54*, 12868–12884. [CrossRef] [PubMed]

55. Lin, L.H.; Ou, H.H.; Zhang, Y.; Wang, X.C. Tri-s-triazine-based crystalline graphitic carbon nitrides for highly efficient hydrogen evolution photocatalysis. *ACS Catal.* **2016**, *6*, 3921–3931. [CrossRef]

56. Yuan, X.; Wang, Y.; Wang, J.; Zhou, C.; Tang, Q.; Rao, X. Calcined graphene/MgAl-layered double hydroxides for enhanced Cr(VI) removal. *Chem. Eng. J.* **2013**, *221*, 204–213. [CrossRef]

57. Lazaridis, N.K.; Asouhidou, D.D. Kinetics of sorptive removal of chromium(VI) from aqueous solutions by calcined Mg-Al-CO_3 hydrotalcite. *Water Res.* **2003**, *37*, 2875–2882. [CrossRef]

Sample Availability: Samples of the sulfur-doped carbon nitride microrods and bulk g-C3N4 are available from the authors.

Section 5:
Modeling and Testing Photocatalytic Reactors for Air Purification

molecules

MDPI

Review

Integral Design Methodology of Photocatalytic Reactors for Air Pollution Remediation

Claudio Passalía [1], Orlando M. Alfano [2],* and Rodolfo J. Brandi [2]

[1] Facultad de Ingeniería y Ciencias Hídricas, Universidad Nacional del Litoral and CONICET, Santa Fe 3000, Argentina; cpassalia@unl.edu.ar

[2] Instituto de Desarrollo Tecnológico para la Industria Química (CONICET-UNL), Santa Fe 3000, Argentina; rbrandi@santafe-conicet.gov.ar

* Correspondence: alfano@santafe-conicet.gob.ar; Tel.: +54-0342-451-1596

Academic Editor: Pierre Pichat
Received: 22 March 2017; Accepted: 31 May 2017; Published: 7 June 2017

Abstract: An integral reactor design methodology was developed to address the optimal design of photocatalytic wall reactors to be used in air pollution control. For a target pollutant to be eliminated from an air stream, the proposed methodology is initiated with a mechanistic derived reaction rate. The determination of intrinsic kinetic parameters is associated with the use of a simple geometry laboratory scale reactor, operation under kinetic control and a uniform incident radiation flux, which allows computing the local superficial rate of photon absorption. Thus, a simple model can describe the mass balance and a solution may be obtained. The kinetic parameters may be estimated by the combination of the mathematical model and the experimental results. The validated intrinsic kinetics obtained may be directly used in the scaling-up of any reactor configuration and size. The bench scale reactor may require the use of complex computational software to obtain the fields of velocity, radiation absorption and species concentration. The complete methodology was successfully applied to the elimination of airborne formaldehyde. The kinetic parameters were determined in a flat plate reactor, whilst a bench scale corrugated wall reactor was used to illustrate the scaling-up methodology. In addition, an optimal folding angle of the corrugated reactor was found using computational fluid dynamics tools.

Keywords: air pollution; photocatalytic reactors; radiation modeling; reactor optimization

1. Introduction and Scope

Indoor air quality became a generalized concern in the last few decades. People spend most of their time in confined environments and may be exposed to a poor air quality for extended periods of time. Such a situation is regarded as a human health concern [1–3]. The effects in human health of a poor air quality in a room can range from headaches to nausea, dizziness, eye and nose irritations, dry cough and tiredness [4], a situation known as sick building syndrome (SBS). This syndrome is usually associated with the presence of volatile organic compounds (VOCs) in very low concentrations [1,5,6].

A poor indoor air pollution quality may result from in situ generation of compounds or from an exchange with the outside. For many pollutants, it is usual to find indoor concentrations larger than outdoors, particularly VOCs, which are emitted from building materials, furniture and equipment [4]. Among indoor VOCs, simple aldehydes such as formaldehyde (HCHO) are typically found in polluted places [7].

The recognition of the health effects of indoor VOCs implies the need to attain their depletion. The reduction of pollutant concentrations in air has been traditionally centered in the source control, the increasing of air renewal rates and the application of air cleaning devices. Conventional control processes usually employed are filtration and adsorption; these technologies present certain drawbacks, but above all they require final disposal because there is a transfer of the pollutant from the gas phase to a solid one. In this context, indoor air quality may be controlled by heterogeneous photocatalysis, an effective alternative to conventional technologies that has been tested to chemically destroy a large variety of airborne pollutants [8–10].

In the photocatalytic process, the compounds present in the air stream may be adsorbed onto the surface of a catalyst, upon which the irradiation starts a series of superficial reactions that can lead to the chemical degradation of pollutants. Although the final products of organic pollutants containing no hetero-atoms are water and carbon dioxide, numerous intermediate products may appear in the reaction steps and their elimination must also be taken into account [11].

Photocatalytic reactors must gather the molecules of the pollutant, the surface of the photocatalyst particles and the radiation energy in the proper wavelength at once. Thus, the design and modeling of such reactors presents the need for solving the radiation field in addition to the classical momentum, energy and species mass balances.

The aim of this work is to present a methodology for an integral design of photocatalytic reactors for the control of air pollution as a synthesis and review, including the development of intrinsic kinetic models [12], reactor scaling-up [13] and optimization [14]. The work is organized in two main sections, being the first a detailed description of the proposed methodology and the second a concrete and successful application.

2. Integral Design Methodology

The proposed methodology for the complete procedure of integral reactor design is schematically represented in Figure 1, and may be summarized in the following steps.

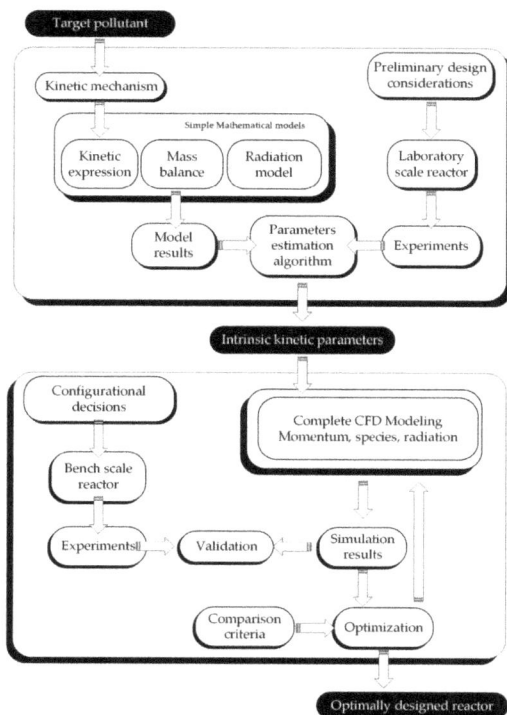

Figure 1. Conceptual framework of the integral design methodology.

2.1. Kinetic Study

There are many chemical compounds usually found in polluted indoor air, among which the family of VOCs can be found. The selection of one or more pollutants to be eliminated is generally the first step in a photocatalytic study. Every compound, with its own physicochemical properties, has its implications regarding the experimental issues, such as generation and analytical determination. Then, a kinetic mechanism for the photocatalytic degradation of the selected target compound can lead to obtaining an analytical reaction rate expression. This rate expression must include the dependence on the pollutant and stable intermediates concentrations and the local superficial rate of photon absorption (LSRPA or $e^{a,s}$) [15].

To carry out experimental tests in photocatalysis, it is important to consider, previously to the reactor design, aspects such as the photocatalyst's characteristics (its optical properties, its chemical stability or its deactivation), the support material in which the catalyst is to be fixed, the radiation source (emission spectrum and power) and the interactive radiation in the interior of the reactor.

The selection of the catalyst is one of the essential steps to be considered in the design and application of a photocatalytic reactor [16]. In general, photocatalysts are metal oxides that behave as semiconductors which absorb radiation in a certain wavelength range. Titanium dioxide (TiO_2) is the most widely used photocatalyst because of its chemical stability, low toxicity and relatively low cost. TiO_2 absorbs radiation in the ultraviolet (UV) range of the spectrum to promote electrons across the energy gap into the conduction band. The vacancies or holes left by those electrons may then form hydroxyl radicals capable of attacking adsorbed organic compounds onto the catalyst's surface.

The depuration of polluted air requires the fixation of TiO_2 over some material acting as a support, given that the catalyst should not be dragged by the gas stream. Among the possibilities of immobilization on the support material one can find photocatalyst coatings, layers and films (but they can also be dispersed in a matrix to build monoliths). The techniques to affix the catalyst and support mainly include variants of dip-coating and sol-gel methods. The purpose is to obtain the largest possible surface area to volume ratio, a large area exposed to radiation and good adherence to an inert substrate. Materials tested to act as inert supports are diverse: glass, metals, fibers, plastics, etc. [17].

The radiation that initiates the superficial phenomena leading to the pollutant elimination may be artificial or natural, i.e., coming from a lamp or from the Sun. According to which source of radiation will be used, the geometry of the reactor may differ greatly. Because of the TiO_2 bandgap, it cannot profit from a large portion of the solar spectrum [18]; thus, efforts are constantly directed towards the doping of TiO_2 for extending the absorption at larger wavelengths (>390 nm) towards the visible range. Doping with C, N, Ce, etc. has been studied with certain success [19,20].

Regarding the size and configuration of the laboratory scale reactor, simplicity is desirable. A simple geometry reactor ensures the application of a simple mathematical model and, under selected operating conditions, the absence of mass transfer limitations. In addition, a uniform radiation flux allows the incorporation of the LSRPA as a constant in the kinetic model. Typically, the radiation model is based on an emission model for the lamps [15] and the determination of spectral optical properties of the catalyst layer and support. In accordance to what has been said above, it is clear why the flat plate reactor may be thought of as a standard for kinetic studies: under the imposed size, geometry and operation, it can be accurately modeled as a plug flow reactor with a straightforward solution.

Thus far, we have a simple reactor model, coupled with the kinetic expressions and mathematically solved. The experimental data from the laboratory reactor together with the mathematical model allow the determination of the unknown kinetic parameters. The procedure for estimation of the kinetic parameters is typically achieved by running a numeric nonlinear algorithm to fit the model predictions to the experimental data. An intrinsic kinetics implies the independence of reactor design variables, including the radiation source and operating variables. In this sense, two important considerations need to be satisfied for the application of the kinetic parameters obtained in the kinetic reactor to the scaling up: (i) to employ the same catalyst and immobilization technique in the larger scale reactor;

and (ii) to ensure that the experimental data in the laboratory reactor are obtained under kinetic control regime to eliminate the effect of mass transfer limitations.

2.2. Scaling and Optimization Methodology

As previously stated, the present work addresses the implementation of a methodology for the scaling-up and optimization of a photocatalytic reactor using experimental data obtained in a simple geometry laboratory scale reactor. As depicted in Figure 1, for scaling-up purposes (changing the reactor shape, size, lamps, and configuration), the kinetic model developed in a previous step is directly applied to the new reactor. The solution of the mass balance and the radiation model allows predicting the performance of the bench scale reactor.

Once the explicit reaction rate and the kinetic parameters are known, their application to any reactor size or configuration is direct, provided that: the mass balances in the reactor can be solved and the radiation field can also be evaluated properly. Thus, the LSRPA must be known in the new reactor which may differ from the laboratory one not only in size or shape but also in its radiative behavior. The obtained field of LSRPA is introduced in the mass balance to simulate the reactor performance and predict its conversion.

When it comes to bench scale reactors, the biggest modifications are related to the inner and outer configuration and size, i.e., the geometrical arrangement of the radiation source with respect to the reaction space; the operation regime (continuous flow, batch, recycle, etc.); and the inner reactor geometry (shape and dimensions).

Photoreactor configuration and geometry are essential because, in addition to the velocity and concentrations profiles, radiation field must be known for a rigorous modeling. As has been previously stated, for the treatment of polluted air streams, the catalyst is immobilized on the support, regardless of the reactor type. In this respect, the best available option is the family of photocatalytic wall reactors. Among the geometries or configurations of photocatalytic wall reactors one can find: the flat plate reactor [12,21], the multi-plate reactor [22], monolith and honeycomb reactors [23,24], mesh reactors [25,26], annular or multi-annular reactor [27,28], optical fiber [29] and corrugated [30,31] reactors, all of which were applied with acceptable to good efficiency in the abatement of innumerable compounds.

The following step is validation, where the data obtained from experimental runs in the bench scale reactor are contrasted against simulation results. Photocatalytic reactors entail simultaneously momentum, mass and radiation transfer. Their design usually needs an optimization step in order to obtain the best global performance possible. The comprehensive mathematical simulation of photochemical reactors is a very helpful and affordable tool nowadays with computational capabilities sufficiently large to provide a detailed approach to all phenomena involved. In particular, computational fluid dynamics (CFD) tools have been increasingly used to model very different kinds of processes, including the modeling of gas phase photocatalytic reactors [31–33].

3. Application: Step-by-Step Design and Optimization of a Photocatalytic Reactor

3.1. Laboratory Scale Reactor

Recalling Figure 1, the determination of kinetic parameters is based on experimental data, in particular, the assays performed in the laboratory scale reactor.

3.1.1. Experimental

A complete experimental set-up was designed and constructed; a scheme is presented in Figure 2. The reaction device itself is a continuous, single-pass flat plate reactor with a photocatalytic wall [12].

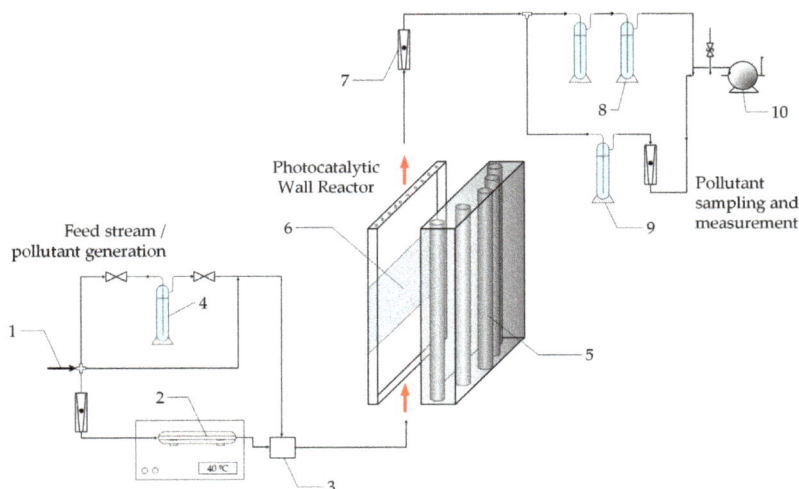

Figure 2. Experimental set-up: (1) Clean dry air; (2) thermostated formaldehyde (HCHO) generator; (3) mixer; (4) water vapor saturator; (5) ultraviolet (UV) lamps; (6) TiO_2 coated stainless steel plate; (7) variable area flowmeter; (8) scrubbers; (9) HCHO collection; and (10) pump.

The reactor consists of an acrylic parallelepiped; in its interior, the photocatalytic wall is a flat stainless steel plate coated with TiO_2. The reactor window is transparent to radiation of wavelengths in the near UV range (300–400 nm). The radiation source is a group of five black light fluorescent lamps (Sylvania F15W T12) that provide a uniform radiation flux over a wide area (Table 1). This is achieved by a convenient arrangement of the lamp positions in relation with the reactor window, as seen in Figure 2.

Table 1. Main characteristics of reactors and operating conditions.

Description	Laboratory Scale Reactor	Bench Scale Reactor
Configuration	Flat plate	Corrugated plate
Main dimensions	8 (W) × 8 (L) × 0.4 (T) cm	20 (W) × 30 (L) × 3 (T) cm
Reactor volume	25.6 cm^3	1800 cm^3
Catalytic surface area	64 cm^2	1843.5 cm^2
Number of lamps	5	10
Flow Rate	1–3.5 L/min	
Inlet HCHO concentration	5–35 ppmv	
Relative Humidity	10–75%	
Maximum radiation	8.94×10^{-5} einstein m^{-2} s^{-1}	
Radiation levels	16%, 26%, 60%, 100%	

The fluid feed to the reactor has a known pollutant concentration (formaldehyde) and relative humidity. The pneumatic circulation within the device is provided by a vacuum pump from its output; the volumetric flow rate is measured using variable area flowmeters. At the reactor outlet, samples are taken to determine the overall pollutant conversion.

3.1.2. Procedures

Catalyst preparation: A flat plate of AISI 316 stainless steel (Acerind S.C., Santa Fe, Argentina) is the support material. TiO_2 was fixed by cycles of impregnation with an aqueous suspension. A solution of methanol in water 25% (*v*/*v*) is prepared and pure TiO$_2$ powder (Aeroxide P25 (Evonik Industries AG, Essen, Germany)) is then added to reach a 45 g/L concentration. Dip coating is performed in a vessel; after extraction, the piece is dried in stove and the cycle is repeated four times. The obtained layers of catalyst presented good uniformity and adherence to the metal support.

Analytical techniques: HCHO was generated online by heating solid paraformaldehyde. For the analytical determination of HCHO concentration in air, the experimental device was adapted to apply a colorimetric method.

Experimental runs: The experimental set-up was designed to collect samples only at the exit of the reactor; thus, the inlet HCHO concentration is established by taking samples over time with the UV lamps occluded. When a steady state inlet concentration is achieved, the lamps are uncovered and samples are taken in this irradiated period until the HCHO concentration reaches a new steady state, the outlet concentration. Different radiation levels were obtained using grey filters, which provided attenuations to give 16%, 26% and 60% of the maximum radiation flux.

3.1.3. Kinetics

The starting point for this study is a simplified kinetic scheme based on a published reaction pathway [34]. After some algebraic work and the application of the micro steady-state approximation (MSSA), an analytical expression for the reaction rate was developed [12]. Table 2 presents the reaction scheme with all the involved steps:

Table 2. Kinetic pathway for formaldehyde (HCHO) photo-oxidation.

Step	Reaction	Reaction
Initiation	1	$TiO_2 + h\nu \rightarrow TiO_2 + e^- + h^+$
Recombination	2	$e^- + h^+ \rightarrow Heat$
$^\bullet OH$ generation	3	$h^+ + OH^- \rightarrow {}^\bullet OH$
Electron trapping	4	$e^- + O_2 \rightarrow {}^\bullet O_2^-$
HCHO Oxidation	5	$HCHO_{ads} + {}^\bullet OH \rightarrow {}^\bullet CHO + H_2O_{ads}$
	6	$^\bullet CHO + {}^\bullet OH \rightarrow HCOOH_{ads}$
Ending reactions	7	$HCOOH_{ads} + {}^\bullet OH \rightarrow Products$
	8	$M + {}^\bullet OH \rightarrow Products$
Adsorption equilibria	9	$HCHO + Site \rightleftarrows HCHO_{ads}$
	10	$H_2O + Site \rightleftarrows H_2O_{ads}$
Site balance	11	$Site|_{Total} = Site|_{HCHO} + Site|_{H_2O} + Site$

Ending reaction 8 in Table 2 is a generic termination step for the active oxidizing species, where M is an inert species or body that can consume hydroxyl radicals. Formic acid (HCOOH) is an intermediate product that has a fast rate of disappearance compared to its formation rate and presents concentrations three orders of magnitude lower than those of HCHO [6]. In addition, a balance for adsorption sites on the catalyst surface for HCHO and water makes it possible to relate surface to bulk concentrations. The final reaction rate expression is:

$$r_F = r_5 = -\alpha_1 \frac{\left(-1 + \sqrt{1 + \alpha_2 r_g}\right) C_F}{1 + \kappa_W C_W + \kappa_F C_F} \tag{1}$$

This four-parameter expression is a function of HCHO concentration (C_F), the water vapor concentration (C_W) and the superficial rate of electron–hole pair generation (r_g). The kinetic parameters α_1 and α_2 are combinations of kinetic constants, concentrations of species that remain inalterable or considered in excess, while κ_F and κ_W are adsorption equilibrium constants that relate bulk to surface concentrations. The local superficial rate of electron–hole pair generation can be defined as follows [15]:

$$r_g(\underline{x}) = \int_\lambda \Phi_\lambda e_\lambda^{a,s}(\underline{x}) d\lambda = \overline{\Phi} \sum_\lambda e_\lambda^{a,s}(\underline{x}) = \overline{\Phi} \cdot e^{a,s}(\underline{x}) \tag{2}$$

where $e_\lambda^{a,s}$ is the LSRPA, a function of position ($\underline{x}$) and wavelength (λ), and $\overline{\Phi}$ is a wavelength averaged primary quantum yield. Experimental evidence of the linear dependence of the reaction rate with the radiation level was also found. Thus, the rate expression was simplified to give:

$$r_F = \frac{-\alpha e^{a,s} C_F}{1 + \kappa_w C_w + \kappa_F C_F} \tag{3}$$

where α is the main kinetic parameter.

3.1.4. Modeling

As stated in Figure 1, the laboratory scale reactor must ensure the simplicity in the mathematical modeling. Considering the advantages of the simple geometry, the mass balance for formaldehyde in the reactor becomes an ordinary differential equation:

$$\langle v_z \rangle_{A_c} \frac{d\langle C_F \rangle_{A_c}}{dz} = a_v \, r_F \tag{4}$$

$$z = 0 \quad \langle C_F \rangle_{A_c} = C_F^{in} \tag{5}$$

where a_v is the external catalytic surface area per unit volume and r_F is the heterogeneous rate of disappearance of HCHO. Equation (4) represents the variation of the pollutant concentration along the reactor length due to the photocatalytic reaction. After inserting Equation (3) into Equation (4) and solving together with its boundary condition (Equation (5)) we have:

$$\kappa_F \left(C_F^{out} - C_F^{in} \right) + (1 + \kappa_w C_w) \ln \left(\frac{C_F^{out}}{C_F^{in}} \right) = -\alpha \frac{A_{cat}}{Q} e^{a,s} \tag{6}$$

This expression is implicit in terms of the cross-section averaged outlet formaldehyde concentration: $C_F^{out} = \langle C_F \rangle_{A_c} (z = L)$. It was solved by means of a non-linear optimization procedure based on the Levenberg–Marquardt algorithm coupled with the set of experimental data to obtain the values of the three kinetic parameters.

Radiation field model. The evaluation of the LSRPA inside the reactor is based on the extense source with superficial emission (ESSE) model for the UV lamps [15] and the ray tracing method. This allows the integration of radiation contributions coming from any point at the lamps surfaces that are visible from the point of incidence considered on the catalytic film. The ESSE model is:

$$I_{\lambda,L_i}(y, z, \phi, \theta) = T_{\lambda,Wi} \, T_{\lambda,Fi} \, I_{\lambda,L_i}^0 = T_{\lambda,Wi} \, T_{\lambda,Fi} \frac{P_{\lambda,L_i}}{2\pi^2 R_{L_i} L_{L_i}} \tag{7}$$

where I_{λ,L_i}^0 is the radiation intensity leaving the source and computed using the superficial lamp emission model of the cylindrical actinic lamp. Here, P_{λ,L_i} is the emission power output of the lamp for a given wavelength provided by the manufacturer, and R_{L_i} and L_{L_i} are the lamp radius and length, respectively. The spectral relative emission power of the lamp and the acrylic window transmittance need to be measured.

Apart from that, for this multilamp system, the additive effect of each lamp "i" on the LSRPA must be taken into account. The expression of the LSRPA ($e^{a,s}$) at each point on the catalytic plate is:

$$e^{a,s}(y, z) = \sum_\lambda \eta_{\lambda,abs} T_{\lambda,Wi} \, T_{\lambda,Fi} \sum_i \int_{\phi_{1,L_i}}^{\phi_{2,L_i}} \int_{\theta_{1,L_i}}^{\theta_{2,L_i}} \frac{P_{\lambda,L_i}}{2\pi^2 R_{L_i} L_{L_i}} \sin^2 \theta \, \sin \phi \, d\theta \, d\phi \tag{8}$$

where $\eta_{\lambda,abs}$ is a radiation absorption fraction that can be defined as the ratio of absorbed to incident radiation; and ϕ_{1,L_i}, ϕ_{2,L_i}, θ_{1,L_i}, θ_{2,L_i} are the limiting angles from which radiation can reach the incidence point for the lamp i. The absorption fraction can be determined from spectral total reflectance measurements. Equation (8) represents the contribution of every lamp i to the superficial radiation absorption over the useful wavelength range of the lamp emission.

3.1.5. Kinetic Parameters Estimation

In order to obtain intrinsic parameters for the degradation kinetics, the experimental data need to be obtained under kinetic control regime, i.e., in the absence of external mass transfer limitations. This was tested for our runs by the application of the following criterion that relates the observed reaction rate r_{obs} and the mass transfer of the pollutant from the bulk to the catalyst surface [35]: $r_{obs}/k_m C_F^{in} < 0.1$; in this expression, k_m is a mass transfer parameter estimated through empirical correlations and C_F^{in} is the inlet HCHO concentration.

The optimization procedure to obtain the values of the kinetic parameters in Equation (6) was a nonlinear algorithm that used the matrix of experimental data containing: (i) inlet HCHO concentration; (ii) relative humidity; (iii) radiation level; and (iv) outlet HCHO concentration. The values and units of the estimated parameters are: $\alpha = 1.34 \times 10^8$ cm^3 einstein^{-1}; $\kappa_F = 7.17 \times 10^9$ cm^3 mol^{-1} and $\kappa_W = 2.97 \times 10^6$ cm^3 mol^{-1}. The comparison between predicted and experimental HCHO conversions proved to be satisfactory as it can be seen in Figure 3.

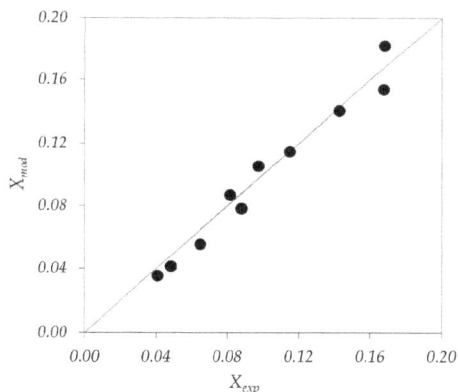

Figure 3. Results of the kinetic parameters estimation. Predicted pollutant conversion vs. experimental conversion (dots). The straight line represents a perfect fit. X_{mod}: Pollutant conversion predicted by the model; X_{exp}: experimental pollutant conversion.

3.2. Bench Scale Reactor: Corrugated Plate Reactor

Following with the procedure schematically shown in Figure 1, at this point we have a validated kinetics for the elimination of HCHO. The kinetic parameters are independent of radiation flux, so they can be applied directly to any reactor size and shape.

3.2.1. Experimental Set-Up and Procedures

The complete experimental set-up and procedures, except for the photoreactor itself, were the same as described in Section 3.1. The support of the bench scale photocatalytic reactor was also made of stainless steel; the stainless steel plate was successively folded into a one radian angle giving a total of 17 complete triangular section channels with their walls coated with TiO$_2$ and placed inside an acrylic frame [31]. The reactor is irradiated by the same arrangement of black light lamps, but in this case from both sides of the reactor (Figure 4).

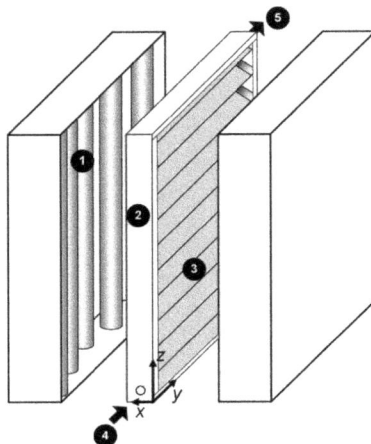

Figure 4. Corrugated wall reactor: (1) UV lamps; (2) acrylic frame; (3) stainless steel corrugated plate coated with TiO_2; (4) reactor inlet; and (5) reactor outlet. Adapted with permission from Passalía, C.; Alfano, O.M.; Brandi, R.J. A methodology for modeling photocatalytic reactors for indoor pollution control using previously estimated kinetic parameters. *J. Hazard. Mater.* **2012**, *211–212*, 357–365. Copyright 2012 Elsevier [13].

The corrugated reactor operates in continuous mode; air flows in a zigzag pattern from one triangular channel to another. Inlet pollutant concentration, water vapor concentration, radiation flux and total flow rate are fixed for each run.

By folding the metal plate, it is possible to obtain a more efficient use of available space and radiation, increasing the ratio between catalytic area per unit volume of reactor compared to a flat plate reactor under the same operating conditions. In addition, due to the possibility of interaction between catalytic walls by radiation reflection, one may expect the existence of an optimum angle: at more closed angles, the greater the reflective interaction between walls, the smaller the incoming radiation from the window.

3.2.2. CFD Modeling

CFD packages allow the numerical, iterative and simultaneous solution of the governing equations of motion and reaction in a certain domain by volume discretization. The fundamentals of modeling the present system are: a reactive isothermal system with heterogeneous reaction; air is considered a Newtonian incompressible fluid with constant physical properties; and the process is under steady state and the flow regime is laminar. According to these conditions, the CFD model implies the resolution of the following set of differential equations:

Continuity equation:

$$\underline{\nabla} \cdot (\rho \underline{v}) = 0 \tag{9}$$

Momentum equations:

$$\underline{\nabla} \cdot (\rho \underline{v}\underline{v}) = -\underline{\nabla}P + \underline{\nabla} \cdot (\underline{\underline{\tau}}) + \rho \underline{g} \tag{10}$$

Conservation equations for species:

$$\underline{\nabla} \cdot (\rho \underline{v} C_i) = -\underline{\nabla} \cdot \underline{J_j} + R_i \tag{11}$$

In Equations (9)–(11), ρ is the density, $\underline{v}$ the velocity, P the pressure, $\underline{\underline{\tau}}$ the viscous stress tensor, g the gravitational acceleration, C_i the molar concentration of species i, $\underline{J_j}$ the diffusion flux vector and R_i is the rate of production or depletion of species i by the homogeneous chemical reaction. The energy balance equation is not included provided that isothermal operation is assumed, with a fixed temperature of 298 K. In the system under study, there are no homogeneous reactions. Consequently, the term R_i is zero for every species i. The calculated Reynolds number within the reactor was 140, so the laminar model was employed. The Newton's law of viscosity is combined with Equation (10) as a constitutive equation for the definition of the stress tensor.

The approach for modeling the radiative interchange between windows and catalytic walls is by the introduction of user defined functions (UDFs). The incident radiation on the reactor windows was evaluated with the same approach to that of the flat plate reactor, i.e., based on the ESSE model for the lamps and the ray tracing method. The radiative interchange between windows and catalytic walls was also modeled externally and introduced as UDFs.

The radiation interchange can be modeled as follows: the reactor window is considered an emitting surface, while the catalytic walls are opaque and therefore can only reflect and absorb radiation. The radiation reflection on the catalyst surface is considered diffuse. At any position over the catalytic surface, radiation may come directly from the window or indirectly by reflection on the opposite catalytic surface.

The approach for modeling this radiative interaction is the surface to surface model by means of view factors [31,36]. Each surface is divided into small rectangular flat elements to compute de LSRPA. Elements in the same plane do not interact; in addition, the absorbed and reflected fractions do not depend on the position. The matrix of interaction between differential surface elements is solved numerically and the complete field of LSRPA over the corrugated plate is obtained.

3.2.3. Simulation Results and Validation

The CFD resolution was performed with Fluent 12.1 (ANSYS Inc., Canonsburg, PA, USA). The pressure based segregated solver was used in a 3D model, under laminar flow regime and time-independent mode. The LSRPA was introduced in the suite as a custom function as well as the heterogeneous reaction rate expression (Equation (3)). In this way, the software solves the species balances subject to the imposed boundary conditions over the folded plate.

The 3D mesh was built with tetrahedrons and wedges and tested for grid independence. After a converged solution was obtained, the complete velocity, radiation and pollutant concentration fields inside the reactor were analyzed. The criteria of convergence were: scaled residuals decreased five orders, stabilized macroscopic monitors and overall mass balance conservation.

Radiation Field: The absorbed radiation on one side of the corrugated plate is presented in Figure 5a. It can be seen that the absorbed radiation is higher at the top than at the bottom of the channels, despite the reflection of the coated surface. In addition, the central channel looks more "illuminated" while the channels towards the extremes receive less radiation, because of the edge effect of the lamps.

Velocity field: Figure 5b shows how the velocity profiles develop to reach a symmetrical distribution, with maximum velocities at the centroid of the triangular section and minimum in the corners and walls, where the flow is subject to a no-slip shear condition. The change in direction of 180 degrees and a slight narrowing of the passage section between channels causes the acceleration and the asymmetry in the contours of velocity at the beginning of each channel.

Concentration field: Figure 5c depicts a typical output of the pollutant concentration inside the reactor. HCHO concentration field shows a decreasing concentration along the reactor pathway. The effect of the catalytic walls along each channel is noticeable: the concentration of pollutant is maximum at the triangle's center, in the bulk, and decreases toward the walls where the reaction takes place. It is also clear that the change in direction from one channel to the other acts as a static mixer, renewing the concentration profile of HCHO. This mixing between each channel renews the concentration profile, smoothing the gradients and allowing the arrival of the pollutant to the reactive walls.

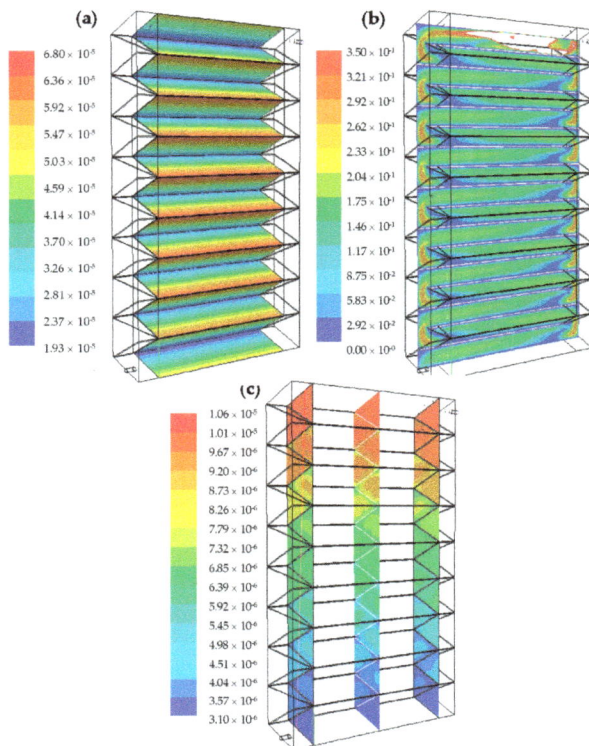

Figure 5. (**a**) LSRPA on the corrugated plate (einstein m^{-2} s^{-1}); (**b**) contours of velocity (m/s) in the central plane of the reactor; and (**c**) contours of molar HCHO fractions.

Five operating condition sets were simulated, using different inlet HCHO concentrations, water vapor concentrations and radiation levels. The five sets of conditions were experimentally tested in the bench scale photocatalytic reactor. Pollutant conversions from the model simulations were contrasted to the experimental results; good agreement was found between simulation results and experimental data. The root mean square error, based on the experimental and predicted HCHO conversions, was lower than 4%.

3.3. Bench Scale Reactor Optimization

The ultimate output of the proposed methodology (Figure 1) is an optimized reactor. The optimization step aims at finding the best configuration given certain restrictions. In order to evaluate different reactor configurations regarding overall conversion of a certain pollutant, some of the parameters that have been employed are: the quantum yields [25,37], the photonic efficiencies [23,38] and the electrical energy per order or per unit mass [39]. A different parameter was also proposed: the photochemical thermodynamic efficiency factor [40].

Here, we use the concept of multiple independent efficiencies in series contributing to the overall reactor performance [14]. This approach allows the identification of reactor design advantages and drawbacks, offering useful tools for subsequent optimization.

The global efficiency of a photocatalytic process can be defined as the product of individual efficiencies, according to:

$$\eta_T = \prod_i \eta_i = \eta_{ele} \cdot \eta_{inc}^{out} \cdot \eta_{inc}^{in} \cdot \eta_{abs} \cdot \eta_{rxn} \tag{12}$$

These efficiencies are linked by the output spectral power of the lamps, the transmittance of the windows material and the spectral absorption of the immobilized catalyst. In fact, the range of wavelengths in which the integration of properties is performed determines the coupling among the different efficiencies.

The electrical efficiency: $\eta_{ele} = P_L/P$ represents the capacity of the radiation source of converting electrical energy (actually consumed power, P) into emitted radiant power (P_L) in the desired wavelength range [41]. This parameter has a fixed value for a given brand and model of lamp.

The outer geometrical incidence efficiency represents the ratio of emitted photons that effectively reach the reactor windows. The possibility of occurrence of such event may be called η_{inc}^{out} and defined by:

$$\eta_{inc}^{out} = \frac{\int_{A_W} \int_\lambda q_{\lambda,W}^{out} d\lambda dA}{\int_\lambda P_{\lambda,L} d\lambda} = \frac{\langle q_w^{out} \rangle A_w}{P_L} \tag{13}$$

It is a relative optical parameter between the radiation source and the reactor exterior. The reactor geometry and the lamp arrangement to irradiate the reactor are the only parameters determining this efficiency. The definition of the denominator in Equation (13) must take into account the number of lamps and the presence of reflectors.

Next, we have the inner incidence efficiency. A fraction of the photons that get to the reactor window is transmitted; once inside, they can either reach a catalytic surface or not. The probability of a radiation ray leaving the inner side of the reactor window to get to a catalytic area may be defined as:

$$\eta_{inc}^{in} = \frac{\langle q_{inc} \rangle A_{cat}}{\langle q_w \rangle A_w} \tag{14}$$

The internal configuration of the reactor may allow the enhancement of this efficiency by surface to surface radiative interaction through reflection. Thus, the optical properties of the support material are essential because transmittance or reflection may have a positive or negative impact depending on the shape and configuration of the reactor.

Ultimately, the radiation that reaches the catalyst surface may be absorbed or reflected, giving place to another parameter, η_{abs}, the absorption efficiency.

$$\eta_{abs} = \frac{\langle e^{a,s} \rangle}{\langle q_{inc} \rangle} \tag{15}$$

which represents the ratio between absorbed and incident energy on the catalytic surface. It is independent of the external reactor geometry, but an intrinsic property of the photocatalyst nature.

Once the catalyst had absorbed radiation, the reaction efficiency is defined as:

$$\eta_{rxn} = \frac{\langle r_{\sup} \rangle}{\langle e^{a,s} \rangle} \tag{16}$$

The reaction efficiency can be regarded as the moles of pollutant units eliminated per absorbed energy (einsteins) over the catalyst surface.

Optimal Folding Angle of the Corrugated Reactor

For the sake of optimization, some design and operative variables were kept constant: the total gas flow rate Q_g and the pollutant concentration $C_{i,0}$ to be reduced; the radiation source; and the reactor volume. The design variable to be optimized in the corrugated wall reactor was then the folding angle.

The influence of the folding angle on the inner incidence efficiency is depicted in Figure 6. It can be seen that this parameter shows an optimum value at angles below 30 degrees. For the case with no reflection (hypothetical), the optimum is lower than that of the real case (with reflection). The fraction of reflected radiation clearly enhances the capabilities of this corrugated wall configuration.

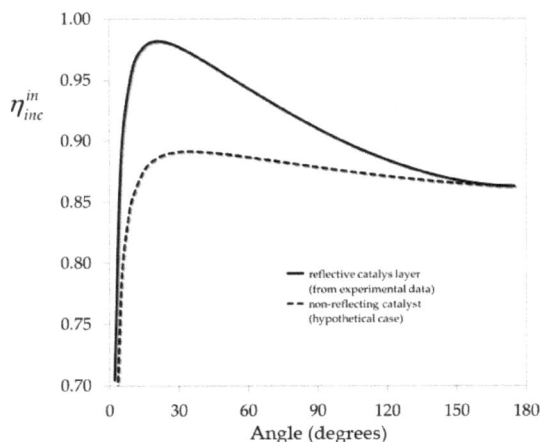

Figure 6. Inner incidence efficiency as a function of folding angle. Reprinted with permission from Passalía, C.; Alfano, O.M.; Brandi, R.J. Optimal design of a corrugated-wall photocatalytic reactor using efficiencies in series and computational fluid dynamics (CFD) modeling. *Ind. Eng. Chem. Res.* **2013**, *52*, 6916–6922. Copyright 2013 American Chemical Society [14].

A study was performed by simulation of four folding angles (φ): 97.2, 59.1, 31.6 and 15.2 degrees. Table 3 summarizes the results for the four folding angles. The simulations were performed in the academic version of a CFD software (Fluent 12.1) and the main output of them was the overall pollutant conversion.

The second column in Table 3 contains the absorbed energy relative to the entering radiation through the reactor window; as can be seen, it gets higher when the catalyst surface is almost normal to the radiation source. The third column shows that the averaged superficial reaction rate increases almost five-fold when changing the angle from 15.2 to 97.2 degrees; this is due to the linear dependence of the reaction rate with the LSRPA. The fourth column in Table 3 presents the reaction efficiency, derived indirectly from the previous columns; as can be seen, η_{rxn} has a smooth trend to decrease when the folding angle is increased. This behavior is based on the simultaneous increase in the relative absorbed energy and the averaged reaction rate.

The results presented in Table 3 indicate the presence of an optimum between the closest angles simulated, i.e., 15 and 30 degrees. This is consistent with the maximum found for the inner incident efficiency (Figure 6).

Table 3. Results of efficiencies for different folding angles. Adapted with permission from Passalía, C.; Alfano, O.M.; Brandi, R.J. Optimal design of a corrugated-wall photocatalytic reactor using efficiencies in series and computational fluid dynamics (CFD) modeling. *Ind. Eng. Chem. Res.* **2013**, *52*, 6916–6922. Copyright 2013 American Chemical Society [14].

Angle	$\langle e^{a,s} \rangle / \langle q_w \rangle$	$\langle r_{sup} \rangle \times 10^{11}$	$\eta_{rxn} \times 10^3$	η_{inc}
(degrees)	-	(mol cm^{-2} s^{-1})	(mol Einstein^{-1})	-
15.2	0.078	1.04	9.24	0.978
31.6	0.155	2.09	9.29	0.976
59.1	0.271	3.50	8.93	0.944
97.2	0.395	4.74	8.28	0.903

4. Conclusions

An integral design methodology has been proposed and applied to scaling-up and optimization of photocatalytic wall reactors employed in air pollution control. Starting from a kinetic study in a laboratory scale reactor and the evaluation of the absorbed radiation, a bench scale reactor could be successfully modeled and optimized. The proposed methodology shows a systematic approach based on engineering concepts and can be used to design and optimize any photocatalytic wall reactor provided that the balances involved, including radiation, can be solved by a mathematical model.

Acknowledgments: The authors are grateful to Universidad Nacional del Litoral (UNL), Consejo Nacional de Investigaciones Científicas y Técnicas (CONICET), and Agencia Nacional de Promoción Científica y Tecnológica (ANPCyT) for the financial support.

Author Contributions: Claudio Passalía, Orlando Alfano and Rodolfo Brandi contributed to the preparation and writing of the review. All the authors read and approved the final version of the manuscript.

Conflicts of Interest: The authors declare no conflict of interest.

Nomenclature

a_v	external catalytic surface area per unit volume (cm^{-1})
A_c	cross sectional area (cm^2)
A_{cat}	photocatalytic area (cm^2)
C	molar concentration (mol cm^{-3})
$e^{a,s}$	local superficial rate of photon absorption, LSRPA (einstein cm^{-2} s^{-1})
H	height (cm)
I	specific radiation intensity (einstein cm^{-2} s^{-1} sr^{-1})
I^0	specific radiation intensity on the lamp surface (einstein cm^{-2} s^{-1} sr^{-1})
k_m	gas phase mass transfer coefficient (cm s^{-1})
L	length (cm)
P	emission power (einstein s^{-1})
q	radiation flux (einstein cm^{-2} s^{-1})
Q	volumetric gas flow rate (cm^3 s^{-1})
R	homogeneous reaction rate (mol cm^{-3} s^{-1})
r	heterogeneous reaction rate (mol cm^{-2} s^{-1})
r_g	local superficial rate of electron-hole pair generation (mol cm^{-2} s^{-1})
R_{Li}	lamp radius (cm)
T	reactor thickness (cm)
T_λ	spectral transmittance (dimensionless)
v_z	axial velocity (cm s^{-1})

W	reactor width (cm)
x, y, z	rectangular coordinate system (cm)
$\underline{x}$	position vector (cm)
X	conversion (dimensionless)
Greek letters	
α	kinetic parameter (cm^3 einstein^{-1})
η	efficiency (dimensionless)
κ	adsorption constant (cm^3 mol^{-1})
λ	wavelength (nm)
ρ	gas density (g cm^{-3})
$\underline{\underline{\tau}}$	viscous stress tensor
ϕ, θ	spherical coordinates (rad)
Φ	primary quantum yield (mol einstein^{-1})
$\overline{\Phi}$	wavelength averaged primary quantum yield (mol einstein^{-1})
Subscripts	
ads	indicates adsorbed species
abs	relative to radiation absorption
F	relative to formaldehyde
F_i	relative to radiation neutral filter
Li	relative to lamp *i*
W	relative to water
W_i	relative to reactor window
λ	indicates wavelength dependence
exp	relative to experimental data
mod	relative to model
Superscripts	
in	relative to reactor inlet
out	relative to reactor outlet
Special symbols	
$\langle \rangle$	denotes an averaged property
$-$	denotes a vector

References

1. Jones, A.P. Indoor air quality and health. *Atmos. Environ.* **1999**, *33*, 4535–4564. [CrossRef]
2. Bruce, N.; Perez-Padilla, R.; Albalak, R. Indoor air pollution in developing countries: A major environmental and public health challenge. *Bull. World Health Organ.* **2000**, *78*, 1078–1092. [PubMed]
3. Wang, S.; Ang, H.M.; Tade, M.O. Volatile organic compounds in indoor environment and photocatalytic oxidation: state of the art. *Environ. Int.* **2007**, *33*, 694–705. [CrossRef] [PubMed]
4. Zhang, Y.; Luo, X.; Wang, X.; Qian, K.; Zhao, R. Influence of temperature on formaldehyde emission parameters of dry building materials. *Atmos. Environ.* **2007**, *41*, 3203–3216. [CrossRef]
5. Lu, Y.; Wang, D.; Ma, C.; Yang, H. The effect of activated carbon adsorption on the photocatalytic removal of formaldehyde. *Build. Environ.* **2010**, *45*, 615–621. [CrossRef]
6. Shiraishi, F.; Toyoda, K.; Miyakawa, H. Decomposition of gaseous formaldehyde in a photocatalytic reactor with a parallel array of light sources: 2. Reactor performance. *Chem. Eng. J.* **2005**, *114*, 145–151. [CrossRef]
7. Hanoune, B.; Lebris, T.; Allou, L.; Marchand, C.; Lecalve, S. Formaldehyde measurements in libraries: Comparison between infrared diode laser spectroscopy and a DNPH-derivatization method. *Atmos. Environ.* **2006**, *40*, 5768–5775. [CrossRef]
8. Hay, S.O.; Obee, T.; Luo, Z.; Jiang, T.; Meng, Y.; He, J.; Murphy, S.C.; Suib, S. The viability of photocatalysis for air purification. *Molecules* **2015**, *20*, 1319–1356. [CrossRef]
9. Dumont, É.; Héquet, V. Determination of the Clean Air Delivery Rate (CADR) of photocatalytic oxidation (PCO) purifiers for indoor air pollutants using a closed-loop reactor. Part I: Theoretical considerations. *Molecules* **2017**, *22*, 407. [CrossRef] [PubMed]

10. Héquet, V.; Batault, F.; Raillard, C.; Thévenet, F.; Le Coq, L.; Dumont, É. Determination of the clean air delivery rate (CADR) of photocatalytic oxidation (PCO) purifiers for indoor air pollutants using a closed-loop reactor. Part II: Experimental results. *Molecules* **2017**, *22*, 408.

11. Pichat, P. Some views about indoor air photocatalytic treatment using TiO$_2$: Conceptualization of humidity effects, active oxygen species, problem of C1-C3 carbonyl pollutants. *Appl. Catal. B Environ.* **2010**, *99*, 428–434. [CrossRef]

12. Passalía, C.; Martínez Retamar, M.E.; Alfano, O.M.; Brandi, R.J. Photocatalytic Degradation of Formaldehyde in Gas Phase on TiO$_2$ Films: A Kinetic Study. *Int. J. Chem. React. Eng.* **2010**, *8*, A161. [CrossRef]

13. Passalía, C.; Alfano, O.M.; Brandi, R.J. A methodology for modeling photocatalytic reactors for indoor pollution control using previously estimated kinetic parameters. *J. Hazard. Mater.* **2012**, *211–212*, 357–365. [CrossRef] [PubMed]

14. Passalía, C.; Alfano, O.M.; Brandi, R.J. Optimal design of a corrugated-wall photocatalytic reactor using efficiencies in series and computational fluid dynamics (CFD) modeling. *Ind. Eng. Chem. Res.* **2013**, *52*, 6916–6922. [CrossRef]

15. Cassano, A.E.; Martin, C.A.; Brandi, R.J.; Alfano, O.M. Photoreactor analysis and design: Fundamentals and applications. *Ind. Eng. Chem. Res.* **1995**, *34*, 2155–2201. [CrossRef]

16. Schneider, J.; Bahnemann, D.W.; Ye, J.; Puma, G.L.; Dionysiou, D.D. *Photocatalysis: Fundamentals and Perspectives*; RSC: London, UK, 2016.

17. Ibhadon, A.O.; Fitzpatrick, P. Heterogeneous Photocatalysis: Recent Advances and Applications. *Catalysts* **2013**, *3*, 189–218. [CrossRef]

18. Pichat, P. *Photocatalysis: Fundamentals, Materials and Potential*; MDPI AG: Basel, Switzerland, 2016.

19. Hernandez-Alonso, M.D.; Fresno, F.; Suarez, S.; Coronado, J.M. Development of alternative photocatalysts to TiO$_2$: Challenges and opportunities. *Energy Environ. Sci.* **2009**, *2*, 1231–1257. [CrossRef]

20. Binas, V.; Venieri, D.; Kotzias, D.; Kiriakidis, G. Modified TiO$_2$ based photocatalysts for improved air and health quality. *J. Mater.* **2017**, *1*, 3–16.

21. Salvadó-Estivill, I.; Brucato, A.; Puma, G.L. Two-Dimensional Modeling of a Flat-Plate Photocatalytic Reactor for Oxidation of Indoor Air Pollutants. *Ind. Eng. Chem. Res.* **2007**, *46*, 7489–7496. [CrossRef]

22. Zazueta, A.L.L.; Destaillats, H.; Puma, G.L. Radiation field modeling and optimization of a compact and modular multi-plate photocatalytic reactor (MPPR) for air/water purification by Monte Carlo method. *Chem. Eng. J.* **2013**, *217*, 475–485. [CrossRef]

23. Singh, M.; Salvado-Estivill, I.; Li Puma, G. Radiation Field Optimization in Photocatalytic Monolith Reactors for Air Treatment. *AIChE J.* **2007**, *53*, 678–686. [CrossRef]

24. Taranto, J.; Frochot, D.; Pichat, P. Photocatalytic air purification: Comparative efficacy and pressure drop of a TiO$_2$-coated thin mesh and a honeycomb monolith at high air velocities using a 0.4 m^3 close-loop reactor. *Sep. Purif. Technol.* **2009**, *67*, 187–193. [CrossRef]

25. Ibrahim, H.; de Lasa, H. Novel photocatalytic reactor for the destruction of airborne pollutants reaction kinetics and quantum yields. *Ind. Eng. Chem. Res.* **1999**, *38*, 3211–3217. [CrossRef]

26. Esterkin, C.R.; Negro, A.C.; Alfano, O.M.; Cassano, A.E. Air pollution remediation in a fixed bed photocatalytic reactor coated with TiO$_2$. *AIChE J.* **2005**, *51*, 2298–2310. [CrossRef]

27. Mohseni, M.; David, A. Gas phase vinyl chloride (VC) oxidation using TiO$_2$-based photocatalysis. *Appl. Catal. B Environ.* **2003**, *46*, 219–228. [CrossRef]

28. Imoberdorf, G.E.; Cassano, A.E.; Irazoqui, H.A.; Alfano, O.M. Simulation of a multi-annular photocatalytic reactor for degradation of perchloroethylene in air: Parametric analysis of radiative energy efficiencies. *Chem. Eng. Sci.* **2007**, *62*, 1138–1154. [CrossRef]

29. Ma, C.; Ku, Y.; Chou, Y.; Jeng, F. Performance of Tubular-Type Optical Fiber Reactor for Decomposition of Vocs in Gaseous Phase. *J. Environ. Eng. Manag.* **2008**, *18*, 363–369.

30. Shang, H.; Zhang, Z.; Anderson, W.A. Nonuniform radiation modeling of a corrugated plate photocatalytic reactor. *AIChE J.* **2005**, *51*, 2024–2033. [CrossRef]

31. Passalia, C.; Alfano, O.M.; Brandi, R.J. Modeling and Experimental Verification of a Corrugated Plate Photocatalytic Reactor Using CFD. *Ind. Eng. Chem. Res.* **2011**, *50*, 9077–9086. [CrossRef]

32. Mohseni, M.; Taghipour, F. Experimental and CFD analysis of photocatalytic gas phase vinyl chloride (VC) oxidation. *Chem. Eng. Sci.* **2004**, *59*, 1601–1609. [CrossRef]

33. Queffeulou, A.; Geron, L.; Schaer, E. Prediction of photocatalytic air purifier apparatus performances with a CFD approach using experimentally determined kinetic parameters. *Chem. Eng. Sci.* **2010**, *65*, 5067–5074. [CrossRef]

34. Yang, J.; Li, D.; Zhang, Z.; Li, Q.; Wang, H. A study of the photocatalytic oxidation of formaldehyde on Pt/Fe$_2$O$_3$/TiO$_2$. *J. Photochem. Photobiol. A Chem.* **2000**, *137*, 197–202. [CrossRef]

35. Walter, S.; Malmberg, S.; Schmidt, B.; Liauw, M.A. Mass transfer limitations in microchannel reactors. *Catal. Today* **2005**, *110*, 15–25. [CrossRef]

36. Siegel, R.; Howell, J. *Thermal Radiation Heat Transfer*, 4th ed.; Taylor & Francis: New York, NY, USA, 2002.

37. Imoberdorf, G.E.; Irazoqui, H.A.; Cassano, A.E.; Alfano, O.M. Quantum efficiencies in a multi-annular photocatalytic reactor. *Water Sci. Technol.* **2007**, *55*, 161. [CrossRef] [PubMed]

38. Serpone, N. Relative photonic efficiencies and quantum yields in heterogeneous photocatalysis. *J. Photochem. Photobiol. A Chem.* **1997**, *104*, 1–12. [CrossRef]

39. Sun, L.; Bolton, J.R. Determination of the Quantum Yield for the Photochemical Generation of Hydroxyl Radicals in TiO$_2$ Suspensions. *J. Phys. Chem.* **1996**, 4127–4134. [CrossRef]

40. Hernandez, J.M.G.; Rosales, B.S.; de Lasa, H. The photochemical thermodynamic efficiency factor (PTEF) in photocatalytic reactors for air treatment. *Chem. Eng. J.* **2010**, *165*, 891–901. [CrossRef]

41. Cerdá, J.; Marchetti, J.L.; Cassano, A.E. Radiation efficiencies in elliptical photoreactors. *Lat. Am. J. Heat Mass Transf.* **1977**, *1*, 33–63.

Sample Availability: Samples of the compounds are not available from the authors.

molecules MDPI

Article

Determination of the Clean Air Delivery Rate (CADR) of Photocatalytic Oxidation (PCO) Purifiers for Indoor Air Pollutants Using a Closed-Loop Reactor. Part I: Theoretical Considerations

Éric Dumont * and Valérie Héquet

UMR CNRS 6144 GEPEA, IMT Atlantique, La Chantrerie, 4 rue Alfred Kastler, CS 20722,
44307 Nantes CEDEX 3, France; valerie.hequet@imt-atlantique.fr
* Correspondence: eric.dumont@imt-atlantique.fr; Tel.: +33-2-51-85-82-66

Academic Editor: Pierre Pichat
Received: 15 December 2016; Accepted: 2 March 2017; Published: 6 March 2017

Abstract: This study demonstrated that a laboratory-scale recirculation closed-loop reactor can be an efficient technique for the determination of the Clean Air Delivery Rate (CADR) of PhotoCatalytic Oxidation (PCO) air purification devices. The recirculation closed-loop reactor was modeled by associating equations related to two ideal reactors: one is a perfectly mixed reservoir and the other is a plug flow system corresponding to the PCO device itself. Based on the assumption that the ratio between the residence time in the PCO device and the residence time in the reservoir τ_P/τ_R tends to 0, the model highlights that a lab closed-loop reactor can be a suitable technique for the determination of the efficiency of PCO devices. Moreover, if the single-pass removal efficiency is lower than 5% of the treated flow rate, the decrease in the pollutant concentration over time can be characterized by a first-order decay model in which the time constant is proportional to the CADR. The limits of the model are examined and reported in terms of operating conditions (experiment duration, ratio of residence times, and flow rate ranges).

Keywords: photocatalysis; Clean Air Delivery Rate (CADR); indoor air quality; volatile organic compounds (VOCs); air cleaner

1. Introduction

Attention to Indoor Air Quality (IAQ) has greatly increased over the last 30 years. The most common gaseous pollutants present in indoor air fall within the concentration range of 1–1000 ppbv. Among them, several hundred volatile organic compounds (VOCs) have been identified. In addition, analysis of the available data demonstrates a statistical association between IAQ conditions and occupants' health [1]. The traditional dilution method of ventilation is not always recommended in current practice due to its limitation in terms of outdoor air quality and energy cost [2] while the control of pollutant emissions is not always possible. Therefore, it appears necessary to develop technologies and effective strategies to improve IAQ.

Several processes can be used to remove VOCs from indoor air. Among them, PhotoCatalytic Oxidation (PCO) air cleaning is considered an efficient technology suitable for the elimination of a broad range of VOCs [3]. In a recent review, Paz [4] noted that the number of scientific publications on indoor air treatment has significantly increased over the last 15 years even though it is still lower than the number dedicated to photocatalytic water treatment. Surprisingly, however, the number of patents related to indoor air has greatly increased over the same period and to date is much higher than the patents dedicated to water treatment (53% for air, 38% for water, and 9% for self-cleaning). This indicates a growing interest in the implementation of photocatalysis for air treatment purposes.

In fact, the current concern about indoor air quality has provided opportunities for manufacturers to develop PCO devices, and many indoor air applications are now available on the market. Thus, the design of reliable methodologies to assess the performances of commercial air purifiers appears vital for manufacturers and consumers, and a careful evaluation and certification of commercial PCO units for consumer safety is needed [3,5]. ISO standards already published propose several methods to assess the performance of photocatalytic materials in the areas of air and water purification, self-cleaning, and photo-sterilization [6]. At the European level, the CEN TC 386 "photocatalysis" has been working on the French AFNOR standard, including the standard XP B44-013 that is dedicated to the evaluation of air cleaners including photocatalytic functions in a closed-chamber test [5,7]. Ideally, the performances of all PCO cleaners that are available on the market or under development should be determined in real conditions, i.e., (i) in a real room; (ii) with a mixture of numerous pollutants; (iii) at pollutant concentrations encountered in indoor air, around a few tens of parts per billion by volume (ppbv). However, such tests can be costly, difficult, and time-consuming [8]. Prior to a final qualification of air cleaners using standard methods, it can be useful to have reliable methodologies to assess the performance of designed photocatalytic systems or to test and optimize the major operating parameters of the systems. In fact, key parameters influencing the performances of PCO air purifiers under realistic indoor conditions still need to be assessed [3,9]. Laboratory procedures and design tools are still lacking with regard to improving PCO techniques and comparing the results obtained by different research teams more objectively [2]. To be compared, the performances of the different tested PCO devices have to be quantified in reproducible conditions. Thus, the best way to achieve this at laboratory scale is to perform experiments either in a continuous mode plug flow reactor, in which the air makes a single pass through the reactor, or in a batch closed-loop reactor operating in recirculation mode [10–14]. Then, there are at least three main descriptors of the efficiency of photocatalytic systems: (i) the VOC degradation rate; (ii) the one-pass removal efficiency; and (iii) the Clean Air Delivery Rate (CADR) [15,16]. The CADR, initially defined by the Association of Home Appliance Manufacturers (AHAM) and well recognized by manufacturers, indicates the clean air volume delivered by the treatment system (usually in $m^3 \cdot h^{-1}$).

In the present work, the approach using a batch closed-loop reactor operating in recirculation mode is investigated to determine theoretically the performances of any PCO device. In this reactor, the PCO apparatus is inserted into a gas-tight chamber with a well-known volume (Figure 1). An amount of pollutant (or a mixture of pollutants) is injected into the chamber to obtain polluted air at the desired concentration. A centrifugal fan provides a controlled flow of the air through the PCO device. Once the initial pollutant concentration at steady-state is reached, the experiment is initiated by illuminating the light source. Samples of the air in the chamber are periodically analyzed with appropriate apparatus to monitor the decrease in the pollutant concentration over time. Then, the experimental data can be analyzed according to an appropriate mathematical model. However, even if different models for PCO devices are available in the literature [17] to fit the experimental data (exponential, linear, and polynomial fits [10]), some can generate bias in the performance determination. Therefore, the aim of the present study is to show that a laboratory closed-loop reactor operating in recirculation mode can be adequately modeled to determine the performances of photocatalytic devices for critical comparison. The rigorous mathematical model describing the decrease in the pollutant concentration over time is simplified by using realistic assumptions, enabling a convenient analysis of the experimental data. The limits of the model are given in terms of operating conditions (experiment duration, ratio of residence times, and flow rate ranges). Moreover, the model highlights that a laboratory closed-loop reactor can be an efficient technique for the determination of the CADR of PCO devices, which is of practical significance in the assessment of the effectiveness of systems for air purification.

Molecules **2017**, *22*, 407

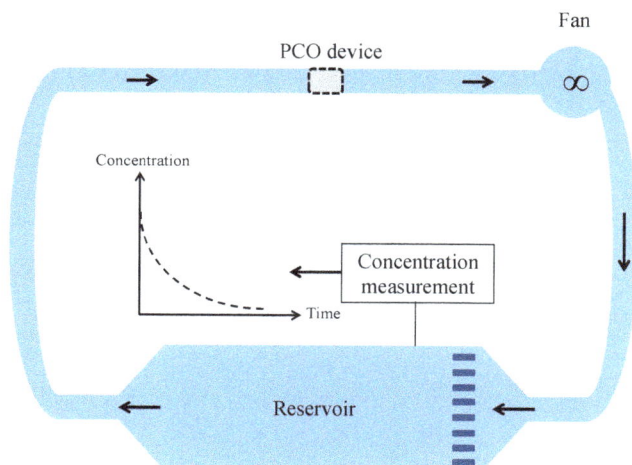

Figure 1. Diagram of a closed-loop reactor operating in recirculation mode (batch).

2. Model

The performances of a PCO device inserted in a recirculation closed-loop reactor can be determined if the decrease in the pollutant concentration in the reactor can be predicted versus time. The recirculation closed-loop reactor (Figure 1) can be modeled by associating equations related to two ideal reactors connected together by pipes of negligible volume: a perfectly mixed reservoir with a volume V_R and a plug flow system, corresponding to the PCO device, with a volume V_P (Figure 2). The mathematical model proposed here has been described by Walker and Wragg [18] for concentration-time relationships established for recirculating electrochemical reactor systems similar in design to the recirculation system used in this study. Although electrochemical reactor systems occur in a liquid phase whereas the gas phase is considered in the PCO device, there is fundamentally no difference between both systems. Indeed, the electrochemical reactor systems described by Walker and Wragg [18] involved the deposition of metal ions leading to a gradual depletion of the concentration of the metal ions in the system, which presents similarity with the gradual degradation of the pollutant in the PCO apparatus. The main difference between both models lies in the fact that in the electrochemical systems, the depletion of the concentration of the metal ions in the plug flow reactor depended on the mass transfer coefficient, whereas in the photocatalytic system, the decrease in the pollutant concentration depends on the overall kinetic rate constant, as it will be shown hereafter. The basic assumptions of the rigorous model are:

(i)   Idealized plug flow occurs in the PCO device;
(ii)   The reservoir is a perfectly-mixed system;
(iii) The mass transfer of pollutant occurs under convective-diffusion control;
(iv)   The kinetic constant does not change during the experiment;
(v)   The flow rate Q is constant with time and with position in the system;
(vi)   Temperature and thus the physical properties of air are constant both in space and time.

Referring to Figure 2, a differential mass balance at the plane x gives the following partial differential equation:

$$A\frac{\partial C}{\partial t}(t,x) = -Q\frac{\partial C}{\partial x}(t,x) - kAC(t,x) \tag{1}$$

In Equation (1), the degradation rate is proportional to the pollutant concentration. Although numerous studies showed that kinetic of organic pollutant removal are well described by the Langmuir-Hinshelwood (LH) expression, at low pollutant concentrations in air (i.e., ppb level) literature reported that the LH expression could be reduced to a first-order reaction [17,19,20]. More than that, in this study, it is considered a global apparent kinetic constant k, including chemical kinetics as well as the reactor dynamics. Walker and Wragg [18] solved Equation (1) with the suitable choice of the boundary condition C(0,x), which corresponds to the concentration at the plane x of the reactor when t = 0. The procedure giving the solution of the equation, fully detailed in the original paper, is mathematically complex:

$$C(t) = C_0 \left[1 + \left(1 - \exp\left(\frac{kV_P}{Q}\right)\right) \int_0^t \sum_{n=0}^{\left(\frac{z}{\beta}\right)-1} \left\{ \frac{1}{n! \left(\frac{V_R}{Q} \exp\left(\frac{kV_P}{Q}\right)\right)^{n+1}} (z - n\beta)^n \exp\left[-\frac{Q}{V_R}(z - n\beta)\right] \right\} dz \right] \qquad (2)$$

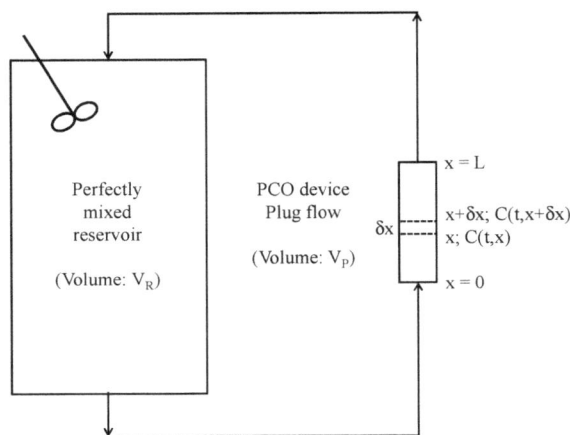

Figure 2. Recirculation closed-loop reactor system: association of two ideal reactors (all connecting lines are of negligible volume).

2.1. First Assumption

As Equation (2) cannot be easily used, Walker and Wragg [18] suggested an alternative approximate model based on the assumption that it is possible to neglect the $\partial C/\partial t$ term in Equation (1). This simplification is valid only if the volume of the PCO device is small in relation to the volume of the reservoir (i.e., if the ratio of the residence times (τ_P/τ_R) tends to 0). In this case, the concentration change with time at any plane x may be regarded as insignificant in comparison with the change in concentration with distance x. Thus, for the closed-loop reactor depicted in Figure 2, the concentration-time relationship is:

$$C = C_0 \exp\left(-t\frac{Q}{V_R}\left[1 - \exp\left(-\frac{k V_P}{Q}\right)\right]\right) \qquad (3)$$

Walker and Wragg [18] compared C/C_0 values for various times computed from both the rigorous and the approximate solutions (Equations (2) and (3), respectively; Table 1). It can be observed that the discrepancy is less than 1% for times less than 1.5 h and reaches 3% at 2.5 h. For longer times, the approximate solution becomes markedly inaccurate. Consequently, it can be concluded that the approximate solution can satisfactorily be used to describe the decrease in the pollutant concentration over time provided that the total time of the experiment does not exceed 2 h.

Table 1. Difference between the rigorous solution (Equation (2)) and the approximate solution (Equation (3)).

Time (s)	Time (h)	Difference (%)
900	0.25	0.04
2700	0.75	0.16
5400	1.50	0.83
9000	2.50	3.00
10800	3.00	7.35
13500	3.75	23.72

In Equation (3), the volume of the PCO device V_P corresponds to the product of the cross-sectional area of the pipe connecting the apparatus (A) and the length of the apparatus (L). Thus, if there is no change in the cross-sectional area along the PCO device, the air residence time in the apparatus is $\tau_P = V_P/Q$. In other words, the term V_P is the real volume of the device; it is also important to note that the configuration of the photocatalytic material inside the device is not taken into account. Considering that the residence time of a molecule in the closed-loop reactor is $\tau_R = V_R/Q$, Equation (3) can be rewritten as follows:

$$C = C_0 \, \exp\left(-\frac{t}{\tau_R}[1 - \exp(-k\,\tau_P)]\right) \tag{4}$$

In Equation (4) the term $\alpha = (k\,\tau_P)$ is the fractional yield of the treated flow rate of the PCO device. In other words, α corresponds to the percentage of the total flow rate treated during the time τ_R (i.e., during one cycle):

$$C = C_0 \, \exp\left(-\frac{t}{\tau_R}[1 - \exp(-\alpha)]\right) \tag{5}$$

The term α can be related to the parameter CADR (Clean Air Delivery Rate) typically used to evaluate the air cleaning capacity of a PCO reactor [16]. Indeed, many manufacturers define the CADR as the product of the device efficiency and the volumetric air flow rate through the apparatus [21,22]. The higher the CADR value, the faster the PCO device cleans the air. Assuming that there is no natural decay rate during experiments in a closed-loop reactor, the CADR is [23]:

$$CADR = \alpha\,Q \tag{6}$$

The CADR ratings were originally developed by the Association of Home Appliance Manufacturers (AHAM) to characterize the ability of air filters to treat particulate matter, not gases. Moreover, CADR measurements must be performed in an 1008-cubic-foot (28.5 m^3) standard room according to a procedure specified by ANSI/AHAM AC-1 [24]. Nonetheless, the CADR concept is now equally used to quantify the performance of air cleaning devices treating polluted gases whatever the size of the test chamber [5,21,25,26].

The degradation rate is a decreasing function of time according to:

$$r = -\frac{dC}{dt} = \frac{C_0}{\tau_R}\,(1 - \exp(-\alpha))\,\exp\left(-t\frac{Q}{V_R}[1 - \exp(-\alpha)]\right) \tag{7}$$

The maximum degradation rate can be determined at the beginning of the experiment, i.e., at $t = 0$:

$$r_{max} = -\frac{dC}{dt}\bigg|_{max} = \frac{C_0}{\tau_R}\,(1 - \exp(-\alpha)) \tag{8}$$

2.2. Second Assumption

As the volume of the PCO device must be small in relation to the volume of the reservoir, it is expected that the volume of air treated during one cycle is small in comparison with the total volume to be treated. In other words, the CADR value should be much smaller than the total air flow rate Q. Consequently, the term α should be small (some %). In this case, Taylor's theorem leads to the approximation that $\exp(-\alpha) \approx (1 - \alpha)$. For $\alpha = 0.05$, the discrepancy between $\exp(-\alpha)$ and $(1 - \alpha)$ is 2.5%. Thus, Equations (5 and 8) can be rewritten as:

$$C = C_0 \exp\left(-t\frac{\alpha}{\tau_R}\right) = C_0 \exp\left(-t\frac{CADR}{V_R}\right) \tag{9}$$

$$r_{max} = C_0\frac{\alpha}{\tau_R} = C_0\frac{CADR}{V_R} \tag{10}$$

According to Equation (9) and due to both the assumptions defined above, the decrease in pollutant concentration follows a first-order decay model. It should be noted that this model is usually employed to determine the overall kinetic constant value of photochemical reactions [27,28]. This finding is in agreement with the first literature data reporting the oxidation of odor compounds in a photocatalytic monolith recirculating batch reactor [12]. A similar closed-loop reactor was also studied by Sauer and Ollis [13] for ethanol treatment. In this case, reactor performance was modeled with the Langmuir-Hinshelwood (LH) expression. However, even if numerous studies showed that Langmuir-Hinshelwood local rate form is successful in correlating much steady state photocatalysis data [29], it was demonstrated for a closed-loop reactor treating low pollutant concentrations in air that the LH expression could be reduced to a first-order reaction [17,19,20]. In this case, the kinetic constant obtained (i.e., k in the present study) is a composite expression consisting of elements originating in chemical kinetics and reactors dynamics [19] that differs from the actual photocatalytic rate constant. Relationship between the actual photocatalytic rate constant and the kinetic constant was discussed between Wolfrum and Turchi [20] and Davis and Hao [19]. Thus, in the present case, the first-order decay model enables an easy and direct determination of the overall kinetic constant, and consequently a direct determination of the CADR of a PCO device. As a result, the effectiveness of different systems for air purification can be assessed and compared irrespective of their geometry and configuration provided that experiments are carried out for the same operating conditions. Rearranging Equation (9), the overall efficiency of the PCO device to treat the air is given by Equation (11) where τ_R / α is the time constant (t_c) of the closed-loop reactor (i.e., the time needed to reach a 63.2% conversion of pollutant) and $1/\alpha$ is the cycle number needed to obtain E = 0.632.

$$E = \frac{C_0 - C}{C_0} = 1 - \exp\left(-t\frac{\alpha}{\tau_R}\right) \tag{11}$$

Experimental results may be drawn for analysis according to either the dimensional form or the dimensionless form of Equation (11) (Figure 3).

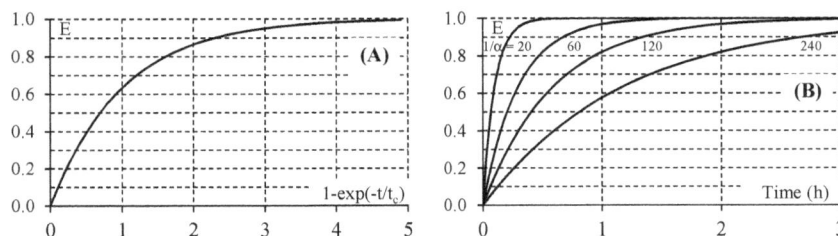

Figure 3. Overall efficiency of the batch recirculation closed-loop reactor system according to Equation (11). (**A**) Dimensionless form (time constant $t_c = \tau_R / \alpha$); (**B**) Example of the dimensional form according to the cycle number $1/\alpha$ ($\tau_R = 17.5$ s).

3. Discussion

Theoretically, the change in the pollutant concentration over time is described by means of a decreasing exponential function (Equation (5) or Equation (9) for the simplest case). According to Moulis et al. [10], who suggested choosing an exponential fit, a linear fit, or a polynomial fit for data analysis, the exponential fit shows deviations for long irradiation times, i.e., when the pollutant concentration tends to zero. Such deviations may be due to the difference that can exist between the rigorous solution of the model (Equation (2)) and the approximate solution for longer times. However, at low concentrations, a change in mass transfer must also be considered as well as the basic assumption of the rigorous model stating that "the kinetic constant does not change during the experiment", which is also questionable. Nevertheless, although deviations can be observed at low pollutant concentrations, the use of the exponential fit is preferable for data analysis rather than the linear fit, which requires the careful selection of a few experimental points (as highlighted in Part II). It is also preferable to the polynomial fit, which can accurately describe the data but has no proven physical basis.

As can be observed in Equations (10) and (11), the maximum degradation rate and the overall efficiency do not depend on the flow rate since $\alpha/\tau_R = k(V_P/V_R)$. Consequently, for a given PCO device, the CADR should be the same whatever the flow rate used in the experiment, provided that all other parameters, mainly the initial concentration C_0 and the irradiance I, are kept constant. In fact, an increase in the flow rate will lead to a decrease in the contact time τ_P in the PCO device, which will reduce the α value, and thus a great number of cycles ($=1/\alpha$) will be required for the total conversion of the pollutant. This result is inherent to the recirculation closed-loop system.

Given that the CADR could be easily determined from experimental concentration-time curves, the knowledge of the volume of the PCO device might be used to determine the value of the kinetic constant k [17,19,20]. This kinetic constant, which is the overall characteristic parameter of the apparatus, depends on the geometry of the apparatus, the surface area and the configuration (plate, pleated, honeycomb) of the photocatalytic material, the irradiance of the lamp, and the initial concentration of the pollutant [29,30]. For a given PCO device to be tested, the latter parameter can be adapted in order to obtain an α value lower than 5%. On the basis that α should not be greater than 5% to satisfy the second assumption, the minimum value of the cycle number is 20. However, if the total conversion of the pollutant is reached for $1/\alpha < 20$, Equation (5) could be used, provided that the first assumption ($\tau_P/\tau_R \to 0$) is valid. Since the duration of the total conversion of the pollutant should not exceed 2 h (first assumption), it can be calculated that the residence time in the reservoir must be less than 72 s ($5t_c < 2$ h; Figure 3). This residence time can be extended to 90 s and 120 s in the case where the total conversion of the pollutant is considered at $t = 4t_c$ and $t = 3t_c$, respectively (E = 98.2% and E = 95.0%, respectively; Figure 3). Consequently, the residence time in the PCO device should not exceed 1 s on the basis that $\tau_P/\tau_R = 1/100$. Indeed, for a ratio $\tau_P/\tau_R = 1/40$, Walker and Wragg [18] calculated that the difference based on the rigorous and the approximate solutions (Equations (2) and (3), respectively) is significant (10.8% at t = 0.5 h). As a result, a ratio $\tau_P/\tau_R \leq 1/100$ seems a reasonable order of magnitude for the determination of the CADR of the PCO device using a recirculation closed-loop reactor. Since the air flow rate through the reservoir and the PCO device is constant, the condition $\tau_P/\tau_R \leq 1/100$ corresponds to $V_P/V_R \leq 1/100$. According to the volume of the PCO device to be tested, the flow rate through the recirculation closed-loop system, or the volume of the reservoir, should be adapted to check this condition. Nonetheless, it should be kept in mind that the actual residence time of a molecule on the surface of the irradiated medium installed inside the PCO device is much lower than τ_P. From the determination of the performance of PCO air purification in realistic indoor conditions, Destaillats et al. calculated the residence time of a parcel of air inside the PCO medium on the basis of the medium volume and the air flow rate. Their results ranged from 0.027 to 0.159 s according to the geometry of the filter (flat or pleated) and the flow rate [3].

Can the Model Be Used in a Large Closed Chamber?

Since the CADR of any small PCO device could be determined using a recirculation closed-loop system characterized by a first-order decay model, provided that the two hypotheses are fulfilled ($\tau_P/\tau_R \to 0$ and $\alpha = CADR/Q < 5\%$), it seems useful to consider the possibility of determining the CADR of any commercial PCO device in a real room irrespective of its size (Figure 4). Indeed, in real conditions, the presence of humidity, of dust, of organic compounds containing heteroatoms (S, N, P, Si), of intermediates of reaction, as well as the presence of catalyst poisoning molecules or the decomposition of the TiO_2 support can significantly decrease the performances of the apparatus. In such a case, the volume of the room corresponds approximately to the volume of the reservoir, assuming that the size of the PCO device is small in comparison with the size of the room (otherwise, the volume of the reservoir is the volume of the room minus the volume of the PCO device). As the total time of the experiment does not exceed 2 h in order to use satisfactorily the approximate solution giving the decrease in the pollutant concentration over time (Equation (3)), the volume of the room should be judiciously adapted to the size of the PCO device, which may clearly be complicated to implement. In fact, the use of a 1 m³ closed chamber is recommended in the XP B44-013 French standard, for which the ratio between the volume of the PCO system and the chamber volume has to be less than 0.25 [31]. In this standard, CADR is classically calculated according to the following Equation:

$$CADR = V\ (k_e - k_n) \tag{12}$$

where V is the volume of the chamber, k_e is the pollutant decay rate with the PCO device in operation and k_n is the natural decay rate. In a recent study, a closed chamber (1.2 m³) was used to compare the performances of several commercial photocatalytic devices according to the French XP B44-013 AFNOR standard [5]. In this case, the CADR values were directly calculated assuming a first-order decay, although both the assumptions described above were not necessarily checked. Thus, the authors indicated that the CADR values could be quite inaccurate, potentially due to the fact that the operating parameters of the systems and the quality of the photocatalytic medium could not be controlled. It could also be due to the methodology of the calculation described in the XP B44-013 standard.

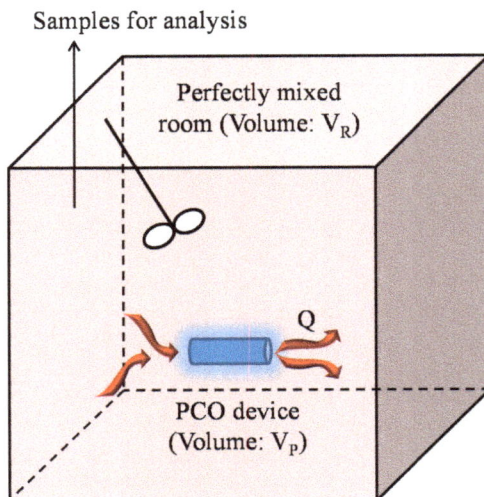

Figure 4. Experimental set-up for the determination of the Clean Air Delivery Rate (CADR) of PhotoCatalytic Oxidation (PCO) devices in real conditions.

Such an experiment was also carried out in a relatively large closed chamber, 4 m^3 in volume, by Kim et al. [25] to determine the CADR of commercial air cleaners (carbon filters for gas removal). In this study, the initial gas concentrations ranged between 8 and 13 ppmv (i.e., thousands of times more than the concentrations treated by PCO devices) and the concentration decreases over time were continuously recorded using an FTIR analyzer. The authors calculated that the single-pass removal efficiency of an air cleaner was equal to [0.83 CADR/Q] against α = CADR/Q in the present study. According to Kim et al. [25], the term 0.83 corresponds to a short-circuit factor taking into account the re-entrainment of part of the cleaned air in the apparatus. Obviously, in spite of the difficulty of carrying out an accurate measurement of the concentration decrease in the pollutant at ppb level over time, it would be interesting to test any PCO device in such real conditions.

In conclusion, it can be said that the recirculation closed-loop reactor considered in this study can provide CADR results similar to the process used in the XP B44-013 standard. It should be noted that by using a closed-loop reactor, the operating parameters can be better controlled and using the model described in this paper leads to more accurate results. Both set-ups (recirculation closed-loop reactor and closed chamber) are actually complementary; the closed-loop reactor can be used to evaluate PCO devices regarding the main operating parameters. This set-up can give results in terms of (i) VOC degradation rate; (ii) one-pass removal efficiency; and (iii) Clean Air Delivery Rate (CADR). This is a reliable methodology to assess the performance of photocatalytic systems and optimize the major operating parameters. This can lead to recommendations for manufacturers in terms of the operating range of flow rate and irradiation for instance, according to the geometry and design of the device. The experimental closed chamber set-up is better adapted for testing any type and geometry of PCO device. Moreover, it refers to the French standard, which enables devices to be compared in the same experimental conditions. It is more dedicated to testing already designed commercial devices for final qualification.

4. Conclusions

Because of recent developments in photocatalytic air cleaning technology, reliable methodologies are needed to assess the performance of commercial PCO devices, as well as standard tests for consumer safety. In addition, rigorous mathematical models describing the decay of the pollutant concentrations over time are required for the convenient analysis of experimental data and to compare the performances of the PCO devices. With these objectives, the performances of PCO devices inserted in a recirculation closed-loop system were investigated using a theoretical approach. The construction of a rigorous model that can be applied to a recirculation closed-loop reactor was explained. It consists of associating the equations describing two ideal reactors: a perfect mixed reservoir and a plug flow reactor. The model was simplified due to both assumptions ($\tau_P/\tau_R \rightarrow 0$ and $\alpha < 5\%$). Once the assumptions are fulfilled, the decrease in pollutant concentration over time can be characterized by a first-order decay model. Thus, it was shown that this model enables the assessment of the overall efficiency of PCO devices. Moreover, it was demonstrated that the model would enable the Clean Air Delivery Rate (CADR) of the PCO devices to be determined experimentally with respect to the low value of the residence time in the closed-loop system. In such conditions, laboratory recirculation closed-loop systems could potentially be used for standardization and for the evaluation of small commercial PCO units. In this case, the experimental CADR could be compared with the efficiency claimed by the manufacturers.

Author Contributions: Éric Dumont and Valérie Héquet wrote the paper.

Conflicts of Interest: The authors declare no conflict of interest.

Appendix A   Nomenclature

A: cross-sectional area of the PCO device (m^2)

C: pollutant concentration (mol·m^{-3})

CADR: Clean Air Delivery Rate $(m^3 \cdot s^{-1})$
E: efficiency (dimensionless)
k: overall degradation rate constant (s^{-1})
k_e: pollutant decay rate constant with the PCO device in operation (s^{-1})
k_n: natural decay rate constant (s^{-1})
L: length of the PCO device (m)
Q: flow rate $(m^3 \cdot s^{-1})$
r: degradation rate $(mol\ m^{-3} \cdot s^{-1})$
t: time (s)
t_c: time constant of the closed-loop reactor (s)
V: chamber volume (m^3)
V_P: PCO device volume (m^3)
V_R: reservoir volume (m^3)

Appendix B Greek letters

τ_P: residence time in the PCO device (s)
τ_R: residence time in the reservoir (s)
α: fractional yield of the treated flow rate (dimensionless)

References

1. Seppanen, O.A.; Fisk, W.J. Some quantitative relations between indoor environmental quality and work performance or health. *Lawrence Berkeley Natl. Lab.* **2006**, *12*, 957–973. [CrossRef]
2. Zhong, L.; Haghighat, F. Photocatalytic air cleaners and materials technologies—Abilities and limitations. *Build. Environ.* **2015**, *91*, 191–203. [CrossRef]
3. Destaillats, H.; Sleiman, M.; Sullivan, D.P.; Jacquiod, C.; Sablayrolles, J.; Molins, L. Key parameters influencing the performance of photocatalytic oxidation (PCO) air purification under realistic indoor conditions. *Appl. Catal. B Environ.* **2012**, *128*, 159–170. [CrossRef]
4. Paz, Y. Application of TiO_2 photocatalysis for air treatment: Patents' overview. *Appl. Catal. B Environ.* **2010**, *99*, 448–460. [CrossRef]
5. Costarramone, N.; Kartheuser, B.; Pecheyran, C.; Pigot, T.; Lacombe, S. Efficiency and harmfulness of air-purifying photocatalytic commercial devices: From standardized chamber tests to nanoparticles release. *Catal. Today* **2015**, *252*, 35–40. [CrossRef]
6. Mills, A.; Hill, C.; Robertson, P. K.J. Overview of the current ISO tests for photocatalytic materials. *J. Photochem. Photobiol. A Chem.* **2012**, *237*, 7–23. [CrossRef]
7. Kartheuser, B.; Costarramone, N.; Pigot, T.; Lacombe, S. NORMACAT project: Normalized closed chamber tests for evaluation of photocatalytic VOC treatment in indoor air and formaldehyde determination. *Environ. Sci. Pollut. Res.* **2012**, *19*, 3763–3771. [CrossRef] [PubMed]
8. Disdier, J.; Pichat, P.; Mas, D. Measuring the effect of photocatalytic purifiers on indoor air hydrocarbons and carbonyl pollutants. *J. Air Waste Manag. Assoc.* **2005**, *55*, 88–96. [CrossRef] [PubMed]
9. Batault, F.; Héquet, V.; Raillard, C.; Thévenet, F.; Locoge, N.; Le Coq, L. How chemical and physical mechanisms enable the influence of the operating conditions in a photocatalytic indoor air treatment device to be modeled. *Chem. Eng. J.* **2017**, *307*, 766–775. [CrossRef]
10. Moulis, F.; Krýsa, J. Photocatalytic degradation of several VOCs (*n*-hexane, *n*-butyl acetate and toluene) on TiO_2 layer in a closed-loop reactor. *Catal. Today* **2013**, *209*, 153–158. [CrossRef]
11. Taranto, J.; Frochot, D.; Pichat, P. Photocatalytic air purification: Comparative efficacy and pressure drop of a TiO_2-coated thin mesh and a honeycomb monolith at high air velocities using a $0.4\ m^3$ close-loop reactor. *Sep. Purif. Technol.* **2009**, *67*, 187–193. [CrossRef]
12. Suzuki, K. *Photocatalytic Air Purification on TiO_2 Coated Honeycomb Support, Proceedings of the First International Conférence on TiO_2 Photocatalytic Purification and Treatment of Air and Water*; Ollis, D.F., Al-Ekabi, H., Eds.; Elsevier Science Ltd.: Amsterdam, The Netherlands, 1993; p. 421.

13. Sauer, M.L.; Ollis, D.F. Photocatalyzed oxidation of ethanol and acetaldehyde in humidified air. *J. Catal.* **1996**, *158*, 570–582. [CrossRef]

14. Nimlos, M.R.; Wolfrum, E.J.; Brewer, M.L.; Fennell, J.A.; Bintner, G. Gas-phase heterogeneous photocatalytic oxidation of ethanol: Pathways and kinetic modeling. *Environ. Sci. Technol.* **1996**, *30*, 3102–3110. [CrossRef]

15. Li Puma, G.; Salvadó-Estivill, I.; Obee, T.N.; Hay, S.O. Kinetics rate model of the photocatalytic oxidation of trichloroethylene in air over TiO$_2$ thin films. *Sep. Purif. Technol.* **2009**, *67*, 226–232. [CrossRef]

16. Mo, J.; Zhang, Y.; Xu, Q.; Lamson, J.J.; Zhao, R. Photocatalytic purification of volatile organic compounds in indoor air: A literature review. *Atmos. Environ.* **2009**, *43*, 2229–2246. [CrossRef]

17. Davis, A.P.; Hao, O.J. Reactor dynamics in the evaluation of photocatalytic oxidation kinetics. *J. Catal.* **1991**, *131*, 285–288. [CrossRef]

18. Walker, A.T.S.; Wragg, A.A. The modelling of concentration-time relationships in recirculating electrochemical reactor systems. *Electrochim. Acta* **1977**, *22*, 1129–1134. [CrossRef]

19. Davis, A.P.; Hao, O.J. Reply to comments on "reactor dynamics in the evaluation of photocatalytic oxidation kinetics". *J. Catal.* **1992**, *136*, 629–630. [CrossRef]

20. Wolfrum, E.J.; Turchi, C.S. Comments on "reactor dynamics in the evaluation of photocatalytic oxidation kinetics". *J. Catal.* **1992**, *136*, 626–628. [CrossRef]

21. Noh, K.-C.; Oh, M.-D. Variation of clean air delivery rate and effective air cleaning ratio of room air cleaning devices. *Build. Environ.* **2015**, *84*, 44–49. [CrossRef]

22. Ginestet, A. Development and Evaluation of a New Test Method for Portable Air Cleaners—CR15_New Test Method for Portable Air Cleaners. Available online: http://www.aivc.org/sites/default/files/members_area/medias/pdf/CR/CR15_New%20test%20method%20for%20portable%20air%20cleaners.pdf (accessed on 17 September 2016).

23. Zhang, Y.; Mo, J.; Li, Y.; Sundell, J.; Wargocki, P.; Zhang, J.; Little, J.C.; Corsi, R.; Deng, Q.; Leung, M.H.K.; et al. Can commonly-used fan-driven air cleaning technologies improve indoor air quality? A literature review. *Atmos. Environ.* **2011**, *45*, 4329–4343. [CrossRef]

24. ANSI/AHAM (American National Standards Institute/Association of Home Appliance Manufacturers) AC-1-2006 Air Cleaners—Portable—CADR; AHAM: Washington, DC, USA, 2006.

25. Kim, H.-J.; Han, B.; Kim, Y.-J.; Yoon, Y.-H.; Oda, T. Efficient test method for evaluating gas removal performance of room air cleaners using FTIR measurement and CADR calculation. *Build. Environ.* **2012**, *47*, 385–393. [CrossRef]

26. Guieysse, B.; Hort, C.; Platel, V.; Munoz, R.; Ondarts, M.; Revah, S. Biological treatment of indoor air for VOC removal: Potential and challenges. *Biotechnol. Adv.* **2008**, *26*, 398–410. [CrossRef] [PubMed]

27. Debono, O.; Thévenet, F.; Gravejat, P.; Héquet, V.; Raillard, C.; Le Coq, L.; Locoge, N. Gas phase photocatalytic oxidation of decane at ppb levels: Removal kinetics, reaction intermediates and carbon mass balance. *J. Photochem. Photobiol. A Chem.* **2013**, *258*, 17–29. [CrossRef]

28. Debono, O.; Thevenet, F.; Gravejat, P.; Hequet, V.; Raillard, C.; Lecoq, L.; Locoge, N. Toluene photocatalytic oxidation at ppbv levels: Kinetic investigation and carbon balance determination. *Appl. Catal. B Environ.* **2011**, *106*, 600–608. [CrossRef]

29. Ollis, D.F. Photocatalytic purification and remediation of contaminated air and water. *Comptes Rendus de l'Académie des Sci. Series IIC Chem.* **2000**, *3*, 405–411. [CrossRef]

30. Peral, J.; Domènech, X.; Ollis, D.F. Heterogeneous photocatalysis for purification, decontamination and deodorization of air. *J. Chem. Technol. Biotechnol.* **1997**, *70*, 117–140. [CrossRef]

31. AFNOR (Association Française de Normalisation) XP B44-013—Photocatalyse—Méthode d'essais et d'analyses pour la mesure d'efficacité de systèmes photocatalytiques pour l'élimination des composés organiques volatils/odeurs dans l'air intérieur en recirculation—Test en enceinte confinée. Available online: http://portailgroupe.afnor.fr/public_espacenormalisation/AFNORB44A/3%20pollutec-bk.pdf (accessed on 17 September 2016).

Sample Availability: Not available.

![molecules](molecules logo) MDPI

Article

Determination of the Clean Air Delivery Rate (CADR) of Photocatalytic Oxidation (PCO) Purifiers for Indoor Air Pollutants Using a Closed-Loop Reactor. Part II: Experimental Results

Valérie Héquet [1], Frédéric Batault [1,2], Cécile Raillard [1], Frédéric Thévenet [2], Laurence Le Coq [1] and Éric Dumont [1,*]

[1] UMR CNRS 6144 GEPEA, IMT Atlantique, La Chantrerie, 4 rue Alfred Kastler, CS 20722, 44307 Nantes CEDEX 3, France; valerie.hequet@imt-atlantique.fr (V.H.); frederic.batault@mines-nantes.fr (F.B.); cecile.raillard@univ-nantes.fr (C.R.); laurence.le-coq@imt-atlantique.fr (L.L.C.)
[2] IMT Lille-Douai, Université de Lille, SAGE, F-59000 Lille, France; frederic.thevenet@imt-lille-douai.fr
* Correspondence: eric.dumont@imt-atlantique.fr; Tel.: +33-251-85-82-66

Academic Editor: Pierre Pichat
Received: 16 December 2016; Accepted: 2 March 2017; Published: 6 March 2017

Abstract: The performances of a laboratory PhotoCatalytic Oxidation (PCO) device were determined using a recirculation closed-loop pilot reactor. The closed-loop system was modeled by associating equations related to two ideal reactors: a perfectly mixed reservoir with a volume of $V_R = 0.42$ m^3 and a plug flow system corresponding to the PCO device with a volume of $V_P = 5.6 \times 10^{-3}$ m^3. The PCO device was composed of a pleated photocatalytic filter (1100 cm^2) and two 18-W UVA fluorescent tubes. The Clean Air Delivery Rate (CADR) of the apparatus was measured under different operating conditions. The influence of three operating parameters was investigated: (i) light irradiance I from 0.10 to 2.0 mW·cm^{-2}; (ii) air velocity v from 0.2 to 1.9 m·s^{-1}; and (iii) initial toluene concentration C_0 (200, 600, 1000 and 4700 ppbv). The results showed that the conditions needed to apply a first-order decay model to the experimental data (described in Part I) were fulfilled. The CADR values, ranging from 0.35 to 3.95 m^3·h^{-1}, were mainly dependent on the light irradiance intensity. A square root influence of the light irradiance was observed. Although the CADR of the PCO device inserted in the closed-loop reactor did not theoretically depend on the flow rate (see Part I), the experimental results did not enable the confirmation of this prediction. The initial concentration was also a parameter influencing the CADR, as well as the toluene degradation rate. The maximum degradation rate r_{max} ranged from 342 to 4894 ppbv/h. Finally, this study evidenced that a recirculation closed-loop pilot could be used to develop a reliable standard test method to assess the effectiveness of PCO devices.

Keywords: photocatalysis; Clean Air Delivery Rate (CADR); indoor air quality; Volatile Organic Compounds (VOCs); air cleaner

1. Introduction

Indoor Air Quality (IAQ) is a current problem because so many people spend time indoors and are exposed to numerous pollutants, such as Volatile Organic Compounds (VOCs). To overcome this issue, PhotoCatalytic Oxidation (PCO) appears as an efficient technology for air cleaning [1,2]. However, the performances of PCO cleaners in development at the laboratory scale or already available on the market have to be evaluated for critical comparison [3]. Consequently, there is a need to develop reliable methodologies to assess the effectiveness of such cleaners and standard test methods prior to their coming onto the market [4,5]. In Part I of the paper, the performance of any PCO device inserted in a closed-loop recirculation system was investigated through a theoretical model. This model allows the Clean Air Delivery Rate (CADR), considered a useful tool to evaluate the air cleaning capacity of a PCO device, to be determined experimentally [6]. The higher the CADR value, the faster the PCO device removes the primary pollutants. As a result, a recirculation closed-loop system can be used (i) to assess the influence of operating parameters and (ii) potentially to compare PCO units. Therefore, the aim of this part of the paper is to check experimentally that the model can be used effectively to determine the CADR of a PCO laboratory device under different operating conditions (pollutant concentration, light irradiance intensity and flow rate). A more accurate analysis of the results using the proposed model leads to a discussion of the main influential parameters. Toluene was chosen as a representative VOC target because it is usually encountered in indoor air at concentrations of around a few tens of parts per billion by volume (ppbv). It is also a common VOC used to study PCO devices found in the literature and in different standard tests.

2. Model

The mathematical model, extensively described in Part I, is based on the one proposed by Walker and Wragg [7] for concentration-time relationships established for recirculating electrochemical reactor systems. The closed-loop reactor operating in recirculation mode (Figure 1) was modeled by associating two ideal reactors: a perfectly mixed reservoir with a volume V_R and a plug flow system corresponding to the PCO device with a volume V_P. Considering that the volume of the PCO device is very small in relation to the volume of the reservoir (i.e., if the ratio of the residence times (τ_P/τ_R) tends to zero), the concentration-time relationship is:

$$C = C_0 \, \exp\left\{ -\frac{t}{\tau_R}[1 - \exp(-\alpha)] \right\} \tag{1}$$

Equation (1) should be able to describe satisfactorily the decrease in the pollutant concentration over time provided that the total time of the experiment does not exceed 2 h. In this equation, the term α (= $k\,\tau_p$) is the fractional yield of the treated flow rate of the PCO device. In other words, α corresponds to the percentage of the total flow rate treated during the time τ_R (i.e., during one cycle). As developed in Part I, the term α can be related to the parameter CADR (Clean Air Delivery Rate) typically used to evaluate the air cleaning capacity of a PCO reactor. The CADR is defined as the product of the device efficiency and the volumetric air flow rate through the apparatus [8,9]:

$$CADR = \alpha\,Q \tag{2}$$

Since the volume of the PCO device is small in relation to the volume of the reservoir, it is expected that the CADR value is small with regard to the air flow rate Q, and consequently, α does not exceed some %. In this case, Equation (1) can be simplified as:

$$C = C_0 \, \exp\left(-t\frac{\alpha}{\tau_R} \right) = C_0 \, \exp\left(-t\frac{CADR}{V_R} \right) \tag{3}$$

Thus, the decrease in pollutant concentration follows a first-order decay model. The value of the parameter α is then directly deduced from the slope of the curve $\ln(C/C_0)$ vs. t. Rearranging Equation (3), the overall efficiency of the PCO device to treat the air is given by Equation (4), where τ_R/α is the time constant (t_c) of the closed-loop reactor (i.e., the time needed to reach a 63.2% conversion of pollutant) and $1/\alpha$ is the cycle number needed to obtain E = 0.632.

$$E = \frac{C_0 - C}{C_0} = 1 - \exp\left(-t\frac{\alpha}{\tau_R} \right) = 1 - \exp\left(-\frac{t}{t_c} \right) \tag{4}$$

The maximum degradation rate determined at the beginning of the experiment (at t = 0) is then:

$$r_{max} = C_0 \frac{\alpha}{\tau_R} = C_0 \frac{CADR}{V_R} \tag{5}$$

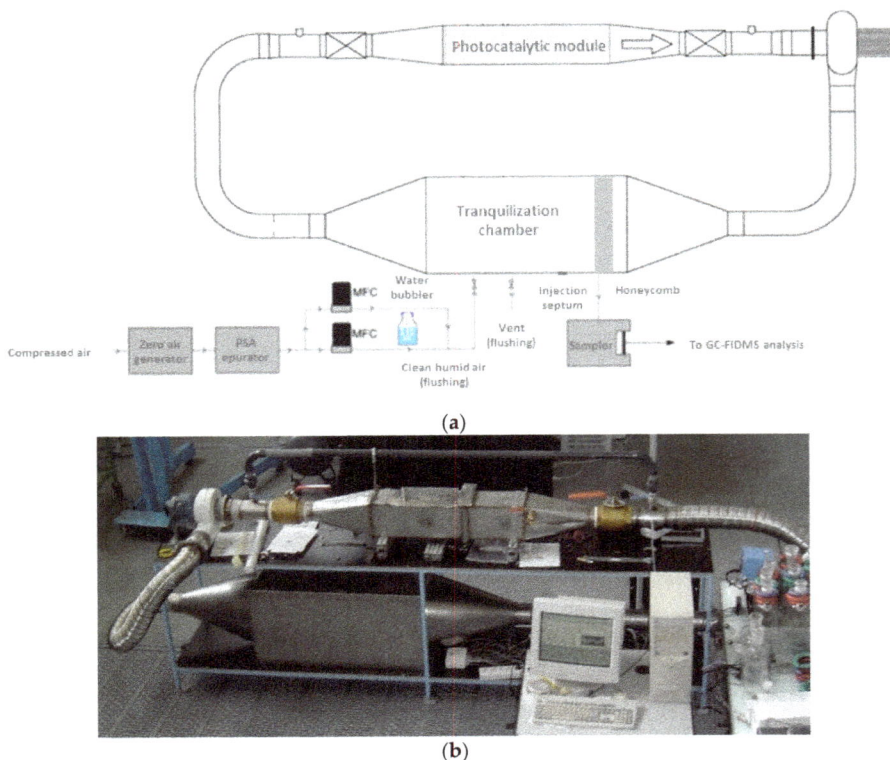

(a)

(b)

Figure 1. Experimental closed-loop reactor operating in recirculation mode. (**a**) Schematic representation; and (**b**) photograph of the whole system.

3. Experimental Methods

3.1. Recirculation Closed-Loop System

The closed-loop (batch) system used in this study is described in Figure 1. As experiments were mainly carried out at concentrations lower than 1 ppmv, the whole pilot (V_R = 0.42 m^3) was constituted of stainless-steel to limit the loss of toluene by adsorption on the walls. The PCO device was installed in an internal duct and operated in recirculation mode. The air flow rate was controlled by a variable speed fan operating inside the closed-loop system. Flow rates were determined from pressure drop measurements between the two sides of a calibrated diaphragm. A tranquilization chamber was used for toluene injection and gas sampling. A honeycomb part located at the inlet of the chamber provided a homogeneous flow distribution in the reservoir. In such conditions, the polluted air could be considered perfectly mixed in the reservoir.

3.2. PCO Device

The PCO device was composed of a photocatalytic medium and UV lamps (Figure 2). The photocatalytic medium (Ahlström Paper Group) consisted of a thin fibrous support (250 μm thick) composed of a mixture of cellulose, polyester and polyamide and coated with P25 TiO_2 in a SiO_2 binder. The TiO_2 load was 17 g·m^{-2}. This medium was specifically produced for these experiments. The BET specific surface area of the medium (31 ± 1 m^2·g^{-1}) was measured by a Micromeritics ASAP 2010 device (Micromeritics Instrument Corp., Norcross, GA, USA). A pleated geometry filter was used in order to increase the surface implemented in the reactor for air treatment. Due to this geometry, the total surface area of the photocatalytic medium was 1100 cm^2, and the corresponding mass of TiO_2 was 1.87 g. Two 18-W UVA UV fluorescent tubes (Philips PL-L series, Koninklijke Philips N.V., Amsterdam, The Netherlands) were placed inside the filter folds as illustrated in Figure 2. Taking into account the pleated configuration of the PCO device, its overall volume V_P was 5.6 L (0.2 × 0.2 × 0.14 = 5.6 10^{-3} m^3), corresponding to 1.3% of the whole pilot. In order to control the irradiance received by the photocatalytic medium, a variable voltage supply was used to change the light intensity provided by the lamps. Before experiments, light irradiance was calibrated using a Vilber Laurmat VLX 3-W radiometer equipped with a calibrated CX-365 sensor (Vilber Lourmat Deutschland GmbH, Eberhardzell, Germany). The calibration procedure is extensively described in Batault et al. [10].

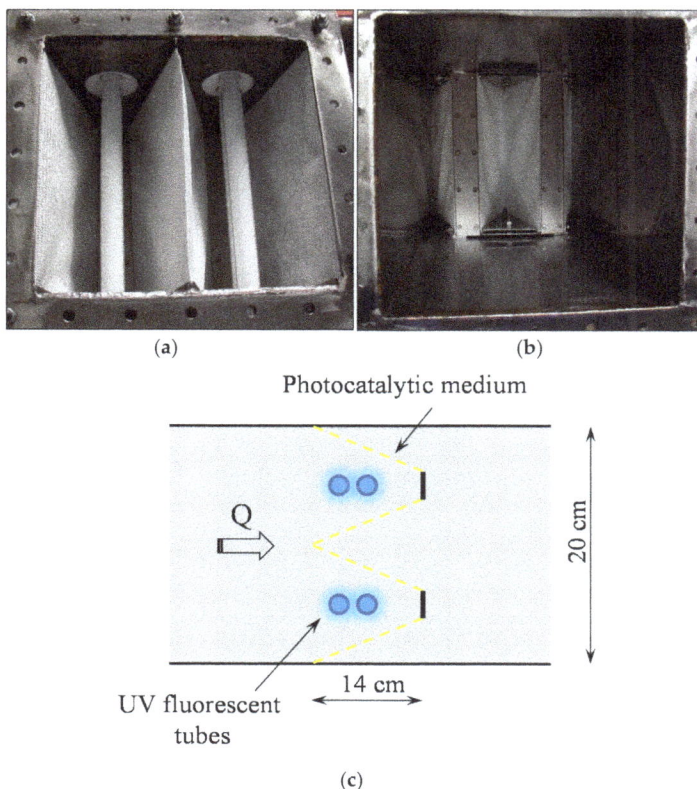

(a)   (b)

Photocatalytic medium

UV fluorescent tubes

20 cm

14 cm

(c)

Figure 2. PhotoCatalytic Oxidation (PCO) device. View of: (**a**) the inlet side; (**b**) the outlet side; and (**c**) diagram of the pleated filter configuration.

3.3. VOC Generation and Analytical Methods

To prepare the polluted gas phase at the desired concentrations, liquid toluene (99.99% purity, purchased from VWR International, Fontenay sous Bois, France) was evaporated in a 1-L glass reactor at room temperature. The toluene gaseous phase was then injected into the tranquilization chamber of the pilot by means of a 50-mL syringe through a septum (Figure 1). The decrease in toluene concentration over time in the photocatalytic reactor was monitored during the experiments using a Markes FLEC Air Pump 1001 sampler (Markes International Ltd., Llantrisant, UK) at room temperature and at a flow of 100 mL/min. It consisted of an off-line sampling on multi-sorbent cartridges containing three beds of Supelco activated carbon: Carbopack B, Carbopack C and Carbopack X. Sorbent tubes were then analyzed using a TD/GC/FID/MS system (Thermal Desorption/Gas Chromatography/Flame Ionization Detector/Mass Spectrometer, Perkin-Elmer, Lyon, France). The gas chromatograph was equipped with a 60-m Rxi-624Sil MS non-polar column from Restek, which was simultaneously connected to two detectors: (i) a Flame Ionization Detector (FID) for the quantification of compounds; and (ii) a Mass Spectrometer (MS) for chromatographic peak identification. A calibration curve for toluene quantification was carried out by diluting a standard toluene cylinder (1040 ppbv in N_2, Prax'air) with clean air. The detection limit was estimated to be lower than 21 ppbv. Three experiments were performed in identical experimental conditions (I = 0.35 mW·cm^{-2}, v = 0.6 m·s^{-1}, C_0 = 600 ppbv) to assess the experimental error in the determination of pollutant concentrations. As evidenced in the paper by Batault et al. [10], the degradation curves of the three experiments indicated that the results were repeatable. The experimental error in the determination of the initial concentration C_0 was 17%, which corresponds to the value given in Destaillats et al. [1] for a pleated filter (around 15%).

3.4. Operating Conditions

Before each experiment, the closed-loop reactor was first flushed with VOC-free air in order to remove traces of VOCs in the system. The photocatalytic material itself is cleaned by irradiation under 50% of relative humidity for several hours. Then, the reactor was flushed again. For experiments, the air entering the reactor was first treated with a zero air generator (Claind 2020, CryoService Limited, Worcester, UK). The CO_2 and humidity were then removed using a suppressor TDGSi PSA device (F-DGSi, Evry, France). Next, the relative humidity of the air stream in the closed-loop reactor was set using a hygro-transmitter measurement, a room temperature bubbler and according to an appropriate gas dilution by means of two Brooks mass flow controllers. The relative humidity was set at the targeted value, 50%, i.e., 13,000 ppm, to be close to real indoor air conditions (Figure 1).

The influence of three operating parameters (light irradiance I, air velocity v and initial toluene concentration C_0) was investigated in order to determine the change in the performance of the PCO device. The light irradiance ranged from 0.1 to 2.0 mW·cm^{-2}. The air velocity was determined as the ratio between the air flow rate, ranging from 28.8 to 273.6 m^3·h^{-1}, and the cross-sectional area of the PCO device (20 cm × 20 cm square-section duct, i.e., 0.04 m^2). Thus, the air velocity was varied from 0.2 to 1.9 m·s^{-1}. In the present work, the photocatalytic module was inserted in a 20 cm × 20 cm square cross-section so that the hydraulic diameter DH was 20 cm. The Reynolds number thus ranged between 2500 and 26,000, depending on the air velocity value (0.2 to 1.9 s^{-1}), and the reaction rate was not limited by mass transfer [11]. In these conditions, the residence time in the reservoir ranged from 52.5 to 5.5 s. Four initial concentrations were selected for experiments: 200, 600, 1000 and 4700 ppbv. The 17 experiments were carried out under the conditions defined in Table 1.

Table 1. Operating conditions used in each experiment and experimental results.

	Experimental Conditions						Experimental Results				
	I (mW·cm^{-2})	v (m·s^{-1})	τ_R (s)	C_0 (ppbv)	α (−)	R^2	CADR (m^3·h^{-1})	$1/\alpha$ (−)	t_c (h)	t_c (s)	r_{max} (ppbv·h^{-1})
Exp 1	0.10	0.2	52.5	600	0.0278	0.988	0.80	36	0.53	1917	1140
Exp 2	0.10	0.6	17.5	200	0.0086	0.973	0.74	117	0.57	2052	342
Exp 3	0.10	0.6	17.5	1000	0.0043	0.956	0.37	232	1.13	4065	890
Exp 4	0.10	1.0	10.5	600	0.0059	0.937	0.85	169	0.49	1781	1183
Exp 5	0.35	0.2	52.5	200	0.0337	0.969	0.97	30	0.44	1584	449
Exp 6	0.35	0.2	52.5	1000	0.0259	0.927	0.75	39	0.57	2055	1780
Exp 7	0.35	0.6	17.5	600	0.0166	0.971	1.43	60	0.30	1062	2045
Exp 8	0.35	0.6	17.5	4700	0.0041	0.964	0.35	244	1.19	4273	3962
Exp 9	0.35	1.0	10.5	200	0.0118	0.961	1.70	85	0.25	901	786
Exp 10	0.35	1.0	10.5	1000	0.0090	0.984	1.30	111	0.33	1171	3097
Exp 11	0.35	1.9	5.5	600	0.0047	0.966	1.29	212	0.33	1173	1841
Exp 12	0.60	0.2	52.5	600	0.0643	0.994	1.85	16	0.23	843	2639
Exp 13	0.60	0.6	17.5	200	0.0457	0.991	3.95	22	0.11	392	1827
Exp 14	0.60	0.6	17.5	1000	0.0170	0.929	1.47	59	0.29	1039	3505
Exp 15	0.60	1.0	10.5	600	0.0167	0.969	2.40	60	0.18	635	3971
Exp 16	1.00	0.6	17.5	600	0.0238	0.979	2.06	42	0.21	743	2934
Exp 17	2.00	0.6	17.5	600	0.0397	0.962	3.43	25	0.12	449	4894

4. Results and Discussion

Figure 3 presents a typical result of changing toluene concentration over irradiation time (corresponding to Exp 10). The experimental C/C_0 measurements were fitted using both the predictive model (Equation (1)) and the simplified model (Equation (3)). Numerical resolutions were carried out using Excel® solver. The procedure is based on the linear least-squares method, which minimizes the Sum of Squared Residuals (SSR) between experimental and predicted values (for i ranging from one to n, which corresponds to the n concentration values measured during the experiment).

$$\mathbf{SSR} = \sum_{i=1}^{n}\left\{\left(\frac{\mathbf{C}}{\mathbf{C_0}}\right)_i^{\mathbf{exp}} - \left(\frac{\mathbf{C}}{\mathbf{C_0}}\right)_i^{\mathbf{model}}\right\}^2 \tag{6}$$

According to Figure 3, it can be observed that the predictive model (Equation (1)) described the experimental data satisfactorily within a total time for the toluene degradation experiment of less than two hours. Moreover, as the percentage of the total flow rate treated during one cycle ($\tau_R = 10.5$ s) was low ($\alpha = 0.9\%$ in the present case; Table 1), the assumption of using the simplified model (Equation (3)) was satisfied, and thus, the first-order decay model could be used for the CADR determination (Figure 3). Nonetheless, for irradiation times longer than 0.7 h, corresponding to a C/C_0 ratio value lower than 0.15, it appears that the model slightly over-predicts the experimental data (this observation can be made for all experiments as soon as $C/C_0 < 0.20$). This result, reported in the literature [12], is clearly highlighted in the curve $\ln(C/C_0)$ vs. time (Figure 3). Such deviations may be due to physical reasons (mass transfer limitation), analytical reasons (accuracy of the measurements when the toluene concentration tends to zero) or mathematical reasons (because $\lim \ln(C/C_0)$ tends to $-\infty$ for C/C_0 tending to zero). Whatever the reasons, which need to be specifically studied, the problem with using the curve $\ln(C/C_0)$ vs. time for the slope determination is choosing the number of points to take into account. As highlighted in Figure 3, the slope is between -2.97 and -3.90 h^{-1} according to the number of points selected, which corresponds to a 35% deviation. These slope values have to be compared with the value obtained by modeling all of the experimental data using Equation (3) and the Excel® solver. For the experiment Exp 10, the term α/τ_R corresponding to the absolute value of the slope is 3.08 h^{-1} ($\alpha/\tau_R = 0.0090/(10.5/3600)$), which indicates that only the first six or seven points should have been chosen for a correct slope determination. Since the Excel® solver is easily usable for data analysis, such a tool should be preferred for CADR determination using Equation (1) or Equation (3) rather than the analysis of the curve $\ln(C/C_0)$ vs. time. Consequently, all of the results given in the present paper (Table 1) came from the analysis of all of the experimental data using Equation (1).

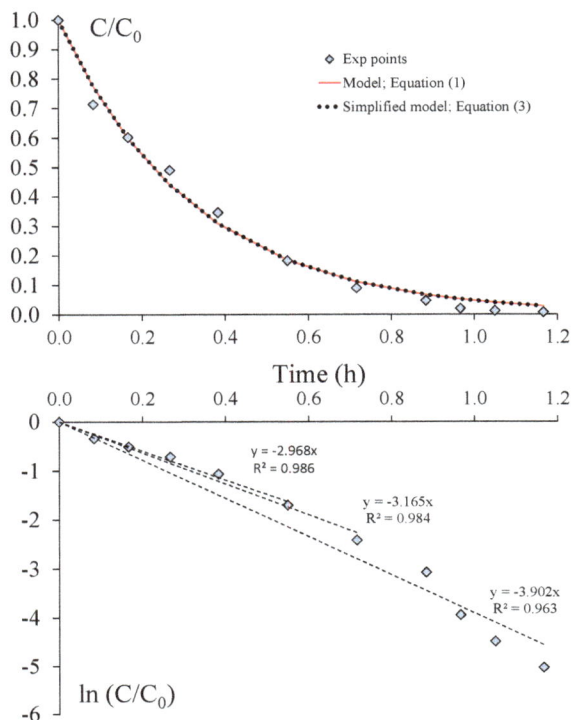

Figure 3. Typical change in toluene concentration over time. Experimental data and models (Exp 10; $I = 0.35$ mW·cm^{-2}; $v = 1.0$ m·s^{-1}; $C_0 = 1000$ ppmv).

From Table 1, it can be concluded that the duration of each experiment (corresponding to 5 t_c) was less than 2 h with the exception of Exp 3 (due to the low value of the light irradiance and high initial concentration) and Exp 8 (because the initial concentration was significantly higher than that of the other experiments, $C_0 = 4700$ ppbv). Consequently, experiments can be satisfactorily modeled using Equation (1). For Exps 3 and 8, the total time of the experiment was longer than 3 h, and the approximate solution given by Equation (1) becomes inaccurate as indicated in Part I. From Table 1, it also appears that the maximum α value was 6.4% for Exp 12 and less than 5% for the other experiments. As a result, the simplified model given by Equation (3) can also be used for the analysis of almost all of the experiments, as already highlighted in Figure 3 for Exp 10. Figures 4 and 5 present the toluene removal efficiency (normalized concentrations) for all of the experiments according to the generalized (dimensionless) form and as a function of the number of cycles, respectively (Equation (4)). Figure 4 confirms that irrespective of the operating conditions, the experiments are satisfactorily described by the first-order decay model. Moreover, Figure 5 shows that the number of cycles needed for toluene degradation is strongly dependent on the operating conditions. Such a graphical representation is useful for a comparison at a glance.

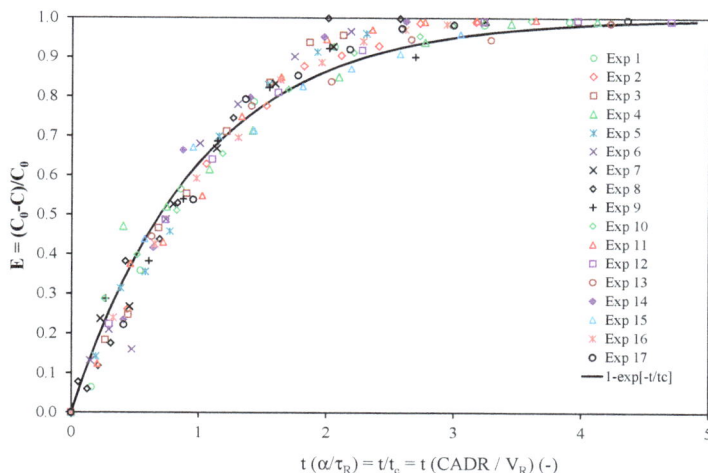

Figure 4. Toluene removal efficiency for all experiments according to the generalized form (Equation (4)).

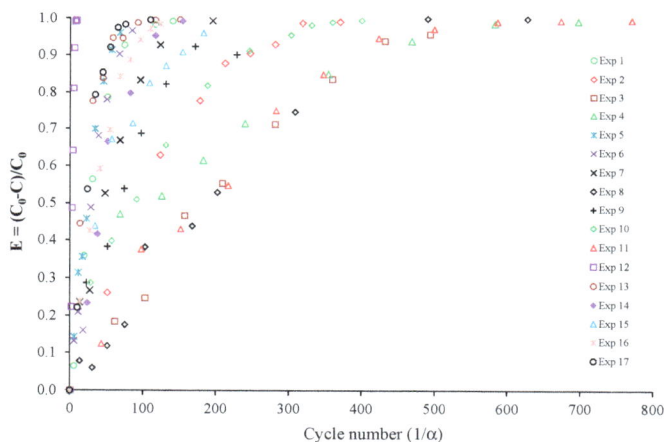

Figure 5. Toluene removal efficiency for all experiments as a function of the number of cycles in the closed-loop reactor.

4.1. CADR Determination

Depending on the operating conditions, the results given in Table 1 show that CADR ranged from 0.35 to 3.95 m$^3 \cdot$h^{-1}, i.e., one order of magnitude for the same PCO device. The time constant t_c of the closed-loop reactor varied accordingly, from 392 to 4065 s. Moreover, the number of cycles ($1/\alpha$) needed to reach a 63.2% conversion of pollutant varied from 16 to 244 corresponding to a ratio of 15. From Table 1, it appears that the performances of the PCO device are strongly dependent on the operating parameters.

Since the experimental data reported in the literature are often obtained under different operating conditions (the design of the reactor or test chamber, flow rate, light intensity, pollutant nature and concentrations), the direct comparison of CADR values given in the literature is therefore not suitable. Nonetheless, the study of CADR values obtained using a similar procedure can be informative. For instance, the curve $\ln(C/C_0)$ vs. t was applied by Costarramone et al. to compare the efficiency of different photocatalytic air purifiers [4]. Studying the toluene degradation performance of eight commercial PCO devices in a 1.2-m^3 closed-loop chamber, these authors reported CADR values from 0.67 to 24.5 m$^3 \cdot$h^{-1} for an initial concentration of 1000 ppbv and from 2.35 to 10.94 m$^3 \cdot$h^{-1} for an initial concentration of 250 ppbv. However, the CADR values were directly calculated assuming a first-order decay, although both assumptions described in Part I of the paper were not necessarily checked. Thus, the authors indicated that the CADR values could be quite inaccurate. Moreover, they reported that the comparison of the PCO devices was not straightforward since so many parameters varied. For instance, the CADR did not follow the order of maximum flow rates of the devices. Consequently, with the aim of developing a standard for the certification of different types of PCO apparatus, there is a need to assess the relative influence of each operating parameter on the performances of the same PCO device.

4.2. Effect of the Light Irradiance Intensity

Figure 6 depicts the influence of the light irradiance intensity on toluene removal at a constant flow rate and constant concentration. The greater the light intensity is, the faster the degradation and the higher the CADR are. The degradation rate is varying with light intensity I as I^n with n approaching one at very low intensities and n approaching zero at very high values. An intermediate case is often described in the literature where n is around 0.5 [13]. The insert in Figure 6 shows that the relationship between CADR and light intensity is: CADR = 2.3 $I^{0.49}$. Thus, the CADR varies as the square root of the light intensity, which corresponds to the trend usually reported in the literature [14]. In fact, it is generally agreed that the reaction rate increases with increasing light intensity, as the heterogeneous photocatalytic reaction depends on the irradiation of the TiO$_2$ surface by UV light to produce electron/hole pairs, even though some of them recombine. Several regimes are defined: a first-order regime in which the electron-hole pairs are consumed more rapidly by chemical reactions than by recombination; and a half-order regime in which the recombination rate dominates [6,15,16]. Nevertheless, it has to be recalled that the CADR is proportional to the overall kinetic rate constant of the apparatus, which can differ from the photocatalytic reaction rate since the reactor dynamics are also taken into account (see Part I). Despite this difference, it can be considered that the CADR determination reflects the actual photocatalytic activity of the PCO device satisfactorily.

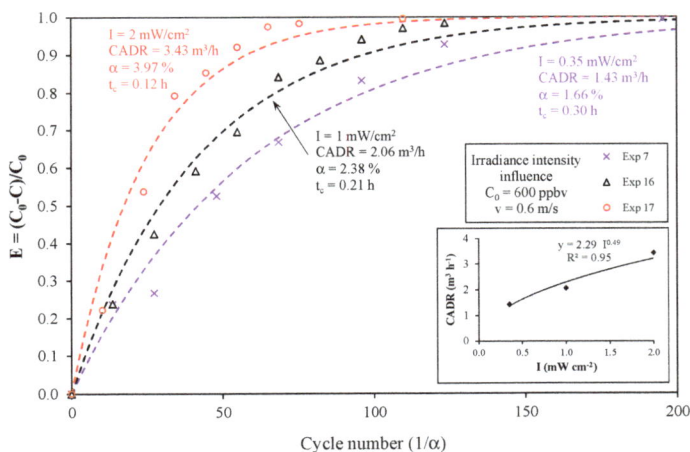

Figure 6. Effect of irradiance intensity on toluene removal efficiency (experimental points and model). Insert: effect of irradiance intensity on the CADR value.

4.3. Effect of the Flow Rate

As indicated in Part I, the overall efficiency, as well as the maximum degradation rate in the closed-loop reactor do not theoretically depend on the flow rate. In fact, an increase in the flow rate leads to a decrease in the residence time τ_P in the PCO device, which reduces the α value, and consequently, many cycles are required for the total conversion of the pollutant. As a result, for a given PCO device, the CADR should be the same irrespective of the flow rate used in the experiment, provided that all other parameters are kept constant. Figure 7 displays five cases illustrating the influence of flow rate for a constant initial concentration and a constant light irradiance. The results are contrasting. For example, cases {**A**} and {**C**} are in agreement with the theory, whereas cases {**B**} and {**D**} indicate that the flow rate had a significant influence on both the CADR and the maximum degradation rate. Moreover, taking into account the accuracy of the experimental measurement, case {**E**} cannot be used to tip the balance. The results from Figure 7 clearly highlight the difficulty of intuitively sensing the effect of the flow rate. It should be noted that a statistical analysis of the results using an experimental design was not sufficient to understand the actual role of the air velocity [10], whereas a previous study carried out at the ppmv level demonstrated a negative influence of the increase in the flow rate on PCO performances [16]. According to Destaillats et al. [1], the contact time of a pollutant on the surface of a PCO medium is one of the most critical parameters for reactor design because a sequence of multiple adsorption/desorption cycles takes place inside the medium. Due to the internal mass transfer of molecules in the inter-fiber and at the surface of the medium fibers [17], the contact time in the medium is likely to be longer than the residence time calculated from geometrical considerations only. One assumption is that the air flow rate may interfere with the sequence of adsorption/desorption cycles and influence the overall performance of the PCO device. In addition, even though all experiments carried out in this study were based on the same PCO device and the same pollutant, it is probable that air flowed differently across the medium according to the operating conditions. As indicated in Figure 2, the presence of the UV lamps upstream of the medium and the configuration of the pleated medium itself could significantly change the air flow distribution along the filter and inside the material. Turbulence effects could be generated at high flow rates. However, from Figure 7, it can be assumed that the order of magnitude of the air velocity would not necessarily be influential because the theory is verified for the couple of velocities (v = 0.6 m/s and v = 1.9 m/s; case {**A**}) and the couple of velocities (v = 0.2 m/s and v = 1.0 m/s; case {**C**}), whereas it is not verified for the latter couple of velocities in the other cases. A numerical study of the air flow around and inside the medium seems necessary to investigate this point. Moreover, future studies using different filter geometries need to be performed to evidence the real influence of the flow rate.

4.4. Effect of the Initial Concentration

For experiments carried out at a constant light irradiance and a constant air velocity, it is expected that the time needed for total toluene degradation should increase with the increase in the initial concentration. Figure 8 shows the effect of C_0 on the determined parameters: CADR, time constant t_c and maximum degradation rate r_{max}. Cases {**F**} {**G**} {**H**} and {**I**} enable a direct comparison between the results obtained since concentrations of 200 and 1000 ppbv were used in these four cases (corresponding to a concentration ratio of five). The results are contrasting. For I = 0.35 mW·cm^{-2} (cases {**F**} and {**G**}), the CADR values and the t_c values are of the same order of magnitude (ratio of 1.3), whereas for I = 0.10 mW·cm^{-2} (case {**H**}) and for I = 0.60 mW·cm^{-2} (case {**I**}), the ratios are equal to 2.0 and 2.7, respectively. Moreover, according to Figure 8, the maximum degradation rate r_{max} was higher for 1000 ppbv than for 200 ppbv, but the ratio $r_{max(1000\ ppbv)}/r_{max(200\ ppbv)}$ was around 4 for cases {**F**} and {**G**}, 2.6 for case {**H**} and only 2 for case {**I**}. In case {**J**} corresponding to a concentration ratio of eight due to the very high concentration used (4700 ppbv), the difference between CADR values was clear (ratio of four), but the maximum degradation rate obtained for this concentration (3962 ppbv/h) was not significantly higher than that obtained for a concentration of 1000 ppbv in other conditions. Actually, at this level of concentration, it is known that the reaction mechanism tends to a saturation reaction well described in the literature by the Langmuir–Hinshelwood kinetic rate law model [16]. It should be noted that cases {**H**} and {**J**} are based on Exp 3 and Exp 8, respectively, and thus, these results must be considered with caution (as discussed above). They confirm that it is difficult to determine experimentally the influence of a given parameter independently of the others [10]. Although these results highlight that the initial concentration is not the main parameter influencing the degradation of the pollutant, in comparison with the light irradiance, they nonetheless demonstrate that it is an important parameter that must be taken into account for the characterization of any PCO device. Consequently, with the aim of developing a reliable standard test method to assess the effectiveness of types of PCO apparatus, the given values of the initial concentration should be selected.

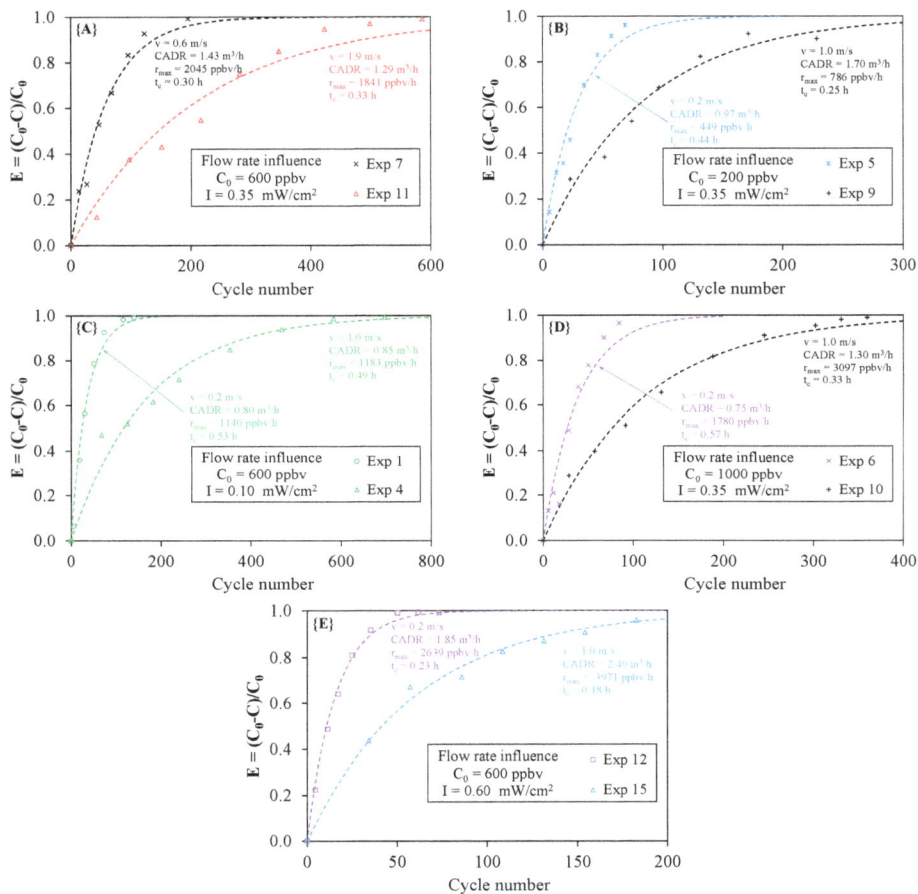

Figure 7. Flow rate influence on toluene removal efficiency according to the initial concentration C_0 and the light irradiance I (cases {**A**}–{**E**}; experimental points and model; cycle number = $1/\alpha$).

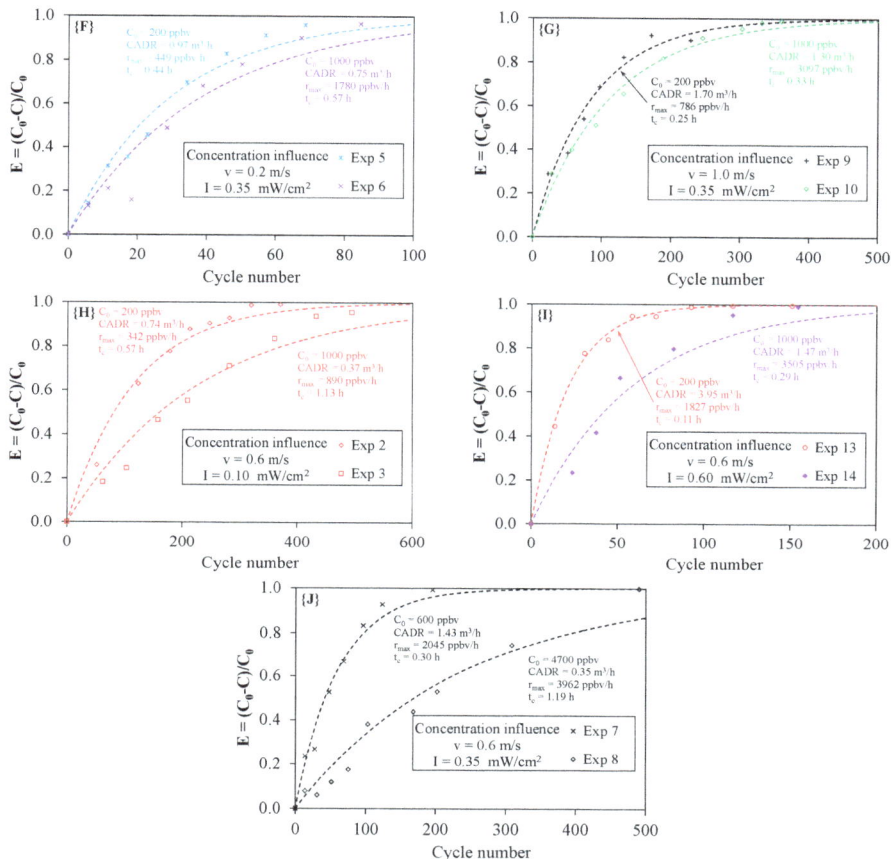

Figure 8. Effect of the initial concentration on toluene removal efficiency according to the light irradiance I and the air velocity v (cases {**F**}–{**J**}; experimental points and model; cycle number = $1/\alpha$).

5. Conclusions

The performances of a laboratory PCO device were determined using a recirculation closed-loop pilot. The Clean Air Delivery Rate (CADR) of the apparatus was calculated under different operating conditions. The main results of this study are:

(i) Since the volume of the PCO device is very small in relation to the volume of the reservoir, the ratio of the residence times (τ_P/τ_R) tends to zero, and consequently, the concentration-time relationship given by Equation (1) can be used to model the experimental points. Moreover, as the term α is usually determined to be lower than 5%, the simplified model characterized by a first-order decay model (Equation (3)) can be used to determine the CADR of the PCO device.

(ii) Since the Excel® solver is easily usable for data analysis, such a tool should be preferred for CADR determination rather than the analysis of the curve $\ln(C/C_0)$ vs. time, which involves selecting a given number of points for a correct calculation.

(iii) According to the operating conditions, the CADR ranged from 0.35 to 3.95 $m^3 \cdot h^{-1}$, i.e., one order of magnitude for the same PCO device. The CADR was mainly dependent on the light irradiance intensity. An increase in the CADR with the square root of the light irradiance was observed.

(iv) Although the CADR of the PCO device inserted in the closed-loop reactor did not theoretically depend on the flow rate (see Part I), the experimental results did not enable the confirmation of this point. The results were contrasting. Some experimental data were in agreement with the theory, whereas others disagreed. Further investigations are therefore needed to explain this ambiguity. Numerical simulations of the air stream line and velocity through the medium may be useful.

(v) The maximum degradation rate r_{max} ranged from 342 to 4894 ppbv·h^{-1}. As the initial concentration is one parameter influencing the degradation rate of the pollutant, tests should be performed at a given value of the initial concentration in order to compare the performances of different types of PCO apparatus.

Finally, this study demonstrated that a recirculation closed-loop pilot can be used to develop a reliable standard test method to assess the effectiveness of any PCO device with regard to the operating parameters. The conditions needed to apply a first-order decay model to the experimental data (described in Part I) were usually fulfilled. However, experimental errors in the determination of the pollutant concentrations still impact the assessment of the influential parameters too much. More accurate results should be obtained using the same operating conditions, and future research is still required to determine pollutant concentrations at low part-per-billion levels.

Acknowledgments: The authors thank the Institut Carnot Mines for its financial support, Yvan Gouriou, François-Xavier Blanchet and Éric Chevrel for their precious technical assistance.

Author Contributions: V.H., C.R., L.L.C. and F.T. supervised the F. Batault PhD work; V.H., F.B. and C.R. conceived and designed the experiments; F.B. performed the experiments; all authors analyzed the data and results; V.H. and E.D. wrote the paper.

Conflicts of Interest: The authors declare no conflict of interest.

Abbreviations

A	cross-sectional area of the PCO device (m^2)
C	pollutant concentration (mol·m^{-3})
CADR	Clean Air Delivery Rate (m^3·s^{-1})
E	efficiency (dimensionless)
k	overall degradation rate constant (s^{-1})
L	length of the PCO device (m)
Q	flow rate (m^3·s^{-1})
r	degradation rate (mol m^{-3}·s^{-1})
t	time (s)
t_c	time constant of the closed-loop reactor (s)
V_P	PCO device volume (m^3)
V_R	reservoir volume (m^3)
Greek letters	
τ_P	residence time in the PCO device (s)
τ_R	residence time in the reservoir (s)
α	fractional yield of the treated flow rate (dimensionless)

References

1. Destaillats, H.; Sleiman, M.; Sullivan, D.P.; Jacquiod, C.; Sablayrolles, J.; Molins, L. Key parameters influencing the performance of photocatalytic oxidation (PCO) air purification under realistic indoor conditions. *Appl. Catal. B Environ.* **2012**, *128*, 159–170. [CrossRef]

2. Pichat, P. Some views about indoor air photocatalytic treatment using TiO$_2$: Conceptualization of humidity effects, active oxygen species, problem of C1–C3 carbonyl pollutants. *Appl. Catal. B Environ.* **2010**, *99*, 428–434. [CrossRef]

3. Zhong, L.; Haghighat, F. Photocatalytic air cleaners and materials technologies—Abilities and limitations. *Build. Environ.* **2015**, *91*, 191–203. [CrossRef]

4. Costarramone, N.; Kartheuser, B.; Pecheyran, C.; Pigot, T.; Lacombe, S. Efficiency and harmfulness of air-purifying photocatalytic commercial devices: From standardized chamber tests to nanoparticles release. *Catal. Today* **2015**, *252*, 35–40. [CrossRef]

5. Mills, A.; Hill, C.; Robertson, P.K.J. Overview of the current ISO tests for photocatalytic materials. *J. Photochem. Photobiol. A Chem.* **2012**, *237*, 7–23. [CrossRef]

6. Mo, J.; Zhang, Y.; Xu, Q.; Lamson, J.J.; Zhao, R. Photocatalytic purification of volatile organic compounds in indoor air: A literature review. *Atmos. Environ.* **2009**, *43*, 2229–2246. [CrossRef]

7. Walker, A.T.S.; Wragg, A.A. The modelling of concentration—time relationships in recirculating electrochemical reactor systems. *Electrochim. Acta* **1977**, *22*, 1129–1134. [CrossRef]

8. Ginestet, A. Development and Evaluation of a New Test Method for Portable air cleaners—CR15_New Test Method for Portable Air Cleaners. Avaiable Online: http://www.aivc.org/sites/default/files/members_area/medias/pdf/CR/CR15_New%20test%20method%20for%20portable%20air%20cleaners.pdf (accessed on 17 September 2016).

9. Zhang, Y.; Mo, J.; Li, Y.; Sundell, J.; Wargocki, P.; Zhang, J.; Little, J.C.; Corsi, R.; Deng, Q.; Leung, M.H.K.; et al. Can commonly-used fan-driven air cleaning technologies improve indoor air quality? A literature review. *Atmos. Environ.* **2011**, *45*, 4329–4343. [CrossRef]

10. Batault, F.; Héquet, V.; Raillard, C.; Thévenet, F.; Locoge, N.; le Coq, L. How chemical and physical mechanisms enable the influence of the operating conditions in a photocatalytic indoor air treatment device to be modeled. *Chem. Eng. J.* **2017**, *307*, 766–775. [CrossRef]

11. Obee, T.N. Photooxidation of Sub-Parts-per-Million Toluene and Formaldehyde Levels on Titania Using a Glass-Plate Reactor. *Environ. Sci. Technol.* **1996**, *30*, 3578–3584. [CrossRef]

12. Moulis, F.; Krýsa, J. Photocatalytic degradation of several VOCs (*n*-hexane, *n*-butyl acetate and toluene) on TiO2 layer in a closed-loop reactor. *Catal. Today* **2013**, *209*, 153–158. [CrossRef]

13. Peral, J.; Ollis, D.F. Heterogeneous photocatalytic oxidation of gas-phase organics for air purification: Acetone, 1-butanol, butyraldehyde, formaldehyde, and *m*-xylene oxidation. *J. Catal.* **1992**, *136*, 554–565. [CrossRef]

14. Herrmann, J.-M. Photocatalysis fundamentals revisited to avoid several misconceptions. *Appl. Catal. B Environ.* **2010**, *99*, 461–468. [CrossRef]

15. Puma, G.L.; Salvadó-Estivill, I.; Obee, T.N.; Hay, S.O. Kinetics rate model of the photocatalytic oxidation of trichloroethylene in air over TiO2 thin films. *Separ. Purif. Technol.* **2009**, *67*, 226–232. [CrossRef]

16. Raillard, C.; Maudhuit, A.; Héquet, V.; le Coq, L.; Sablayrolles, J.; Molins, L. Use of Experimental Designs to Establish a Kinetic Law for a Gas Phase Photocatalytic Process. *Int. J. Chem. React. Eng.* **2014**, *12*, 113–122. [CrossRef]

17. Zhong, L.; Haghighat, F. Modeling and validation of a photocatalytic oxidation reactor for indoor environment applications. *Chem. Eng. Sci.* **2011**, *66*, 5945–5954. [CrossRef]

Sample Availability: Not available.

MDPI AG

St. Alban-Anlage 66

4052 Basel, Switzerland

Tel. +41 61 683 77 34

Fax +41 61 302 89 18

http://www.mdpi.com

Molecules Editorial Office

E-mail: molecules@mdpi.com

http://www.mdpi.com/journal/molecules

www.ingramcontent.com/pod-product-compliance
Lightning Source LLC
Chambersburg PA
CBHW051714210326
41597CB00032B/5476